PRINCIPLES OF
DEVELOPMENT

Lewis Wolpert is Professor of Biology as Applied to Medicine at University College London, UK. He is the author of *The Triumph of the Embryo*, *A Passion for Science*, and *The Unnatural Nature of Science* and is a well-known broadcaster and journalist.

Rosa Beddington is Head of the Division of Mammalian Development at the National Institute for Medical Research, London, UK.

Jeremy Brockes is Professor in the Department of Biochemistry and Molecular Biology at University College London, UK.

Thomas Jessell is Professor of Biochemistry and Molecular Biophysics, a member of the Center for Neurobiology and Behavior, and a Howard Hughes Medical Institute investigator at the College of Physicians and Surgeons of Columbia University, New York, USA. He is an author of *Principles of Neural Science* and *Essentials of Neural Science and Behavior*.

Peter Lawrence is in the Cell Biology Division at the Medical Research Council Laboratory of Molecular Biology, Cambridge, UK. He is the author of *The Making of a Fly*.

Elliot Meyerowitz is Professor in the Division of Biology at the California Institute of Technology, Pasadena, USA.

PRINCIPLES OF
DEVELOPMENT

Lewis Wolpert

Rosa Beddington

•

Jeremy Brockes

•

Thomas Jessell

•

Peter Lawrence

•

Elliot Meyerowitz

CURRENT BIOLOGY LTD

London • New York

1998

OXFORD UNIVERSITY PRESS

Oxford • New York • Tokyo

Text editor: Eleanor Lawrence
Project editor: Giles Montier
Copy editor: Hazel Richardson
Indexer: Liza Weinkove

Illustrator: Matthew McClements
Design and layout: Huw Woodman
Production: Zoe Edwards

Published by:
Current Biology Ltd.,
Middlesex House,
34–42 Cleveland Street,
London W1P 6LB, UK.

Oxford University Press,
Great Clarendon Street,
Oxford OX2 6DP, UK.

Oxford, New York
Athens, Auckland, Bangkok, Bogota, Bombay,
Buenos Aires, Calcutta, Cape Town, Dar es Salaam,
Delhi, Florence, Hong Kong, Istanbul, Karachi,
Kuala Lumpur, Madras, Madrid, Melbourne,
Mexico City, Nairobi, Paris, Singapore, Taipei,
Tokyo, Toronto, Warsaw, and associated
companies in Berlin, Ibadan.

Oxford is a trade mark of Oxford University Press.

Published in the United States
by Oxford University Press, Inc. New York.
198 Madison Avenue,
New York NY 10016, USA.

The cover represents cells communi-
cating positional information and cell
identity by production of a diffusion
gradient in a signaling molecule. At
a threshold concentration, cells are
triggered to differentiate, thus expressing a new phenotype. Design by
Matthew McClements.

A catalog record for this book is available from the British Library.

**Library of Congress Cataloging in Publication Data
(data available)**

ISBN 0-19-850263-X.

This book was produced using Corel VENTURA Publisher, CorelDRAW,
Adobe Photoshop, and Strata Vision3d. Editorial assistance was provided by
Mary Devane, Clare Rathbone, and Daniel Kirschner.

Printed in Singapore by Stamford Press.

Preface

Developmental biology is at the core of all biology. It deals with the process by which the genes in the fertilized egg control cell behavior in the embryo and so determine its pattern, its form, and much of its behavior. The progress in developmental biology in recent years with the applications of advances in cell and molecular biology, has been remarkable and an enormous amount of information is now available.

Principles of Development is designed for undergraduates as well as graduates, and the emphasis is on principles and key concepts. Central to our approach is that development can be best understood by understanding how genes control cell behavior. We have assumed that the students have some very basic familiarity with cell biology and genetics, but all key concepts, like the control of gene activity, are explained in the text.

Conscious of the pressures on students, we have tried to make the principles as clear as possible and to provide numerous summaries, both in words and in pictures. The illustrations in the book are a special feature and have been carefully designed and chosen to illuminate both experiments and mechanisms.

We have resisted the temptation to cover every aspect of development and have, instead, focused on those systems that best illuminate common principles. Indeed a theme that runs throughout the book is that universal principles govern the process of development. At all stages, what we included has been guided by what we believe undergraduates should know about development.

We have thus concentrated our attention on vertebrates and *Drosophila*, but not to the exclusion of the other systems, such as nematodes and sea urchins, where they best illustrate a concept. An important feature of our book is the inclusion of the development of plants, which is usually neglected in text books. There have been striking advances in plant developmental biology in recent times, and some unique and important features have emerged. As knowing the basic features of the embryology of the main organisms used to study development is essential for an understanding of molecular mechanisms, we have introduced embryology at an early stage.

Whereas our emphasis has been on the laying down of the body plans and organ systems, such as limbs and the nervous system, we have also included later aspects of development, including growth and regeneration. The book concludes with a consideration of evolution and development.

In providing references, our prime concern has been to guide the students to helpful papers rather than giving credit to the scientists who have made major contributions: to those whom we have neglected, we apologize.

The way the book was written was rather special. Although I was in continual consultation with my co-authors, I did all the writing—and I mean writing, which was skillfully typed by Maureen Moloney. Each chapter was also reviewed by a number of experts (see page xv), to whom we give thanks. The text was initially edited, and often re-written, by Eleanor Lawrence, whose expertise and influence pervades the book. Further critical editing was carried out by Hazel Richardson. And Huw Woodman magically turned the whole text into finished pages.

Central to the book are the illustrations, which were brilliantly created or adapted by Matthew McClements. The whole complex project was masterfully managed by Giles Montier. Particularly to Giles and Matthew, I offer my thanks for their patient dealing with my impatience and incompetence. The complete team was a pleasure, even fun, to work with.

Finally my thanks to Peter Newmark, the head of Current Biology Ltd., and to Vitek Tracz, the head of the Current Science Group; without them, the book would never have been started, let alone completed.

Lewis Wolpert

Contents

List of headings

Chapter 4: Patterning the Vertebrate Body Plan II: The Mesoderm and Early Nervous System

Chapter 5: Development of the *Drosophila* Body Plan

Chapter 11: Development of the Nervous System

Chapter 12: Germ Cells and Sex

Chapter 13: Regeneration

Chapter 14: Growth and Post-Embryonic Development

Chapter 15: Evolution and Development

Text acknowledgments

Many thanks to the following who kindly reviewed various parts of the book.

Chapter 1

Vernon French: University of Edinburgh.
Sir John Gurdon: Wellcome CRC Institute, Cambridge.
Andrew Murray: University of California, San Francisco.
Daniel St Johnston: Wellcome CRC Institute, Cambridge.

Chapter 2

Susan Darling: University College London.
Brigid Hogan: Vanderbilt University School of Medicine.
Nigel Holder: King's College London.
Andrew Murray: University of California, San Francisco.
Jonathan Slack: University of Bath.
James Smith: National Institute for Medical Research, London.
Daniel St. Johnston: Wellcome CRC Institute, Cambridge.

Chapter 3

Richard Harland: University of California, Berkeley.
Brigid Hogan: Vanderbilt University School of Medicine.
Nigel Holder: King's College London.
Janet Rossant: Samuel Lunenfield Research Institute, Toronto.
James Smith: National Institute for Medical Research, London.

Chapter 4

Richard Harland: University of California, Berkeley.
Brigid Hogan: Vanderbilt University School of Medicine.
Nigel Holder: King's College London.
Janet Rossant: Samuel Lunenfield Research Institute, Toronto.

Chapter 5

Kathryn Anderson: University of California, Berkeley.
Peter Bryant: University of California, Irvine.
Stephen Cohen: European Molecular Biology Laboratory, Heidelburg.
Herbert Jackle: Max-Planck Institute für Biophysikalische Chemie, Gottingen.
Daniel St Johnston: Wellcome CRC Institute, Cambridge.

Chapter 6

Jeff Hardin: University of Wisconsin, Madison.
James Priess: Fred Hutchinson Cancer Center, Seattle.
Paul Sternberg: California Institute of Technology, Pasadena.
Jeff Williams: University College London.

Chapter 7

David Meinke: Oklahoma State University.
Scott Poethig: University of Pennsylvania.
Ian Sussex: University of California, Berkeley.

Chapter 8

Marianne Bronner-Fraser: University of California, Irvine.
Charles Ettensohn: Carnegie Mellon University, Pittsburgh.
Raymond Keller: University of California, Berkeley.
Malcolm Steinberg: Princeton University.

Chapter 9

Margaret Buckingham: Institut Pasteur, France.
David Latchman: University College London Medical School.

Nicole Le Douarin: Institut d'Embryologie Cellulaire et Moléculaire du CNRS et du Collège de France, Nogent-sur-Marne.
Roger Patient: The Randall Institute, King's College London.
Fiona Watt: Imperial Cancer Research Fund, London.

Chapter 10

Stephen Cohen: European Molecular Biology Laboratory, Heidelburg.
Uma Thesleff: Institute of Biotechnology, University of Helsinki.
Cheryl Tickle: University College London.
Andrew Tomlinson: College of Physicians and Surgeons of Columbia University, New York.

Chapter 11

Michael Bate: University of Cambridge.
Steven Easter: University of Michigan.
Andrew Lumsden: United Medical and Dental School, Guy's Hospital, London.
Jack Price: SmithKline Beecham, Harlow, Essex.

Chapter 12

Denise Barlow: The Netherlands Cancer Institute, Amsterdam.
Tom Cline: University of California, Berkeley.
Jonathan Hodgkin: MRC Laboratory of Molecular Biology, Cambridge.
Anne McLaren: Wellcome CRC Institute, Cambridge.
Michael Whitaker: Newcastle University.

Chapter 13

Hans Bode: University of California, Irvine.
Susan Bryant: University of California, Irvine.
Vernon French: University of Edinburgh.
David Stocum: Indiana University-Purdue University, Indianapolis.

Chapter 14

Michael Ashburner: University of Cambridge.
Charles Brook: University College London.
Peter Bryant: University of California, Irvine.
Tom Kirkwood: University of Manchester.
Stuart Reynolds: University of Bath.
Jonathan Slack: University of Bath.
J. R. Tata: National Institute for Medical Research, London.

Chapter 15

Michael Akam: Wellcome CRC Institute, Cambridge.
Michael Coates: University College London.
John Gerhart: University of California, Berkeley.
Jonathan Slack: University of Bath.
Cliff Tabin: Harvard Medical School, Boston.

General

Brigid Hogan: Vanderbilt University, School of Medicine,
David Kimelman: University of Washington, Seattle.
Charles Rutherford: Virginia Polytechnic Institute and State University, Blacksburg.
Nancy Wall: Lawrence University,

Figure acknowledgments

Chapter 1

Fig. 1.15, illustration after Moore, K.L.: *Before we are Born: Basic Embryology and Birth Defects, 2nd edition*. Philadelphia: W.B. Saunders, 1983.

Chapter 2

Fig. 2.3, top photograph reproduced with permission from Alberts, B., Bray, D., Lewis, J., Raff, M., Roberts, K., Watson, J.D.: *Molecular Biology of the Cell, 3rd edition*. New York: Garland Publishing, 1994.

Fig. 2.5, photographs reproduced with permission from Kessel, R.G., Shih, C.Y.: *Scanning Electron Microscopy in Biology: A Student's Atlas of Biological Organization*. London, Springer-Verlag, 1974. © 1974 Springer-Verlag GmbH & Co. KG.

Fig. 2.6, illustration after Balinsky, B.I.: *An Introduction to Embryology*. Fourth edition. Philadelphia, W.B. Saunders, 1975.

Fig. 2.8, photograph reproduced with permission from Hausen, P., Riebesell, M.: *The Early Development of Xenopus laevis*. Berlin: Springer-Verlag, 1991.

Fig. 2.11, top photograph reproduced with permission from Kispert, A., Ortner, H., Cooke, J., Herrmann, B.G.: **The chick *Brachyury* gene: developmental expression pattern and response to axial induction by localized activin**. *Dev. Biol.* 1995, **168**:406–415.

Fig. 2.13, illustration adapted, with permission, from Balinsky, B.I.: *An Introduction to Embryology, 4th edition*. Philadelphia, W.B. Saunders, 1975. © 1975 Saunders College Publishing.

Fig. 2.17, illustration after Patten, B.M.: *Early Embryology of the Chick*. New York, Mc Graw-Hill, 1971.

Fig. 2.18, photographs reproduced with permission from Kispert, A., Ortner, H., Cooke, J., Herrmann, B.G.: **The chick *Brachyury* gene: developmental expression pattern and response to axial induction by localized activin**. *Dev. Biol.* 1995, **168**:406–415.

Fig. 2.19, illustration after Patten, B.M.: **The first heart beat and the beginning of embryonic circulation**. *American Scientist* 1951, **39**:225–243.

Fig. 2.20, top photograph reproduced with permission from Bloom, T.L.: **The effects of phorbol ester on mouse blastomeres: a role for protein kinase C in compaction?** *Development* 1989, **106**:159–171 Published by permission of The Company of Biologists Ltd.

Fig. 2.22, illustration after Hogan, B., Beddington, R., Costantini, F., Lacy, E.: *Manipulating the Mouse Embryo: A Laboratory Manual, 2nd edition*. New York: Cold Spring Harbor Laboratory Press, 1994.

Fig. 2.23, illustration adapted, with permission, from McMahon, A.P.: **Mouse development. Winged-helix in axial patterning**. *Curr. Biol.* 1994, **4**:903–906.

Fig. 2.24, illustration after Hogan, B., Beddington, R., Costantini, F., Lacy, E.: *Manipulating the Mouse Embryo: A Laboratory Manual, 2nd edition*. New York: Cold Spring Harbor Laboratory Press, 1994.

Fig. 2.25, illustration after Kaufman, M.H.: *The Atlas of Mouse Development*. London: Academic Press, 1992.

Fig. 2.26, top photograph reproduced with permission from Kimmel, C.B., Ballard, W.W., Kimmel, S.R., Ullmann, B., Schilling, T.F.: **Stages of embryonic development of the zebrafish**. *Dev. Dynamics* 1995. **203**:253–310. © 1995 Wiley-Liss, Inc.

Fig. 2.27, photographs reproduced with permission from Kessel, R.G., Shih, C.Y.: *Scanning Electron Microscopy in Biology: A Student's Atlas of Biological Organization*. London, Springer-Verlag, 1974. © 1974 Springer-Verlag GmbH & Co. KG.

Fig. 2.29, top photograph reproduced with permission from Turner, F.R., Mahowald, A.P.: **Scanning electron microscopy of *Drosophila* embryogenesis. I. The structure of the egg envelopes and the formation of the cellular blastoderm**. *Dev. Biol.* 1976, **50**:95–108. Middle photograph reproduced with permission from Turner, F.R., Mahowald, A.P.: **Scanning electron microscopy of *Drosophila* melanogaster embryogenesis. III. Formation of the head and caudal segments**. *Dev. Biol.* 1979, **68**:96–109.

Fig. 2.38, illustration after Sulston, J.E., Schierenberg, E., White, J.G., Thomson, J.N.: **The embryonic cell lineage of the nematode *Caenorhabditis elegans*.** *Dev. Biol.* 1983, **100**:64–119.

Fig. 2.32, left photograph reproduced with permission from Turner, F.R., Mahowald, A.P.: **Scanning electron microscopy of *Drosophila* melanogaster embryogenesis. II. Gastrulation and segmentation**. *Dev. Biol.* 1977, **57**:403–416. Center photograph reproduced with permission from Alberts, B., Bray, D., Lewis, J., Raff, M., Roberts, K., Watson, J.D.: *Molecular Biology of the Cell, 3rd edition*. New York, Garland Publishing, 1994.

Fig. 2.41, photographs reproduced with permission from Meinke, D.W.: **Seed development in *Arabidopsis thaliana*.** In *Arabidopsis*. Edited by Meyerowitz, E.M., & Somerville, C.: New York, Cold Spring Harbor Laboratory Press, 1991:pp253–295.

Chapter 3

Fig. 3.31, photograph reproduced with permission from Smith W.C., Harland, R.M.: **Expression cloning of noggin, a new dorsalizing factor localized to the Spemann organizer in *Xenopus* embryos.** *Cell* 1992, **70**:829–840. © 1992 Cell Press.

Chapter 4

Fig. 4.6, illustration after Johnson, R.L., Laufer, E., Riddle, R.D., Tabin, C.: **Ectopic expression of *Sonic hedgehog* alters dorsal-ventral patterning of somites**. *Cell* 1994, **79**:1165–1173.

Box 4A, illustration after Coletta, P.L., Shimeld, S.M., Sharpe, P.T.: **The molecular anatomy of Hox gene expression**. *J. Anat.* 1994, **184**:15–22.

Fig. 4.10, illustration after Burke, A.C., Nelson, C.E., Morgan, B.A., Tabin, C.: **Hox genes and the evolution of vertebrate axial morphology**. *Development* 1995, **121**: 333–346.

Fig. 4.16, illustration after Mangold, O.: **Über die induktionsfahigkeit der verschiedenen bezirke der neurula von urodelen**. *Naturwissenschaften* 1933, **21**:761–766.

Fig. 4.17, illustration after Kelly, O.G., Melton, D.A.: **Induction and patterning of the vertebrate nervous system**. *Tr. Genet.* 1995, **11**:273–278.

Fig. 4.18, illustration after Kintner, C.R., Dodd, J.: **Hensen's node induces neural tissue in *Xenopus* ectoderm. Implications for the action of the organizer in neural induction**. *Development* 1991, **113**:1495–1505.

Fig. 4.19, illustration after Holtfreter, J., Hamburger, V.: **Amphibians**. In *Analysis of Development,* Edited by Willier, B.H., Weiss, P.A., Hamburger, V. Philadelphia: Saunders, 1955; pp230–295.

Fig. 4.20, illustration after Doniach, T., Phillips, C.R., Gerhart, J.C.: **Planar induction of antero-posterior pattern in the developing central nervous system of *Xenopus laevis**. Science* 1992, **257**: 542–545.

Fig. 4.21, illustration adapted, with permission, from Lumsden, A.: **Cell lineage restrictions in the chick embryo hindbrain**. *Phil. Trans. Roy. Soc. Lond.* B 1991, **331**:281–286.

Fig. 4.22, illustration adapted, with permission, from Lumsden, A.: **Cell lineage restrictions in the chick embryo hindbrain**. *Phil. Trans. Roy. Soc. Lond.* B 1991, **331**:281–286.

Fig. 4.23, illustration after Krumlauf, R.: **Hox genes and pattern formation in the branchial region of the vertebrate head**. *Tr. Genet.* 1993, **9**:106–112.

Fig. 4.24, photograph reproduced with permission from Lumsden, A., Krumlauf, R.: **Patterning the vertebrate neuraxis**. *Science* 1996, **274**:1109–1115 (image on front cover). © 1996 American Association for the Advancement of Science.

Chapter 5

Fig. 5.5, photograph reproduced with permission from Griffiths, A.J.H., Miller, J.H., Suzuki, D.T., Lewontin, R.C., Gelbart, W.M.: *An Introduction to Genetic Analysis, 6th edition*. New York: W.H. Freeman & Co., 1996.

Fig. 5.6, photograph reproduced with permission from Griffiths, A.J.H., Miller, J.H., Suzuki, D.T., Lewontin, R.C., Gelbart, W.M.: *An Introduction to Genetic Analysis, 6th edition*. New York: W.H. Freeman & Co., 1996.

Fig. 5.12, illustration after González-Reyes, A., Elliott, H., St. Johnston, D.: **Polarization of both major body axes in *Drosophila* by *gurken-torpedo* signalling**. *Nature* 1995, **375**:654–658.

Fig. 5.20, illustration after Lawrence, P.: *The Making of a Fly*. Oxford: Blackwell Scientific Publications, 1992.

Fig. 5.24, photograph reproduced with permission from Lawrence, P.: *The Making of a Fly*. Oxford: Blackwell Scientific Publications, 1992.

Fig. 5.25, illustration after Lawrence, P.: *The Making of a Fly*. Oxford: Blackwell Scientific Publications, 1992.

Box 5B, illustration after Lawrence, P.: *The Making of a Fly*. Oxford: Blackwell Scientific Publications, 1992.

Fig. 5.30, illustration after Lawrence, P.: *The Making of a Fly*. Oxford: Blackwell Scientific Publications, 1992.

Fig. 5.36, illustration after Lawrence, P.: *The Making of a Fly*. Oxford: Blackwell Scientific Publications, 1992. Photograph reproduced with permission from Bender, W., Akam, A., Karch, F., Beachy, P.A., Peifer, M., Spierer, P., Lewis, E.B., Hogness, D.S.: **Molecular genetics of the bithorax complex in *Drosophila melanogaster**. Science* 1983, **221**:23–29 (image on front cover). © 1983 American Association for the Advancement of Science.

Chapter 6

Fig. 6.2, illustration after Sulston, J.E., Schierenberg, E., White, J.G., Thompson, J.N.: **The embryonic cell lineage of the nematode *Caenorhabditis elegans**. Dev. Biol.* 1983, **100**:69–119.

Fig. 6.3, photograph reproduced with permission from Strome, S., Wood, W.B.: **Generation of asymmetry and segregation of germline granules in early *C. elegans* embryos**. *Cell* 1983 **35**:15-25. © 1983 Cell Press.

Fig. 6.4, illustration after Sulston, J.E., Schierenberg, E., White, J.G., Thompson, J.N.: **The embryonic cell lineage of the nematode *C. elegans**. Dev. Biol.* 1983, **100**:69–119.

Fig. 6.5, photographs reproduced with permission from Wood W.B: **Evidence from reversal of handedness in C. elegans embryos for early cell interactions determining cell fates**. *Nature* 1991, **349**:536–538. © 1991 Macmillan Magazines Ltd.

Fig. 6.6, illustration after Mello, C.C., Draper, B.W., Priess, J.R.: **The maternal genes *apx-1* and *glp-1* and establishment of dorsal-ventral polarity in the early *C. elegans* embryo**. *Cell* 1994, **77**:95–106.

Fig. 6.7, illustration after Bürglin, T.R., Ruvkun, G.: **The *Caenorhabditis elegans* homeobox gene cluster**. *Curr. Opin. Gen. Devel.* 1993, **3**:615–620.

Fig. 6.10, illustration after Morgan, T.H. *Experimental Embryology*. New York: Columbia University Press, 1927.

Fig. 6.11, illustration after Wilmer, P. *Invertebrate Relationships*. Cambridge: Cambridge University Press, 1990.

Fig. 6.13, illustration after van den Biggelaar, J.A.M.: **Asymmetries during molluscan embryogenesis**. In *Biological Asymmetry and Handedness. Ciba Foundation Symp*. Chichester: J. Wiley, 1991; **162**:128–142.

Fig. 6.15, illustration after Bissen, S.T., Smith, C.M.: **Unequal cleavage in leech embryos: zygotic transcription is required for correct spindle orientation in a subset of early blastomeres**. *Development* 1996, **122**:599–606.

Fig. 6.17, illustration after Wedeen, C.J., Weisblat, D.A.: **Segmental expression of an engrailed-class gene during early development and neurogenesis in an annelid**. *Development* 1991, **113**:805–814.

Fig. 6.18, illustration after Shankland, M.: **Leech segmentation: a molecular perspective**. *BioEssays* 1994, **16**:801–808.

Fig. 6.22, illustration after Ransick, A., Davidson, E.H.: **A complete second gut induced by transplanted micomeres in the sea urchin embryos**. *Science* 1993, **259**:1134–1138.

Fig. 6.26 photograph reproduced with permission from Corbo, J.C., Levine, M., Zeller, R.W.: **Characterization of a notochord-specific enhancer from the Brachyury promoter region of the ascidian, *Ciona intestinalis**. Development* 1997, **124**:589–602. Published by permission of The Company of Biologists Ltd.

Fig. 6.28, illustration after Conklin, E.G.: **The organization and cell lineage of the ascidian egg**. *J. Acad. Nat. Sci. Philadelphia* 1905, **13**:1–119.

Fig. 6.29, illustration after Nakatani, Y., Yasuo, H., Satoh, N., Nishida, H.: **Basic fibroblast growth factor induces notochord formation and the expression of *As-T*, a *Brachyury* homolog, during ascidian embryogenesis**. *Development* 1996, **122**:2023–2031.

Fig. 6.30, top photograph reproduced with permission from Early, A.E., Gaskell, M.J., Traynor, D., Williams, J.G.: **Two distinct populations of prestalk cells within the tip of the migratory Dictyostelium slug with differing fates at culmination**. *Development* 1993, **118**:353–362. Published by permission of The Company of Biologists Ltd. Middle panel, photograph reproduced with permission from Jermyn, K., Traynor, D., Williams, J.: **The initiation of basal disc formation in *Dictyostelium discoideum* is an early event in culmination**. *Development* 1996, **122**:753–760. Published by permission of The Company of Biologists Ltd.

Chapter 7

Fig. 7.2, illustration after Alberts, B., Bray, D., Lewis, J., Raff, M., Roberts, K., Watson, J.D.: *Molecular Biology of the Cell, 2nd edition*. New York: Garland Publishing, 1989.

Fig. 7.5, illustration after Scheres, B., Wolkenfelt, H., Willemsen, V., Terlouw, M., Lawson, E., Dean, C., Weisbeek, P.: **Embryonic origin of the *Arabidopsis* primary root and root meristem initials.** *Development* 1994, **120**:2475–2487.

Fig. 7.6, illustration after Mayer, U., Torres-Ruiz, R.A., Berleth, T., Misera, S., Jurgens, G.: **Mutations affecting body organization in the *Arabidopsis* embryo.** *Nature* 1991, **353**:402–407.

Fig. 7.8, illustration after Alberts, B., Bray, D., Lewis, J., Raff, M., Roberts, K., Watson, J.D.: *Molecular Biology of the Cell, 2nd edition.* New York: Garland Publishing, 1989.

Fig. 7.9, photograph reproduced with permission from Bowman, J. (ed.).: *Arabidopsis: an Atlas of Morphology and Development.* New York: Springer-Verlag, 1994. © 1994 Springer-Verlag GmBH & Co.

Fig. 7.10, illustration after Steeves, T.A., Sussex, I.M.: *Patterning in Plant Development.* Cambridge: Cambridge University Press, 1989.

Fig. 7.12, illustration after McDaniel, C.N., Poethig, R.S.: **Cell lineage patterns in the shoot apical meristem of the germinating maize embryo.** *Planta* 1988, **175**:13–22.

Fig. 7.13, illustration after Irish, V.F.: **Cell lineage in plant development.** *Curr. Opin. Gen. Devel.* 1991, **1**:169–173.

Fig. 7.14, illustration after Sachs, T.: *Pattern Formation in Plant Tissues.* Cambridge: Cambridge University Press, 1994; p.133.

Fig. 7.15, top panel, illustration after Poethig, R.S., Sussex, I.M.: **The cellular parameters of leaf development in tobacco: a clonal analysis.** *Planta* 1985, **165**: 170–184. Bottom panel, illustration after Sachs, T.: *Pattern Formation in Plant Tissues.* Cambridge: Cambridge University Press, 1994.

Fig. 7.18, illustration after Scheres, B., Wolkenfelt, H., Willemsen, V., Terlouw, M., Lawson, E., Dean, C., Weisbeek, P.: **Embryonic origin of the *Arabidopsis* primary root and root meristem initials.** *Development* 1994, **120**:2475–2487.

Fig. 7.19, illustration after Coen, E.S., Meyerowitz, E.M.: **The war of the whorls: genetic interactions controlling flower development.** *Nature* 1991, **353**:31–37.

Fig. 7.20, photograph reproduced with permission from Meyerowitz, E.M., Bowman, J.L., Brockman, L.L., Drews, G.N., Jack, T., Sieburth, L.E., Weigel, D.: **A genetic and molecular model for flower development in *Arabidopsis thaliana*.** *Development Suppl.* 1991, pp157–167. Published by permission of The Company of Biologists Ltd.

Fig. 7.21, photographs reproduced with permission from Meyerowitz, E.M., Bowman, J.L., Brockman, L.L., Drews, G.N., Jack, T., Sieburth, L.E., Weigel, D.: **A genetic and molecular model for flower development in *Arabidopsis thaliana*.** *Development Suppl.* 1991, pp157–167. Published by permission of The Company of Biologists Ltd. (left panel); center panel from Bowman, J.L., Smyth, D.R., Meyerowitz, E.M.: **Genes directing flower development in *Arabidopsis*.** *Plant Cell* 1989, **1**:37–52. Published by permission of The American Society of Plant Physiologists.

Fig. 7.23, illustration after Dennis, E., Bowman, J.L.: **Manipulating floral identity.** *Curr. Biol.* 1993, **3**:90–93.

Fig. 7.24, illustration after Meyerowitz, E.M., Bowman, J.L., Brockman, L.L., Drews, G.N., Jack, T., Sieburth, L.E., Weigel, D.: **A genetic and molecular model for flower development in *Arabidopsis thaliana*.** *Development Suppl.* 1991, pp157–167.

Fig. 7.25, illustration after Coen, E.S., Meyerowitz, E.M.: **The war of the whorls: genetic interactions controlling flower development.** *Nature* 1991, **353**:31–37.

Fig. 7.28, photograph reproduced with permission from Coen, E.S., Meyerowitz, E.M.: **The war of the whorls: genetic interactions controlling flower development.** *Nature* 1991, **353**:31–37. © 1991 Macmillan Magazines Ltd.

Fig. 7.29, illustration after Drews, G.N., Goldberg, R.B.: **Genetic control of flower development.** *Trends Genet.* 1989 **5**:256-261.

Chapter 8

Fig. 8.3, photographs reproduced with permission from Steinberg, M.S., Takeichi, M.: **Experimental specification of cell sorting, tissue spreading, and specific spatial patterning by quantitative differences in cadherin expression.** *Proc. Natl. Acad. Sci./Dev. Biol.* 1994, **91**:206–209. © 1994 National Academy of Sciences.

Fig. 8.6, illustration after Strome, S.: **Determination of cleavage planes.** *Cell* 1993, **72**:3–6.

Fig. 8.9, photograph reproduced with permission from Bloom T.L.: **The effects of phorbol ester on mouse blastomeres: a role for protein kinase C in compaction?** *Development* 1989, **106**:159–71.

Fig. 8.12, illustration after Coucouvanis, E., Martin, G.R.: **Signals for death and survival: a two-step mechanism for cavitation in the vertebrate embryo.** *Cell* 1995, **83**:279–287.

Fig. 8.16, illustration after Odell, G.M., Oster, G., Alberch, P., Burnside, B.: **The mechanical basis of morphogenesis. I. Epithelial folding and invagination.** *Dev. Biol.* 1981, **85**:446–462.

Fig. 8.18, photographs reproduced with permission from Leptin, M., Casal, J., Grunewald, B., Reuter, R.: **Mechanisms of early *Drosophila* mesoderm formation.** *Development Suppl.* 1992, pp23–31. Published by permission of The Company of Biologists Ltd.

Fig. 8.19, illustration after Balinsky, B.I.: *An Introduction to Embryology, 4th edition.* Philadelphia, W.B. Saunders, 1975.

Fig. 8.20, illustration after Hardin, J., Keller, R.: **The behavior and function of bottle cells during gastrulation of *Xenopus laevis*.** *Development* 1988, **103**:211–230.

Fig. 8.23, photograph reproduced with permission from Smith, J.C., Cunliffe, V., O'Reilly, M-A.J., Schulte-Merker, S., Umbhauer, M.: ***Xenopus Brachyury*.** *Semin. Dev. Biol.* 1995, **6**:405–410. © 1995 by permission of the publisher, Academic Press Ltd., London.

Fig. 8.30, illustration after Schoenwolf, G.C., Smith, J.L.: **Mechanisms of neurulation: traditional viewpoint and recent advances.** *Development* 1990 **109**:243–270.

Fig. 8.33, photograph reproduced with permission from Morrill, J.B., Santos, L.L.: **A scanning electron micrographical overview of cellular and extracellular patterns during blastulation and gastrulation in the sea urchin, *Lytechinus variegatus*.** In *The Cellular and Molecular Biology of Invertebrate Development.* Edited by Sawyer, R.H. and Showman, R.M. University of South Carolina Press, 1985; pp3–33.

Fig. 8.38, illustration after Alberts, B., Bray, D., Lewis, J., Raff, M., Roberts, K., Watson, J.D.: *Molecular Biology of the Cell, 2nd edition.* New York: Garland Publishing, 1989.

Fig. 8.41, photographs reproduced with permission from Priess, J.R., Hirsh, D.I.: ***Caenorhabditis elegans morphogenesis*: the role of the cytoskeleton in elongation of the embryo.** *Dev. Biol.* 1986, **117**:156–173. © 1986 Academic Press.

Fig. 8.43, photographs reproduced with permission from Tsuge, T., Tsukaya, H., Uchimaya, H.: **Two independent and polarized processes of cell elongation regulate leaf blade expansion in *Arabidopsis thaliana* (L.) Heynh.** *Development* 1996, **122**:1589–1600. Published by permission of The Company of Biologists Ltd.

Chapter 9

Fig. 9.1, illustration after Friederich, E., Prignault, E., Arpin, M., Louvard, D.: **From the structure to the function of villin, an actin-binding protein of the brush border.** *BioEssays* 1990, **12**:403–408.

Fig. 9.5, illustration after Okada, T.S.: *Transdifferentiation.* Oxford: Clarendon Press, 1992.

Fig. 9.6, illustration after Doupe, A.J., Landis, S.C., Patterson, P.H.: **Environmental influences in the development of neural crest derivatives: glucocorticoids, growth factors, and chromaffin cell plasticity.** *J. Neurosci.* 1985, **5**:2119–2142.

Fig. 9.7, illustration after Janeway, C.A., Travers, P.: *Immunobiology: The Immune System in Health and Disease, 3rd edition.* London: Current Biology/Garland, 1997.

Fig. 9.8, illustration after Janeway, C.A., Travers, P.: *Immunobiology: The Immune System in Health and Disease, 3rd edition.* London: Current Biology/Garland Publishing, 1997.

Fig. 9.9, illustration after Alberts, B., Bray, D., Lewis, J., Raff, M., Roberts, K., Watson, J.D.: *Molecular Biology of the Cell, 2nd edition.* New York: Garland Publishing, 1989.

Fig. 9.11, illustration after Alberts, B., Bray, D., Lewis, J., Raff, M., Roberts, K., Watson, J.D.: *Molecular Biology of the Cell, 2nd edition.* New York: Garland Publishing, 1989.

Fig. 9.13, illustration after Alberts, B., Bray, D., Lewis, J., Raff, M., Roberts, K., Watson, J.D.: *Molecular Biology of the Cell, 2nd edition.* New York: Garland Publishing, 1989.

Fig. 9.15, illustration after Tijian, R.: **Molecular machines that control genes.** *Sci. Amer.* 1995, **272**:54–61.

Fig. 9.23, illustration after Metcalf, D.: **Control of granulocytes and macrophages: molecular, cellular, and clinical aspects.** *Science* 1991, **254**:529–533.

Fig. 9.26, illustration after Crossley, M., Orkin, S.H.: **Regulation of the β-globin locus.** *Curr. Opin. Genet. Dev.* 1993, **3**:232–237.

Fig. 9.29, illustration after Doupe, A.J., Landis, S.C., Patterson, P.H.: **Environmental influences in the development of neural crest derivatives: glucocorticoids, growth factors, and chromaffin cell plasticity.** *J. Neurosci.* 1985, **5**:2119–2142.

Chapter 10

Box 10A, top illustration after Meinhardt, H., Gierer, A.: **Applications of a theory of biological pattern formation based on lateral inhibition.** *J. Cell Sci.* 1974, **15**:321–346.

Fig. 10.12, photograph reproduced with permission from Cohn, M.J., Izpisúa-Belmonte, J.C., Abud, H., Heath, J.K., Tickle, C.: **Fibroblast growth factors induce additional limb development from the flank of chick embryos.** *Cell* 1995, **80**:739–746. © 1995 Cell Press.

Fig. 10.20, photograph reproduced with permission from Garcia-Martinez, V., Macias, D., Gañan, Y., Garcia-Lobo, J.M., Francia, M.V., Fernandez-Teran, M.A., Hurle, J.M.: **Internucleosomal DNA fragmentation and programmed cell death (apoptosis) in the interdigital tissue of embryonic chick leg bud.** *J. Cell Sci.* 1993, **106**:201–208. Published by permission of The Company of Biologists Ltd.

Fig. 10.22, illustration after French, V., Daniels, G.: **Pattern formation: the beginning and the end of insect limbs.** *Curr. Biol.* 1994, **4**:35–37.

Fig. 10.25, photograph reproduced with permission from Nellen, D., Burke, R., Struhl, G., Basler, K.: **Direct and long-range action of a dpp morphogen gradient.** *Cell* 1996, **85**:357–368. © 1996, Cell Press.

Fig. 10.28, illustration after Diaz-Benjumea, F.J., Cohen, S.M.: **Interaction between dorsal and ventral cells in the imaginal disc directs wing development in *Drosophila*.** *Cell* 1993, **75**:741–752.

Fig. 10.29, photograph reproduced with permission from Zecca, M., Basler, K., Struhl, G.: **Direct and long-range action of a wingless morphogen gradient.** *Cell* 1996, **87**:833–844. © 1996 Cell Press.

Fig. 10.31, illustration after Bryant, P.J.: **The polar coordinate model goes molecular.** *Science* 1993, **259**:471–472.

Fig. 10.39, illustration after Lawrence, P.: *The Making of a Fly.* Oxford: Blackwell Scientific Publications, 1992.

Fig. 10.42, illustration after Horvitz, H.R., Sternberg, P.W.: **Multiple intercellular signalling systems control the development of the *Caenorhabditis elegans* vulva.** *Nature* 1991, **351**:535–541.

Chapter 11

Fig. 11.3, photograph reproduced with permission from Skeath, J.B., Doe, C.Q.: **The achaete-scute complex proneural genes contribute to neural precursor specification in the *Drosophila* CNS.** *Curr. Biol.* 1996, **6**:1146–1152.

Fig. 11.5, illustration after Campuzano, S., Modolell, J.: **Patterning of the *Drosophila* nervous system: the achaete-scute gene complex.** *Trends Genet.* 1992, **8**:202–208.

Fig. 11.6, illustration after Jan, Y.N., Jan, L.Y.: **Genes required for specifying cell fates in Drosophila embryonic sensory nervous system.** *Trends Neurosci.* 1990, **13**:493–498.

Fig. 11.8, illustration after Guo, M., Jan, L.Y., Jan, Y.N.: **Control of daughter cell fates during asymmetric division: interaction of Numb and Notch.** *Neuron* 1996, **17**:27–41.

Fig. 11.15, illustration after Rakic, P.: **Mode of cell migration to the superficial layers of fetal monkey neocortex.** *J. Comp. Neurol.* 1972, **145**:61–83.

Fig. 11.18, illustration after Alberts, B., Bray, D., Lewis, J., Raff, M., Roberts, K., Watson, J.D.: *Molecular Biology of the Cell, 2nd edition.* New York: Garland Publishing, 1989.

Fig. 11.22, illustration after O'Connor, T.P., Duerr, J.S., Bentley, D.: **Pioneer growth cone steering decisions mediated by single filopodial contacts *in situ*.** *J. Neurosci.* 1990, **10**:3935–3946.

Fig. 11.29, illustration after Tessier-Lavigne, M., Placzek, M.: **Target attraction: are developing axons guided by chemotropism?** *Trends Neurosci.* 1991, **14**:303–310.

Fig. 11.30, photograph reproduced with permission from Serafini, T., Colamarino, S.A., Leonardo, E.D., Wang, H., Beddington, R., Skarnes, W.C., Tessier-Lavigne, M.: **Netrin-1 is required for commissural axon guidance in the developing vertebrate nervous system.** *Cell* 1996, **87**:1001–1014. © 1996 Cell Press.

Fig. 11.32, illustration after Davies, A.M.: **Neurotrophic factors: switching neurotrophin dependence.** *Curr. Biol.* 1994, **4**:273–276.

Fig. 11.33, illustration after Kandell, E.R., Schwartz, J.H., Jessell, T.M.: *Principles of Neural Science, 3rd edition.* New York: Elsevier Science Publishing Co., Inc., 1991.

Fig. 11.37, illustration after Goodman, C.S., Shatz, C.J.: **Developmental mechanisms that generate precise patterns of neuronal connectivity.** *Cell Suppl.* 1993, **72**:77–98

Fig. 11.38, illustration after Goodman, C.S., Shatz, C.J.: **Developmental mechanisms that generate precise patterns of neuronal connectivity.** *Cell Suppl.* 1993, **72**:77–98

Fig. 11.39 illustration after Kandell, E.R., Schwartz, J.H., Jessell, T.M.: *Essentials of Neural Science and Behavior*. Norwalk, Connecticut: Appleton & Lange, 1991.

Fig. 11.40, illustration after Goodman, C.S., Shatz, C.J.**: Developmental mechanisms that generate precise patterns of neuronal connectivity**. *Cell Suppl.* 1993, **72**:77–98

Chapter 12

Fig. 12.2, illustration after Goodfellow, P.N., Lovell-Badge, R.: ***SRY and sex determination in mammals***. *Ann. Rev. Genet.* 1993, **27**:71–92.

Fig. 12.3, illustration after Higgins, S.J., Young, P., Cunha, G.R.: **Induction of functional cytodifferentiation in the epithelium of tissue recombinants II. Instructive induction of Wolffian duct epithelia by neonatal seminal vesicle mesenchyme**. *Development* 1989, **106**:235–250.

Fig. 12.7, illustration after Cline, T.W.: **The *Drosophila* sex determination signal: how do flies count to two?** *Trends Genet.* 1993 **9**:385–390.

Fig. 12.11, illustration after Clifford, R., Francis, R., Schedl, T.: **Somatic control of germ cell development**. *Semin. Dev. Biol.* 1994, **5**:21–30.

Fig. 12.16, illustration after Wylie, C.C., Heasman, J.: **Migration, proliferation, and potency of primordial germ cells**. *Semin. Dev. Biol.* 1993, **4**:161–170.

Fig. 12.22, illustration after Alberts, B., Bray, D., Lewis, J., Raff, M., Roberts, K., Watson, J.D.: *Molecular Biology of the Cell, 2nd edition*. New York: Garland Publishing, 1989.

Chapter 13

Fig. 13.5, photograph reproduced with permission from Müller, W.A.: **Diacylglycerol-induced multihead formation in *Hydra***. *Development* 1989, **105**:309–316. Published by permission of The Company of Biologists Ltd.

Fig. 13.11, photograph reproduced with permission from Müller, W.A.: **Diacylglycerol-induced multihead formation in *Hydra***. *Development* 1989, **105**:309–316. Published by permission of The Company of Biologists Ltd.

Fig. 13.18, photographs reproduced with permission from Pecorino, L.T. Entwistle A., Brockes, J.P.: **Activation of a single retinoic acid receptor isoform mediates proximodistal respecification** *Curr. Biol.* 1996, **6**:563–569.

Fig. 13.20, illustration after French, V., Bryant, P.J., Bryant, S.V.: **Pattern regulation in epimorphic fields**. *Science* 1976, **193**:969–981.

Fig. 13.21 illustration after French, V., Bryant, P.J., Bryant, S.V.: **Pattern regulation in epimorphic fields**. *Science* 1976, **193**:969–981.

Chapter 14

Fig. 14.3, illustration after Edgar, B.A., Lehman, D.A., O'Farrell, P.H.: **Transcriptional regulation of *string* (*cdc25*): a link between developmental programming and the cell cycle**. *Development* 1994, **120**:3131–3143.

Fig. 14.5, illustration after Gray, H.: *Gray's Anatomy*. Edinburgh: Churchill-Livingstone, 1995.

Fig. 14.8, photograph reproduced with permission from Harrison, R.G.: *Organization and Development of the Embryo*. New Haven: Yale University Press, 1969. © 1969 Yale University Press.

Fig. 14.9, illustration after Wallis, G.A.: **Here today, bone tomorrow**. *Curr. Biol.* 1993, **3**:687–689.

Fig. 14.12, illustration after Alberts, B., Bray, D., Lewis, J., Raff, M., Roberts, K., Watson, J.D.: *Molecular Biology of the Cell, 2nd edition*. New York: Garland Publishing, 1989.

Fig. 14.18, illustration after Tata, J.R.: **Gene expression during metamorphosis: an ideal model for post-embryonic development**. *BioEssays* 1993, **15**: 239–248.

Fig. 14.19, illustration after Tata, J.R.: **Gene expression during metamorphosis: an ideal model for post-embryonic development**. *BioEssays* 1993, **15**: 239–248.

Chapter 15

Fig. 15.2, illustration after Larsen, W.J.: *Human Embryology*. New York: Churchill Livingstone, 1993.

Fig. 15.3, illustration after Romer, A.S.: *The Vertebrate Body*. Philadelphia: W.B. Saunders, 1949.

Fig. 15.5, photographs reproduced with permission from Sordino, P., van der Hoeven, F., Duboule, D.: **Hox gene expression in teleost fins and the origin of vertebrate digits**. *Nature* 1995, **375**:678–681. © 1995 Macmillan Magazines Ltd.

Fig. 15.6, illustration after Coates, M.I.: **Limb evolution: fish fins or tetrapod limbs—a simple twist of fate?** *Curr. Biol.* 1995, **5**:844–848.

Fig. 15.9, illustration after Garcia-Fernández, J., Holland, P.W.: **Archetypal organization of the amphioxus Hox gene cluster**. *Nature* 1994 **370**: 563–566.

Fig. 15.10, illustration after Akam, M.: **Hox genes and the evolution of diverse body plans**. *Phil. Trans. Roy. Soc. Lond. B* 1995, **349**:313–319.

Fig. 15.11, illustration after Ferguson, E.L.: **Conservation of dorsal-ventral patterning in arthropods and chordates**. *Curr. Opin. Genet. Dev.* 1996, **6**:424–431.

Fig. 15.12, illustration after Gregory, W.K.: *Evolution Emerging*. New York: Macmillan, 1957.

History and Basic Concepts

1

The origins of developmental biology.

A conceptual tool kit.

"Ideas about us have changed a lot; but we now have some reliable rules."

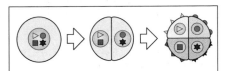

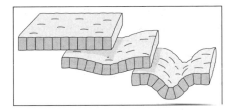

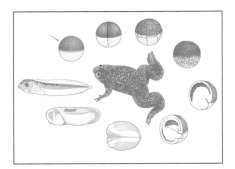

The development of multicellular organisms from a single cell—the fertilized egg—is a brilliant triumph of evolution. During embryonic development the egg divides to give rise to many millions of cells, which form structures as complex and varied as eyes, arms, the heart, and the brain. This amazing achievement raises a multitude of questions. How do the cells arising from division of the fertilized egg become different from each other? How do they become organized into structures such as limbs and brains? What controls the behavior of individual cells so that such highly organized patterns emerge? How are the organizing principles of development embedded within the egg and in particular within the genetic material—DNA? Much of the excitement in developmental biology today comes from our growing understanding of how genes direct these developmental processes, and genetic control is one of the main themes of this book.

One of the main tasks of early embryogenesis is to lay down the overall body plan of the organism and we shall see that different organisms solve this fundamental problem in several ways. The focus of this book is mainly on animal development—that of vertebrates such as frogs, birds, fish and mammals, and that of a selection of invertebrates, such as the sea urchin, ascidians, the leech, the nematode worm, and above all the fruit fly, *Drosophila melanogaster* (Fig. 1.1). It is with that small fly that our understanding of the genetic control of development is most advanced. We also look at some aspects of plant development, which differs crucially in some respects from that of animals.

The development of individual organs such as the vertebrate limb, the insect eye, and the nervous system illustrates multicellular organization and tissue differentiation at later stages in embryogenesis, and we consider some of these systems in detail. We also deal with the development of sexual characteristics. The study of developmental biology, however, goes well beyond the development of the embryo. We also need to understand how some animals can regenerate lost organs (Fig. 1.2), and how post-embryonic growth of the organism is controlled, a process which includes metamorphosis and aging. Taking a wider view, we finally consider how developmental mechanisms have evolved and how they constrain the very process of evolution itself.

One might ask whether it is necessary to cover so many different organisms and developmental systems in order to understand the basic features of development? The answer at present is yes. Developmental biologists do indeed believe that there are general principles of development that apply to all animals, but that life is too wonderfully diverse to find all the answers in a single organism. As it is, developmental biologists have tended to focus their efforts on a relatively small number of animals, chosen originally because they were convenient to study and amenable to experimental manipulation or genetic analysis. This is why some creatures, such as the frog *Xenopus laevis* (Fig. 1.3), the nematode *Caenorhabditis elegans*, and the fruit fly *Drosophila*, have such a dominant place in developmental biology, and are encountered again and again in this book.

One of the most exciting and satisfying aspects of developmental biology is that understanding a developmental process in one organism can help to illuminate similar processes elsewhere, for example in organisms much more like ourselves. Nothing illustrates this more dramatically than the influence our understanding of *Drosophila* development, and especially its genetic basis, has had throughout developmental biology. In particular, the identification of genes controlling early embryogenesis in *Drosophila* has led to the discovery of related genes being used in similar ways in the development of mammals and other vertebrates. Such discoveries encourage us to believe in the emergence of general developmental principles.

Frogs have long been a favorite organism for studying development because their eggs are large, and their embryos are robust, easy to grow in a simple culture medium, and relatively easy to experiment on. The South African claw-toed frog *Xenopus* (sometimes mistakenly called the 'South African clawed toad'), is the model organism for many aspects of vertebrate development, and the main features of its development (Box 1A) can serve to illustrate some of the basic stages of development in all animals.

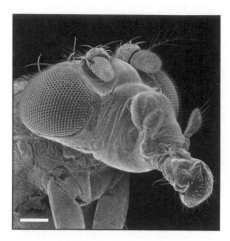

Fig. 1.1 Scanning electron micrograph of the head of an adult *Drosophila melanogaster*. Scale bar = 0.1 mm. Photograph by D. Scharfe, from Science Photo Library.

Fig. 1.2 Photograph of a lizard, the South Eastern five-lined skink, after it has released its tail in defense. This species can deliberately shed its tail as a technique to avoid capture by predators; and then regenerate it. A piece of discarded tail can be seen below the skink. Photograph from Oxford Scientific Films.

In the rest of this chapter we first look at the history of embryology—as the study of developmental biology has been called for most of its existence. (The term developmental biology itself is of rather recent origin.) We then introduce some key concepts that are used repeatedly in studying and understanding development.

The origins of developmental biology.

Many questions in embryology were first posed hundreds, and in some cases thousands, of years ago. Appreciating the history of these ideas helps us to understand why we approach developmental problems in the way that we do today.

1-1   Aristotle first defined the problem of epigenesis and preformation.

A scientific approach to explaining development started with Hippocrates in Greece in the 5th century BC. Using ideas that were current at the time, he tried to explain development in terms of the principles of heat, wetness, and solidification. The study of embryology advanced about a century later when the Greek philosopher Aristotle formulated a question that was to dominate much thinking about development until the end of the 19th century. Aristotle addressed the problem of how the different parts of the embryo were formed. He considered two possibilities: one was that everything in the embryo was preformed from the very beginning and simply got bigger during development; the other, which Aristotle supported, was that new structures arose progressively, a process he termed epigenesis (which means 'upon formation') and likened metaphorically to the 'knitting of a net'. Aristotle favored epigenesis and his conjecture was correct.

Aristotle's influence on European thought was enormous and his ideas remained dominant well into the 17th century. The contrary view to epigenesis, namely that the embryo was preformed from the beginning, was championed anew in the late 17th century. Many could not believe that physical or chemical forces could mold a living entity like the embryo. Against the contemporaneous background of belief in the divine creation of the world and all living things, they believed that all embryos had existed from the beginning of the world and that the first embryo of a species must contain all future embryos.

Even the brilliant 17th century Italian embryologist Marcello Malpighi could not free himself from preformationist ideas. While he provided a remarkably accurate description of the development of the chick embryo, he remained convinced, against the evidence of his own observations, that the embryo was already present from the very beginning (Fig. 1.4). He argued that at very early stages the parts were so small that they could not be observed, even with his best microscope. Yet other preformationists believed that the sperm contained the embryo and some even claimed to be able to see a tiny human—a homunculus—in the head of each human sperm (Fig. 1.5).

The preformation/epigenesis issue was the subject of vigorous debate throughout the 18th century. But the problem could not be resolved until one of the great advances in biology had taken place—the recognition that living things, including embryos, were composed of cells.

Fig. 1.3 Photograph of the adult South African claw-toed frog, *Xenopus laevis.* Scale bar = 1 cm. Photograph courtesy of J. Smith.

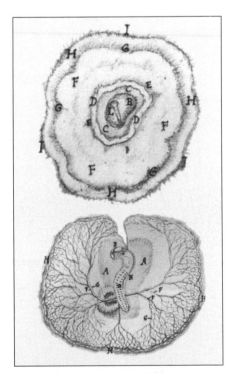

Fig. 1.4 Malpighi's description of the chick embryo. The figure shows Malpighi's drawings, made in 1673, depicting the early embryo (top), and at 2 days' incubation (bottom). His drawings accurately illustrate the shape and blood supply of the embryo. Reprinted by permission of the President and Council of the Royal Society.

Box 1A Basic stages of *Xenopus laevis* development.

Stage 2

Stage 6

Stage 8

Blastula

Stage 1

Egg

Animal pole

Sperm

Vegetal pole

Fertilization

Cleavage

Stage 10

Gastrula (section)

blastocoel

blastopore

Gastrulation

mesoderm

future notochord

endoderm

ectoderm

future gut

Stage 12

(section)

Adult

Stage 66

Metamorphosis

Free-swimming tadpole

Dorsal

Stage 45

Ventral

Organogenesis

Neurulation

notochord

Anterior

neural folds

Neurula

Stage 13

(surface removed to reveal notochord)

brain

spinal cord

notochord

Ventral

Dorsal

somites

yolk mass

Posterior

Anterior

Posterior

Stage 15

(dorsal view)

Tailbud stage embryo (dorsal view with surface removed)

Tailbud stage embryo (lateral view)

Stage 26

Although vertebrate development is very varied there are a number of basic stages that can be illustrated by following the development of the frog *Xenopus laevis*, which is a favorite organism for experimental embryology. The unfertilized egg is a large cell. It has a pigmented upper surface (the **animal pole**) and a lower region (the **vegetal pole**) characterized by an accumulation of yolk granules. So even at the beginning, the egg is not uniform; in subsequent development, cells from the animal half become the anterior (head) end of the embryo.

After fertilization of the egg by a sperm, and the fusion of male and female nuclei, **cleavage** begins. Cleavages are mitotic divisions in which cells do not grow between each division, and so with successive cleavages the cells become smaller. After about 12 division cycles the embryo, now known as a **blastula**, consists of many small cells surrounding a fluid-filled cavity (the **blastocoel**) above the larger yolky cells. Already, changes have occurred within the cells and they have interacted with each other so that some future tissue types—the **germ layers**—have become partly specified. The future **mesoderm** for example, which gives rise to muscle, cartilage, bone, and other internal organs like the heart, blood, and kidney is present in the blastula as an equatorial band. Adjacent to it is the future **endoderm**, which gives rise to the gut, lungs, and liver. The animal region will give rise to **ectoderm**, which forms both the epidermis and the nervous system. The future endoderm and mesoderm, which are destined to form internal organs, are still on the surface of the embryo. During the next stage—**gastrulation**—there is a dramatic rearrangement of cells; the endoderm and mesoderm move inside, and the basic body plan of the tadpole is established. Internally, the mesoderm gives rise to a rod-like structure (the **notochord**), which runs from the head to the tail, and lies centrally beneath the future nervous system. On either side of the notochord are segmented blocks of mesoderm called **somites**, which will give rise to the muscles and vertebral column, as well as the dermis of the skin.

Shortly after gastrulation the ectoderm above the notochord folds to form a tube (the **neural tube**), which gives rise to the brain and spinal cord—a process known as **neurulation**. By this time other organs, such as limbs, eyes, and gills, are specified at their future location, but only develop a little later during organogenesis. During organogenesis, specialized cells such as muscle, cartilage, and neurons differentiate. Within 48 hours the embryo has become a feeding tadpole with typical vertebrate features. Because the timing of each stage can vary, depending on environmental conditions, developmental stages in *Xenopus* and other embryos are often denoted by numbers rather than by hours of development.

1-2   Cell theory changed the conception of embryonic development and heredity.

The cell theory put forward by the German botanist Mathais Schleiden and the physiologist Theodor Schwann between 1838 and 1839 was one of the most illuminating advances in biology, and had an enormous impact. It was at last recognized that all living organisms consist of cells, which are the basic units of life, and which arise only by division from other cells. Multicellular organisms such as animals and plants could then be viewed as communities of cells. Development could not therefore be based on preformation but must be epigenetic because during development many new cells are generated by division from the egg, and new types of cell are formed. A crucial step forward in understanding development was the recognition in the 1840s that the egg itself is but a single, albeit specialized, cell.

An important advance was the proposal by the 19th century German biologist August Weismann that the offspring does not inherit its characteristics from the body (the soma) of the parent but only from the **germ cells**—egg and sperm—and that the germ cells are not influenced by the body that bears them. Weismann thus drew a fundamental distinction between germ cells and **somatic cells** or body cells (Fig. 1.6). Characteristics acquired by the body during an animal's life cannot be transmitted to the germ cells. As far as heredity is concerned the body is merely a carrier of germ cells. As the English novelist and essayist Samuel Butler put it: "A hen is only an egg's way of making another egg".

Work on sea urchin eggs showed that after fertilization the egg contains two nuclei, which eventually fuse; one of these nuclei belongs to the egg while the other comes from the sperm. Fertilization therefore results in

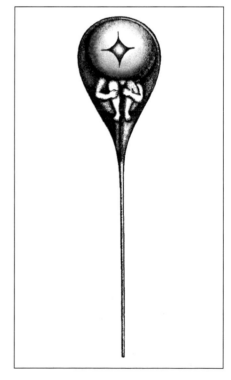

Fig. 1.5 Some preformationists believed that a homunculus was curled up in the head of each sperm. An imaginative drawing, after Nicholas Hartsoeker (1694).

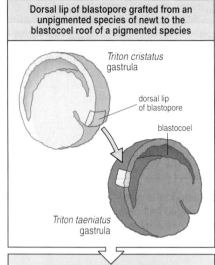

Dorsal lip of blastopore grafted from an unpigmented species of newt to the blastocoel roof of a pigmented species

Triton cristatus gastrula

dorsal lip of blastopore

blastocoel

Triton taeniatus gastrula

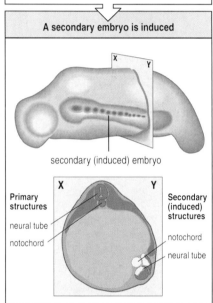

A secondary embryo is induced

X   Y

secondary (induced) embryo

X   Y

Primary structures

neural tube

notochord

Secondary (induced) structures

notochord

neural tube

Fig. 1.10 The dramatic demonstration by Spemann and Mangold of induction of a new main body axis by the organizer region in the early amphibian gastrula. A piece of tissue (yellow) from the dorsal lip of the blastopore of a newt (*Triton cristatus*) gastrula is grafted to the opposite side of a gastrula from another, pigmented, newt species (*Triton taeniatus*, pink). The grafted tissue induces a new body axis containing neural tube and somites. The unpigmented graft tissue forms a notochord at its new site (see section in lower panel) but the neural tube and the other structures of the new axis have been induced from the pigmented host tissue.

Driesch had completely separated the cells at the two-cell stage and obtained a normal but small larva. That was just the opposite of Roux's result and was the first clear demonstration of the developmental process known as **regulation**: the ability of the embryo to develop normally even when some portions are removed or rearranged. An explanation of Roux's experiment is given in Section 3-2.

1-4   The discovery of induction.

Although the concept of regulation implied that cells must interact with each other, the central importance of cell–cell interactions in embryonic development was not really established until the discovery of the phenomenon of **induction**, in which one tissue directs the development of another, neighboring, tissue.

The importance of induction and other cell–cell interactions in development was proved dramatically in 1924 when Spemann and his assistant Hilde Mangold carried out their famous organizer transplant experiment in amphibian embryos. They showed that a partial second embryo could be induced by grafting one small region of a newt embryo onto a new site on another embryo (Fig. 1.10). The grafted tissue was taken from the dorsal lip of the **blastopore**—the slit-like invagination that forms where gastrulation begins on the dorsal surface of the amphibian embryo (see Box 1A). This small region they called the **organizer**, since it seemed to be ultimately responsible for controlling the organization of a complete embryonic body. For their discovery, Spemann received the Nobel Prize for Physiology or Medicine in 1935, one of only two ever given for embryological research. Sadly, Hilde Mangold had died earlier, in an accident, and so could not be honored.

1-5   The coming together of genetics and development.

During much of the early part of this century there was little connection between embryology and genetics. When Mendel's laws were rediscovered in 1900 there was a great surge of interest in mechanisms of inheritance, particularly in relation to evolution, but less so in relation to development. Genetics was seen as the study of the transmission of hereditary elements from generation to generation, whereas embryology was the study of how an individual organism develops and in particular how cells in the early embryo became different from each other. Genetics seemed, in this respect, to be irrelevant to development.

An important concept that eventually helped to link genetics and embryology was the distinction between **genotype** and **phenotype**. This was first put forward by the Danish botanist Wilhelm Johannsen in 1909. The genetic endowment of an organism—the genetic information it acquires from its parents—is the genotype. Its visible appearance, internal structure, and biochemistry at any stage of development is the phenotype. While the genotype certainly controls development, environmental factors interacting with the genotype influence the phenotype. Despite having identical genotypes, identical twins can develop considerable differences in their phenotypes as they grow up (Fig. 1.11), and these tend to become more evident with age. The problem of development could now be posed in terms of the relationship between genotype and phenotype; how the genetic endowment becomes 'translated' or 'expressed' during development to give rise to a functioning organism.

The coming together of genetics and embryology was a slow and tortuous process. Little progress was made until the nature and function of the genes were much better understood. The discovery, in the 1940s, that genes encode proteins was a major turning point. Since it was already clear that the properties of a cell are determined by the proteins it contains, the fundamental role of genes in development could at last be appreciated. By controlling which proteins were made in a cell, genes could control the changes in cell properties and behavior that occurred during development.

Summary.

The science of embryonic development started with the Greeks more than 2000 years ago. Aristotle put forward the idea that embryos were not contained completely preformed in miniature within the egg, but that form and structure emerged gradually, as the embryo developed. This idea was challenged in the 17th and 18th centuries by those who believed in preformation, the idea that all embryos that had been, or ever would be, had existed from the beginning of the world. The emergence of the cell theory in the 1880s finally settled the issue in favor of epigenesis, and it was realized that the sperm and egg are single, albeit highly specialized, cells. Some of the earliest experiments showed that very early sea urchin embryos are able to regulate, that is to develop normally even if cells are removed or killed. This established the important principle that development must depend at least in part on communication between the cells of the embryo. Direct evidence for the importance of cell–cell interactions came from the organizer graft experiment carried out by Spemann and Mangold in 1924, showing that the cells of the amphibian organizer region could induce a new partial embryo from host tissue when transplanted into another embryo. The role of the genes in controlling development has only been fully appreciated in the past 30 years and the study of the genetic basis of development has been made much easier in recent times by the techniques of molecular biology.

Fig. 1.11 The difference between genotype and phenotype. These 'identical twins' have the same genotype because one fertilized egg split into two during development. Their slight difference in appearance is due to nongenetic factors, such as environmental influences. Photograph courtesy of José and Jaime Pascual.

A conceptual tool kit.

Development into a multicellular organism is the most complicated fate a single living cell can undergo; in this lies both the fascination and the challenge of developmental biology. Yet only a few basic principles are needed to start to make sense of developmental processes. The rest of this chapter is devoted to introducing these key concepts. These principles are encountered repeatedly throughout the book, as we look at different organisms and developmental systems, and should be regarded as a conceptual tool kit, essential for embarking on a study of development.

1-6   Development involves cell division, the emergence of pattern, change in form, cell differentiation, and growth.

Development is essentially the emergence of organized structures from an initially very simple group of cells. It is convenient to distinguish five main developmental processes, even though in reality they overlap with, and influence, one another considerably.

Fertilization is followed by a period of rapid cell division where the egg divides into a number of smaller cells (Fig. 1.12). These divisions are known as cleavage divisions and, unlike the cell divisions that take place

Fig. 1.12 Light micrograph of *Xenopus* eggs after four cell divisions. Scale bar = 1 mm. Photograph courtesy of J. Slack.

during cell proliferation and growth of a tissue, there is no increase in cell mass between each division. The cell cycle during cleavage consists simply of phases of DNA replication, mitosis, and cell division with no intervening stage of cell growth. The cleavage stage of embryogenesis thus rapidly divides the embryo into a number of cells, each containing a copy of the genome.

Pattern formation is the process by which a spatial and temporal pattern of cell activities is organized within the embryo so that a well-ordered structure develops. In the developing arm, for example, pattern formation is the process that enables cells to 'know' whether to make the upper arm or fingers, and where the muscles should form. There is no single universal strategy or mechanism of patterning; rather, it is achieved by a variety of cellular and molecular mechanisms in different organisms and at different stages of development.

Pattern formation initially involves laying down the overall **body plan** —defining the main **axes** of the embryo so that the anterior and posterior ends, and the dorsal and ventral sides of the body are specified. One can distinguish at least one main body axis in all multicellular organisms. In animals this refers to the axis that runs from 'head' to 'tail' (antero-posterior) and in plants from the growing tip to the roots. Many animals also have a distinguishable front and back, which defines another axis (dorso-ventral). A striking feature of these axes is that they are almost always at right angles to one another. The two axes may therefore be thought of as making up a coordinate system (Fig. 1.13).

The next stage in pattern formation in animal embryos is allocation of cells to the different **germ layers**—the ectoderm, mesoderm, and endoderm (Box 1B). During further pattern formation, cells of these germ layers acquire different identities so that organized spatial patterns of cell differentiation emerge, such as the arrangement of skin, muscle, and cartilage in developing limbs, and the arrangement of neurons in the nervous system. In the earliest stages of pattern formation, differences between cells are not easily detected and probably consist of subtle chemical differences caused by a change in activity of a very few genes.

The third important developmental process is change in form or morphogenesis. Embryos undergo remarkable changes in three-dimensional form—you need only look at your hands and feet. At certain stages in development, there are characteristic and dramatic changes in form, of which **gastrulation** is the most striking. Almost all animal embryos undergo gastrulation, during which the gut is formed and the main body plan emerges. During gastrulation, cells on the outside of the

Fig. 1.13 The main axes of a developing embryo. The antero-posterior axis and the dorso-ventral axis are at right angles to one another as in a coordinate system.

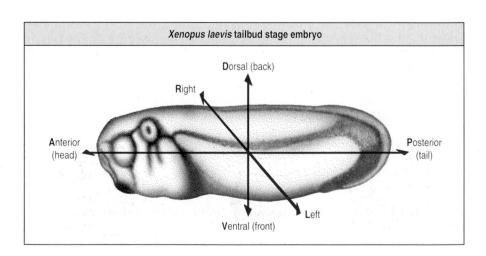

Box 1B Germ layers.

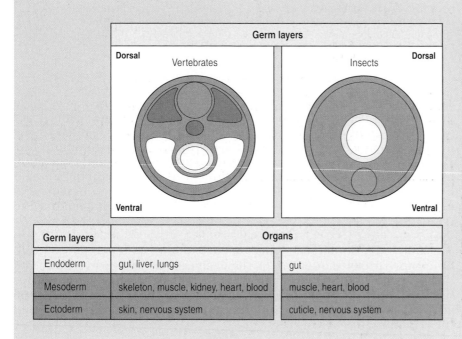

Germ layers	Organs	
Endoderm	gut, liver, lungs	gut
Mesoderm	skeleton, muscle, kidney, heart, blood	muscle, heart, blood
Ectoderm	skin, nervous system	cuticle, nervous system

The concept of germ layers is useful to distinguish between regions of the early embryo that give rise to quite distinct types of tissues. It applies to both vertebrates and invertebrates. All the animals considered in this book have three germ layers: the endoderm, which gives rise to the gut and its derivatives, such as the liver and lungs in vertebrates; the mesoderm, which gives rise to the skeleto-muscular system, connective tissues, and other internal organs such as the kidney and heart; and the ectoderm, which gives rise to the epidermis and nervous system. These are specified early in development. The boundaries between the different layers can be fuzzy and there are notable exceptions. The neural crest in vertebrates, for example, is ectodermal in origin but gives rise both to neural tissue and to some skeletal elements, which would normally be considered mesodermal in origin.

embryo move inwards and, in animals such as the sea urchin, gastrulation even transforms a hollow spherical blastula into a gastrula with a hole through the middle—the gut (Fig. 1.14). Morphogenesis in animal embryos can also involve extensive cell migration. Most of the cells of the human face, for example, are derived from cells that migrated from the **neural crest**, which originate on the back of the embryo.

The fourth developmental process we must consider here is **cell differentiation**, in which cells become structurally and functionally different from each other, ending up as distinct cell types, such as blood, muscle, or skin cells. Differentiation is a gradual process, cells often going through several divisions between the time they start differentiating and the time they are fully differentiated (when some cell types stop dividing altogether). In humans, the fertilized egg gives rise to at least 250 clearly distinguishable types of cell.

Pattern formation and cell differentiation are very closely interrelated, as we can see by considering the difference between human arms and legs. Both contain exactly the same types of cell—muscle, cartilage, bone, skin, and so on—yet the pattern in which they are arranged is clearly different. It is essentially pattern formation that makes us different from elephants and chimpanzees.

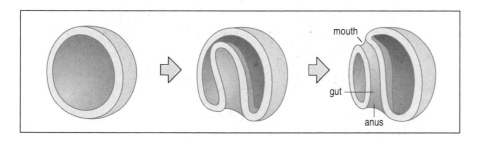

Fig. 1.14 Gastrulation in the sea urchin. Gastrulation transforms the spherical blastula into a structure with a hole through the middle, the gut. The left-hand side of the embryo has been removed.

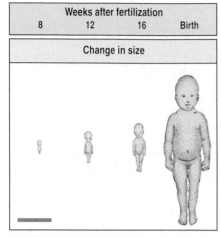

Weeks after fertilization

| 8 | 12 | 16 | Birth |

Change in size

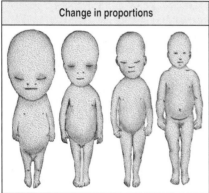

Change in proportions

Fig. 1.15 The human embryo changes shape as it grows. From the time the body plan is well established at 8 weeks until birth the embryo increases in length some ten-fold (upper panel), while the relative proportion of the head to the rest of the body decreases (lower panel). As a result, the shape of the embryo changes. Scale bar = 10 cm. After Moore, K.L.: 1983.

The fifth process is growth—the increase in size. In general there is little growth during early embryonic development and the basic pattern and form of the embryo is laid down on a small scale, always less than a few millimeters in extent. Subsequent growth can be brought about in a variety of ways: cell multiplication, increase in cell size, and deposition of extracellular materials such as bone and shell. Growth can also be morphogenetic. Differences in growth rates between organs, and parts of the body, can generate changes in the overall shape of the embryo (Fig. 1.15).

These five developmental processes are neither independent of each other nor strictly sequential. In very general terms, however, one can think of pattern formation in early development specifying differences between cells that lead to changes in form, cell differentiation, and growth. But in any real developing system there will be many twists and turns in this sequence of events.

1-7   Cell behavior provides the link between gene action and developmental processes.

Gene expression within cells is translated into embryonic development through the consequent properties and behavior of those cells. The main categories of cell behavior that concern us are changes in cell state (i.e. the pattern of gene activity), cell-to-cell signaling, changes in cell shape and cell movement, cell proliferation, and cell death.

Changing patterns of gene activity during early development are essential for pattern formation. They give cells identities that determine their future behavior and lead eventually to their final differentiation and, as we saw in the example of induction by the Spemann organizer, the capacity of cells to influence each other's fate by producing and responding to signals is crucial for development. By their movement or change in shape, cells generate the physical forces that bring about morphogenesis. The curvature of a sheet of cells into a tube, as happens in *Xenopus* and other vertebrates during formation of the neural tube, is the result of contractile forces generated by cells changing shape at certain positions within the cell layer (Fig. 1.16). Gastrulation in sea urchins, for example, first becomes visible as a small dimple on the surface of the embryo, which is due to a small group of cells contracting and pulling the surface inward. Cells also carry adhesion molecules on their surface, which hold them together as tissues and can also guide the migration of cells such as the neural crest cells of vertebrates, which leave the neural tube to form many structures elsewhere in the body.

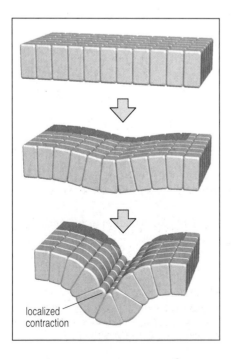

localized contraction

Fig. 1.16 Localized contraction of particular cells can cause a whole sheet of cells to fold. Contraction of a line of cells at their apices due to the contraction of cytoskeletal elements causes a furrow to form in a sheet of epidermis.

Later in development, growth involves cell proliferation, which can also influence final form, as parts of the body grow at different rates. Programmed cell death (**apoptosis**) is also an intrinsic part of the developmental process; cell death in developing hands and feet helps to form the fingers and toes from a continuous sheet of tissues. We can therefore describe and explain developmental processes in terms of how individual cells and groups of cells behave. Because the final structures generated by development are themselves composed of cells, explanations and descriptions at the cellular level can provide an account of how these adult structures are formed.

Since development can be understood at the cellular level, we can pose the question of how genes control development in a more precise form. We can now ask how genes are controlling cell behavior. The many possible ways in which a cell can behave therefore provides the link between gene activity and the morphology of the adult animal—the final outcome of development. Cell biology provides the means by which the genotype becomes translated into the phenotype.

1-8	**Genes control cell behavior by controlling which proteins are made by a cell.**

What a cell can do is determined very largely by the proteins present within it. The hemoglobin in red blood cells enables them to transport oxygen; skeletal muscle cells are able to contract because they contain an arrangement of the contractile apparatus of myosin, actin, tropomyosin, and other muscle-specific proteins. All these are very special or luxury proteins that are not involved in what are considered to be the housekeeping activities of the cell that keep it alive and functioning. Housekeeping activities are common to all cells and include the production of energy and all the intermediate biochemical pathways involved in the breakdown and synthesis of molecules necessary for the life of the cell. Although there are some variations, both qualitative and quantitative in housekeeping proteins in different cells, they are not important players in development. In development we are concerned primarily with those luxury or tissue-specific proteins that make cells different from one another.

Genes control development mainly by determining which proteins are made in which cells and when. In this sense they are passive participants in the process, compared with the proteins for which they code, which directly determine cell behavior. Whether a particular protein is synthesized in a cell requires that its gene is switched on—that the gene is being **transcribed** into RNA. The RNA is eventually **translated** into protein, but this does not necessarily follow automatically, because protein production can be controlled at some later stage in gene expression. Fig. 1.17 shows the main stages in gene expression at which the production of a protein can be controlled. For example, **messenger RNA (mRNA)** may be degraded before it can be exported from the nucleus. Even if it reaches the cytoplasm its translation may be inhibited there. In the eggs of many animals, pre-formed mRNA is prevented from being translated until after fertilization. Even if a gene has been transcribed and the mRNA translated, the protein may not yet be able to function. Many newly synthesized proteins require further **post-translational modification** before they acquire biological activity, as in the digestive enzymes trypsin and chymotrypsin, where the protein chain is cut and fragments of the protein removed. But, as a rough guide, if a gene is being transcribed,

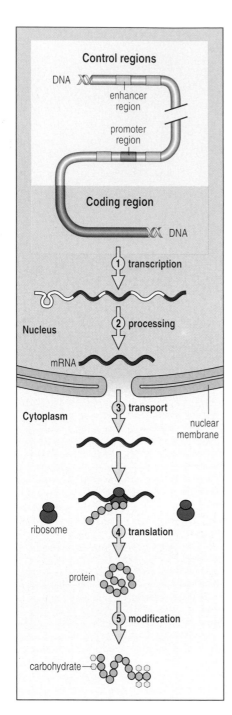

Fig. 1.17 Gene expression and protein synthesis. A protein-coding gene comprises a stretch of DNA that contains a coding region, which contains the instructions for making the protein, and adjacent control regions—promoter and enhancer regions—at which the gene is switched on or off. The promoter region is the site at which RNA polymerase binds and starts transcribing. The enhancer regions may be thousands of base pairs distant from the promoter. Transcription of the gene into RNA (1) may be either stimulated or inhibited by transcription factors that bind to promoter and enhancer regions. The RNA formed by transcription is spliced to remove introns (yellow) and processed within the nucleus (2) to produce mRNA that is exported to the cytoplasm (3) for translation into protein at the ribosomes (4). Control of gene expression and protein synthesis occurs mainly at the level of transcription but can also occur at later stages. For example, mRNA may be degraded before it can be translated. If it is not translated immediately it may be stored in inactive form in the cytoplasm for translation at some later stage. Some proteins require post-translational modification (5) to become biologically active.

it is reasonable to expect the protein for which it codes to be present in the cell. Untranslated sequences in RNA can also play an important regulatory role. Also, some genes, such as those for ribosomal RNA, do not code for proteins.

Both housekeeping and luxury genes encode proteins that are involved directly in cell function, i.e. enzymes, receptor proteins, growth factors, and the structural components of cells. Of particular importance in development are genes that encode **transcription factors** or gene regulatory proteins, proteins that are involved in activating or repressing transcription. Transcription factors act by binding to the control regions of genes (see Fig. 1.17) or by interacting with other DNA-binding proteins.

An intriguing question is how many genes out of the total genome are involved in the development of an embryo? This is not an easy estimate to make. In a few particularly well-studied cases we have rough estimates of the number of genes involved in a particular aspect of development. In the early development of *Drosophila* about 60 genes are known to be involved in pattern formation up to the time of segmentation; that is, when the embryo becomes divided into segments. In the nematode worm *Caenorhabditis*, some 50 genes are need to specify a small reproductive structure known as the vulva. Indirect evidence from mutations affecting limb development in humans and mice suggests that perhaps 100 genes may be involved in development of a vertebrate limb. All these are quite small numbers compared to the thousands of housekeeping genes that are active at the same time; these are essential to development in that they are necessary for maintaining life, but provide no information for influencing pattern formation. For flies and vertebrates, a guess at the total number of developmental genes could range from 1000 to 50,000, respectively. This number can be compared to the 80,000 genes thought to be present in the mammalian genome.

| 1-9 | **Differential gene activity controls development.** |

All the somatic cells in an embryo are derived from the fertilized egg by successive rounds of mitotic cell division. With rare exceptions, they all therefore contain identical genetic information, the same as the zygote. The differences between cells must therefore be generated by differences in gene activity. Turning the correct genes on or off in the correct cells at the correct time becomes the central issue in development.

As all the key steps in development reflect changes in gene activity, one might be tempted to think of development simply in terms of mechanisms for controlling gene expression. But this would be highly misleading. For gene expression is only the first step in a cascade of cellular processes that change cell behavior and so direct the course of embryonic development. To think only in terms of genes is to ignore crucial aspects of cell biology, such as change in cell shape, that may be initiated at several stages removed from gene activity. In fact, there are very few cases established where the complete sequence of events from gene expression to altered cell behavior has been worked out. The route leading from the genes to a structure such as the five-fingered hand may be tortuous.

| 1-10 | **Development is progressive and the fate of cells becomes determined at different times.** |

As embryonic development proceeds, the organizational complexity of the embryo becomes vastly increased over that of the fertilized egg. Many different cell types are formed, spatial patterns emerge, and there are major changes in shape. All this occurs more or less gradually, depending on the particular organism. But, in general, the embryo is first divided up into a few broad regions, such as the future germ layers (mesoderm, ectoderm, and endoderm). Subsequently the cells within these regions have their fates more and more finely determined. Mesoderm, for example, becomes differentiated into muscle cells, cartilage cells, bone cells, the fibroblasts of connective tissue, and the cells of the dermis. **Determination** implies a stable change in the internal state of a cell, and an alteration in the pattern of gene activity is assumed to be the initial step.

It is important to understand clearly the distinction between the normal fate of a cell at any particular stage, and its state of determination. The **fate** of a group of cells merely describes what they will normally develop into. By marking cells of the early embryo one can find out, for example, which ectodermal cells will normally give rise to the nervous system, and of those, which to the retina in particular. However, that in no way implies that those cells can only develop into a retina or are already determined or committed to doing so.

A group of cells is called **specified** if, when isolated and cultured in the neutral environment of a simple culture medium away from the embryo, they develop more or less according to their normal fate (Fig. 1.18). For example, cells at the animal pole of the amphibian blastula are specified to form ectoderm, and will form epidermis when isolated. Cells that are specified in this technical sense need not yet be determined, for influences from other cells can change their normal fate; if tissue from the animal pole is put in contact with cells from the vegetal pole it will form mesoderm instead of epidermis. At a later stage of development, however, the cells in the animal region have become determined as ectoderm and their fate cannot then be altered. Tests for specification rely on culturing the tissue in a 'neutral' environment lacking any inducing signals, and this is often difficult to achieve.

The state of determination of cells at any particular stage can be demonstrated by transplantation experiments. At the blastula stage of the amphibian embryo, one can graft the ectodermal cells that give rise to the eye into the side of the body and show that the cells develop

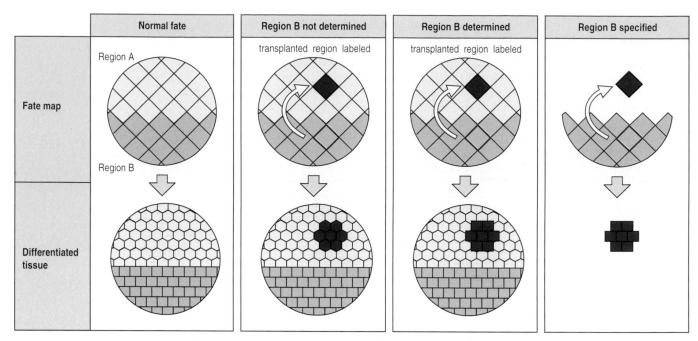

Normal fate	Region B not determined	Region B determined	Region B specified

Fate map

Region A

Region B

transplanted region labeled

transplanted region labeled

Differentiated tissue

Fig. 1.18 The distinction between cell fate, determination, and specification. In this idealized system, regions A and B differentiate into two different sorts of cell, depicted as hexagons and squares. The fate map (first panel) shows how they would normally develop. If cells from region B are grafted into region A and now develop as A-type cells, the fate of region B has not yet been determined (second panel). By contrast, if region B cells are already determined when they are grafted to region A, they will develop as B cells (third panel). Even if B cells are not determined, they may be specified, in that they will form B cells when cultured in isolation from the rest of the embryo (fourth panel).

according to their new position; that is, into mesodermal cells like those of the notochord and somites (Fig. 1.19). At this early stage, their potential for development is much greater than their normal fate. However, if the same operation is done at a later stage, then the future eye region will form structures typical of an eye. At the earlier stage the cells were not yet determined as eye cells, whereas later they had become so.

It is a general feature of development that cells in the early embryo are less narrowly determined than those at later stages; with time, cells become more and more restricted in their developmental potential. We assume that determination involves a change in which genes are being expressed by the cell and that this change fixes or restricts the cell's fate, thus reducing its developmental options.

We have already seen how, even at the two-cell stage, the cells of the sea urchin embryo do not seem to be determined. Each thus has the potential to generate a whole new larva (see Section 1-3). Embryos like these, where the potential of cells is much greater than that indicated by their normal fate, are termed **regulative**. Vertebrate embryos are among those capable of considerable regulation. In contrast, those embryos where, from a very early stage, the cells can develop only according to their early fate are termed mosaic. As we saw in Section 1-3, this terminology has a long history. It describes eggs and embryos that develop as if their pattern of future development is laid down very early, even in the egg, as a 'mosaic' of different molecules. The different parts of the embryo then develop quite independently of each other. In such embryos, cell interactions may be quite limited. The demarcation between these two strategies of development is not always sharp, and in part reflects the time when determination occurs; it occurs much earlier in mosaic systems.

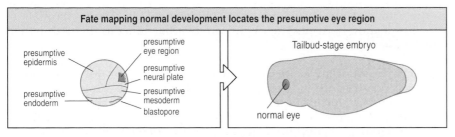

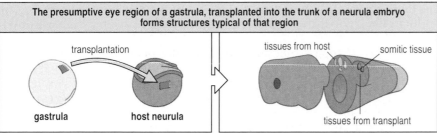

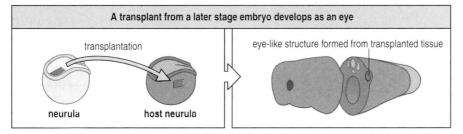

Fig. 1.19 Determination of the eye region with time in amphibian development. If the region of the gastrula that will normally give rise to an eye is grafted into the trunk region of a neurula (middle panel), the graft forms structures typical of its new location, such as notochord and somites. If, however, the eye region from a neurula is grafted into the same site (bottom panel), it develops as an eye-like structure, since at this later stage it has become determined.

The difference between regulative and mosaic embryos reflects the relative importance of cell–cell interactions in each system. The occurrence of regulation absolutely requires interactions between cells, for how else could normal development take place and deficiencies be recognized and restored? A truly mosaic embryo, however, would in principle not require such interactions. No purely mosaic embryos are known to exist.

1-11 Inductive interactions can make cells different from each other.

Making cells different from one another is central to development. There are numerous examples in development where a signal from one group of cells influences the development of an adjacent group of cells. This is known as **induction**, and the classic example is the action of the Spemann organizer in amphibians (see Section 1-4). Inducing signals may be propagated over several or even many cells, or be highly localized. The inducing signal from the amphibian organizer affects many cells, whereas other inducing signals may pass from one cell to its immediate neighbor.

There are three main ways in which inducing signals may be passed between cells (Fig. 1.20). First, the signal is transmitted through the extracellular space, usually by means of a secreted diffusible molecule. Second, cells may interact directly with each other by means of molecules located on their surface. In both these cases, the signal is generally received by receptor proteins in the cell membrane and is subsequently relayed through intracellular signaling systems to produce the eventual cellular response. Third, the signal may pass from cell to cell directly through gap junctions. These are specialized protein pores

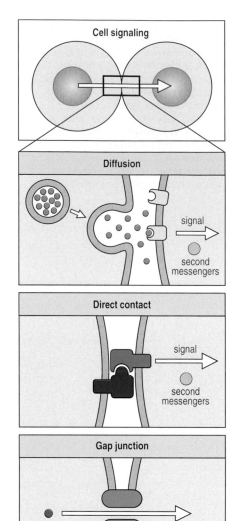

Fig. 1.20 An inducing signal can be transmitted from one cell to another in three main ways. The signal can be a diffusible molecule, which interacts with a receptor on the target cell surface (top panels), or the signal can be produced by direct contact between two complementary proteins at the cell surfaces (middle panel). If the signal involves small molecules it may pass directly from cell to cell through gap junctions in the plasma membrane (bottom panel).

in the apposed plasma membranes, which provide direct channels of communication between the cytoplasm of adjacent cells through which small molecules can pass.

One further important feature of induction is whether or not the responding cell is **competent** to respond to the inducing signal. This competence may depend on, for example, the presence of the appropriate receptor and transducing mechanism, or on the presence of the particular transcription factors needed for gene activation. A cell's competence for a particular response can change with time; the Spemann organizer can induce changes in the cells it affects only during a restricted time window.

In embryos, it seems that small is generally beautiful where signaling and pattern formation are concerned. Whenever a pattern is being specified, the size of the group of cells involved is barely, if ever, greater than 0.5 mm in any direction; that is, some 50 cell diameters. Many patterns are specified on a much smaller scale and involve just tens or a few hundred cells. This means that the inducing signals involved in pattern formation reach over distances of the order of only ten times a cell diameter. The final organism may be very big, but this is almost entirely due to growth of the basic pattern.

1-12   The response to inductive signals depends on the state of the cell.

Inductive signals can alter how the induced cells develop. They can thus be regarded as providing the cells with instructions as how to behave. It is important to realize that the response to inductive signals is entirely dependent on the current state of the cell. It not only has to be competent to respond but the number of possible responses is usually very limited. An inductive signal can only select one response from a small number of possible cellular responses. All inductions and signals are not really instructive but are essentially selective. A truly instructive signal would be one that provided the cell with entirely new information and capabilities, by providing it with, for example, new DNA or proteins, which is not thought to occur in development.

As an analogy, consider a juke-box containing a hundred records. When one selects a record to be played, the machine has not been given any new information; rather, one of its repertoire of records has simply been selected. It would be quite different if a new record were added. That would be to provide new information, and would be equivalent to introducing a completely new gene or protein into a cell, which rarely occurs in development. Like the juke-box, a cell's behavior can be changed only by external signals within the constraints provided by its current state.

Because signals are essentially selective and depend on the state of the cell, different signals can activate a particular gene at different stages of development. Genes can be turned on and off repeatedly during development.

The emphasis on instructive signals acting by selection has a further important implication for biological economy. The same signal can be used to elicit different responses in different cells. A particular signaling molecule, for example, can act on several types of cell, evoking a characteristic and different response from each, depending on their developmental history. As we see in future chapters, evolution has been quite lazy with respect to this aspect of development; once a suitable set of signaling molecules has been assembled, its members are used again and again.

1-13 Patterning can involve the interpretation of positional information.

One general mode of pattern formation can be illustrated by considering the patterning of a simple nonbiological model—the French flag (Fig. 1.21). The French flag has a simple pattern: one third blue, one third white, and one third red, along just one axis. Moreover, the flag comes in many sizes but always the same pattern, and thus can be thought of as mimicking the capacity of an embryo to regulate. Given a line of cells, any one of which can be blue, white, or red, and given also that the line of cells can be of variable length, what sort of mechanism is required for the line to develop the pattern of a French flag?

One solution is for the cells to acquire **positional information**. That is, the cells acquire an identity or **positional value** that is related to their position along the line with respect to the boundaries at either end. After they have acquired their positional values, the cells interpret this information by differentiating according to their genetic program. Those in the left-hand third of the line will become blue, those in the middle third white, and so on.

Pattern formation using positional information implies at least two distinct stages: first the positional value has to be specified with respect to some boundary and then it has to be interpreted. The separation of these two processes has an important implication: it means that there need be no set relation between the positional values and how they are interpreted. In different circumstances, the same set of positional values could be used to generate the Italian flag or another pattern. How positional values will be interpreted will depend on the particular genetic instructions active in the group of cells and will be influenced by their developmental history.

Cells could have their position specified by a variety of mechanisms. The simplest is based on a gradient of some substance. If the concentration of some chemical decreases from one end of a line of cells to the other, then the concentration of that chemical in any cell along the line effectively specifies the position of the cell with respect to the boundary (Fig. 1.22). A chemical whose concentration varies, and which is involved in pattern formation, is called a **morphogen**. In the case of the French flag, we assume a source of morphogen at one end and a sink at the other, and that the concentrations of morphogen at both ends are kept constant but are different from each other. Then, as the morphogen diffuses down the line, its concentration at any point effectively provides positional information. If the cells can respond to **threshold concentrations** of the morphogen—for example, above a particular concentration the cells develop as blue, while below this concentration they become white, and at yet another

Fig. 1.21 The French flag.

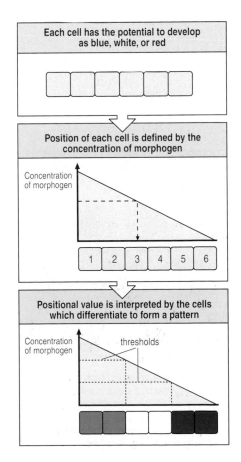

Fig. 1.22 The French flag model of pattern formation. Each cell in a line of cells has the potential to develop as blue, white, or red. The line of cells is exposed to a concentration gradient of some substance and each cell acquires a positional value defined by the concentration at that point. Each cell then interprets the positional value it has acquired and differentiates into blue, white, or red, according to a predetermined genetic program, thus forming the French flag pattern. Substances that can direct the development of cells in this way are known as morphogens. The basic requirements of such a system are that the concentration of substance at either end of the gradient must remain different from each other but constant, thus fixing boundaries to the system. Each cell must also contain the necessary information to interpret the positional values. Interpretation of the positional value is based upon different threshold responses to different concentrations of morphogen.

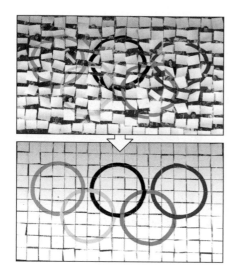

Fig. 1.23 Positional information could be used to generate an enormous variety of patterns. A good example, as shown here, is where the people seated in a stadium each have a position defined by their row and seat numbers. Each position has an instruction about which colored card to hold up, and this makes the pattern. If the instructions were changed, a different pattern would be formed.

lower concentration they become red—the line of cells will develop as a French flag (see Fig. 1.22). Thresholds can represent the amount of morphogen that must bind to receptors to activate an intracellular signaling system, or concentrations of transcription factors required to activate particular genes. The use of threshold concentrations of transcription factors to specify position is most beautifully illustrated in the early development of *Drosophila*, as we shall see in Chapter 5.

The French flag model, simple though it is, illustrates two important features of development in the real world. The first is that, even if the length of the line varies, the pattern will still form correctly, given that the boundaries of the system are properly defined by keeping the different concentrations of morphogen constant at either end. The second is that the system could also regenerate the complete original pattern if it were cut in half, provided that the boundary concentrations were reestablished. It is therefore truly regulative. We have discussed it here as a one-dimensional patterning problem, but the model can easily be extended to provide two-dimensional patterning (Fig. 1.23).

1-14   Lateral inhibition can generate spacing patterns.

Many structures, such as the feathers on the skin of a bird, are found to be more or less regularly spaced with respect to one another. Such spacing could occur by the mechanism of **lateral inhibition** (Fig. 1.24). Given a group of cells that all have the potential to differentiate in a particular way, for example as feathers, it is possible to regularly space the cells that will form feathers by a mechanism in which cells that begin to form feathers inhibit the adjacent cells from doing so. This is reminiscent of the spacing of trees in a forest being caused by competition for sunlight and nutrients. In embryos, lateral inhibition is often the result of the differentiating cell secreting an inhibitory molecule that acts locally on the nearest cell neighbors to prevent them developing in a similar way.

1-15   Localization of cytoplasmic determinants and asymmetric cell division can make cells different from each other.

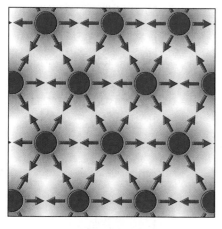

Fig. 1.24 Lateral inhibition can give a spacing pattern. If developing structures produce an inhibitor which prevents the formation of any similar structures in the area adjacent to them, the structures may become evenly spaced.

Positional specification is just one way in which cells can be given a particular identity. A separate mechanism is based on **cytoplasmic localization** and **asymmetric cell division** (Fig. 1.25). Asymmetric divisions are so called because they result in daughter cells having properties different from each other, independently of any environmental influence. The properties of such cells therefore depend on their **lineage** or line of descent, and not on environmental cues. Although some asymmetric cell divisions are unequal divisions in that they produce cells of different sizes, this is not usually the most important feature in animals; it is the unequal distribution of cytoplasmic factors that makes the division asymmetric. An alternative way of making the French flag pattern from the egg would be to have chemical differences (representing blue, white, and red) distributed in the egg in the form of determinants that foreshadowed the French flag. When the egg underwent cleavage, these cytoplasmic determinants would become distributed amongst the cells in a particular way and a French flag would develop. This would require no interactions between the cells, which would have their fates determined from the beginning.

Although such extreme examples of mosaic development are not known in nature, there are many cases where eggs or cells divide so that some cytoplasmic determinant becomes unequally distributed between the two daughter cells and they develop differently. This happens at the first cleavage of the nematode egg, and defines the antero-posterior axis of the embryo. The germ cells of *Drosophila* are also specified by cytoplasmic determinants, in this case contained in the cytoplasm located at the posterior end of the egg. However, daughter cells most often become different because of signals from other cells or from their extracellular environment, rather than because of the unequal distribution of cytoplasmic determinants.

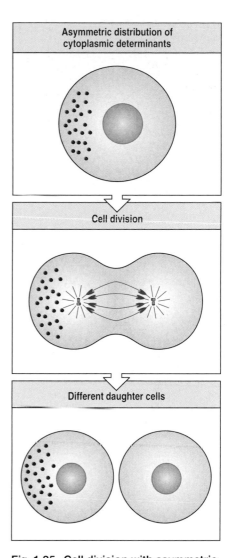

Fig. 1.25 Cell division with asymmetric distribution of cytoplasmic determinants. If a particular molecule is distributed unevenly in the parent cell, cell division will result in its being shared unequally between the cytoplasm of the two daughter cells. The more localized the cytoplasmic determinant is in the parental cell, the more likely it will be that one daughter cell will receive all of it and the other none, thus producing a distinct difference between them.

| 1-16 | **The embryo contains a generative rather than a descriptive program.** |

All the information for embryonic development is contained within the fertilized egg. So how is this information interpreted to give rise to an embryo? One possibility is that the structure of the organism is somehow encoded as a descriptive program in the genome. Does the DNA contain a full description of the organism to which it will give rise? The answer is no. The genome contains instead a program of instructions for making the organism—a generative program—in which the cytoplasmic constituents of eggs and cells are essential players along with the genes like the DNA coding for the sequence of amino acids in a protein.

A descriptive program, like a blueprint or a plan, describes an object in some detail, whereas a generative program describes how to make an object. For the same object the programs are very different. Consider origami, the art of paper folding. By folding a piece of paper in various directions it is quite easy to make a paper hat or a bird from a single sheet. To describe in any detail the final form of the paper with the complex relationships between its parts is really very difficult, and not of much help in explaining how to achieve it. Much more useful and easier to formulate are instructions on how to fold the paper. The reason for this is that simple instructions about folding have complex spatial consequences. In development, gene action similarly sets in motion a sequence of events that can bring about profound changes in the embryo. One can thus think of the genetic information in the fertilized egg as equivalent to the folding instructions in origami; both contain a generative program for making a particular structure.

A further distinction can be made between the embryo's genetic program and the developmental program of a particular cell or group of cells. The genetic program refers to the totality of information provided by the genes, whereas a developmental program may refer only to that part of the genetic program that is controlling a particular group of cells. As the embryo develops, different parts acquire their own developmental programs as a result of cell–cell interactions and the activities of selected sets of genes. Each cell in the embryo thus has its own developmental program, which may change as development proceeds.

| | **Summary.** |

Development results from the coordinated behavior of cells. The major processes involved in development are cell division, pattern formation, morphogenesis or change in form, cell differentiation, cell migration, cell death, and growth. Genes control cell behavior, and thus cell biology provides the link between gene action and developmental processes. During development, cells undergo changes in the genes they express, in their shape, in the signals they produce and respond to, in their rate of proliferation, and in their migratory behavior. All these aspects of cell behavior are controlled largely by the presence of specific proteins;

gene activity controls which proteins are made. Since the somatic cells in the embryo generally contain the same genetic information, the changes that occur in development are controlled by the differential activity of selected sets of genes in different groups of cells. Development is progressive and the fate of cells becomes determined at different times. The potential for development of cells in the early embryo is usually much greater than their normal fate, but this potential becomes more restricted as development proceeds. Inducing interactions, involving signals from one tissue or cell to another, are one of the main ways of changing cell fate and directing development. Asymmetric cell divisions, in which cytoplasmic components are unequally distributed to daughter cells, can also make cells different. One widespread means of pattern generation is through positional information; cells first acquire a positional value with respect to boundaries and then interpret their positional values by behaving in different ways. Developmental signals are more selective than instructive, choosing one or other of the developmental pathways open to the cell at that time. The embryo contains a generative not a descriptive program—it is more like the instructions for making a structure by paper-folding than a blueprint.

Summary to Chapter 1.

All the information for embryonic development is contained within the fertilized egg—the diploid zygote. The genome of the zygote contains a program of instructions for making the organism. In the working out of this developmental program, the cytoplasmic constituents of the egg and of the cells it gives rise to are essential players along with the genes. Highly regulated gene activity directs a sequence of cellular events that bring about the profound changes that occur in the embryo during development. The major processes involved in development are cell division, pattern formation, morphogenesis, cell differentiation, cell migration, cell death, and growth. All of these processes can be influenced by communication between the cells of the embryo. Development is progressive, with the fate of cells becoming more precisely specified as development proceeds. Cells in the early embryo usually have a much greater potential for development than is evident from their normal fate, and this enables embryos to develop normally even if cells are removed, added, or transplanted to different positions. Cells' developmental potential becomes much more restricted as development proceeds.

References.

The origins of developmental biology.

Cole, F.J.: *Early Theories of Sexual Generation.* Oxford: Clarendon Press, 1930.

Hamburger, V.: *The Heritage of Experimental Embryology: Hans Spemann and the Organizer.* New York: Oxford University Press, 1988.

Needham, J.: *A History of Embryology.* Cambridge: Cambridge University Press, 1959.

Sander, K.: **"Mosaic work" and "assimilating effects" in embryogenesis: Wilhelm Roux's conclusions after disabling frog blastomeres.** *Roux's Arch. Dev. Biol.* 1991, **200**:237–239.

Sander, K.: **Shaking a concept: Hans Driesch and the varied fates of sea urchin blastomeres.** *Roux's Arch. Dev. Biol.* 1992, **201**:265–267.

Wilson, E.B.: *The Cell in Development and Heredity.* New York: Macmillan, 1896.

Wolpert, L.: **Evolution of the cell theory.** *Phil. Trans. Roy. Soc. Lond. B* 1995, **349**:227–233.

A conceptual tool kit.

Roberts, K., *et al.*: *Essential Cell Biology: An Introduction to Molecular Biology of the Cell.* New York: Garland Publishing, 1998.

Wolpert, L.: **Do we understand development?** *Science* 1994, **266**:571–572.

Wolpert, L.: **One hundred years of positional information.** *Trends Genet.* 1996, **12**:359–364.

Wolpert, L.: *The Triumph of the Embryo.* Oxford: Oxford University Press, 1991.

Model Systems

2

Model organisms: vertebrates.

Model organisms: invertebrates.

Model systems: plants.

Identifying developmental genes.

*"In order to understand how we build things,
you must first just watch us do it ."*

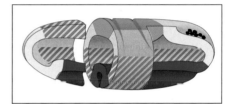

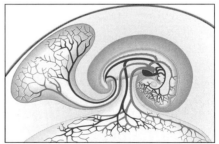

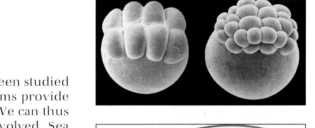

Although the development of a wide variety of species has been studied at one time or another, a relatively small number of organisms provide most of our knowledge about developmental mechanisms. We can thus regard them as models for understanding the processes involved. Sea urchins and frogs were the main animals used for the first experimental investigations at the beginning of the century (see Chapter 1) because their developing embryos are both easy to obtain and, in the case of the frog, sufficiently large and robust for relatively easy experimental manipulation, even at quite late stages. Among vertebrates, the frog *Xenopus*, the mouse, the chick, and more recently the zebrafish, are the main model systems now studied. Among invertebrates, the fruit fly *Drosophila* and the nematode worm *Caenorhabditis elegans* are presently the focus of most attention because a great deal is known about their developmental genetics and they can also be deliberately genetically modified.

The reasons for these choices are partly historical—once a certain amount of research has been done on one animal it is more efficient to continue to study it rather than start at the beginning again with another species—and partly a question of ease of study and biological interest. Each species has its advantages and disadvantages as a developmental model. The chick embryo, for example, has long been studied as an example of vertebrate development because fertile eggs are easily available, the embryo withstands experimental microsurgical manipulation very well, and it can be cultured outside the egg. A disadvantage, however, is that little is known about the chick's developmental genetics. However, we know a great deal about the genetics of the mouse, although the mouse is more difficult to study in some ways, as development takes place entirely within the mother. Many developmental mutations have been identified in the mouse, and it is also amenable to genetic modification by transgenic techniques (see Box 3A, page 64). It is also the best experimental model we have for studying mammalian development, including that of humans. The zebrafish (*Brachydanio rerio*) is a very recent addition to the select list of vertebrate model systems; it is easy to breed in large numbers, the embryos are transparent and so cell divisions and tissue movements can be followed visually, and it has great potential for genetic investigations.

Although the genetics of the fruit fly *Drosophila melanogaster* have been studied since the beginning of this century, research on its development has come into its own only in relatively recent times, with the advent of molecular biological techniques that can exploit the immense fund of knowledge of fruit fly genetics. The nematode worm's prominence as a model developmental system is even more recent; it relies on the simplicity of the organism—the embryo has fewer than 1000 cells (not counting the germ cells)—the ability to follow development cell by cell in the transparent embryo, and its invariant cell lineages. The ancestry of every cell in a nematode can be traced back to the zygote through a series of invariant cell divisions. In addition, the nematode is amenable to genetic analysis and genetic modification.

Among plants, the small crucifer *Arabidopsis thaliana* is playing an increasingly important role as a developmental model for flowering plants, especially in relation to the genetic basis of plant development.

Although this book focuses mainly on the organisms covered in this chapter, and mainly on animal models, there are many other animals and plants whose development is of great interest and has been studied in some detail; some of these, for example mollusks and tunicates, are considered in later chapters. By comparing developmental mechanisms in a variety of organisms it becomes possible to identify those mechanisms that are conserved; that is, mechanisms that are used in different groups of organisms. Comparative studies suggest that, despite immense differences in the details, it is likely that the most basic mechanisms of development are similar in all animals and are derived from the earliest animal ancestors. Thus, elucidation of a developmental process in one animal is often of great help in understanding development in another.

Before embarking on a consideration of the mechanisms of embryonic development, one must first be familiar with the stages that embryos pass through. It is essential to understand clearly how their structure, or morphology, changes during development to give rise to the larval or adult form. This chapter takes you through the main developmental stages of a number of model systems, and introduces some essential terminology. Although this aspect of development is largely descriptive, its importance cannot be underestimated.

The developmental cycles of the main model organisms are dealt with only in outline here; the details and mechanisms of developmental processes will be covered in later chapters. Here, you will find background information and terminology that will be needed when developmental mechanisms are considered later, so regard this chapter also as one to refer back to. Since a major aim of developmental biology is to identify those genes that control development, we finally look briefly at some basic strategies for obtaining and screening developmental mutants.

Model organisms: vertebrates.

All vertebrate embryos pass through a similar set of developmental stages. After fertilization, the zygote undergoes cleavage, during which the embryo divides into a number of smaller cells without any increase in overall mass. This is followed by gastrulation, in which cell movements result in the germ layers (see Box 1B, page 11) moving into the correct places for further development. At the end of gastrulation, the **ectoderm** covers the embryo, and the **mesoderm** and **endoderm** have moved inside. One of the earliest mesodermal structures that can be recognized is is the rod-shaped notochord, which forms along the antero-posterior axis of the body. This later becomes incorporated into the vertebral column. The vertebral column and the muscles of the trunk and limbs develop from blocks of mesodermal tissue—somites—that form in an antero-posterior sequence on either side of the notochord. All vertebrates form these structures, although the details of their earlier development differ between the vertebrate classes. The brain and spinal cord are derived from ectoderm directly above the notochord that forms the neural tube. The vertebrate 'body plan' is illustrated in Fig. 2.1.

Fig. 2.2 shows the differences in form and shape of a range of early vertebrate embryos. Most of the apparent differences in the development of different vertebrates up to and including gastrulation are related to the nutrition of the embryo. The overall form of the embryo is affected by the amount of yolk in the egg, and the structures the embryo has to develop to utilize this nutrient source. In the case of mammals, whose eggs have no yolk, the extra-embryonic structures of the placenta have to develop. But after gastrulation, all vertebrate embryos pass through a **phylotypic stage**—at which they all more or less resemble each other (see Fig. 2.2) and show the specific features of chordate embryos such as the notochord, somites, and neural tube.

Before describing the development of some model vertebrates, a general point must be made in relation to the staging of development. Amphibians,

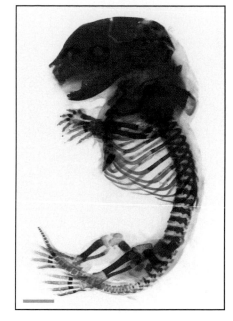

Fig. 2.1 The skeleton of a mouse embryo illustrates the vertebrate body plan. The skeletal elements in this embryo have been stained with a dye. The vertebral column, which develops from blocks of somites, is divided up into cervical (neck), thoracic (chest), lumbar (waist), and sacral (hip and lower) regions. The paired limbs can also be seen. Scale bar = 1mm. Photograph courtesy of M. Maden.

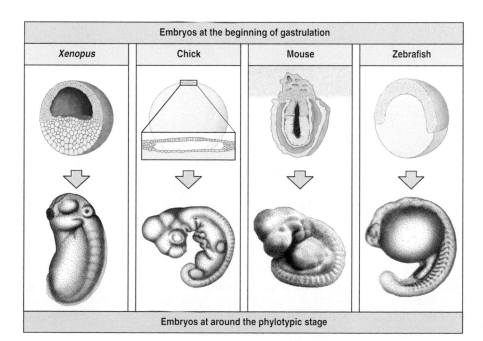

Embryos at the beginning of gastrulation

| Xenopus | Chick | Mouse | Zebrafish |

Embryos at around the phylotypic stage

Fig. 2.2 Vertebrate embryos go through a similar phylotypic stage, but the embryos show considerable differences in form before gastrulation. The top row shows representative vertebrate embryos, in cross-section, at the stage corresponding roughly to the *Xenopus* blastula (left panel) just before gastrulation commences. The main determinant of tissue arrangement is the amount of yolk (yellow) in the egg. The mouse embryo (third from left) at this stage has implanted into the uterine wall and thus has already developed some extra-embryonic tissues required for implantation. The mouse embryo proper is a small cup-shaped blastoderm at the center of these structures, seen here in cross-section as a U-shaped epithelial layer. The bottom row shows all of these embryos after gastrulation, at around the phylotypic stage, when they all more-or-less resemble each other and show the characteristic vertebrate features.

for example, can develop quite normally over a range of temperatures, but the rate of development changes considerably at different temperatures. Therefore one needs a way of charting development by stages rather than by the time after fertilization. For amphibians, this is provided by tables describing normal development, in which different developmental stages are identified by their main features and given a number. A stage 10 *Xenopus* embryo, for example, refers to an embryo at a very early stage of gastrulation. Similar tables of normal development describe stages in the chick embryo; they are required here since one does not always know the time of fertilization or when the fertilized egg has been placed in the incubator. For mouse embryos, which develop in a much more constant environment, staging is again by reference to the structure of the embryo; somite number is often used as an indication of developmental stage. For earlier stages of mouse development, before the somites have formed, time is often expressed as days *post coitum*, that is, days after mating.

2-1   Amphibians: *Xenopus laevis.*

The amphibian most commonly used for developmental work nowadays is the African claw-toed frog *Xenopus laevis*, which is able to develop normally in tap water. The life cycle of *Xenopus* is shown in Fig. 2.3. Much classical embryology was, however, done on embryos of newts and

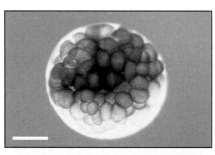
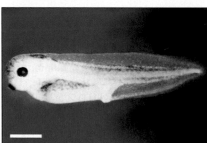

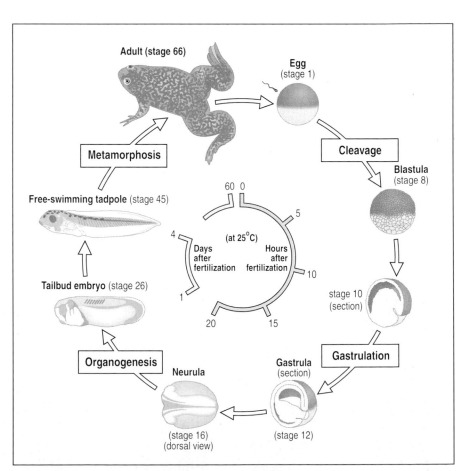

Fig. 2.3 Life cycle of the African claw-toed frog *Xenopus laevis*. The numbered stages refer to standardized stages of *Xenopus* development. More stages can be seen in the larger figure in Box 1A, page 4. The photographs show: an embryo at the blastula stage (top, scale bar = 0.5 mm); a tadpole at stage 45 (middle, scale bar = 1 mm); and an adult frog (bottom, scale bar = 1 cm). Photographs courtesy of J. Slack (top, from Alberts, B., *et al.*: 1994) and J. Smith (middle and bottom).

salamanders. These belong to an order of amphibians different from that of *Xenopus*, and their development is not the same in some details. A great advantage of *Xenopus* is that its fertilized eggs are easy to obtain; females and males need only be injected with the human hormone chorionic gonadotropin and put together overnight. Eggs can also be fertilized in a dish by adding sperm to eggs released after hormonal stimulation of the female. The embryos of *Xenopus* are extremely hardy and, like those of many other amphibians, are highly resistant to infection after microsurgery. The description of amphibian development in this chapter will focus almost exclusively on *Xenopus*. The large eggs of amphibians —1.to 2 mm in diameter—are invaluable for experimental manipulation. It is also easy to culture fragments of early *Xenopus* embryos in a simple, chemically defined solution.

The mature *Xenopus* egg has a dark, pigmented **animal region** and a pale, yolky, and heavier **vegetal region** (Fig. 2.4). Before fertilization, the egg is enclosed in a protective **vitelline membrane**, which is embedded in a gelatinous coat. Meiosis is not yet complete and, while the first meiotic division has resulted in a small cell—a **polar body**—forming at the animal pole, the second meiotic division is completed only after fertilization, when the second polar body also forms at the animal pole (Box 2A).

At fertilization, one sperm enters the egg in the animal region. The egg completes meiosis and the egg and sperm nuclei then fuse to form the diploid zygote nucleus. The vitelline membrane lifts off the egg surface and within about 15 minutes the egg has rotated within it under the influence of gravity so that the heavier, yolky, vegetal region is now downward. About an hour after fertilization, there is a rotation of the egg cortex—a gel-like superficial layer beneath the plasma membrane —which moves relative to the cytoplasm beneath it. As we shall see in Chapter 3, cortical rotation determines the future dorsal side of the *Xenopus* embryo, which develops opposite the site of sperm entry.

The first cleavage occurs along the animal-vegetal axis within 90 minutes of fertilization, and divides the embryo into equal halves (Fig. 2.5). Further cleavages follow rapidly at intervals of about 20 minutes. The second cleavage is also along the animal-vegetal axis but at right angles to the first. The

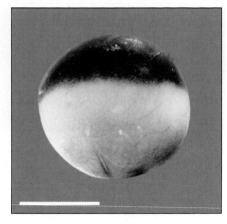

Fig. 2.4 The unfertilized egg of *Xenopus*. The surface of the animal half (top) is pigmented and the paler, vegetal half of the egg is heavy with yolk. Scale bar = 1mm. Photograph courtesy of J. Smith.

Box 2A Polar body formation.

Polar bodies are small cells formed by meiosis during the development of an oocyte into an egg. In this highly schematic illustration, the segregation of only one pair of chromosomes is shown for simplicity. There are two cell divisions associated with meiosis, and one daughter from each division is almost always very small compared with the other, which becomes the egg—hence the term polar bodies for these smaller cells.

The timing of meiosis in relation to the development of the oocyte varies in different animals, and in some species meiosis is completed and the second polar body formed only after fertilization. In general, polar body formation is of little importance for later development, but in some animals the site of formation is a useful marker for the embryonic axes.

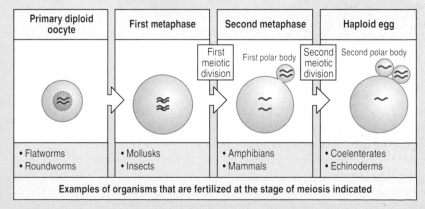

Primary diploid oocyte	First metaphase	Second metaphase	Haploid egg
• Flatworms • Roundworms	• Mollusks • Insects	• Amphibians • Mammals	• Coelenterates • Echinoderms

Examples of organisms that are fertilized at the stage of meiosis indicated

Fig. 2.5 Cleavage of the *Xenopus* embryo. The *Xenopus* embryo undergoes successive cleavages at intervals of about 20 minutes.   Photographs courtesy of R. Kessel, from Kessel, R.G, *et al.*: 1974.

third cleavage is equatorial, at right angles to the first two, and divides the embryo into four animal cells and four larger vegetal cells. The cells deriving from cleavage divisions are often called **blastomeres**. Continued cleavage results in formation of smaller and smaller blastomeres, since there is no growth between cell divisions. Cells at the vegetal pole are larger than those at the animal pole. Inside this spherical mass of cells, a fluid-filled cavity—the blastocoel—develops in the animal region and the embryo is now called a blastula.

At the end of blastula formation the *Xenopus* embryo has gone through about 12 cell divisions and is made up of several thousand cells. At the blastula stage, the mesodermal and endodermal germ layers that give rise to internal structures are located in the equatorial and vegetal regions and are essentially on the outside of the embryo, while the ectoderm, which will eventually cover the whole of the embryo, is still confined to the animal region (Fig. 2.6, first panel). The belt of tissue around the equator which, as we shall see, plays a crucial part in future development, is known as the **marginal zone**. At this stage, the blastula is in the form of a hollow sphere with radial symmetry. The cell movements of gastrulation convert this into a three-layered structure with clearly recognizable antero-posterior and dorso-ventral axes and bilateral symmetry.

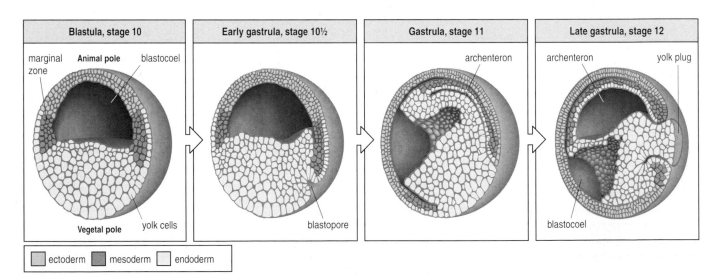

Fig. 2.6 Gastrulation in amphibians. The blastula (first panel) contains several thousand cells and there is a fluid-filled cavity, the blastocoel, beneath the cells at the animal pole. Gastrulation begins (second panel) at the blastopore, which forms on the dorsal side of the embryo. Future mesoderm and endoderm of the marginal zone move inside at this site through the dorsal lip of the blastopore, the mesoderm ending up sandwiched between the endoderm and ectoderm in the animal region (third panel). The tissue movements create a new internal cavity—the archenteron—which will become the gut. Endoderm in the ventral region also moves inside through the ventral lip of the blastopore (fourth panel) and will eventually completely line the archenteron. At the end of gastrulation the blastocoel has considerably reduced in size. After Balinsky, B.I.: 1975.

Gastrulation involves extensive cell movements and rearrangement of the tissues of the blastula so that they become located in their proper positions in relation to the overall body plan of the animal. Because it involves changes in form in three dimensions, gastrulation can be quite difficult to visualize. Gastrulation is initiated by a small slit-like infolding —the **blastopore**—that forms on the surface of the blastula on the dorsal side (see Fig. 2.6, second panel). Once gastrulation has started, the embryo is known as a **gastrula**. The layers of future endoderm and mesoderm in the marginal zone move inside the gastrula through the dorsal lip of the blastopore and converge and extend along the antero-posterior axis beneath the ectoderm, while the ectoderm spreads downward to cover the whole embryo. The layer of dorsal endoderm is closely applied to the mesoderm; the space between it and the yolky vegetal cells is known as the **archenteron** (see Fig. 2.6, third panel) and is the precursor of the gut cavity. The inward movement of endoderm and mesoderm eventually spreads to form a complete circle around the blastopore.

By the end of gastrulation, the blastopore has closed, the dorsal mesoderm lies beneath the dorsal ectoderm, and the lateral mesoderm begins to spread in a ventral direction on either side; the inner surface of the archenteron becomes completely covered by a layer of endoderm, forming the gut. At the same time, the ectoderm has spread to cover the whole embryo by a process known as **epiboly**. There is still a large amount of yolk present, which provides nutrients until the larva—the tadpole—starts feeding.

During gastrulation, the mesoderm in the dorsal region develops into two main structures, the notochord and the somites. The notochord is a stiff, rod-like structure that forms along the dorsal midline and eventually becomes incorporated into the vertebrae. The somites form by segmentation of the mesoderm lying immediately either side of the notochord. Somites are formed in pairs, and segmentation proceeds in an antero-posterior direction.

Gastrulation is succeeded by neurulation—the formation of the neural tube, the early embryonic precursor of the central nervous system. While the notochord and somites are developing, the neural plate ectoderm above them begins to develop into the neural tube and the embryo is then called a **neurula**. The early sign of neural development is the formation of the **neural folds**, which form on the edges of the **neural plate**. These rise up, fold toward the midline and fuse together to form the **neural tube**, which sinks beneath the epidermis (Fig. 2.7). The anterior neural tube gives rise to the brain; further back, the neural tube overlying the notochord will develop into the spinal cord.

A section taken across the middle of the body shows the internal structure of the *Xenopus* embryo just after the end of neurulation (Fig. 2.8). Now that the germ layers are in place they begin to develop into specific tissues. The main structures that can be recognized at this stage are the neural tube, the notochord, the somites, the **lateral plate mesoderm**, and the endoderm lining the gut. By this stage, different parts of the somites can be distinguished: the most dorsal region has formed the **dermatome**, which will give rise to the dermis. The rest of the somite gives rise to the vertebrae and to the muscles of the trunk. The unsegmented lateral plate mesoderm, lying lateral and ventral to the somites, gives rise to tissues of the heart and kidney, as well as to the gonads and gut muscle, while the most ventral mesoderm gives rise to the blood-forming tissues. The endoderm lining the gut will bud off organs such as the liver and lungs.

Fig. 2.7 Neurulation in amphibians.
Top row: the notochord begins to form in the midline. At the same time the neural plate develops neural folds. Middle and bottom rows: the neural folds come together in the midline to form the neural tube, from which the brain and spinal cord will develop. During neurulation, the embryo elongates along the antero-posterior axis. The left panel shows cross-sections through the embryo in the planes indicated by the red dotted lines in the center panel. The center panel shows dorsal surface views of the amphibian embryo. The right panel shows cross-sections through the embryo in the planes indicated by the blue dotted lines in the center panel.

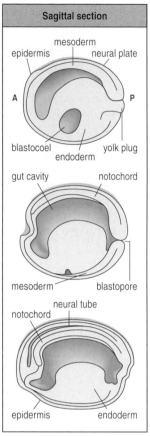

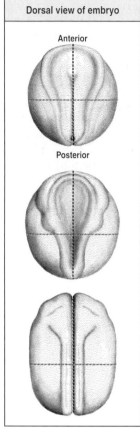

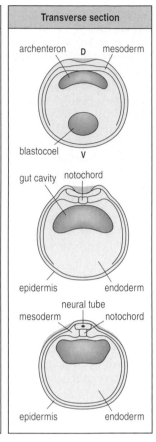

The embryo now begins to look something like a tadpole and we can recognize the main vertebrate features (Fig. 2.9). At the anterior end the brain is already divided up into a number of regions, and the eye and ear have begun to develop. There are also three branchial arches, of which the most anterior one will form the lower jaw. More posteriorly, the somites and notochord are well developed. The post-anal tail of the tadpole is formed last. It develops from the tailbud which, at the dorsal lip of the blastopore, gives rise to the continuation of notochord, somites, and neural tube.

Fig. 2.8 A cross-section through a stage 22 *Xenopus* embryo just after gastrulation and neurulation are completed. The germ layers are now all in place for future development and organogenesis. The most dorsal parts of the somites have already begun to differentiate into the dermatome, which will give rise to the dermis. Scale bar = 0.2 mm. Photograph from Hausen, P., and Riebesell, M.: 1991.

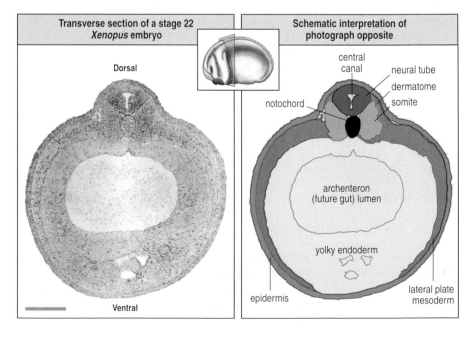

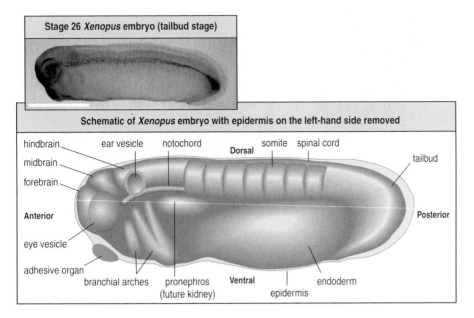

Stage 26 *Xenopus* embryo (tailbud stage)

Schematic of *Xenopus* embryo with epidermis on the left-hand side removed

hindbrain   ear vesicle   notochord   **Dorsal**   somite   spinal cord

midbrain   tailbud

forebrain

Anterior   **Posterior**

eye vesicle

adhesive organ

branchial arches   pronephros   **Ventral**   endoderm
                   (future kidney)   epidermis

Fig. 2.9 The early tailbud stage (stage 26) of a *Xenopus* embryo. At the anterior end, in the head region, the future eye is prominent and an ear vesicle has formed. The brain is divided into forebrain, midbrain, and hindbrain. Just posterior to the site at which the mouth will form are the branchial arches, the first of which will form the lower jaw. More posteriorly, a succession of somites lies on either side of the notochord. The embryonic kidney (pronephros) is beginning to form from lateral mesoderm. Ventral to these structures is the gut (not visible in this picture). The tailbud will give rise to the tail of the tadpole, forming a continuation of somites, neural tube, and notochord. Scale bar = 1mm. Photograph courtesy of B. Herrmann.

Many other internal structures in vertebrates are formed from **neural crest cells**. They come from tissue at the tip of the neural folds, and after neural tube fusion detach and migrate as single cells between the mesodermal tissues. Neural crest gives rise to a remarkable variety of tissues, including the sensory and autonomic nervous systems, the bones of the skull, and pigment cells. They provide the exception to the general rule that ectodermal cells form either nervous system or epidermis, since they also give rise to cartilage.

After organogenesis is completed, the mature tadpole hatches out of its jelly covering and begins to swim and feed. Later, the tadpole larva will undergo metamorphosis to give rise to the adult frog; the tail regresses and the limbs form.

2-2   Birds: the chicken.

Avian embryos are very similar to those of mammals in the morphological complexity of the embryo and the general course of embryonic development, but are easier to obtain and observe. Many observations and manipulations can be carried out simply by opening the egg, but the embryo can also be cultured outside the egg. This is particularly convenient for some experimental microsurgical manipulations and investigation of the effects of chemical compounds. The later development of a chick embryo is similar to that of a mouse embryo, and so provides a valuable complement to studies of mouse embryology.

The egg is fertilized and begins to undergo cleavage while still in the hen's oviduct. The cytoplasm and nucleus of the fertilized egg is confined to a small patch, several millimeters in diameter, lying on a large mass of yolk. Cleavage in the oviduct results in the formation of a disc of cells called a **blastodisc** or **blastoderm**. During a 20 hour passage down the oviduct, the egg becomes surrounded by albumen (egg white), the shell membranes, and the shell (Fig. 2.10). At the time of laying, the blastoderm, which is analogous to the amphibian blastula, is composed of some 60,000 cells. The chick developmental cycle is shown in Fig. 2.11.

After the egg is laid, cleavage continues with the formation of the furrows. The early cleavage furrows extend downward from the surface of the cytoplasm but do not completely separate the cells, whose ventral faces initially remain open to the yolk. Cleavage results in a circular blastoderm several cells thick. Its central region, which overlies a cavity, is translucent and is

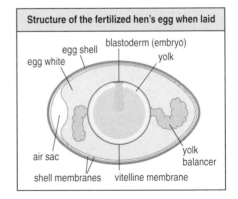

Structure of the fertilized hen's egg when laid

egg shell   blastoderm (embryo)

egg white   yolk

air sac   yolk
                                  balancer

shell membranes   vitelline membrane

Fig. 2.10 The development of the hen egg at the time of laying. Cleavage begins after fertilization while the egg is still in the oviduct. The albumen (egg white) and shell are added during the egg's passage down the oviduct. At the time of laying the embryo is a disc-shaped cellular blastoderm lying on top of a massive yolk, which is surrounded by the egg white and shell.

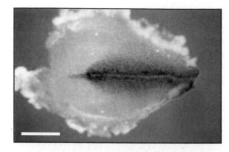

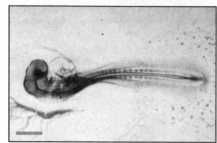

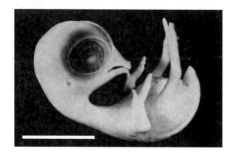

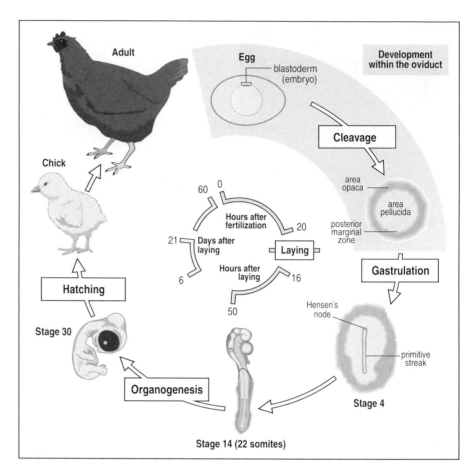

Fig. 2.11 Life cycle of the chicken. The egg is fertilized in the hen and by the time it is laid, cleavage is complete and a cellular blastoderm lies on the yolk. After gastrulation, the primitive streak forms. Regression of Hensen's node is associated with somite formation. The photographs show: the primitive streak surrounded by the area pellucida (top, scale bar = 1 mm); a stage 14 embryo (50–53 hours after laying) with 22 somites (middle, scale bar = 1mm). The head region is well-defined. The transparent organ adjacent to it is the ventricular loop of the heart; a stage 35 embryo, about 8½–9 days after laying, with a well developed eye and beak (bottom, scale bar = 10mm). Top photograph courtesy of B. Herrmann, from Kispert. A., *et al*.: 1994.

known as the **area pellucida**, in contrast to the outer region, which is in the darker **area opaca** (Fig. 2.12). Between the area pellucida and the yolk is a cavity—the subgerminal space—and a layer of cells called the **hypoblast** now develops over the yolk. The hypoblast cells come from two sources: the **posterior marginal zone**, which lies at the junction between the area opaca and area pellucida at the posterior of the embryo, and the overlying cells of the blastoderm. The hypoblast gives rise to extra-embryonic structures such as the stalk of the yolk sac, whereas the embryo proper is formed from the remaining blastoderm cells, known as the **epiblast**.

The posterior marginal zone is a slightly thickened region of the epiblast and defines both the dorsal side and posterior end of the embryo. The onset of gastrulation is marked by the development of the **primitive streak**. This is the forerunner of the antero-posterior axis; it develops from the posterior marginal zone and is fully extended by 16 hours' after laying. The primitive streak first becomes visible as a denser strip extending from the posterior marginal zone to just over half-way across the area pellucida. It represents a region where cells of the epiblast are proliferating and moving inward beneath the upper layer (Fig. 2.13) and is thus similar in some respects to the blastopore region of amphibians. Unlike in amphibians however, cell proliferation and growth in size occurs during gastrulation in birds (and in mammals).

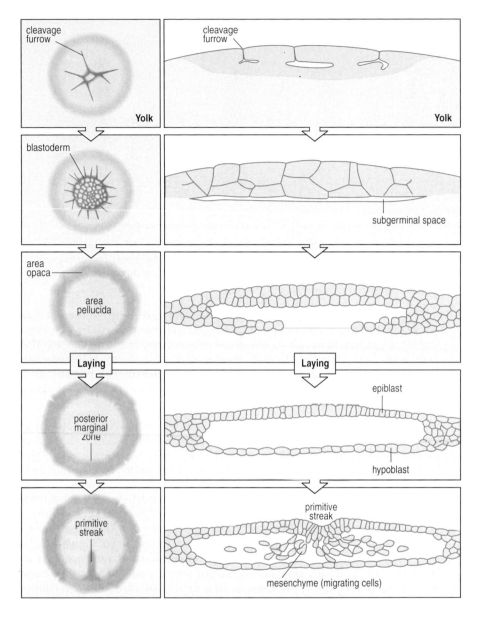

Fig. 2.12 Cleavage and epiblast formation in the chick embryo. By the time the egg is laid, cleavage has divided the small area of egg cytoplasm free from yolk into a disc-shaped cellular blastoderm. The first cleavage furrows extend downward from the surface of the egg cytoplasm and initially do not separate the blastoderm completely from the yolk. In the cellular blastoderm the central area overlying the subgerminal space is called the area pellucida and the marginal region the area opaca. The hypoblast forms as a layer of cells overlying the yolk and will give rise to extra-embryonic structures, while the upper layers of the blastoderm —the epiblast—give rise to the embryo proper.

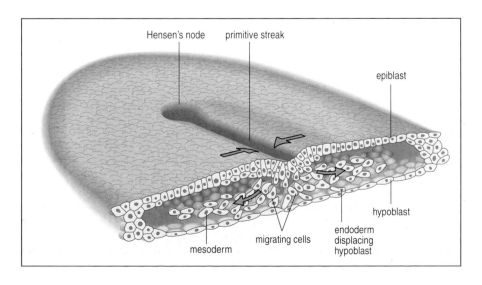

Fig. 2.13 Ingression of mesoderm and endoderm during gastrulation in the chick embryo. Gastrulation begins with the formation of the primitive steak, a region of proliferating and migrating cells, which elongates from the posterior marginal zone. Future mesodermal and endodermal cells migrate through the primitive streak into the interior of the blastoderm. During gastrulation, the primitive streak extends about halfway across the area pellucida (see Fig. 2.12). At its anterior end an aggregation of cells known as Hensen's node forms. As the streak extends, cells of the epiblast move toward the primitive streak (arrows), move through it, and then outward again underneath the surface to give rise to the mesoderm and endoderm internally, the latter displacing the hypoblast. Adapted from Balinsky, B.I., *et al.*: 1975.

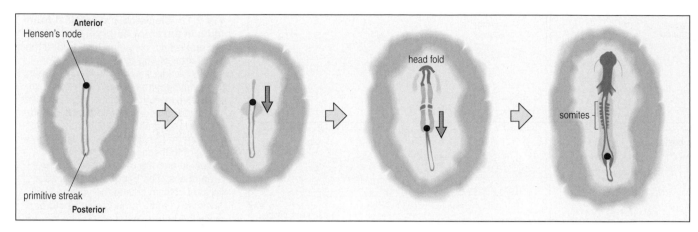

Fig. 2.14 Regression of Hensen's node. After extending about half-way across the blastoderm, the primitive streak begins to regress, with Hensen's node moving in a posterior direction as the head fold and neural plate begin to form. As the node moves backward, the notochord develops in the area anterior to it and somites begin to form on either side of the notochord.

The cells in the epiblast of the posterior marginal zone move forward as the streak extends across the area pellucida. Those cells that converge on the streak move through it, and then move away from it beneath the surface layer will give rise to mesoderm and endoderm, whereas the surface layer, of the epiblast gives rise to the ectoderm. The prospective endoderm displaces the hypoblast and the mesoderm forms a layer between ectoderm and endoderm.

During gastrulation, the area pellucida changes from circular to pear shaped, and at the anterior end of the primitive streak a condensation of cells known as **Hensen's node** is formed. Once most of the mesoderm and endoderm has moved inward, the primitive streak begins to regress, Hensen's node moving toward the posterior end of the embryo (Fig. 2.14). The head region of the embryo is demarcated anterior to the node by the head fold, an infolding of the blastoderm composed of both ectoderm and endoderm. Cells from Hensen's node give rise to the notochord and contribute to the somites during node regression (Fig. 2.15). As the node regresses in a posterior direction, notochord and somites form immediately anterior to it (Fig. 2.16): by 25 hours after laying, about seven pairs of somites have formed. Somite formation progresses in a posterior direction at a rate of about one pair of somites per hour in the anterior region of the pre-somitic mesoderm, which lies between Hensen's node and the last-formed somite.

As the notochord is formed, the neural tube begins to develop as a pair of folds on either side of the midline of the neural plate ectoderm above it. Unlike in *Xenopus*, where the neural tube forms at the same time along the whole length of the midline, folding and closure of the chick embryo neural tube starts at the anterior end and proceeds in a posterior direction (Fig. 2.17). The folds fuse in the dorsal midline and cells detach from the neural crest

Fig. 2.15 Head fold and notochord formation during node regression in the chick embryo. The diagram shows a sagittal section through the chick embryo (inset, dorsal view) at the stage of head fold formation as Hensen's node starts to regress. As the node regresses, the notochord starts to form anterior to it; the undifferentiated mesenchyme on either side of the notochord will form somites.

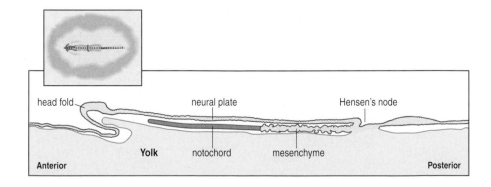

Fig. 2.16 Scanning electron micrograph of chick early somites and neural tube. There are blocks of somites adjacent to the neural tube and the notochord lies beneath it. The lateral plate mesoderm flanks the somites. Scale bar = 0.1mm. Photograph courtesy of J. Wilting.

on either side of the site of fusion. At the same time the head fold develops, separating the head from the surface of the epiblast. Accompanying neurulation and development of the head fold, the embryo also folds on the ventral side to form the gut. This brings the two heart rudiments together to form one organ lying ventral to the gut. The further development of the mesoderm is rather similar to that of *Xenopus*, the somites giving rise to

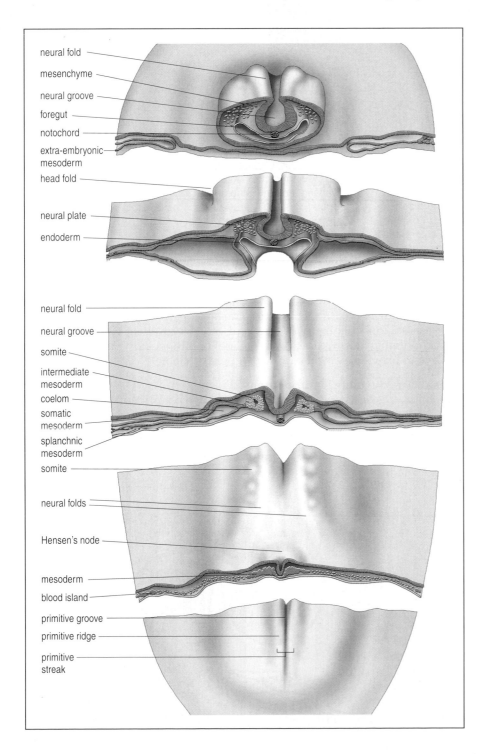

Fig. 2.17 Development of the neural tube and mesoderm in the chick embryo. Once the notochord has formed, neurulation begins, following notochord formation in an anterior to posterior direction. The figure shows a series of sections along the antero-posterior axis of a chick embryo. Neural tube formation is well advanced at the anterior end (top two sections), where the head fold has already separated the future head from the rest of the blastoderm and the ventral body fold has brought endoderm from both sides of the body together to form the gut. During neurulation, the neural plate changes shape: neural folds rise up on either side and form a tube when they meet in the midline. The mesenchymal mesoderm in this region will give rise to head structures. Further back (middle section), in the future trunk region of the embryo, notochord and somites have formed and neurulation is starting. At the posterior end, behind Hensen's node (bottom section), notochord formation, somite formation, and neurulation have not yet begun. The mesoderm internalized through the primitive streak starts to form structures appropriate to its position along the antero-posterior and dorso-ventral axes. For example, in the future trunk region, the intermediate mesoderm will form the mesodermal parts of the kidney, and the splanchnic mesoderm will give rise to the heart. The body fold will continue down the length of the embryo, forming the gut and also bringing paired organ rudiments that initially form on each side of the midline (e.g. those of the heart and dorsal aorta) together to form the final organs lying ventral to the gut. Blood islands, from which the first blood cells are produced, form from the ventral-most part of the lateral mesoderm. After Patten, B.M.: 1971.

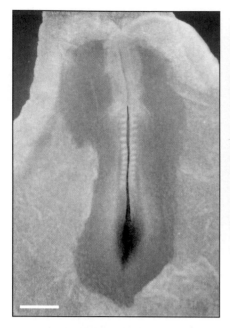
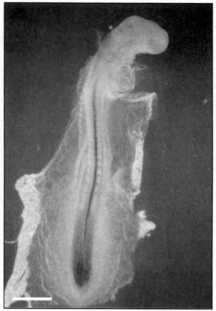
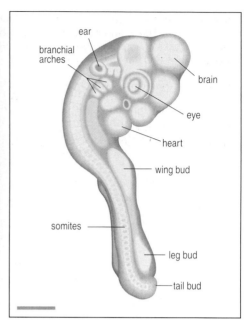

Fig. 2.18 Development of the chick embryo. Left panel: at the 13-somite stage. At the anterior end (top), the head fold has formed. The dark region at the posterior end is Hensen's node. The somites can be seen on either side of the notochord as blocks of white tissue. Between the node and the last formed somite is mesoderm that will segment into somites. Center panel: at the 20-somite stage. Right panel: at the 40-somite stage. Development of the head region and the heart are quite well advanced, and the wing and leg buds are present as small protrusions. Scale bars = 1 mm. Photographs courtesy of B. Herrmann, from Kispert, A., *et al.*: 1994.

the vertebrae, the axial and limb muscles, and the dermis. By 2 days after laying, the embryo has reached the 20-somite stage (Fig. 2.18).

By 3 days after laying, 40 somites have formed, the head is well developed, the heart is formed, and the limbs are beginning to develop. Blood islands, where hematopoesis is occurring, and blood vessels have developed in the extra-embryonic tissues; these vessels connect up with those of the embryo to provide a circulation with a beating heart.

At this stage, the embryo turns on its side and the head is strongly flexed. The embryo gets its nourishment through extra-embryonic membranes (Fig. 2.19), which also provide protection. The fluid-filled **amniotic sac**

Fig. 2.19 The extra-embryonic structures and circulation of the chick embryo. A chick embryo at the same stage as that shown in Fig. 2.18 is depicted *in situ*. The embryo has turned on its side, and its heart is beating. The yolk is surrounded by the yolk sac membrane. The vitelline vein takes nutrients from the yolk sac to the embryo and the blood is returned to the yolk sac via the vitelline artery. The umbilical artery takes waste products to the allantois and the umbilical vein brings oxygen to the embryo. The amnion and fluid-filled amniotic cavity provide a protective chamber for the embryo. After Patten, B.M.: 1951.

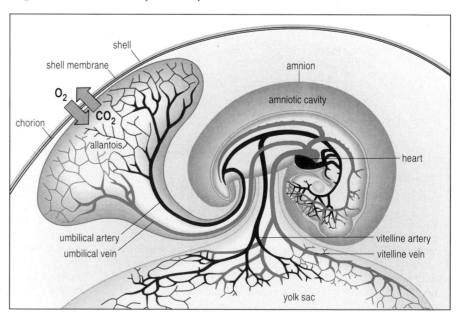

provides mechanical protection; a **chorion** surrounds the whole embryo and lies just beneath the shell; an **allantois** both receives excretory products and provides the site of oxygen and carbon dioxide exchange; and a **yolk sac** surrounds the yolk.

In the remaining time before hatching, eyes develop from the optic vesicles, and the inner ear develops from the otic vesicles. The embryo grows in size, the internal organs develop, the wings, legs, and beak are formed, and down feathers grow on the wings and body. The chick hatches 21 days after the laying of the egg.

2-3   Mammals: the mouse.

The mouse has a life cycle of 9 weeks, from fertilization to mature adult (Fig. 2.20), which is relatively short for a mammal, and is one of the reasons that the mouse has become a model organism for vertebrate development. Another is that it is amenable to both classical genetic analysis and to the generation of mutants by genetic modification. But like all mammals, the mouse embryo develops inside the mother, and so is not easily accessible for experimental manipulation or continuous observation, although it can

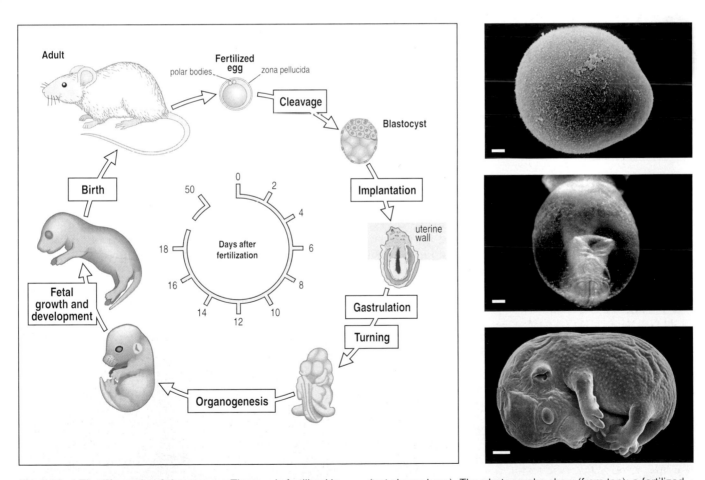

Fig. 2.20   The life cycle of the mouse. The egg is fertilized in the oviduct, where cleavage also takes place before implantation of the blastocyst in the uterine wall at 5 days after fertilization. Gastrulation and organogenesis then take place over a period of around 7 days and the remaining 6 days before birth are largely a time of overall growth. After gastrulation the mouse embryo undergoes a complicated movement known as 'turning' in which it becomes surrounded by its extra-embryonic membranes (not shown here). The photographs show (from top): a fertilized mouse egg just before the first cleavage (scale bar = 10 μm); anterior view of a mouse embryo at 8 days after fertilization (scale bar = 0.1 mm); and a mouse embryo at 14 days after fertilization (scale bar = 1 mm). Photographs courtesy of: T. Bloom (top, from Bloom, T.L.: 1989); N. Brown (middle); and J. Wilting (bottom).

be cultured outside the mother for short periods. The mouse is the mammalian model system that is most often used to help us to understand human development.

Fertilization of the egg takes place internally in the oviduct; meiosis is then completed and the second polar body forms. The egg is small, about 100 μm in diameter. It is surrounded by a protective external coat, the **zona pellucida**, composed of mucopolysaccharides and glycoproteins. Mammalian embryos rely on nutrients obtained from the mother via the placenta.

Cleavage takes place in the oviduct and only after 4½ days does the embryo implant into the uterine wall after being released from the zona pellucida. Gastrulation takes place over the next few days, and by 10 days after fertilization all the organs have begun to develop. During the next nine days before birth, organogenesis continues, as in the chick, and the embryo grows in size.

Early cleavages are very slow compared with *Xenopus* and chick, the first occurring about 24 hours after fertilization and subsequent cleavages at about 12-hour intervals. They produce a solid ball of cells, the **morula** (Fig. 2.21). At the eight-cell stage the blastomeres increase the area of cell surface in contact with each other in a process called compaction. After compaction, the cells are polarized; their exterior surfaces carry microvilli whereas their inner surfaces are smooth. Further cleavages are somewhat variable and are both radial and tangential, so that by the equivalent of the 32-cell stage the morula contains about 10 internal cells and more than 20 outer cells.

A special feature of mammalian development is that the early cleavages give rise to two groups of cells—the **trophectoderm** and the **inner cell mass**. The internal cells of the morula give rise to the inner cell mass and the outer cells to the trophectoderm. The trophectoderm will give rise to extra-embryonic structures such as the placenta, which provides a pathway for nutrition of the embryo via the mother, while the embryo proper develops from a small set of cells in the inner cell mass. At this stage (3½ days' gestation) the embryo is known as a **blastocyst** (see Fig. 2.21). Fluid is pumped by the trophectoderm into the interior of the blastocyst, which causes the trophectoderm to expand and form a fluid-filled vesicle containing the inner cell mass at one end.

From 3½ to 4½ days' gestation the inner cell mass becomes divided into two regions. The surface layer in contact with the fluid-filled cavity of

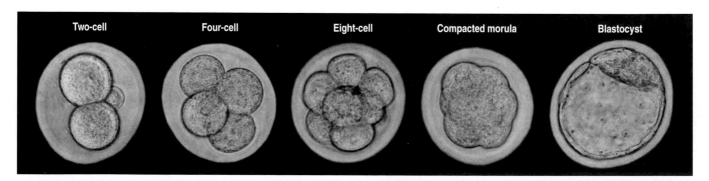

| Two-cell | Four-cell | Eight-cell | Compacted morula | Blastocyst |

Fig. 2.21 Cleavage in the mouse embryo. The photographs show the cleavage of a fertilized mouse egg from the two-cell stage through to the formation of the blastocyst. After the eight-cell stage compaction occurs, forming a solid ball of cells, the morula, in which individual cell outlines can no longer be discerned. The internal cells of the morula give rise to the inner cell mass, which can be seen as the compact clump of the top of the blastocyst. It is from this that the embryo proper forms. The outer layer of the hollow blastocyst—the trophectoderm—gives rise to the extra-embryonic structures. Photographs courtesy of T. Fleming.

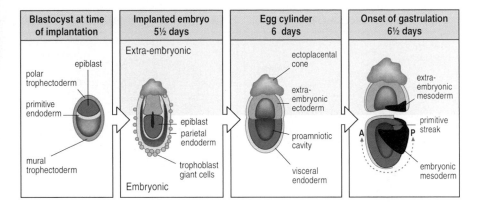

Blastocyst at time of implantation	Implanted embryo 5½ days	Egg cylinder 6 days	Onset of gastrulation 6½ days

Fig. 2.22 Early post-implantation development of the mouse embryo. First panel: before implantation, the fertilized egg has undergone cleavage to form a hollow blastocyst, in which a small group of cells, the inner cell mass, will give rise to the embryo, while the rest of the blastocyst forms the trophectoderm, which will develop into extra-embryonic structures. At the time of implantation the inner cell mass divides into two regions: the primitive ectoderm or epiblast, which will develop into the embryo proper, and the primitive endoderm, which will contribute to extra-embryonic structures. The polar trophectoderm in contact with the epiblast forms extra-embryonic tissues, the ectoplacental cone, and extra-embryonic ectoderm which contributes to the placenta. The mural trophectoderm gives rise to the trophoblast giant cells. Second panel: the epiblast elongates and develops an internal cavity (proamniotic cavity) which gives it a cup-shaped form. Third panel: the cylindrical structure containing both the epiblast and the extra-embryonic tissue derived from the polar trophectoderm is known as the egg cylinder. The parietal endoderm and trophoblast giant cells are not shown in this or any subsequent figures. Fourth panel: the primitive streak appears at the posterior of the epiblast (P) and extends anteriorly (A) to the tip of the cylinder. Epiblast cells passing through the streak will become mesoderm and endoderm. After Hogan, B, *et al.*: 1994.

the blastocyst becomes the **primitive endoderm**, and will contribute to extra-embryonic membranes, while the remainder of the inner cell mass—the **primitive ectoderm** or **epiblast**—will develop into the embryo proper as well as some extra-embryonic membranes. At this stage the embryo releases itself from the zona pellucida still surrounding it, and implants into the uterine wall.

The course of early post-implantation development of the mouse embryo from around 4½ days to 8½ days appears more complicated than that of the chick, partly because of the need to produce a larger variety of extra-embryonic membranes, and partly because the epiblast from which the embryo will develop is distinctly cup shaped in the early stages. In essence, however, the development of the embryo proper is very similar to that of the chick.

The first two days of post-implantation development are shown in Fig. 2.22. At implantation, the cells of the mural trophectoderm (not the region in contact with the inner cell mass) replicate their DNA without cell division (endo-reduplication), giving rise to trophoblast giant cells which invade the uterus during implantation. The rest of the trophectoderm grows to form the ectoplacental cone and the **extra-embryonic ectoderm**, which both contribute to the formation of the placenta. Some cells from the primitive endoderm migrate to cover the whole inner surface of the mural trophectoderm. They become the parietal endoderm. The remaining primitive endoderm cells form the **visceral endoderm** that covers the elongating egg cylinder containing the epiblast.

By 6 days after fertilization, an internal cavity has formed inside the epiblast, which becomes cup shaped—U-shaped when seen in cross-section (see Fig. 2.22, third panel). The embryo proper develops from this curved layer of epithelium, which at this stage contains about 1000 cells. The first visible sign of its future axis is at about 6½ days, when gastrulation begins with the formation of the primitive streak. The streak starts as a localized thickening at a point on the circumference of the cup; this is the future posterior end of the embryo. The inside of the cup is the future dorsal side of the embryo. Proliferating epiblast cells migrate through the primitive streak, and spread out laterally and anteriorly between the ectoderm and the visceral endoderm to form a mesodermal layer (Fig. 2.23). Some (epiblast derived) cells—the definitive endoderm which will later give rise to the gut—enter the visceral endoderm layer and gradually displace it.

The development of the primitive streak in the mouse is similar to that in the chick; first it elongates towards the future anterior end of the embryo, with a condensation of cells at its anterior end corresponding to Hensen's

Fig. 2.23   Gastrulation in the mouse embryo. At the beginning of gastrulation, epiblast cells move through the primitive streak to give rise to the mesoderm and definitive endoderm (definitive endoderm is not shown in this diagram).

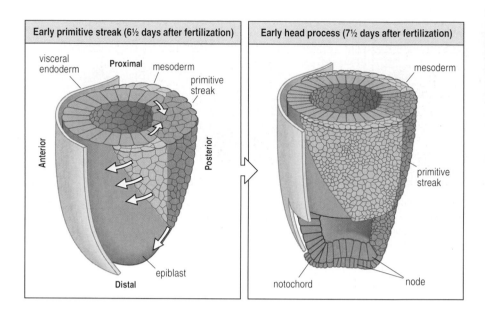

node (Fig. 2.24). Cells migrating anteriorly through the node will form the notochord, and both the notochord and somites form anterior to it. Some migrating cells pass through the mesoderm to form an embryonic endodermal layer (the future gut), eventually displacing the visceral endoderm cells completely.

At around 8½ days, neural folds have started to form at the anterior end on the dorsal side of the embryo. In these final stages of gastrulation the embryo also undergoes an episode of complex folding, in which the embryonic endoderm—initially on the ventral surface of the embryo—becomes internalized to form the gut, while the heart and liver move into their final positions relative to the gut, and the head becomes distinct. The embryo then turns so that it becomes surrounded by its extra-embryonic membranes (Fig. 2.25). By 9 days, gastrulation is complete: the embryo has a distinct head and the forelimb buds are starting to develop. Organogenesis proceeds very much as in the chick embryo, at least in the initial stages.

Fig. 2.24   Early post-implantation development of the mouse embryo. Left panel: the primitive streak continues to extend. Further development of extra-embryonic structures involves the production of extra-embryonic mesoderm at the posterior end of the primitive streak. This eventually contributes to the amnion, the visceral yolk sac, and the allantois and chorion, which are important components of the placenta. Right panel: during the final stages of gastrulation, organogenesis commences in the anterior part of the embryo with the formation of the heart, the cranial neural folds, and the appearance of somites. After Hogan, B., et al.: 1994.

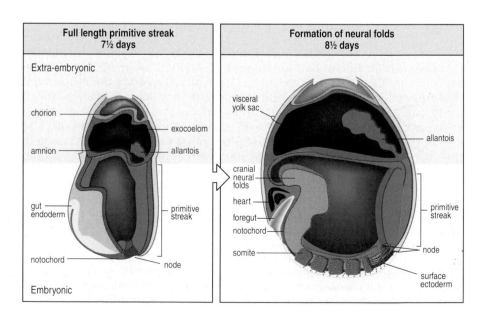

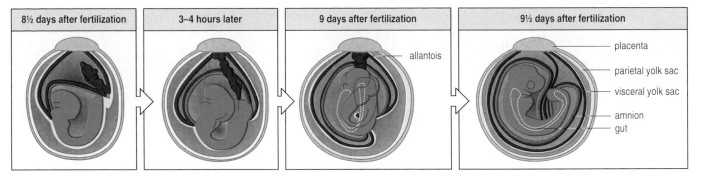

| 8½ days after fertilization | 3–4 hours later | 9 days after fertilization | 9½ days after fertilization |

Labels: allantois; placenta; parietal yolk sac; visceral yolk sac; amnion; gut

Fig. 2.25 Turning in the mouse embryo. Between 8½ and 9½ days the mouse embryo undergoes a complicated change in conformation, in which it 'turns' so that it becomes entirely enclosed in the protective amnion and amniotic fluid. The visceral yolk sac, a major source of nutrition, surrounds the amnion and the allantois connects the embryo to the placenta. After Kaufman, M.H.: 1992.

2-4   Fishes: the zebrafish.

The zebrafish is receiving increasing attention as a model for vertebrate development. Its two great advantages are its short life cycle of approximately 12 weeks (Fig. 2.26), which makes genetic analysis so much easier;

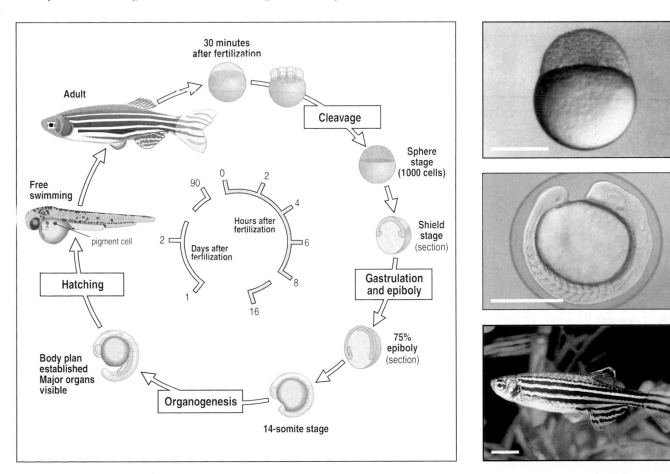

Fig. 2.26 Life cycle of the zebrafish. The zebrafish embryo develops as a cup-shaped blastoderm sitting on top of a large yolk cell. It develops rapidly and by 2 days after fertilization the tiny fish, still attached to the remains of its yolk, hatches out of the egg. The top photograph shows a zebrafish embryo at the sphere stage of development, with the embryo sitting on top of the large yolk cell (scale bar = 0.5 mm). The middle photograph shows an embryo at the 14-somite stage, showing developing organ systems. Its transparency is useful for observing cell behavior (scale bar = 0.5 mm). The bottom photograph shows an adult zebrafish (scale bar = 1 cm). Photographs courtesy of C. Kimmel (top, from Kimmel, C.B., *et al.*: 1995), N. Holder (middle), and M. Westerfield (bottom).

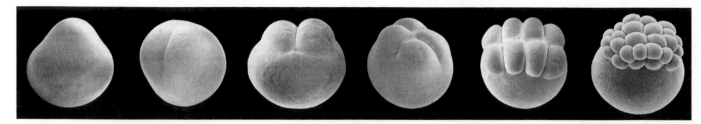

Fig. 2.27   Cleavage of the zebrafish embryo is initially confined to the animal (top) half of the embryo.

Photographs courtesy of R. Kessel, from Kessel, R.G., *et al.*: 1974.

and the transparency of the embryo, so that the fate of individual cells during development can be observed (see Fig. 2.26). The zebrafish egg is about 0.7 mm in diameter, with the cytoplasm and nucleus at the animal pole sitting upon a large mass of yolk. After fertilization, the zygote undergoes cleavage but, as in the chick, cleavage does not extend into the yolk and results in a mound of blastomeres perched above the yolk. The first five cleavages are all vertical, and the first horizontal cleavage gives rise to the 64-cell stage about 2 hours after fertilization (Fig. 2.27).

Further cleavage results in a blastoderm with a single outer layer of flattened cells known as the outer enveloping layer and a deep layer of more rounded cells, overlying the yolk (Fig. 2.28). The blastoderms expand in a vegetal direction by the spreading process known as epiboly, which we have met already in the *Xenopus* gastrula, to cover the yolk cell. By about 5½ hours after fertilization they have spread half-way to the vegetal pole. Gastrulation then begins with the prospective endodermal and mesodermal cells of the deep layer turning inward at the margin of the blastoderm, a process known as involution. These cells migrate toward the future dorsal side, the tissue converging toward the midline of the embryo and extending at the same time as the embryo elongates in an antero-posterior direction. The future mesoderm and endoderm come to lie beneath the ectoderm. Gastrulation in the zebrafish has many features in common with gastrulation in *Xenopus*, but one difference is that involution occurs all around the periphery of the blastoderm at about the same time. By 9 hours the notochord becomes distinct, and gastrulation is complete by 10 hours. Neurulation and somite formation then follow.

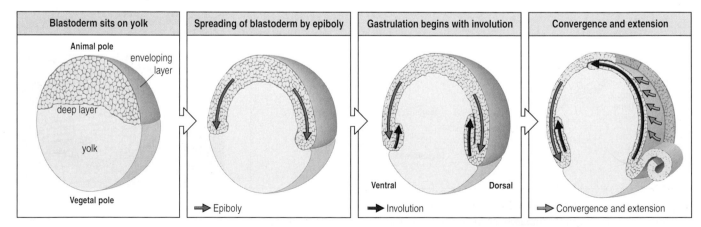

Fig. 2.28   Epiboly and gastrulation in the zebrafish.
At the end of the first stage of cleavage the zebrafish embryo is composed of a cluster of blastomeres sitting on top of the yolk. With further cleavage and spreading out of the layers of cells (epiboly), the upper half of the yolk becomes covered by a cup-shaped blastoderm. Gastrulation occurs by involution of cells in a ring around the edge of the blastoderm. The involuting cells converge on the dorsal midline to form the body of the embryo encircling the yolk.

Over the next 12 hours the embryo elongates, and the rudiments of the primary organ systems become recognizable. Somites appear anteriorly at about 10 hours, and new ones are formed at intervals of initially 2, then 3 hours; by 18 hours, 18 somites are present. The nervous system develops rapidly. Optic vesicles, which give rise to the eyes, can be distinguished at 12 hours as bulges from the brain and, by 18 hours, the body starts to twitch. At 48 hours the embryo hatches, and the young fish begins to swim and feed.

Summary.

While the early development of different vertebrates can differ considerably, they all first undergo cleavage to form a blastula-like structure. The mouse and other mammals are special, since so much of the early mammalian embryo gives rise to extra-embryonic structures whereas the embryo proper derives from just a few cells of the inner cell mass. In all vertebrate embryos, gastrulation is followed, or is accompanied in its late stages, by neurulation—the formation of the neural tube. During gastrulation there is extensive cell movement so that the three germ layers—ectoderm, mesoderm, and endoderm—take up their appropriate positions. The mesoderm immediately on either side of the notochord forms somites, while the ectoderm lying above the notochord forms the neural tube, which develops into the brain and spinal cord. In the chick and mouse, complex extra-embryonic structures involved in nutrition, gas exchange, secretion, and mechanical protection are formed.

Model organisms: invertebrates.

Although the development of the many different kinds of invertebrates is very diverse, certain features are not only common to the development of most invertebrates but are also seen in vertebrate development. These include cleavage, formation of a blastula or blastoderm, and gastrulation. Compared with vertebrate embryos, the number of cells in some invertebrate embryos is small and, as in the nematode worm for example, they have a stereotyped pattern of cleavage in which the fate of each cell can be identified.

2-5   The fruit fly *Drosophila melanogaster*.

The wealth of genetic studies on *Drosophila* development, together with the feasibility of combining genetic and microsurgical manipulation, have made this small fly one of the best understood developmental systems. The life cycle of *Drosophila* is shown in Fig. 2.29.

The *Drosophila* egg is sausage shaped and the future anterior end is easily recognizable by the micropyle, a nipple-shaped structure in the tough external coat surrounding the egg. Sperm enter the anterior end of the egg through the micropyle. After fertilization and fusion of the sperm and egg nuclei, the zygote nucleus undergoes a series of rapid mitotic divisions, one about every 9 minutes but, unlike most animal embryos, there is no cleavage of the cytoplasm. The result is a **syncytium** in which many nuclei are present in a

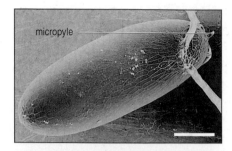

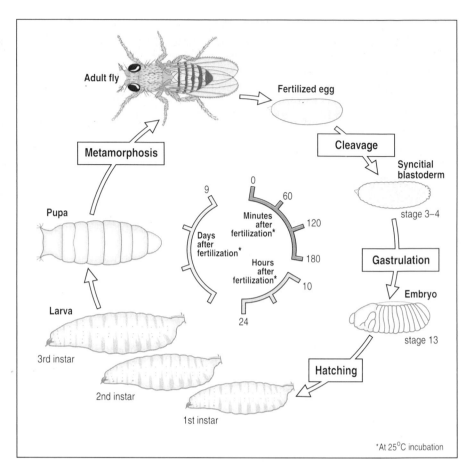

Fig. 2.29 Life cycle of *Drosophila melanogaster*. After cleavage and gastrulation the embryo becomes segmented and hatches out as a feeding larva. The larva grows and goes through two molts (instars), eventually forming a pupa that will metamorphose into the adult fly. The photographs show scanning electron micrographs of: a *Drosophila* egg before fertilization (top). The sperm enters through the micropile. The dorsal filaments are extra-embryonic structures; a *Drosophila*, 2nd instar larva (middle); and a *Drosophila* pupa (bottom). Scale bars = 0.1 mm. Photographs courtesy of F. R. Turner (top, from Turner, F.R., *et al.*: 1976. Middle, from Turner, F.R., *et al.*: 1979).

common cytoplasm (Fig. 2.30); the embryo essentially remains a single cell during its early development. After nine divisions the nuclei move to the periphery to form the **syncytial blastoderm**. This is equivalent to the blastula or blastoderm stage of other animals. Shortly afterwards, membranes grow in from the surface to enclose a nucleus and form cells, and the blastoderm becomes truly cellular after about 13 mitoses. Not all the nuclei give rise to the cells of the cellular blastoderm; some 15 or so end up at the posterior end of the embryo and develop into **pole cells**, which later give rise to germ cells, that is, sperm or eggs. Because of the formation of a syncytium, even large molecules such as proteins can diffuse between nuclei during the first 3 hours of development and, as we shall see in Chapter 5, this is of great importance to early *Drosophila* development.

All the future tissues are derived from the single epithelial layer of the cellular blastoderm. For example, prospective mesoderm is located in the most ventral region, while the future midgut derives from two regions of prospective endoderm, one at the anterior and the other at the posterior end of the embryo. Endodermal and mesodermal tissues move to their future positions inside the embryo during gastrulation, leaving ectoderm as the outer layer (Fig. 2.31). Gastrulation starts at about 3 hours after fertilization when the future mesoderm in the

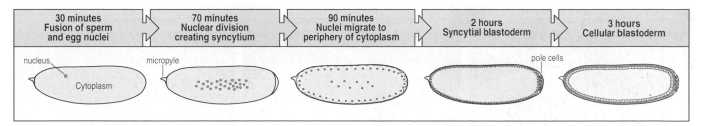

Fig. 2.30 Cleavage of the *Drosophila* embryo. After fusion of the sperm and egg nuclei, there is rapid nuclear division but no cell walls form, the result being a syncytium of many nuclei in a common cytoplasm. After the ninth division the nuclei move to the periphery to form the syncytial blastoderm. After about 3 hours, cell walls develop, giving rise to the cellular blastoderm. About 15 pole cells, which will give rise to germ cells, form a separate group at the posterior end of the embryo. Times given are for incubation at 25°C.

ventral region invaginates to form a furrow along the ventral midline. The mesodermal cells are initially internalized by formation of a mesodermal tube in a process rather similar to neural tube formation in vertebrates. The mesoderm cells then separate from the surface layer of the tube and migrate under the ectoderm to internal locations, where they later give rise to muscle and other connective tissues.

In insects, as in all arthropods, the main nerve cord lies ventrally, rather than dorsally as in vertebrates. Shortly after the mesoderm has invaginated, ectodermal cells of the ventral region that will give rise to the nervous system leave the surface individually and form a layer of **neuroblasts** between the mesoderm and the outer ectoderm. At the same time, two tube-like invaginations develop at the sites of the future anterior and posterior midgut. These grow inward and eventually fuse to form the endoderm of the midgut, while ectoderm is dragged inward behind them at each end to form the foregut and the hindgut. The outer ectoderm layer develops into the epidermis. There are no cell divisions during gastrulation, but once it is completed, cells start to divide again. The cells of the epidermis only divide twice before they secrete a cuticle.

Also during gastrulation, the ventral blastoderm or **germ band**, which comprises the main trunk region, undergoes germ band extension, which drives the posterior trunk regions round the posterior end and onto what was the dorsal side (Fig. 2.32). The germ band later retracts as embryonic development is completed. At the time of germ band extension the first external signs of **segmentation** can be seen. A series of evenly spaced grooves form more or less at the same time and these demarcate **parasegments**, which later give rise to the **segments** of the larva and adult. Parasegments and segments are out of register, so that a segment is formed by the posterior region of one parasegment and the anterior region of the next. There are 14 parasegments: three of which contribute to mouth parts of the head, followed by three thoracic segments, and eight abdominal segments.

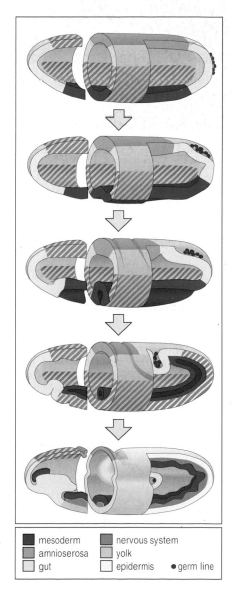

Fig. 2.31 Gastrulation in *Drosophila*. Gastrulation begins when the future mesoderm invaginates in the ventral region, first forming a furrow and then an internalized tube. The cells then leave the tube and migrate internally under the ectoderm. The nervous system comes from cells which leave the surface of the ventral blastoderm and form a layer between the ventral ectoderm and mesoderm. The gut forms from two invaginations at the anterior and posterior end that fuse in the middle. The midgut region is endoderm, while the foregut and hindgut are of ectodermal origin.

■ mesoderm	■ nervous system
■ amnioserosa	□ yolk
□ gut	□ epidermis • germ line

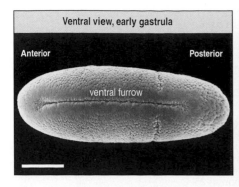

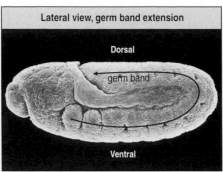

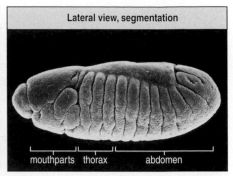

Fig. 2.32 Gastrulation, germ band extension, and segmentation in the *Drosophila* embryo. Gastrulation involves the future mesoderm moving inside through the ventral furrow. During gastrulation, the ventral blastoderm (the germ band) extends, driving the posterior trunk region onto the dorsal side and segmentation now takes place. Later the germ band shortens. Scale bar = 0.1mm. Photographs courtesy of F. Turner (left from Turner, F.R., *et al.*: 1977, middle from Alberts, B., *et al.*: 1994).

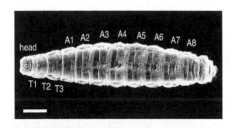

Fig. 2.33 Ventral view of a *Drosophila* larva. T1 to T3 are the thoracic segments and A1 to A8, the abdominal segments. The characteristic pattern of dentricles can be seen in the posterior region of each abdominal segment. Scale bar = 0.1mm. Photograph courtesy of F.R. Turner.

The larva (Fig. 2.33) hatches about 24 hours after fertilization, but the different regions of the larval body are well defined several hours before that. The head is a complex structure, largely hidden from view before the larva hatches. Structures associated with the most anterior region of the head are called the acron. At the posterior end, the most terminal structures are called the telson. Between these extremities, three thoracic segments and eight abdominal segments can be distinguished by specializations in the cuticle secreted by the epidermis. On the ventral side of each segment are denticle belts and other cuticular structures characteristic of each segment. As the larva feeds and grows, it molts, shedding its cuticle. This occurs twice, each stage being called an **instar**.

The *Drosophila* larva has neither wings nor legs; these and other organs emerge when the larva undergoes **metamorphosis** after the third instar, which is driven under the influence of hormones. These structures are, however, already present in the larva as **imaginal discs**, small sheets of prospective epidermal cells derived from the cellular blastoderm and usually containing about 40 cells each. These discs grow throughout larval life and form folded sacs of epithelia to accommodate their increase in size. There are imaginal discs for each of the six legs, two wings, and the two halteres (balancing organs), and for the genital apparatus, eyes, antennae, and other adult head structures (Fig. 2.34). In the segments of the abdomen there

Fig. 2.34 Imaginal discs give rise to adult structures at metamorphosis. The imaginal discs in the *Drosophila* larva are small sheets of epithelial cells; at metamorphosis they give rise to a variety of adult structures. The abdominal cuticle comes from groups of histoblasts located in each larval abdominal segment.

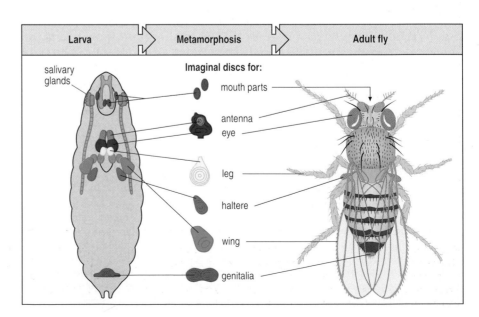

are groups of about ten histoblast cells. These cells do not divide but will contribute to the epidermis of the adult at metamorphosis.

2-6   The nematode *Caenorhabditis elegans*.

The free-living soil nematode *Caenorhabditis elegans*, whose life cycle is shown in Fig. 2.35, is now used as a major model organism in developmental biology. Its advantages are its suitability for genetic analysis, its small number of cells (558 in the first larval stage) and their invariant lineage, and the transparency of the embryo that allows the formation of each cell to be observed. *Caenorhabditis elegans* has a simple anatomy and the adults are about 1 mm long and just 70 µm in diameter. Nematodes can be grown on agar plates in large numbers and early larval stages can be stored frozen and later resuscitated. This nematode reproduces primarily by self-fertilization of adult hermaphrodites, although males can develop under special conditions. Embryonic development is rapid, the larva hatching after 15 hours at 20°C, though maturation through larval stages to adulthood takes about 50 hours.

The nematode egg is small, only 50 µm in diameter. Polar bodies are formed after fertilization. Before the male and female nuclei fuse, there is what appears to be an abortive cleavage, but after fusion of the nuclei

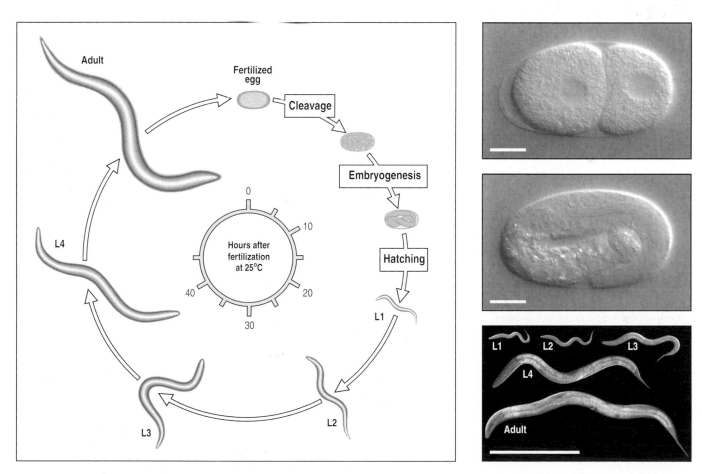

Fig. 2.35 Life cycle of the nematode *Caenorhabditis elegans*. After cleavage and embryogenesis there are four larval stages (L1–L4) before the sexually mature adult develops. Adults of *C. elegans* are usually hermaphrodite, although males can develop. The photographs show: the two-cell stage, an embryo after gastrulation (top, scale bar = 10 µm); with the larva curled up (middle, scale bar = 10 µm); and the four larval stages and adult (bottom, scale bar = 0.5 mm). Photographs courtesy of J. Ahringer.

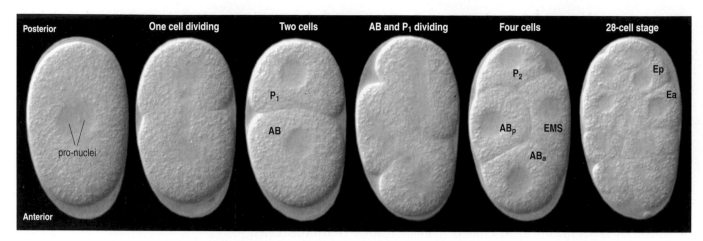

Fig. 2.36 Cleavage of the *C. elegans* embryo. After fertilization the pronuclei of the sperm and egg fuse. The egg then divides into a large anterior AB cell and a smaller, posterior P₁ cell. At the next cell division, AB divides into ABₐ and ABₚ, while P₁ divides into P₂ and EMS. The EMS cell divides into the E cells which give rise to the intestine, and MS cells (not labeled here). Each of the cells will continue to divide within these groups. Photographs courtesy of J. Ahringer.

true cleavage begins (Fig. 2.36). The first cleavage is asymmetric and generates an anterior AB cell and a smaller posterior P₁ cell. At the second cleavage, AB divides to give ABₐ anteriorly and ABₚ posteriorly, while P₁ divides to give P₂ and EMS. At this stage the main axes can already be identified, since P₂ is posterior and ABₚ dorsal. Further cleavage of the AB cells gives rise mainly to hypodermis (the outer layers of the worm), neurons, and muscle. We concentrate here on the fates of the P₂ and EMS cells (Fig. 2.37). EMS divides into E and MS. E gives rise to the gut, whereas MS gives rise to muscle, glands, and neurons. P₂ divides to give P₃ and C. C forms muscle, hypodermis, and neurons, and P₃ divides into P₄ and D. D gives rise to muscle and P₄ gives rise to the germ cells. All these cells undergo a further well-defined pattern of cell divisions. Gastrulation starts at the 28-cell stage, when the descendants of the E cell that will form the gut

Fig. 2.37 Cell lineage and cell fate in the early *C. elegans* embryo. The fertilized egg divides into an anterior AB cell and a posterior P₁ cell. The AB cell gives rise to hypodermis (the outer layers of the embryo), neurons, and some muscle. The P₁ cell divides to give EMS and P₂. The EMS cell then divides to give cells MS and E which develop respectively into muscle, glands, and coelomocytes, and into the gut. Further divisions of the P lineage are rather like stem cell divisions, with one daughter of each division (C and D) giving rise to a variety of tissues while the other (P₂ and P₃) continues to act as a stem cell. Eventually, P₄ gives rise to the germ cells.

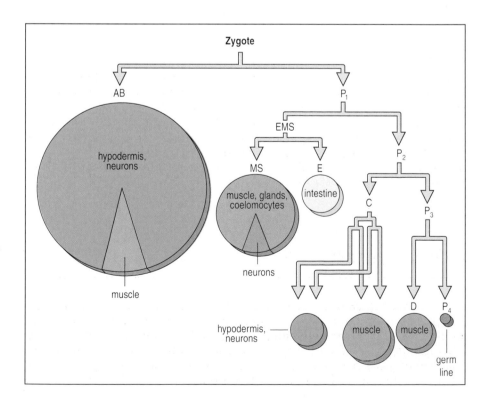

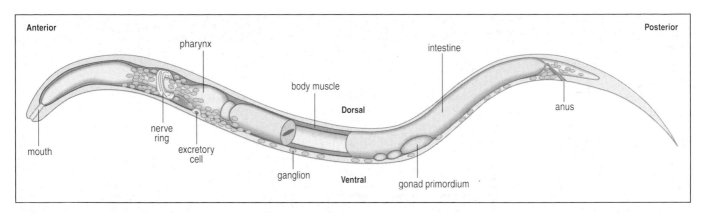

Fig. 2.38 *Caenorhabditis elegans* **larva at the L1 stage (20 hours after fertilization)**. The vulva will form from the gonad primordium.

move inside. Not all cells that are formed during embryonic development survive; **programmed death** of specific cells is an integral feature
of nematode development.

The newly hatched larva (Fig. 2.38), while similar in overall organization
to the mature adult, is sexually immature and lacks a gonad and its associated structures, such as the vulva, which are required for reproduction.
Post-embryonic development takes place during a series of four molts.
While the newly hatched larva contains 558 nuclei, the adult hermaphrodite contains 959 somatic cell nuclei in addition to a variable number
of germ cells. The emphasis is on nuclei rather than cells because some
cells are syncytial and have several nuclei. The additional cells in the
adult are derived largely from precursor blast cells (P cells) that are distributed along the body axis. Each of these blast cells founds an invariant
lineage involving between one and eight cell divisions. The vulva, for
example, is derived from blast cells P_5, P_6, and P_7. One may think of
post-embryonic development in the nematode as the adding of adult
structures to the basic larval plan.

Summary.

Like vertebrates, invertebrate embryos undergo cleavage, form a blastula-
like structure, and then gastrulate so that the endoderm and mesoderm
migrate into the embryo to take up their correct location, with the ectoderm on the outside. There can be considerable differences in these
processes. In insects, the early embryo is a synctium with several thousand nuclei, forming a superficial layer on the outside of the embryo.
This layer later forms a cellular blastoderm of several thousand cells.
The nematode is an example of an embryo with an invariant cell lineage,
and the total number of cells is small.

Model systems: plants.

The study of developmental biology has been dominated by animals,
but plant development is now receiving increasing attention. There are
important differences between plant and animal development, the most
obvious being the absence of cell migration and tissue movements, so
that cell division and cell expansion play a large role in morphogenesis.
There is nothing in plants that corresponds to gastrulation. A particular

feature of plants is the development of all adult structures from the **meristems**—groups of undifferentiated cells set aside in shoot and root tips.

2-7   *Arabidopsis thaliana.*

The equivalent of *Drosophila* in the study of plant development is the small annual crucifer *Arabidopsis thaliana*, often called wall cress, which is well suited to genetic studies. Its life cycle is shown in Fig. 2.39. This annual flowering plant develops as a small ground-hugging rosette of leaves, from which a branched flowering stem is produced with an inflorescence at the end of each branch.

Each flower (Fig. 2.40) consists of four sepals surrounding four white petals; inside the petals are six stamens, which contain the male pollen, and a central ovary of two carpels which contain ovules. Each ovule contains an egg cell. Following fertilization, the embryo develops inside the ovule, taking about 2 weeks to form a mature seed. Flower buds are usually visible on a young plant 3–4 weeks after seed germination. The complete life cycle is thus about 6–8 weeks. In the ovule, the fertilized egg cell is surrounded by specialized nutritive tissue, the **endosperm**, which provides the food source for embryonic development.

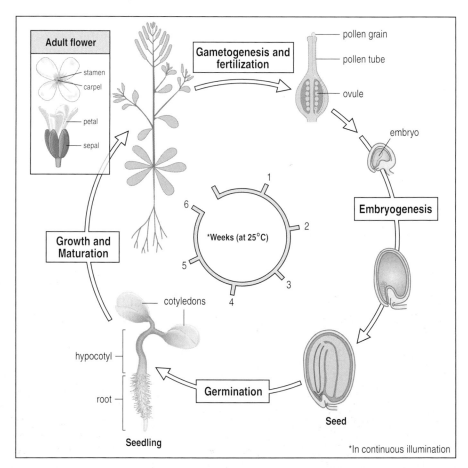

Fig. 2.39 Life cycle of *Arabidopsis*. In flowering plants, egg cells are contained separately in ovules inside the carpels. Fertilization of an egg cell by a male nucleus from a pollen grain takes place inside the ovary. The egg then develops into an embryo contained within the ovule coat, forming a seed. *Arabidopsis* is a dicotyledon, and the mature embryo has two wing-like cotyledons (storage organs) at the apical (shoot) end of the main axis—the hypocotyl—which contains a shoot meristem at one end and a root meristem at the other. Following germination, the seedling develops into a plant with roots, a stem, leaves, and flowers. The photograph shows a mature *Arabidopsis* plant.

The early embryo is a mass of small, undifferentiated cells that becomes differentiated into three main tissues: the outer epidermis; prospective vascular tissue, which runs through the center of the main axis and cotyledons; and the ground tissue that surrounds it. The **cotyledons** (seed leaves) are storage organs, developed by the embryo, that will provide nutrition for the germinating seedling. Monocotyledons, such as maize, have a single cotyledon; dicotyledons, such as *Arabidopsis*, have two. **Apical meristems** composed of undifferentiated cells capable of continued division develop at each end of the main axis and give rise to the root and shoot of the seedling.

The ovule containing the embryo matures into a seed, which remains dormant until suitable external conditions trigger germination. The early stages of germination and seedling growth rely on food supplies stored in the cotyledons. The shoot and root elongate and emerge from the seed. Once the shoot emerges above ground it starts to photosynthesize and forms the first true leaves at the shoot apex. About 4 days after germination the seedling is a self-supporting plant. All the adult structures such as leaves, stem, flowers, and roots are derived from the apical meristems.

The egg cell is fertilized by pollen from the male sex organs, the stamens. A pollen grain deposited on the carpel surface grows a tube that penetrates the carpel and delivers two haploid pollen nuclei to an ovule. One nucleus fertilizes the egg cell while the other fuses with the two so-called polar nuclei, which go on to develop into the triploid endosperm. Early cell division produces an embryo composed of two parts; the embryo proper, and the suspensor, which attaches the embryo to maternal tissue and is a source of nutrients (Fig. 2.41). Initial patterns of cleavage up to the 16-cell stage are highly reproducible. At the octant (eight-cell) stage the embryo proper and suspensor are clearly distinct. Even at this early stage it is possible to make a fate map for the major regions of the seedling. The upper tier of cells gives rise to the cotyledons, which will provide nutrition, the next tier is the origin of the hypocotyl, and the region of the suspensor where it joins the embryo gives rise to the root. At the 16-cell stage the epidermal layer or dermatogen is established. Slightly later, at the globular stage, the future vascular and ground tissue can be identified. With further cell divisions the heart stage is reached, in which the two cotyledons appear as wing-like structures. Further expansion of the cotyledons and hypocotyl takes place to

Fig. 2.40　An individual flower of **Arabidopsis**. Scale bar = 1 mm.

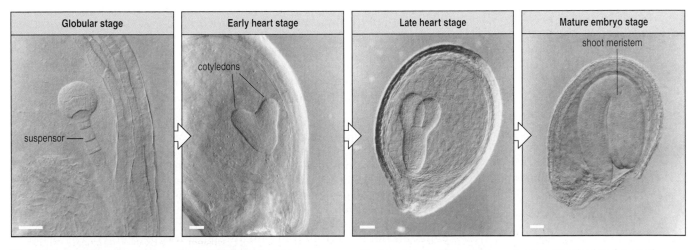

Fig. 2.41　Arabidopsis embryonic development. Light micrographs (Nomarski optics) of cleared wild-type seeds of *Arabidopsis thaliana*. The cotyledons can already be seen at the heart stage. The embryo proper is attached to the seed coat through a filamentous suspensor. Scale bar = 20 μm. Photographs courtesy of D. Meinke, from Meinke, D.W.: 1994.

give the embryo its form in the mature seed. The future shoot apical meristem is a small cluster of cells between the cotyledons and remains quiescent until germination. The embryo is now enclosed in a seed coat and awaits germination.

Meristem activity is entirely post-embryonic. Cell division in the shoot meristem results in both apical growth and the initiation of leaves. Leaf initiation is first indicated by a localized swelling—a leaf primordium—on the apical meristem; this gradually forms a protrusion that gives rise to the leaf. Flowering involves the transformation of a vegetative shoot meristem into reproductive mode. In *Arabidopsis*, the apical shoot meristem becomes an **inflorescence meristem**, throwing off individual **floral meristems**, each of which develops into a single flower. Instead of leaves, a floral meristem gives rise in sequence to sepals, petals, stamens, and carpels.

Summary.

The annual wall cress, *Arabidopsis thaliana*, has a complete life cycle of about 6–8 weeks. Post-embryonic growth is entirely from meristems composed of undifferentiated cells capable of continued division, which develop at each end of the main axis and give rise to the root and shoot of the seedling, and all structures of the adult plant. Early patterns of cleavage in the embryo are highly reproducible and it is possible to make a fate map of the main regions of the seedling at a very early stage. In the adult plant, the shoot apical meristem gives rise to stem and leaves before becoming transformed into an inflorescence meristem. This produces floral meristems, each of which develops into a single flower. A floral meristem gives rise in sequence to sepals, petals, stamens, and carpels.

Identifying developmental genes.

A major goal of developmental biology is to understand how genes control embryonic development, and to do this one must first identify those genes critically involved in controlling development. This task can be approached in a variety of ways, depending on the organism involved, but the general starting point is the identification of mutations that alter development in some specific and informative way. In the following sections general strategies are outlined for obtaining, and screening for, such mutations; other methods will be described in later chapters. Techniques for identifying genes that control development and detecting their expression in the organism will also be described elsewhere, along with techniques for genetic manipulation.

Only some of our model organisms are suitable for genetic analysis. In spite of their importance, amphibian embryos cannot be used for genetic studies, as their breeding period is far too long and their genetics are virtually unknown. The situation with birds is only marginally better, but using techniques of direct DNA analysis, developmental genes are beginning to be identified in these organisms by their sequence similarity to well-characterized developmental genes from organisms such as *Drosophila* and mice. In general, when an important developmental gene has been identified in one animal, it has proved very rewarding to consider whether a **homologous gene** (that is, one with a certain minimum degree of nucleotide sequence similarity that indicates descent from a common ancestral gene) is present and is acting in some developmental capacity in other animals. As we shall see in Chapter 4, this approach has, for example, identified a hitherto unsuspected class of genes in vertebrates that affect segmental patterning along part of the antero-posterior axis. These were identified by their homology with genes that control antero-posterior patterning in early *Drosophila* development.

2-8 **Developmental genes can be identified by rare spontaneous mutation.**

All the organisms dealt with in this book are sexually reproducing diploids so that their **somatic cells** contain two copies of each gene (with the exception of those on sex chromosomes). *Xenopus* is however different, as it is tetraploid, so has double the number of chromosomes compared to diploids. One copy (**allele**) of each gene is contributed by the male parent and the other by the female. For many genes there are several different 'normal' alleles present in the population, which leads to the variation in normal phenotype one sees within any sexually reproducing species. Occasionally, however, a mutation will occur spontaneously in a gene and there will be marked change, usually deleterious, in the phenotype of the organism.

Many of the genes that affect development have been identified by spontaneous mutations that disrupt their function and produce an abnormal phenotype. Such mutations are rare, however, and to produce more it has been necessary to carry out large-scale mutagenesis experiments using chemical mutagens or X-irradiation and to screen for developmental mutations, as described in Section 2-9. Mutations are classified broadly according to whether they are dominant or recessive (Fig. 2.42). **Dominant** and **semi-dominant** mutations are those that produce a distinctive phenotype when the mutation is present in just one of the alleles of a pair; that is, it exerts an effect in the **heterozygous** state. By contrast, **recessive** mutations such as *white* in *Drosophila* alter the phenotype only when both alleles of a pair carry the mutation; that is, when they are **homozygous**.

In general, dominant mutations are more easily recognized, particularly if they affect gross anatomy or coloration, provided that they do not cause the early death of the embryo in the heterozygous state. However, truly

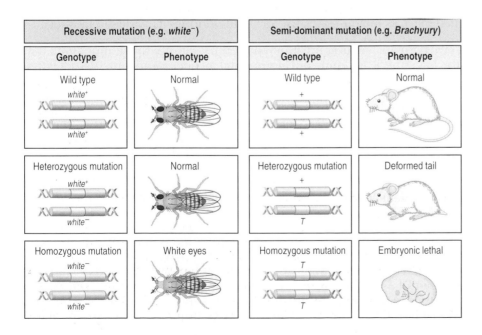

Fig. 2.42 Types of mutations. Left: a mutation is recessive when it only has an effect in the homozygous state, that is, when both copies of the gene carry the mutation. Right: by contrast, a dominant or semi-dominant mutation produces an effect on the phenotype in the heterozygous state; that is, when just one copy of the mutant gene is present. A plus sign denotes wild type, and a minus sign recessive, *T* is the mutant form of the gene *Brachyury*.

Fig. 2.43 Genetics of the semi-dominant mutation in *Brachyury (T)* in the mouse.
A male heterozygote carrying the *T* mutation merely has a short tail. When mated with a normal (wild type, +) female some of the offspring will also be heterozygotes and have short tails. Mating two heterozygotes together will result in some of the offspring being homozygous *(T/T)* for the mutation, resulting in a severe and lethal developmental abnormality in which the posterior mesoderm does not develop.

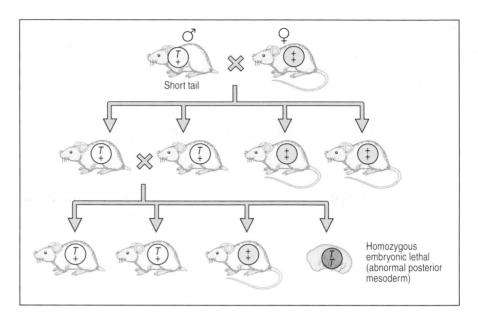

dominant mutations are rare. A mutation in the mouse gene *Brachyury*, which is involved in mesoderm formation, is a classic example of a semi-dominant mutation and was originally identified because mice heterozygous for this mutation (symbolized by *T*) have short tails. When the mutation is homozygous it has a much greater effect, and embryos die at an early stage, indicating that the gene is required for normal embryonic development (Fig. 2.43). In the *Brachyury* mutation, the development of posterior mesoderm is affected; the anatomical defect seen in heterozygotes is relatively minor because the normal copy of the gene also present is able to compensate to some extent. Once breeding studies had confirmed the *Brachyury* mutation as being due to a single gene, the gene could be mapped to a location on a particular chromosome by classical genetic mapping techniques. It has now been cloned; that is, isolated in a pure form that can be sequenced or used for other molecular studies. Identifying recessive mutations is more laborious, as the heterozygote has a phenotype identical to a normal wild-type animal, and a carefully worked-out breeding program is required to obtain homozygotes. Identifying potentially lethal recessive developmental mutations requires careful observation and analysis in mammals, as the homozygotes may die unnoticed inside the mother.

Very rigorous criteria must be applied to identify those mutations that are affecting a genuine developmental process and not just affecting some vital but routine housekeeping function without which the animal cannot survive. One simple criterion for a developmental mutation is embryonic lethality, but this also catches mutations in genes involved in housekeeping functions. Mutations that produce abnormal patterns of embryonic development are much more promising candidates for true developmental mutations.

2-9   Identification of developmental genes by induced mutation and screening.

Valuable though spontaneous mutations have been in the study of development, suitable mutations are rare. Many more developmental genes have been identified by inducing random mutations in a large number of organisms by chemical treatments or irradiation with X-rays, and then screening for mutants of developmental interest. The aim, where possible, is to treat a large enough population so that, in total, a mutation is induced in every gene in the genome. This sort of approach can best be used in organisms that breed rapidly and can be obtained and treated conveniently in very large numbers.

Zebrafish offer a potentially very valuable vertebrate system for large-scale mutagenesis because large numbers can be handled, and the transparency and large size of the embryos makes it easier to identify developmental abnormalities. However, unlike the case of *Drosophila* described later, there are as yet no genetic means of eliminating unaffected individuals automatically. This means that all the progeny of a cross have to be examined visually.

A screening program using zebrafish involves breeding for three generations (Fig. 2.44). Male fish treated with a chemical mutagen are crossed with wild-type females; their F_1 male offspring are crossed again with wild-type females, and the female and male siblings from each of these crosses are themselves crossed. The offspring from each of these pairs are examined separately for homozygous mutant phenotypes. If the F_1 fish carry a mutation, then in 25% of the F_2 matings two heterozygotes will mate and 25% of their offspring will be homozygous for the mutation. Zebrafish can also be made to develop as haploids by fertilizing the egg with sperm heavily irradiated with ultraviolet light. This allows one to detect early-acting recessive mutations without having to breed the fish to obtain homozygous embryos.

Many of the developmental mutations that have led to our present understanding of early *Drosophila* development came from a brilliantly successful screening program that searched the *Drosophila* genome

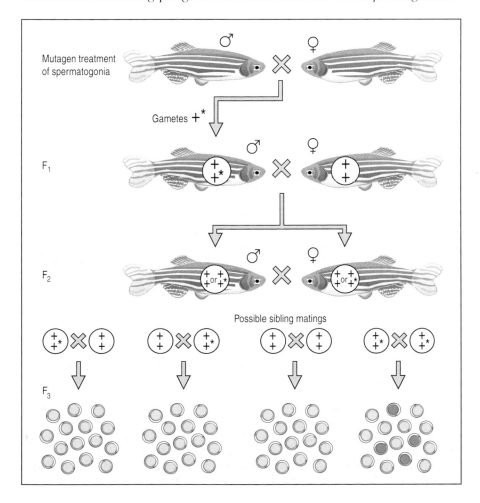

Fig. 2.44 Genetic screening to produce homozygous mutant zebrafish embryos. Male fish are treated with a mutagen and mated with wild-type females. In the F_1 offspring each individual is heterozygous for a different mutated gene. Further mating with wild-type females results in 50% of each F_2 family carrying the same mutation. The siblings within each family are mated and the embryos examined for developmental abnormalities. Embryos homozygous for the induced mutation will be found in the offspring of 25% of the matings. A plus sign with an asterisk indicates the induced mutation.

systematically for mutations affecting the patterning of the early embryo. Its success was recognized with the award of a Nobel Prize for Physiology or Medicine to Edward Lewis, Christiane Nüsslein-Volhard, and Eric Wieschaus in 1995, only the second given for developmental biology.

In this screening program thousands of flies were mutagenized with a chemical mutagen, and were bred and screened according to the strategy described in Box 2B. Given the number of progeny involved, it was

Box 2B Mutagenesis and genetic screening strategy for identifying developmental mutants in *Drosophila*.

The mutagen ethyl methane sulfonate (EMS) was applied to large numbers of male flies homozygous for a recessive mutation on the selected chromosome. The chromosome marked in this way is designated 'a' in the figure. A recessive mutation is chosen that gives adult flies an easily distinguishable but viable phenotype when homozygous (such as the mutation *white⁻*, see Fig. 2.42)

The treated males, who now produce sperm with a variety of induced mutations on the 'a' chromosomes (*a**), were crossed with untreated females that carried different mutations (*DTS* and *b*) on their two 'a' chromosomes, but were otherwise wild type. These mutations track the untreated female-derived chromosomes and automatically eliminate all embryos carrying two female-derived chromosomes in subsequent generations. *DTS* is a dominant temperature-sensitive mutation that causes death of the fly when the incubation temperature is raised to 29°C. *b* is a non-developmental lethal recessive so that any flies homozygous for this female-derived chromosome will die as normal-looking embryos and be automatically eliminated.

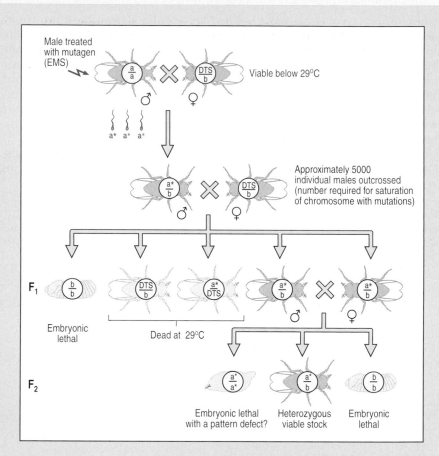

The female flies also carried a balancer chromosome (not shown) that prevented recombination at meiosis; this is to prevent recombination between male-derived and female-derived chromosomes in females. There is no recombination at meiosis in male *Drosophila*.

To identify the new recessive mutations (*a**) caused by the chemical treatment, a large number of the heterozygous males arising from this first cross were again outcrossed to *DTS/b* females. Of the offspring of each individual cross, only *a*/b* flies survive when placed at 29°C; all other combinations die. The surviving siblings were then intercrossed and the offspring of each cross screened for patterning mutants. There are three possible outcomes: flies homozygous for the induced mutation *a** (which will also be homozygous for the original mutation marking chromosome a in males); heterozygous *a** flies; and homozygous *b* flies (which die as embryos).

If *a** is indeed a patterning mutation which is lethal in the larva, then the culture tube in which the cross is made will contain no adult flies with the original male phenotype, say white-eyed. Therefore tubes containing white-eyed flies can be discarded immediately, as the induced mutation *a** they carry must have allowed them to develop to adulthood even when homozygous and is therefore likely to be of little interest. If no white-eyed flies are present in a tube, then the homozygous *a*/a** embryos may have died or been arrested in their development as a result of abnormal development, and the mutation is of potential interest. The embryos of larvae from this cross can then be examined for pattern defects, as in the example illustrated here. The phenotypically wild-type adults in this tube are heterozygous for *a** and are used as breeding stock to study the mutant further. This whole program has to be repeated for each of the four chromosome pairs of *Drosophila*.

important to devise a strategy that would reduce the number of flies that had to be examined to find a mutation. So the search was restricted to mutations on just one chromosome at a time. As described in Box 2B, the program also incorporated a means of identifying flies homozygous for the male-derived chromosomes carrying the induced mutations and, most importantly, also included a way of automatically deleting from the population flies that could not be carrying a mutated chromosome.

A phenotypic character of particular value in screening for patterning mutants in *Drosophila* is the stereotyped denticle pattern of the larval segments, irregularities in which enable alterations in pattern to be quickly recognized visually. In this way the key genes involved in patterning the early *Drosophila* embryo were first identified. They were subsequently mapped and many have now been cloned. The success of this screening was also made possible because the mutations that were inevitably produced in housekeeping genes were mostly rescued by the actions of maternal housekeeping genes; that is, those genes acting in the mother that provide many housekeeping functions to the egg. Indeed, it is necessary to do special screens for maternal-effect genes; that is, genes acting in the mother that are involved in patterning the egg during oogenesis.

The nematode is also amenable to this sort of large-scale screening, and that is why its developmental genetics are quite advanced despite the fact that it has only been studied for a relatively short time. *Caenorhabditis* is the subject of an intensive program of genetic analysis and the complete DNA sequence of its genome should soon be known. A similar approach has been used to induce developmental mutants in mice, but screening is much more difficult and is limited by both the enormous numbers of mice needed and the difficulty of screening for embryonic phenotype while the embryo is still inside the mother.

In *Arabidopsis*, mutagenesis is most frequently performed not on the gametes but on the seed. If the mutant cell ends up in the apical meristem it has a chance of contributing to the sex cells when the flower is formed, and can thus give rise to gametes containing the mutation. Since *Arabidopsis* is self-fertilizing, both homozygotes and heterozygotes for the mutation can be obtained in the next generation.

Summary.

Genes controlling development can be identified by mutations that affect embryonic development. It is easier to recognize dominant as distinct from recessive mutations, as the latter have a distinguishable phenotype only in homozygotes. In some animals, such as *Drosophila*, recessive mutations in many genes controlling early development have been identified by chemically inducing mutations in large populations of flies, breeding them, and screening the embryos. Zebrafish offer the possibility of similar large-scale screening for developmental genes in a vertebrate.

Summary to Chapter 2.

Our knowledge and understanding of development depends on a rather small number of intensively studied model organisms, most of which are animals. These model systems represent quite a wide range of organisms, each with their own advantages and disadvantages for studying development. In spite of the differences between various animals, there are broad similarities in their early development. In all the animals studied here, cleavage of the fertilized egg leads to a multicellular blastula or blastoderm stage; this is followed by gastrulation, during which cells of the three germ layers—endoderm, mesoderm, and ectoderm—become located in the correct position for future development of the animal body. In most embryos, future endoderm and mesoderm are located initially on the outside of the embryo and move inside during gastrulation.

In vertebrates, gastrulation is followed by a distinct episode of neurulation, in which the future central nervous system is formed. During gastrulation and neurulation the overall shape of the embryo is altered from a sphere or disc of cells into that of a recognizable embryo, with antero-posterior and dorso-ventral axes. One of the main differences between embryos relates to their rate of development and the number of cells present at different stages.

In plants, there is virtually no cell movement and nothing that corresponds to gastrulation. The shape of the early plant embryo is molded by patterns of cell division. Unlike animals, in which the late embryo can be thought of as a miniature of the larva or adult, all adult plant structures are derived at a later stage from the shoot and root meristems.

A variety of techniques can be used for identifying developmental genes. One powerful and successful strategy has been to induce large numbers of mutations in a population by chemical treatment, and then to screen the offspring for developmental mutations. Such programs have contributed many of the known developmental mutations in *Drosophila, Caenorhabditis*, and zebrafish.

General references.

Bard, J.B.L.: *Embryos. Color Atlas of Development*. London: Wolfe, 1994.

Carlson, B.M.: *Pattern's Foundations of Embryology*. New York: McGraw-Hill, Inc., 1996.

Slack, J.M.W.: *From Egg to Embryo*. Cambridge: Cambridge University Press, 1991.

Section references.

2-1   **Amphibians: *Xenopus laevis*.**

Hausen, P., Riebesell, H.: *The Early Development of Xenopus Laevis*. Berlin: Springer-Verlag, 1991.

Nieuwkoop, P.D., Faber, J.: *Normal Tables of Xenopus Laevis*. Amsterdam: North Holland, 1967.

2-2   **Birds: the chicken.**

Hamburger, V., Hamilton, H.L.: **A series of normal stages in the development of a chick**. *J. Morph*. 1951, **88**:49–92.

Lillie, F.R.: *Development of the Chick: An Introduction to Embryology*. New York: Holt, 1952.

Patten, B.M.: *The Early Embryology of the Chick*. New York: McGraw-Hill, 1971.

2-3   **Mammals: the mouse.**

Hogan, H., Beddington, R., Costantini, F., Lacy, E.: *Manipulating the Mouse Embryo. A Laboratory Manual (2nd edn)*. New York: Cold Spring Harbor Laboratory Press, 1994.

Kaufman, M.H.: *The Atlas of Mouse Development*. London: Academic Press, 1992.

2-4   **Fishes: the zebrafish.**

Kimmel, C.B., Ballard, W.W., Kimmel, S.R., Ullmann, B., Schilling, T.F.: **Stages of embryonic development of the zebrafish**. *Dev. Dynam*. 1995, **203**:253–310.

Westerfield, M. (ed.) *The Zebrafish Book; A Guide for the Laboratory Use of Zebrafish (Brachydanio Rerio)*. Eugene, Oregon: University of Oregon Press, 1989.

2-5   **The fruit fly *Drosophila melanogaster*.**

Ashburner, M.: *Drosophila. A Laboratory Handbook*. New York: Cold Spring Harbor Laboratory Press, 1989.

Lawrence, P.: *The Making of a Fly*. Oxford: Blackwell Scientific Publications, 1992.

2-6   The nematode *Caenorhabditis elegans.*

Sulston, J.E., Scherienberg, E., White, J.G., Thompson, J.N.: **The embryonic cell lineage of the nematode** *Caenorhabditis elegans*. *Dev. Biol.* 1983, **100**:64–119.

Sulston, J.: **Cell lineage**. In *The Nematode Caenorhabditis elegans*. Edited by Wood, W.B. New York: Cold Spring Harbor Laboratory Press, 1988:123–156.

Wood, W.B.: **Embryology**. In The *Nematode Caenorhabditis elegans*. Edited by Wood, W.B. New York: Cold Spring Harbor Laboratory Press, 1988: 215–242.

2-7   *Arabidopsis thaliana.*

Lyndon, R.F.: *Plant Development*. London: Unwin Hymas, 1990.

Mansfield. S.G., Briarty, L.G.: **Early embryogenesis in** *Arabidopsis thaliana.* II. **The developing embryo**. *Can. J. Bot*. 1991, **69**:461–476.

Meyerowitz, E.M.: *Arabidopsis*—**a useful weed**. *Cell*. 1989: **56**:263–269.

2-8   Developmental genes can be identified by rare spontaneous mutation.

&

2-9   Identification of developmental genes by induced mutation and screening.

Driever, W., Solnica-Krezel, L., Schier, A.F., Neuhauss, S.C.F., Malicki, J., Stemple, D.L., Stainier, D.Y.R., Zwartkruis, F., Abdelilah, S., Rangini, Z., Belak, J., Boggs,C.: **A genetic screen for mutations affecting embryogenesis in zebrafish**. *Development*, 1996: **123**:37–46.

Gelbart, W.M., Griffiths, A.J.F., Lewontin, R.C., Miller, J.H., Suzuki, D.T.: *An Introduction to Genetic Analysis (5th edn)*. New York: W.H. Freeman and Co., 1995.

Haffter, P., Granato, M., Brand, M., Mullins, M.C., Hammerschmidt, M., Jiang, Y-J., Heisenberg, C-P., Kelsh, R.N., Fabian, C., Nüsslein-Volhard, C.: **The identification of genes with unique and essential functions in the development of the zebrafish,** *Danio rerio*. *Development*, 1996: **123**:1–36.

Mullins, M.C., Hammerschmidt, M., Haffter, P., Nüsslein-Volhard, C.: **Large scale mutagenesis in the zebrafish: in search of genes controlling development in a vertebrate**. *Curr. Biol*. 1994, **4**:189–202.

Patterning the Vertebrate Body Plan I: Axes and Germ Layers

3

Setting up the body axes.

The origin and specification of the germ layers.

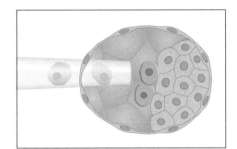

"When young, we needed to decide who would be at the front and who at the back. Thus, we were assigned to these main groups. All this required many conversations and even some messages from the outside."

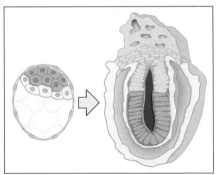

All vertebrates, despite their many outward differences, have a very similar basic body plan. The defining vertebrate structures comprise the segmented vertebral column surrounding the spinal cord, with the brain at the anterior end enclosed in a bony or cartilaginous skull. These prominent structures define the **antero-posterior axis**, the main body axis of a vertebrate. The head is at the anterior end of this axis, followed by the trunk with its paired appendages—limbs in land vertebrates and fins in fish—and terminating in many vertebrates in a post-anal tail. In addition, the vertebrate body has a distinct **dorso-ventral** (back-belly) polarity, with the mouth defining the ventral side. The antero-posterior and dorso-ventral axes define the left and right sides of the animal, and internal organs such as the heart and liver are asymmetrically arranged.

The overall similarity of the body plan in all vertebrates suggests that the developmental processes that establish it are also similar in different animals. All vertebrate embryos do indeed pass through a common stage known as the **phylotypic stage**, in which the head is distinct and a neural tube runs along the dorsal midline, under which runs the notochord, flanked on either side by the mesodermal somites. At this stage, most vertebrate embryos are very similar, and features special to different groups, such as beaks, wings, and fins, appear later (Fig. 3.1).

There are, however, considerable differences between vertebrate embryos at earlier stages of development, particularly in relation to how and when the axes are set up, and how early patterning is established. These differences are related mainly to the very different modes of reproduction amongst vertebrates: yolk provides all the nutrients for amphibian, bird, and fish development. Mammalian eggs, by contrast, are small and non yolky, and the embryo is nourished initially by fluids in the oviduct and uterus and then through the placenta; this requires the development of specialized extra-embryonic structures at a very early stage.

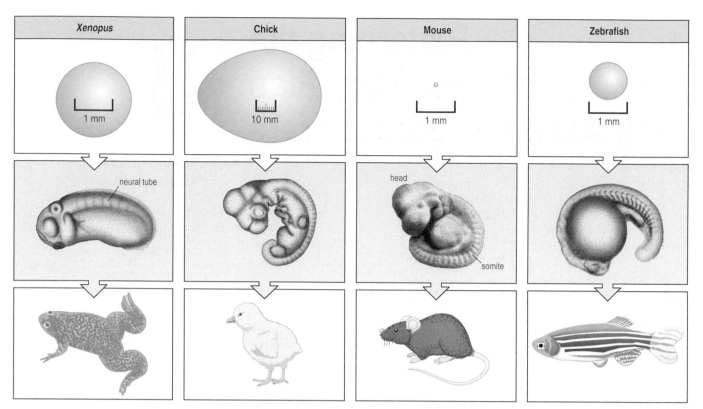

Fig. 3.1 All vertebrate embryos pass through a common phylotypic stage. The eggs of the frog (*Xenopus*), chick, mouse, and zebrafish are very different in size (top row), and their early development (not shown) is rather different, yet they all pass through an embryonic stage at which they look rather similar (middle row). This is the phylotypic stage, when the body axis has developed, and neural tube, somites, notochord, and head structures are present. After this stage their development diverges again. Paired appendages, for example, develop into fins in fish, and wings and legs in the chick (bottom row).

These reproductive differences affect the mechanisms used to specify the axes, the time at which they are established, and what proportion of the earliest 'embryo' forms the embryo proper rather than extra-embryonic structures.

In this and the following chapter, we concentrate on the setting-up of the main body axes, the induction and patterning of the mesoderm, and the induction and early patterning of structures along the antero-posterior axis, including the nervous system. This takes our vertebrate embryo up to the stage at which it has become recognizably vertebrate, with somites, notochord, and neural tube. More detailed consideration of the development of particular organs and structures, such as the limbs, the head, and the nervous system itself, is left to later chapters.

We deal here mainly with three vertebrates whose early development has been particularly well studied: amphibians are represented by the frog *Xenopus*, birds by the chicken, and mammals by the mouse. We also touch briefly on the zebrafish, which is becoming an increasingly important model organism for developmental studies. The development of *Xenopus* inevitably provides the main focus, as this is the organism in which setting-up of the body plan is best understood.

To consider early vertebrate development in an integrated way, we divide it into three main stages and the similarities and differences between organisms at each stage are discussed.

First, we consider the setting-up of the main body axes—the antero-posterior and the dorso-ventral axes—and to what extent these axes are already present in the egg or are specified by external signals. A central issue in the early development of any animal is the role of **maternal factors** in the egg. To what extent is the very early embryo already patterned due to maternal factors laid down in the egg during its development in the mother's ovary? We must draw a distinction between **maternal genes**, which act in the mother during the development of the egg, and which affect subsequent embryonic development through maternal factors—proteins and mRNAs—laid down in the egg during **oogenesis**, and **zygotic genes**, which are expressed in the developing embryo itself. Maternal genes can control not only which proteins and mRNAs are put into the egg but also how they are distributed within the egg. In the second stage, we look at the specification of the three germ layers: the endoderm, which gives rise to the gut, and its derivatives such as liver and lungs; the mesoderm, which forms the notochord, skeletal structures, muscle, connective tissue, kidney, and blood, as well as some other tissues; and the ectoderm, which gives rise to the epidermis, the brain and spinal cord, and neural crest. The third stage to be considered is the patterning of the germ layers, particularly the mesoderm, and the early patterning of the nervous system.

Although these three stages follow each other roughly in developmental time, there are no sharp boundaries between them, and the later stages, in particular, overlap considerably. The first two stages, along with the early patterning of the mesoderm along the dorso-ventral axis, will be the subject of this chapter, while Chapter 4 deals with patterning along the antero-posterior axis and the final emergence of the characteristic vertebrate body plan.

Setting up the body axes.

Different vertebrates use quite different strategies to set up the primary embryonic axes. We begin with amphibians, in which the establishment of the axes is by far the best understood, and then compare this strategy with those of birds and mammals. As well as antero-posterior and dorso-ventral polarity, vertebrates also have bilateral symmetry, with many structures occurring in pairs on either side of the midline, and we shall see how this bilateral symmetry is set very early on in *Xenopus*. We then briefly consider the intriguing question of how the left–right asymmetry or 'handedness' of a number of internal organs might be determined.

3-1   The animal-vegetal axis of *Xenopus* is maternally determined.

The *Xenopus* egg possesses a distinct polarity even before it is fertilized, and this polarity influences the pattern of cleavage. One end of the egg, the **animal pole**, which sits uppermost, has a heavily pigmented surface, while most of the yolk is located toward the opposite end, the **vegetal pole**, which is unpigmented (see Fig. 2.4). (The pigment itself has no role in development but is a useful marker for the developmental differences between the animal and vegetal halves of the egg.) The yolk is confined mainly to the vegetal half, while the animal half contains the nucleus, which is located close to the animal pole. The planes of early cleavages are related to the animal-vegetal axis. The first plane of cleavage is parallel with the axis and often defines a plane of midline symmetry, whereas the third cleavage is at right angles to the axis and divides the embryo into animal and vegetal halves (see Fig. 2.5).

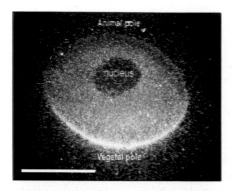

Fig. 3.2 Distribution of mRNA for the growth factor Vg-1 in the amphibian egg. *In situ* hybridization with a radio-active probe for maternal Vg-1 mRNA shows its localization (yellow) in the vegetal region. Scale bar = 1 mm. Photograph courtesy of D. Melton.

Subtle differences in the localization of maternally provided mRNAs and proteins are present along the animal-vegetal axis of the egg before cleavage. Messenger RNAs for housekeeping proteins such as histones are abundant throughout the unfertilized egg, but there are also smaller amounts of localized mRNAs that are presumed to encode proteins with specifically developmental roles. At least nine classes of these rare maternal mRNAs have been identified along the animal-vegetal axis.

The proteins encoded by several of these mRNAs belong to developmentally important families of signaling molecules, and are strong candidates for signals involved in specifying early polarity and inducing the mesoderm.

One such maternal mRNA encodes the signaling protein Vg-1, which is a member of the transforming growth factor-β (TGF-β) family (Box 3A). Vg-1 mRNA is localized at the vegetal pole of the fertilized egg (Fig. 3.2), where its presence can be detected by *in situ* hybridization and autoradiography (Box 3B). Vg-1 mRNA is synthesized during early oogenesis and becomes localized in the vegetal cortex of fully grown oocytes.

Box 3A Protein intercellular signaling molecules.

Proteins that are known to act as signals between cells during development belong to six main families. Some of these families, like that of the fibroblast growth factor (FGF), were originally identified because they were essential for the survival and proliferation of mammalian cells in tissue culture. The members of the six families are either secreted or expressed on cell membranes and they provide intercellular signals in both vertebrates and invertebrates at many stages of development.

These protein factors act by binding to receptors at the cell surface. This produces a signal that is passed across the cell membrane, by the receptor, to the intracellular biochemical signaling pathways, and which eventually results in specific genes being switched on or off. For each type of factor there is a set of corresponding

receptors, and only cells with the appropriate receptors on their surface can respond. Some factors, such as those of the transforming growth factor-β family (TGF-β), act as dimers—two molecules forming a complex that activates the receptor, which is itself a dimer. In some cases, the active form is a heterodimer, made up of two different members of the same family. Delta protein is expressed in the membrane and interacts directly with the Notch receptor protein on an adjacent cell.

Binding of the ligand to the receptor results in each case to membrane transduction and intracellular signaling. The intracellular signaling pathway can be complex and involve a variety of different proteins. For example, the Wnt pathway involves catenin in vertebrates and its homolog, armadillo in *Drosophila*.

Family	Receptors	Examples of roles in development
Fibroblast growth factor (FGF) Ten mammalian FGFs; FGF-1 to FGF-10 and eFGF	Receptor tyrosine kinases	Induction of spinal cord; signal from apical ridge in vertebrate limb
Transforming growth factor-β (TGF-β) Large family, which includes activin, Vg-1, bone morphogenetic proteins (BMPs), Nodal (mouse), decapentaplegic (*Drosophila*)	Receptors associated with a cytoplasmic serine-threonine protein kinase. Receptors act as dimers	Mesoderm induction in *Xenopus*; patterning of dorso-ventral axis and imaginal discs in *Drosophila*
hedgehog hedgehog in insects, Sonic hedgehog and Indian hedgehog in vertebrates	Patched	Positional signal in vertebrate limb and neural tube, and insect wing and leg discs
Wingless Wingless in insect, various Wnt proteins in vertebrates	frizzled	Dorso-ventral axis specification in *Xenopus*; insect segment and imaginal disc specification
Delta and Serrate	Notch	Inhibitory signal in nervous system
Ephrins	Receptor tyrosine kinases	Vertebrate nervous system

Box 3B *In situ* detection of gene expression.

In order to understand how gene expression is guiding development it is essential to know exactly where and when particular genes are active. Genes are switched on and off during development and patterns of gene expression are continually changing. Several powerful techniques are available that show where a gene is being expressed within a tissue or within a whole early embryo.

One set of techniques use ***in situ* hybridization** to detect the mRNA that is being transcribed from a gene. If an anti-sense RNA probe shares complementary sequence with a length of mRNA being transcribed in the cell, it will hybridize (pair tightly) to the mRNA. The RNA probe of the appropriate sequence can therefore be used to locate its complementary mRNA in a tissue slice or a whole embryo. The probe may be labeled in various ways—with a radioactive isotope, a fluorescent dye, or an enzyme for histochemical localization—to enable it to be detected. Radioactively labeled probes are detected by autoradiography, as illustrated in the bottom panels, whereas probes labeled with colored dyes are observed directly. Enzyme-labeled probes are mixed with a substrate that produces a localized colored product. The probes can be applied to both tissue sections and whole mount preparations.

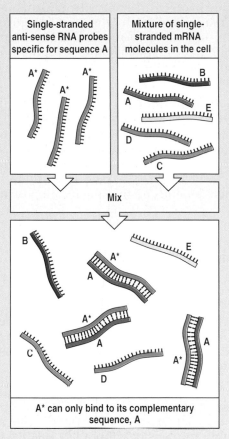

Single-stranded anti-sense RNA probes specific for sequence A

Mixture of single-stranded mRNA molecules in the cell

Mix

A* can only bind to its complementary sequence, A

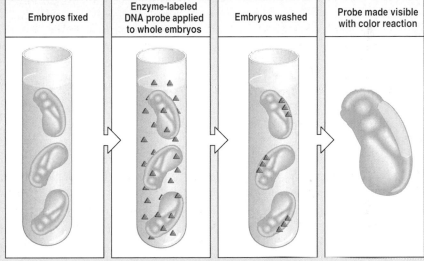

| Embryos fixed | Enzyme-labeled DNA probe applied to whole embryos | Embryos washed | Probe made visible with color reaction |

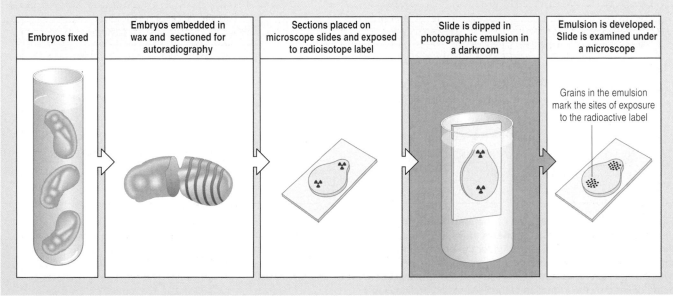

| Embryos fixed | Embryos embedded in wax and sectioned for autoradiography | Sections placed on microscope slides and exposed to radioisotope label | Slide is dipped in photographic emulsion in a darkroom | Emulsion is developed. Slide is examined under a microscope |

Grains in the emulsion mark the sites of exposure to the radioactive label

It then moves into the vegetal cytoplasm prior to fertilization. An mRNA encoding another signaling protein, Xwnt-11, is also located at the vegetal pole. The Wnt family of vertebrate proteins are related to, and named after, a protein found in *Drosophila*, coded for by the gene *wingless*. This is an important signal protein in pattern formation in the fly and other organisms. Since many of the genes controlling development in different organisms are related, they are commonly identified by their similarities in nucleotide sequence.

It should be emphasized that the axes of the tadpole are not directly comparable to those of the fertilized egg. The animal-vegetal axis of the egg is certainly related to the antero-posterior axis of the tadpole, as the head forms from the animal region. However, just where the head will form is not determined until after the second main body axis—the dorso-ventral axis—is fixed, after fertilization. The precise position of the future antero-posterior axis thus depends on the specification of the dorso-ventral axis.

| 3-2 | **The dorso-ventral axis of amphibian embryos is determined by the site of sperm entry.** |

The spherical unfertilized egg of *Xenopus* is radially symmetric about the animal-vegetal axis, and this symmetry is broken only when the egg is fertilized. Sperm entry sets in motion a series of events that defines the dorso-ventral axis of the embryo, with the dorsal side forming more or less opposite the sperm entry point.

Within 90 minutes of fertilization, changes in the egg become distinguishable opposite the site of sperm entry. The plasma membrane and the cortex—a gel-like layer of actin filaments and associated material about 5 μm thick beneath the membrane—rotates about 30° relative to the rest of the cytoplasm, which remains stationary. This **cortical rotation** is toward the site of sperm entry, the vegetal cortex opposite the entry site moving toward the animal pole (Fig. 3.3).

The crucial developmental consequence of cortical rotation is the formation of a signaling center in the vegetal region on the side opposite sperm entry, which defines that side as the future dorsal side of the

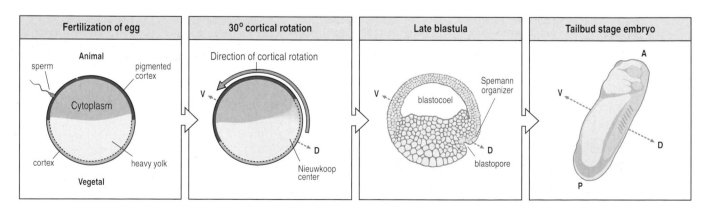

Fig. 3.3 The future dorsal side of the amphibian embryo develops opposite the site of sperm entry. After fertilization (first panel), the cortical layer just under the cell membrane rotates toward the site of sperm entry, moving over the cytoplasm within. This movement results in the dorsal side being defined by the formation of the Nieuwkoop or signaling center (second panel). Later, the Spemann organizer and blastopore will develop in a region just above it (third panel). The fourth panel shows a tailbud stage embryo after gastrulation and neurulation. V = ventral; D = dorsal; A = anterior; P = posterior.

embryo. A signaling center is a localized region of the embryo that exerts a special influence on surrounding tissues and can thus determine how they will develop. This signaling center is known as the **Nieuwkoop center**, after the Dutch embryologist Pieter Nieuwkoop, who discovered it. The Nieuwkoop center sets the initial dorso-ventral polarity in the blastula.

The Nieuwkoop center exerts its influence on dorso-ventrality from a very early stage. The first cleavage usually passes through the point of sperm entry and so divides the egg, and also the Nieuwkoop center, into left and right halves, defining the plane of bilateral symmetry of the animal body. The importance of the center is shown by experiments that divide the embryo into two at the four-cell stage in such a way that one half contains the Nieuwkoop center and the other does not. The half containing the center will develop most structures, although the resulting embryo will lack some ventral regions. The half without the Nieuwkoop center will develop much more abnormally, producing a distorted radially symmetric ventralized embryo that lacks all dorsal and anterior structures (Fig. 3.4).

The influence of the center can also be seen dramatically in experiments in which cells from the region of the Nieuwkoop center of a 32-cell *Xenopus* embryo are grafted into the ventral side of another embryo. This gives rise to a twinned embryo with two dorsal sides (Fig. 3.5), whereas grafting ventral cells to the dorsal side has no effect. Signals from the Nieuwkoop center must therefore be required for the future development of all dorsal and anterior structures.

We can now understand the puzzling result of Roux's classic experiment (see Fig. 1.8) in which he destroyed one cell of a frog embryo at the two-cell stage and got a half-embryo rather than a half-sized whole embryo. The crucial feature of his experiment, it turns out, was that the killed cell remained attached, but the embryo did not 'know' it was dead. The plane of first cleavage had already passed through the Nieuwkoop center. The remaining living cell therefore developed as just one half embryo and had just one half of the center. If the killed cell had been separated from the living blastomere the embryo could than have regulated and developed into a small-sized whole embryo. We now look at the relation between cortical rotation and the specification of the Nieuwkoop center.

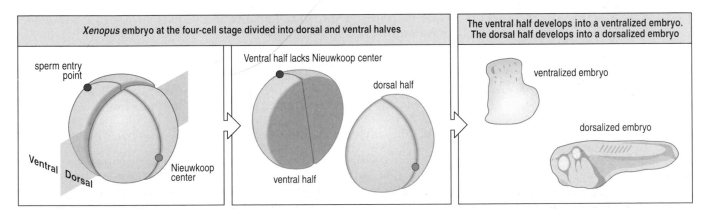

Fig. 3.4 The Nieuwkoop center is essential for normal development. If a *Xenopus* embryo is divided into a dorsal and a ventral half at the four-cell stage, the dorsal half containing the Nieuwkoop center develops as a dorsalized embryo lacking a gut, while the ventral half, which has no Nieuwkoop center, is ventralized and lacks both dorsal and anterior structures.

Fig. 3.5 The Nieuwkoop center can specify a new dorsal side. Grafting vegetal cells containing the Nieuwkoop center from the dorsal side to the ventral side of a 32-cell *Xenopus* blastula results in the formation of a second axis and the development of a twinned embryo.

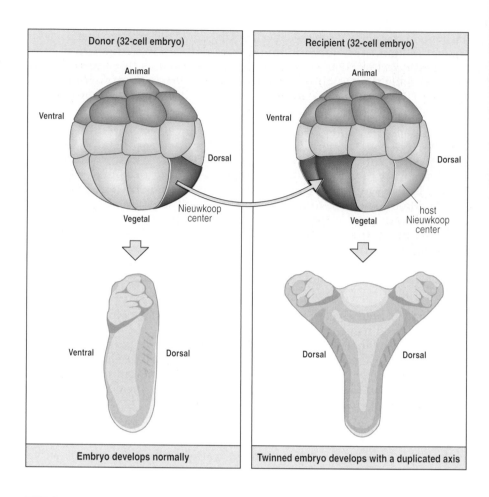

Donor (32-cell embryo)	Recipient (32-cell embryo)
Animal	Animal
Ventral / Dorsal	Ventral / Dorsal
Nieuwkoop center	host Nieuwkoop center
Vegetal	Vegetal
Ventral Dorsal	Dorsal Dorsal
Embryo develops normally	**Twinned embryo develops with a duplicated axis**

3-3   **The Nieuwkoop center is specified by cortical rotation.**

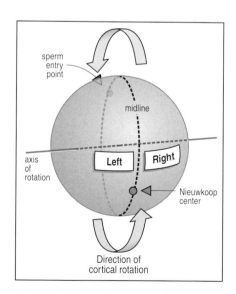

Fig. 3.6 Bilateral symmetry in amphibians results from cortical rotation. Cortical rotation toward the point of sperm entry defines the midline of the embryo, since the Nieuwkoop center lies on this midline and defines left and right sides.

How sperm entry triggers cortical rotation is as yet unknown, but it probably causes a signal to be generated and passed to the egg's cytoskeleton. During cortical rotation the cortex slides with respect to the rest of the cytoplasm; this involves the interaction of the cortex with parallel arrays of microtubules located in the cytoplasm of the vegetal region. The rotation brings about a localized change in the vegetal region on the side of the egg opposite the point of sperm entry. This change specifies the Nieuwkoop center in the vegetal region on that side, just below the equator, possibly because the vegetal cortex can now interact with more animal cytoplasm.

As well as specifying the dorsal side, cortical rotation can also establish the plane of bilateral symmetry of the future embryo. The rotation toward the point of sperm entry is greatest in the plane defined by the point of sperm entry. This is the future midline and contains the Nieuwkoop center (Fig. 3.6). The first cleavage usually passes through the midline, dividing the egg into two initially symmetrical halves.

Blocking cortical rotation, and thus the formation of the Nieuwkoop center, leads to highly abnormal development. Cortical rotation can be prevented by irradiating the vegetal side of the egg with ultraviolet (UV) light, which disrupts the microtubule array responsible for the movement. Embryos developing from such treated eggs are **ventralized**; they are deficient in structures normally formed on the dorsal side (Fig. 3.7), and develop excessive amounts of blood-forming mesoderm, a tissue normally only present at the embryo's ventral midline. With increasing

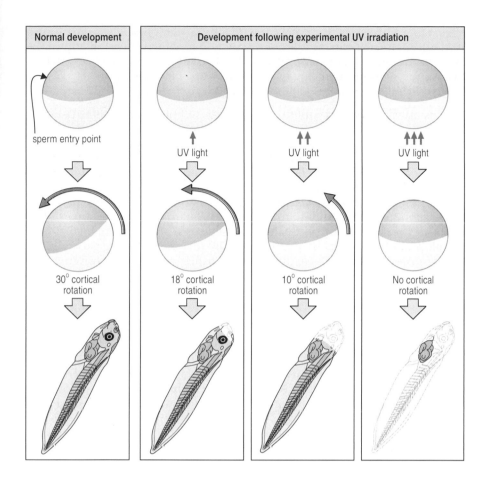

Normal development	Development following experimental UV irradiation

sperm entry point

UV light

UV light

UV light

30° cortical rotation

18° cortical rotation

10° cortical rotation

No cortical rotation

Fig. 3.7 The completeness of the body axis is related to the amount of cortical rotation. Rotation can be prevented by UV irradiation of the vegetal region of the egg. The smaller the degree of rotation, the more defects there are in anterior and dorsal regions of the embryo. The shaded areas, shown in the embryos at the bottom, represent the structures that still develop after the degree of cortical rotation shown.

doses of radiation, both dorsal and anterior structures are lost and the ventralized embryo appears as little more than a small, distorted cylinder. It has developed in the same way as an isolated ventral half of a four-cell embryo that lacks a Nieuwkoop center (see Fig. 3.4).

UV-irradiated eggs can be rescued by establishing a new Nieuwkoop center. This can be done by re-orienting the eggs (and thus mimicking cortical rotation) after irradiation. They can also be rescued at a slightly later stage by a direct graft of dorsal cells containing the Nieuwkoop center from another embryo at a 32-cell stage, showing that it is not the cortical rotation itself but the specification of a Nieuwkoop center that is crucial. By establishing a Nieuwkoop center, both these treatments specify a dorsal side and enable the embryo to continue its development normally.

Whereas UV irradiation ventralizes the embryo, treatment with lithium chloride **dorsalizes** it, promoting the formation of dorsal and anterior structures at the expense of ventral and posterior structures. UV irradiation and lithium chloride probably produce their effects by interfering in some way either with the proteins involved in establishing the dorsoventral axis or with their distribution. Some of these proteins are considered in the next section.

3-4   Maternal proteins with dorsalizing and ventralizing effects have been identified.

How cortical movement establishes the Nieuwkoop center is not yet known, but some localized maternal proteins can both act like an additional Nieuwkoop center in a normal embryo, and rescue UV-irradiated

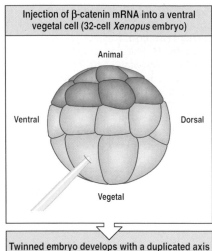

Injection of β-catenin mRNA into a ventral vegetal cell (32-cell *Xenopus* embryo)

Animal

Ventral   Dorsal

Vegetal

Twinned embryo develops with a duplicated axis

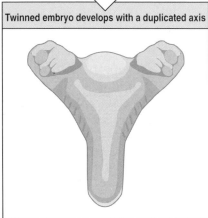

Fig. 3.8 Induction of a new dorsal side by injection of β-catenin mRNA. Injection of mRNA encoding β-catenin into ventral vegetal cells can specify a new Nieuwkoop center at the site of injection, leading to a twinned embryo. Some other vegetally localized maternal mRNAs, like *Vg-1*, also have similar effects.

eggs. One of these is the mRNA for β-catenin, which plays a part in the Wnt intracellular transduction pathway and is also a cell adhesion molecule. Some other maternal mRNAs localized at the vegetal pole, such as that for the secreted signaling molecules Xwnt-11 and Vg-1, can also rescue UV-irradiated eggs to a certain extent (Fig. 3.8).

The establishment and function of the Nieuwkoop center is caused by the suppression of ventralizing signals in the future dorsal region. One such ventralizing signal is the protein kinase GSK-3 (glycogen synthase kinase-3), which is translated from maternal mRNA and is uniformly expressed in the blastula. The fact that it is required for ventral development is shown by blocking its activity. This leads to an almost fully dorsalized embryoid, which in the most extreme cases has greatly overdeveloped head structures and no ventral or posterior structures. The fact that the activity of GSK-3 has to be suppressed for normal dorsal development is shown by experimentally forcing its expression on the future dorsal side. This results in a ventralized embryo, showing that GSK-3 is able to suppress either the formation or the signaling activity of the Nieuwkoop center. Exactly what normally suppresses GSK-3 activity on the dorsal side is uncertain but there is some evidence that proteins of the Wnt pathway of secreted signaling molecules, like β-catenin, are involved and are activated by cortical rotation. Lithium probably acts by inhibiting the action of GSK-3.

One of the main roles of the Nieuwkoop center is to specify another key dorsal signaling center—the **Spemann organizer**—which arises just above the Nieuwkoop center at the late blastula–early gastrula stage (see Fig. A3-03). As we shall see, signals originating in the Spemann organizer are involved in further patterning along both the antero-posterior and dorso-ventral axes of the embryo, and in inducing the central nervous system.

In *Xenopus* therefore, an external signal—sperm entry—sets the dorso-ventral axis, while the antero-posterior axis is related to the animal-vegetal axis, which is already laid down in the egg. We now look at axis development in the chick embryo.

3-5   **The dorso-ventral axis of the chick blastoderm is specified in relation to the yolk and the antero-posterior axis is set by gravity.**

Because there is so much yolk in a hen's egg, cleavage is restricted to a thin layer of cytoplasm and the chick embryo starts off as a disc of cells—the **blastoderm**—sitting on the top of the yolk. This positioning defines the dorso-ventral axis; the dorsal side of the blastoderm faces away from the yolk and the ventral side is adjacent to it. Like the amphibian blastula, the chick blastoderm is initially radially symmetric. This symmetry is broken when the posterior end of the embryo is specified. The future posterior end becomes evident soon after the egg is laid, when a denser area of cells appears at one side of the blastoderm; this is the **posterior marginal zone** (see Fig. 2.13). From this region, the primitive streak will develop. The streak defines the position of the antero-posterior axis within the blastoderm.

The position of the posterior marginal zone and thus of the posterior end of the antero-posterior axis is specified by gravity. During its passage through the hen's uterus, which takes about 20 hours, the fertilized egg moves pointed end first and rotates slowly around its long axis, each revolution taking about 6 minutes. Cleavage has already started at this stage, so the blastoderm contains thousands of cells when the egg is laid. The egg is obliquely tilted in the gravitational field, although the

Fig. 3.9 Gravity defines the antero-posterior axis of the chick. Rotation of the egg in the oviduct of the mother results in the blastoderm being tilted in the direction of rotation, though it tends to remain uppermost. The posterior marginal zone (P) develops at that side of the blastoderm which is uppermost and initiates the primitive streak. A = anterior.

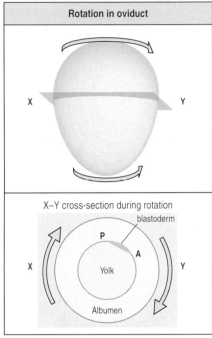

Rotation in oviduct

X–Y cross-section during rotation

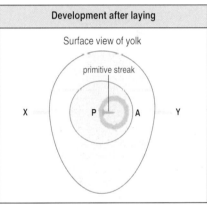

Development after laying

blastoderm tends to remain uppermost. The future posterior marginal zone will develop at the uppermost side of the blastoderm. As the shell and albumen (egg white) rotate, the embryo and its weight of yolk tend to return to the vertical. Thus, the future blastoderm region becomes tipped in the direction of rotation of the egg (Fig. 3.9).

The posterior marginal zone can be thought of as an organizing center analogous in some way to the Nieuwkoop center in *Xenopus* since it too can induce a new axis: if a fragment of the posterior marginal zone is grafted to another part of the marginal zone, it may induce a new primitive streak (Fig. 3.10). In general, however, only one axis develops in the grafted embryos—either the host's normal axis or one induced by the graft. This suggests that the more advanced of the two organizing centers inhibits streak formation elsewhere.

A chick gene related to Vg-1 of *Xenopus* is expressed at the site of primitive streak formation. If cells expressing this Vg-1 protein are grafted to another part of the marginal zone, they can induce a complete new primitive streak, thus simulating grafts from the posterior marginal zone.

3-6 | The axes of the mouse embryo are specified by cell–cell interactions.

There is no sign of polarity in the mouse egg and no evidence for localized maternal factors affecting later development. The very earliest stages of development of a mammalian egg differ considerably from those of either *Xenopus* or the chick, since the mammalian egg contains no yolk. Thus, the extra-embryonic structures of the placenta, which nourishes the embryo, must develop (see Chapter 2).

Early development of the mouse embryo involves segregation of the inner cell mass, which will form the embryo proper, from the trophectoderm, which gives rise to extra-embryonic structures associated with implantation and formation of the placenta. The early cleavages do not follow a well-ordered pattern, and some cleavages are parallel to the egg's surface so that a solid ball of cells (the morula) is formed with outer and inner cell populations. By the 32-cell stage, the mouse embryo has developed into a blastocyst, a hollow sphere of epithelium containing a

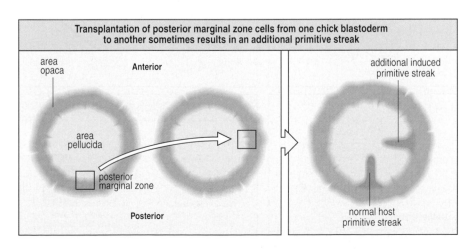

Fig. 3.10 The posterior marginal zone of the chick specifies the posterior end of the antero-posterior axis. Grafting posterior marginal zone cells to another site in the marginal zone can result in the formation of an extra primitive streak, which defines a new antero-posterior axis. This does not always occur. Usually the more advanced of the streaks is the only one to develop because it inhibits the development of the other.

small mass of some 10–15 cells attached at one end (Fig. 3.11). These cells are the inner cell mass, from which the embryo itself will develop, whereas the outer epithelium will form the trophectoderm, which forms only extra-embryonic structures.

The specification of cells as inner cell mass or trophectoderm depends on their relative positions in the cleaving embryo. Determination of their fate occurs only after the 32-cell stage, and during earlier stages all the cells seem to be equivalent in their ability to give rise to either tissue. The most direct evidence for the effect of position comes from taking individual cells (blastomeres) from disaggregated four-cell or eight-cell embryos, labeling them, and then combining the labeled blastomeres in different positions with respect to unlabeled blastomeres from another embryo. If the labeled cells are placed on the outside of a group of unlabeled cells they usually give rise to trophectoderm; if they are placed inside, so that they are surrounded by unlabeled cells, then they more often give rise to inner cell mass (see Fig. 3.11). If a whole embryo is surrounded by other blastomeres, it too can become part of the inner cell mass of a giant embryo. Aggregates composed entirely of either 'outside' or 'inside' cells of early embryos can also develop into normal blastocysts, showing that there is no specification of these cells, other than by their position, at this stage.

The mechanism by which the dorso-ventral axis is established in mammals is related to the position of the inner cell mass. The mouse embryo is a spheroidal ball of cells up to the blastocyst stage when, following the specification of the inner cell mass and trophectoderm, a blastocoel cavity develops asymmetrically within the blastocyst, leaving the inner cell mass attached to one part of the trophectoderm. The blastocyst now has a distinct axis running from the site where the inner cell mass is attached—the embryonic pole—to the opposite end. It is generally assumed that this axis, which corresponds to the dorso-ventral axis, persists through implantation until the beginning of gastrulation (see Section 2-3).

By this time (about 4½ days after fertilization) the inner cell mass has become differentiated into two tissues: primary or primitive endoderm on its blastocoelic surface (which will form extra-embryonic structures);

Fig. 3.11 The specification of the inner cell mass of a mouse embryo depends on the relative position of the cells with respect to the inside and outside of the embryo. If labeled blastomeres from a four-cell mouse embryo are separated and combined with unlabeled blastomeres from another embryo, it can be seen that blastomeres on the outside of the aggregate more often give rise to trophectoderm, 97% of the labeled cells ending up in that layer. The reverse is true for the origin of the inner cell mass, blastomeres on the inside giving rise most often to this.

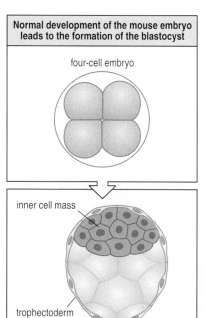

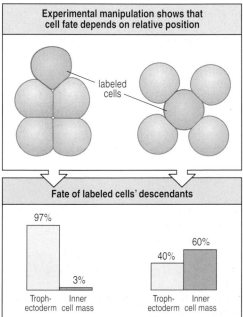

and the **epiblast** within, from which the embryo and some extra-embryonic structures will develop. The blastocyst implants in the uterine wall, and the trophectoderm at the embryonic pole proliferates to form the ectoplacental cone, producing extra-embryonic ectoderm that pushes the inner cell mass across the blastocoel. A cavity—the proam-niotic cavity—is then formed within the epiblast as its cells proliferate. The epiblast (embryonic ectoderm), which is now an epithelial sheet, is formed into a cup shape (U-shaped in cross-section, Fig. 3.12). During this process there is considerable cell mixing, so it is not possible at this early stage to identify individual cells that will become dorsal or ventral.

How the antero-posterior axis of mammalian embryos is established is also unknown. Since it is unrelated to any maternal determinants, it must involve cell–cell interactions. It is possible that signals from the uterus at the time of implantation are involved, since the antero-posterior axis of the embryo is roughly perpendicular to the long axis of the uterus.

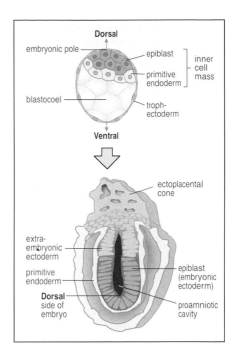

Fig. 3.12 The specification of the dorso-ventral axis of the mouse embryo. The epiblast extends ventrally into the blastocoel and forms a cup, seen as U-shaped in the cross-section above. However, the cells that will form the dorsal side are not yet specified as the epiblast is still a single layer of cells.

3-7 Specification of left-right handedness of internal organs requires special mechanisms.

Vertebrates are bilaterally symmetric about the midline of the body for many structures, such as eyes, ears, and limbs. But, while the verte-brate body is outwardly symmetric, most internal organs are in fact asymmetric with respect to the left and right sides. In mice, for example, the heart is on the left side, the right lung has more lobes than the left, the stomach and spleen lie to the left, and the liver has a single left lobe. The handedness of organs is remarkably consistent, but there are rare individuals, about one in 10,000 in humans, who have the condition known as *situs inversus*, a complete mirror-image reversal of handed-ness. Such people are generally asymptomatic even though all their organs are reversed.

Specification of left and right is fundamentally different from specifying the other axes of the embryo, as left and right have meaning only after the antero-posterior and dorso-ventral axes have been established. If one of these axes is reversed then so too will be the left-right axis (it is for this reason that handedness is reversed when you look in a mirror: your dorso-ventral axis is reversed, and hence left becomes right and *vice versa*). While the molecular mechanism whereby organ handed-ness is specified remains an intriguing mystery, one suggestion is that it requires asymmetry at the molecular level to be converted to an asymmetry at the cellular and multicellular level. If that were so, the asymmetric molecules or molecular structure would need to be oriented with respect to both the antero-posterior and dorso-ventral axes.

3-8 Organ handedness in vertebrates is under genetic control.

In mice, the *iv* gene is involved in specifying organ handedness. In animals homozygous for a mutant *iv* allele, the handedness of the organs is reversed in half of the animals (Fig. 3.13), implying that the specification of handedness has become random in these mutants. The mutation therefore affects not the generation of asymmetry itself but the mecha-nism by which it is normally consistently biased to one side. These mutant mice quite often show heterotaxis, the condition where organs of normal and inverted asymmetry are present in the same animal. This suggests that the generation of asymmetry in different organs may occur independently.

Fig. 3.13 Left-right asymmetry of the mouse heart is under genetic control. Each photograph shows a mouse heart viewed anteriorly after the loops have formed. The normal asymmetry of the heart results in it looping to the left, as indicated by the arrow (left panel). 50% of mice that are homozygous for the mutation in the *iv* gene have hearts that loop to the right (right panel). Scale bar = 0.1 mm. Photographs courtesy of N. Brown.

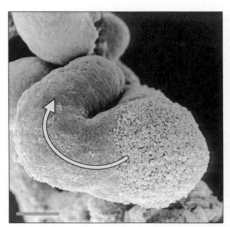
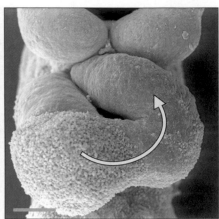

In humans, *situs inversus* sometimes occurs with Kartagener's syndrome, which is a recessive defect. As in the *iv* mice, handedness becomes random; that is 50% of those with the syndrome have altered asymmetry. In individuals with this syndrome, the cilia that line the surface of respiratory organs, such as the lungs, are nonfunctional and do not beat, and so these individuals suffer from respiratory problems. The immotile cilia lack the motor protein dynein, which is essential for their movement. Dynein has other roles in cells, where it is associated with microtubules, and so microtubules and other cytoskeletal structures may play a role in generating asymmetry, since they themselves are asymmetric structures.

The phenotype of the *iv* mouse suggests that the mechanism for generating asymmetry is basically random, but is normally biased by some unknown mechanism toward one side only. However, another, and still unidentified, mutation causes the complete reversal of handedness in mice. Identification of this mutated gene is awaited with great interest.

In the early chick embryo, several genes are expressed asymmetrically with respect to Hensen's node (see Section 2-3), which is at the anterior end of the primitive streak, and may thus be involved in setting the pattern for organ asymmetry. The gene *Sonic hedgehog*, whose protein product is implicated in a variety of developmental processes (see Section 4-2), is expressed on the left side only of Hensen's node. Activin and its receptor are expressed on the right side and repress *Sonic hedgehog* expression on this side. On the left side, *Sonic hedgehog* induces the expression of *Nodal*, another TGF-β family member. If this pattern of expression is made symmetric by placing a pellet of cells secreting Sonic hedgehog protein on the right-hand side, then organ asymmetry is randomized, as in the mouse *iv* mutant. *Nodal* is also asymmetrically expressed in the mouse.

In *Xenopus*, localized injection of processed Vg-1 protein into the right side of an early embryo randomizes organ asymmetry, suggesting that asymmetrical distribution of Vg-1 in the Nieuwkoop center could underlie left/right asymmetry. The generation of asymmetry may well involve orientation of an asymmetric molecule, which specifies a small differential between left and right sides.

Summary.

Setting-up the body axes in vertebrates involves maternal factors, external influences, and cell–cell interactions. In the amphibian embryo, maternal factors determine the animal-vegetal axis, which approximately corresponds to the antero-posterior axis, whereas the dorso-ventral axis is specified by the site of sperm entry and the resulting cortical rotation, which leads to the establishment of the Nieuwkoop center. In chick embryos, the dorso-ventral axis is specified at cleavage in relation to the yolk while the setting of the antero-posterior axis involves gravity, which determines the side of the blastoderm at which the posterior marginal zone, and thus the primitive streak, will form. Specification of the axes in the mouse embryo does not involve any maternal component. They are established in the cells of the inner cell mass from which the embryo proper develops. The dorso-ventral axis is related to the position of the inner cell mass on the trophectoderm, while the antero-posterior axis may be set only at implantation. The generation of the consistent left-right organ asymmetry found in vertebrates is under genetic control.

Summary: vertebrate axis determination		
	Dorso-ventral axis	**Antero-posterior axis**
Xenopus	sperm entry point and cortical rotation. Dorsal side and Nieuwkoop center form on side opposite to sperm entry. GSK-3 suppressed on dorsal side	related to maternal determinants. Future anterior end develops from animal region
Chick	cellularization of blastoderm on top of yolk	gravity
Mouse	interaction between inner cell mass and trophectoderm	intercellular interactions?

The origin and specification of the germ layers.

We have seen in the preceding sections how the main axes are laid down in various vertebrate embryos. We now focus on the earliest patterning of the embryo with respect to these axes: the specification of the three germ layers—endoderm, mesoderm, and ectoderm—and their further diversification.

All the tissues of the body are derived from these three germ layers. The mesoderm becomes subdivided into cells that give rise to notochord, to muscle, to heart and kidney, and to blood-forming tissues, amongst others. The ectoderm becomes subdivided into cells that give rise to the epidermis and that develop into the nervous system. The endoderm gives rise to the gut and organs such as the lungs. We first look at the **fate maps** (see Section 1-10) of early embryos of different vertebrates, which tell us which tissues the different regions of the embryo give rise to. We then consider how the germ layers are specified and subdivided, with the main focus on *Xenopus*, in which these processes are best understood and where some of the genes and proteins involved have been identified. We will see that much of the early patterning can be accounted for by a four-signal model.

3-9 | **A fate map of the amphibian blastula is constructed by following the fate of labeled cells.**

Examination of the *Xenopus* blastula at the 32-cell stage gives no indication of how the different regions will develop, but individual cells can be identified. By following the fate of individual cells, or groups of cells, we can make a map on the blastula surface showing the regions that will give rise to, for example, somites, brain, spinal cord, and gut. The fate map shows where the tissues of each germ layer normally come from, but it indicates neither the full potential of each region for development nor to what extent its fate is already specified or determined in the blastula. Early vertebrate embryos have considerable capacity for regulation when pieces are removed or transplanted to different parts of the same embryo (see Section 1-10). This implies considerable developmental plasticity at this early stage and also that the actual fate of cells is heavily dependent on the signals they receive from neighboring cells.

One way of making a fate map is to stain various parts of the surface of the early embryo with a lipophilic dye such as diI, and observe where the labeled region ends up. Individual cells can also be labeled by injection of stable high molecular weight molecules such as rhodamine-labeled dextran, which cannot pass through cell membranes and so are restricted to the injected cell and its progeny; as rhodamine fluoresces red in UV light, the rhodamine dextran can be easily detected under a UV microscope. Fig. 3.14 shows a *Xenopus* embryo labeled for fate mapping with the green-fluorescing dye fluorescein-dextran-amine.

The fate map of the late *Xenopus* blastula (Fig. 3.15) shows that the yolky vegetal region, which occupies the lower third of the spherical blastula, gives rise to most of the endoderm. The yolk provides all the nutrition for the developing embryo, and is gradually used up as development proceeds. At the other pole, the animal hemisphere becomes ectoderm, which becomes further diversified into epidermis and the future nervous tissue. The mesoderm forms a belt-like region, known as the **marginal zone**, around the equator of the blastula. In *Xenopus*, but not in all amphibians, a thin outer layer of presumptive endoderm overlies the presumptive mesoderm in the marginal zone.

The fate map of the blastula makes it clear why gastrulation is necessary. At the blastula stage, mesodermal tissues that will form internal structures such as the gut, muscle, and internal organs, are on the outside of the embryo. During gastrulation, the marginal zone moves into the interior through the dorsal blastopore, which lies above the Nieuwkoop center. The fate map of the mesoderm (see Fig. 3.15) shows that it becomes subdivided along the dorso-ventral axis of the blastula. The most dorsal mesoderm gives rise to the notochord, followed, as we move ventrally, by somites (which give rise to muscle tissue), lateral plate (which contains heart and kidney mesoderm), and blood islands (tissue where hematopoiesis first occurs in the embryo). There are also important differences between the future dorsal and ventral sides of the animal

Fig. 3.14 Fate mapping of the early *Xenopus* embryo. Left panel: a single cell in the embryo, C3, is labeled by injecting fluorescein-dextran-amine, which fluoresces green under UV light. Right panel: a cross-section of the embryo, made at the tailbud stage, shows that the labeled cell has given rise to mesoderm cells on one side of the embryo. Scale bar = 0.5 mm. Photograph courtesy of L. Dale.

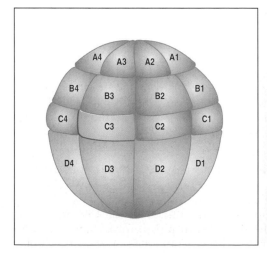

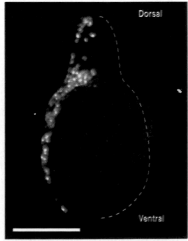

Fig. 3.15 Fate map of a late *Xenopus* blastula. The ectoderm gives rise to the epidermis and nervous system. Along the dorso-ventral axis the mesoderm gives rise to notochord, somites, heart, kidneys, and blood. In *Xenopus*, although not in all amphibians, there is also endoderm (not shown here) overlying the mesoderm in the marginal zone.

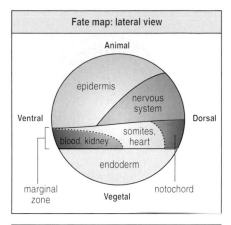

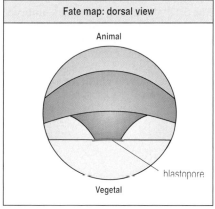

hemisphere: the epidermis comes mainly from the ventral side of the animal hemisphere, whereas the nervous system comes from the dorsal side. The epidermis spreads to cover the whole of the embryo after neural tube formation.

The terms dorsal and ventral in relation to the fate map can be somewhat confusing because the fate map does not correspond exactly to a neat set of axes at right angles to each other. As a result of cell movements during gastrulation, cells from the dorsal side of the blastula give rise to some ventral parts of the anterior end of the embryo, such as the head, as well as to dorsal structures, and will also form some other ventral structures, such as the heart. The ventral region gives rise to ventral structures in the anterior part of the embryo but will also form some dorsal structures posteriorly. This is why the 'dorsalized' embryos described in Section 3-4 have overdeveloped anterior structures and lack posterior regions.

The fate map of *Xenopus* has reasonably well-defined boundaries. However, there is some local cell movement and mixing of cells as the embryo develops and deeper cells move to the surface. The fate map that one can obtain in *Xenopus* in no way implies that the fate of the cells in the early embryo is fixed, but rather reflects the stereotyped pattern of tissue movements that carry cells to their positions in the later embryo.

In contrast to the fate map, which gives no indication of the actual differences between cells at the time they are labeled, a **specification map** gives some indication of such differences (see Section 1-10). A specification map of the blastula is constructed by culturing small pieces of the blastula in a simple culture medium and observing what tissues they form. The specification map of the *Xenopus* blastula corresponds quite well to some features of the fate map, but there are important differences, particularly in the ectodermal and mesodermal regions (Fig. 3.16). No neural tissue develops from explants of cells from the animal half of the blastula, and no muscle develops from most mesodermal fragments. This shows that the ectoderm has not yet become differentiated into prospective neural cells and prospective epidermal cells, and that prospective muscle has not yet been specified within the mesoderm. Nevertheless, the specification map shows that there are already important regional differences in cell states at the blastula stage.

3-10   The fate maps of vertebrates are variations on a basic plan.

Fate maps of the early embryos of chick, mouse, and zebrafish have been prepared using techniques essentially similar to those used for *Xenopus*: cells in the early embryo are labeled and their fate followed.

A fate map of the chick embryo cannot be made at the early blastoderm stage that corresponds roughly to the *Xenopus* blastula. This is partly because so much of the chick embryo comes from the posterior marginal zone, which is still a very small region of the total blastoderm at this stage. Unlike *Xenopus*, there is considerable cell proliferation and growth in the chick embryo during primitive streak formation and

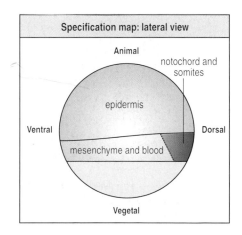

Fig. 3.16 Specification map of a *Xenopus* late blastula. The specification map is constructed from the results of experiments showing how isolated fragments of blastula develop in a simple culture medium.

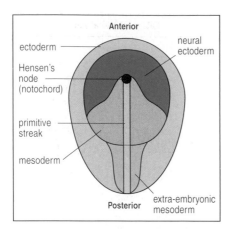

Fig. 3.17 Fate map of a chick embryo when the primitive streak has fully formed. The diagram shows a view of the dorsal surface of the embryo. Almost all the endoderm has already moved through the streak to form a lower layer, so is not shown.

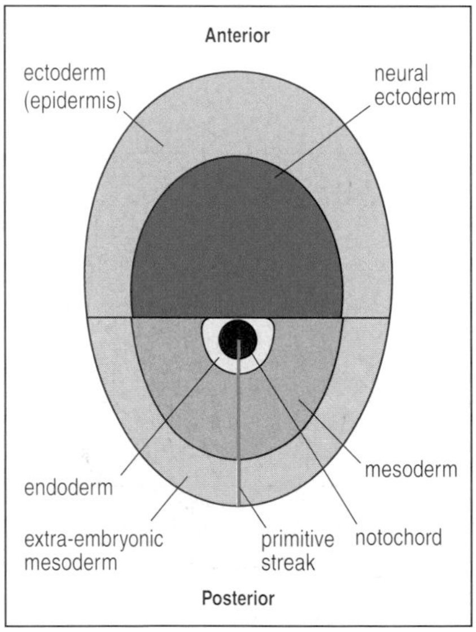

gastrulation. There are also extensive cell movements both before and during the emergence of the primitive streak (see Fig. 2.13) and gastrulation. Once the primitive streak has formed, the picture becomes clearer, and presumptive endoderm, mesoderm, and ectoderm can be mapped (Fig. 3.17).

At the stage shown, the blastoderm has become a three-layered structure. Cells have ingressed through the primitive streak into the interior to form mesodermal and endodermal layers. Most of the cells that now form the outer surface of the blastoderm are prospective ectoderm and will form neural tube and epidermis, but an area of cells has still to move inside the streak; these cells will give rise to mesoderm. Hensen's node, an aggregation of cells at the anterior end of the streak, is prospective mesoderm; as the node regresses it leaves cells behind that will form the notochord and also contribute to the somites. In the mesodermal layers of the blastoderm, the mesoderm lying along the antero-posterior midline will give rise to somites and is surrounded by cells that will form the lateral plate mesoderm and structures such as the heart and kidney. In the lowest layer of the embryo, closest to the yolk, the presumptive endoderm is surrounded by cells that will form extra-embryonic structures.

It is not possible to construct a fate map for the very early mouse embryo, as cells from the inner cell mass of embryos that are less than 3½ days old will give rise to many different embryonic tissues, as well as to some extra-embryonic structures, such as the visceral and parietal endoderm. At about 4–4½ days, the inner cell mass gives rise to an outer layer of cells, the primitive endoderm (see Fig. 3.12). The cells lying between the primitive endoderm and the polar trophectoderm comprise the primitive ectoderm or epiblast. Although the primitive endoderm gives rise to extra-embryonic structures only, the primitive ectoderm gives rise to all of the embryo proper and all the extra-embryonic mesodermal structures.

At 6–7 days gestation the mouse epiblast becomes transformed into the three germ layers by the formation of a primitive streak and gastrulation. Gastrulation in the mouse is essentially very similar to gastrulation in the chick, but the mouse epiblast is at this stage folded into a cup, which makes the process more difficult to follow. A reasonably detailed fate map of this stage has been established by tracing the descendants of single cells that have been labeled by injection with a dye. There is, however, extensive cell mixing and cell proliferation in the epiblast. Descendants of a single cell may spread widely and give rise to cells of different germ layers, so that only about 50% of the labeled clones have progeny in only one germ layer.

Nevertheless, the fate map obtained is basically similar to that of the chick at the primitive streak stage, making allowances for the fact that the mouse epiblast is cup shaped, in contrast to the sheet-like chick epiblast (Fig. 3.18). The node forms at the anterior end of the primitive streak in the mouse embryo, and gives rise to the notochord and part of the somites. The middle part of the streak gives rise mainly to lateral plate mesoderm, while the posterior part of the streak provides the extra-embryonic mesoderm of the amnion, visceral yolk sac, and allantois.

Because there is extensive cell mixing during the transition from blastula to gastrula in the zebrafish embryo, it is not possible to construct a reproducible fate map at cleavage stages. In this the zebrafish resembles the mouse. The zebrafish late blastula comprises a cup-shaped blastoderm of deep cells and a thin overlying layer, sitting on top of a large yolk cell. The overlying layer is largely protective and is eventually lost. At the beginning of gastrulation, the fate of deep layer cells, from which all the cells of the embryo itself will come, is correlated with their position in

Fig. 3.18 Fate map of a mouse at the late gastrula stage. The embryo is depicted as if the 'cup' has been flattened and is viewed from the dorsal side. At this stage the primitive streak is at its full length.

respect of the animal pole. Cells at the margin of the blastoderm give rise to the endoderm, cells slightly further toward the animal pole to mesoderm, while ectodermal cells come from the blastoderm nearest the animal pole (Fig. 3.19).

A fate map for each of the germ layers has also been constructed: in the ectoderm, for example, forebrain structures come from a region near the animal pole, while hindbrain structures come from nearer the margin. In the mesoderm, the future notochord is located on the dorsal side whereas presumptive blood-forming tissue is located ventrally. In general terms, the fate map of the zebrafish is rather similar to that of an amphibian, if one imagines the vegetal region of the amphibian blastula being replaced by one large yolk cell.

The fate maps of the different vertebrates are rather similar when one looks at the relationship between the germ layers and the site of ingression of cells at gastrulation (Fig. 3.20). The differences are due mainly to the yolkiness of the egg, which determines the pattern of cleavage and influences the shape of the early embryo. The similarity in relationship between the germ layers implies that similar mechanisms must be involved in their specification. Fate maps are not specification maps, and they do not reflect the full potential for development of the cells of these early embryos. At the late blastula and early gastrula stages, when these maps are made, vertebrate embryos are still capable of considerable regulation.

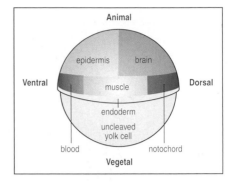

Fig. 3.19 Fate map of zebrafish at the early gastrula stage. The three germ layers come from the blastoderm which sits on the lower hemisphere of an uncleaved yolk cell. The endoderm comes from the margin of the blastoderm and some has already moved inside.

3-11 Cells of early vertebrate embryos do not yet have their fates determined.

All early vertebrate embryos have considerable powers of regulation when parts of the embryo are removed or rearranged. For example, fragments of a fertilized *Xenopus* egg that are only one fourth of the normal volume develop into more or less normally proportioned but small embryos. There must therefore be a patterning mechanism involving cell interactions that can cope with such differences in size. Such experiments show that the fate of the cells can be altered. Cells at this stage are not yet determined (see Section 1-10), and their potential for development is greater than their position on the fate map suggests.

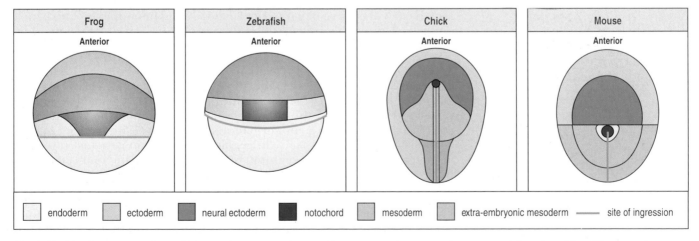

Fig. 3.20 The fate maps of vertebrate embryos at comparable developmental stages. In spite of all the differences in early development, the fate maps of vertebrate embryos at stages equivalent to a late blastula or early gastrula show similarities. All maps are shown in a dorsal view. The future notochord

mesoderm occupies a central dorsal position. The neural ectoderm lies adjacent to the notochord, with the rest of the ectoderm anterior to it. The mouse fate map depicts the late gastrula stage. The future ectoderm of the zebrafish is on its ventral side.

However, there are limitations to the capacity for regulation. Isolated animal and vegetal halves of an eight-cell *Xenopus* embryo do not develop normally; while dorsal halves of eight-cell embryos regulate to produce a reasonably normal embryo, early ventral halves do not, making an abnormal embryo lacking anterior and dorsal structures (see Fig. 3.4) and with much less muscle than the fate map would suggest. As we saw earlier, these results are linked to the presence or absence of a Nieuwkoop center in the fragments.

The early embryo's powers of regulation reflect the state of determination of individual cells. The state of determination of cells, or small regions of an embryo, can be studied by transplanting them to a different region of a host embryo, and seeing how they develop (see Section 1-10). If they are already determined, they will develop according to their original position. If they are not yet determined, they will develop in line with their new position. This can be shown experimentally by introducing a single labeled cell from a *Xenopus* blastula into the blastocoel of a later-stage host embryo and following its fate. The transferred cell divides, and during gastrulation its progeny become distributed to different parts of the embryo. The results, in general, show that cell transplants made from early blastulas are not yet determined; their progeny will differentiate according to the signals they receive at their new location. So cells from the vegetal pole, which would normally form endoderm, can contribute to a wide variety of other tissues such as muscle or nervous system, when grafted at an early stage. Similarly, early animal pole cells, whose normal fate is epidermis or nervous tissue, can form endoderm or mesoderm. With time, cells gradually become determined, so that similar cells taken from later stage blastulas and early gastrulas develop according to their fate at the time of transplantation.

The cells in the inner cell mass of the mouse embryo are not yet determined. We have already seen this in relation to early mouse development, where cells of the inner cell mass and the trophectoderm are specified purely by their relative position on the inside or the outside of the embryo (see Fig. 3.11). The cells of the inner cell mass itself are pluripotent up to 4½ days after fertilization—they can give rise to many cell types. If injected into the inner cell mass of another blastocyst of similar age they can contribute to all tissues of the embryo, including the germ cells. This property enables the creation of **chimeric** mice—that is mice with cells of two different genetic constitutions. Lines of embryonic stem cells (ES cells) can be derived from cells of the inner cell mass and these will behave like inner cell mass cells when injected into a host embryo. This enables the creation of **transgenic** mice by the introduction of embryonic stem cells carrying particular mutations (Box 3C).

Early mouse embryos can regulate to achieve the correct size. Giant embryos formed by aggregation of several embryos in early cleavage stages can achieve normal size within about 6 days by reducing cell proliferation. One can thus combine embryos of differing genetic constitution to give rise to a chimeric mouse (Fig. 3.21). The mouse embryo retains considerable capacity to regulate until late in gastrulation. Even at the primitive streak stage, up to 80% of the cells of the epiblast can be destroyed with the drug mitomycin C and the embryo can still recover and develop with relatively minor abnormalities.

Further evidence for regulation in mammals comes from twinning. Twins can result from a separation of the cells at the two-cell stage, but twinning can also occur in humans at a stage as late as 7 days of gestation, when the primitive streak has already started to form. Since early vertebrate embryos show considerable capacity for regulation and many of the cells are not determined, this implies that cell–cell communication must determine cell fate.

Box 3C Transgenic mice.

ES cells (carrying a mutation in a single gene) in culture

ES cells injected into inner cell mass of normal blastocyst

Chimeric animal produces sperm carrying the mutation

When studying the role of a particular gene in development, it is an enormous advantage to be able to study the effects of a mutation in that gene. One way of obtaining an animal with the desired mutation is simply to wait for it to turn up in the population, but in vertebrates the wait may be very long indeed. Developmental mutations, in particular, are rarely identified. In mice, however, it is possible to produce animals with a particular mutant genetic constitution using transgenic techniques.

One means of producing transgenic mice with a desired mutation is by introducing **embryonic stem cells** (**ES cells**) carrying the mutation into the blastocyst. ES cells are cultured cells derived from the inner cell mass; they can be maintained in culture indefinitely and grown in large numbers. Inner cell mass cells introduced into the inner cell mass of another embryo will populate all of the mouse's tissues and will contribute to the germ cells.

ES cells can be genetically manipulated in culture to produce mutant cells in which a certain gene or genes have been inactivated or new genes introduced. This technique is particularly powerful for creating loss-of-function mutations to ascertain the role of particular genes in development (see Box 4B, page 108). Mutations that lead to the complete absence of the function of the gene are known as gene **knock-outs**. Some mutations do not lead to loss of function, but to a change in function.

Since the initial transgenic animals are a chimeric mixture of mutant and normal cells they may show few if any effects of the mutation, but if they carry the mutant gene—the transgene—in their germ cells, it is possible by inter-breeding to produce a non-chimeric transgenic animal in which the mutation is present in either the heterozygous or the homozygous state. Almost all transmission of transgenes is via sperm.

We now examine mechanisms for specifying the germ layers, focusing particularly on the induction of mesoderm in *Xenopus*, which is the best understood.

| 3-12 | **In *Xenopus* the mesoderm is induced by signals from the vegetal region.** |

At the time the *Xenopus* egg is laid, there are already differences along the animal-vegetal axis (see Section 3-1). When explants from different regions of the early blastula are cultured in a simple medium containing the necessary salts for ion balance, tissue from the animal region will form a ball of epidermal cells, while explants from the vegetal region are endodermal in their development. These results are in line with the normal fates of these regions. It is thus generally accepted that all the ectoderm and most endoderm are specified by maternal factors in the

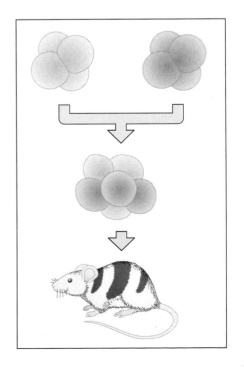

Fig. 3.21 Fusion of mouse embryos gives rise to a chimera. If an eight-cell stage embryo of an unpigmented strain of mouse is fused with a similar embryo of a pigmented strain, the resulting embryo will give rise to a chimeric animal, with a mixture of 'pigmented' and 'unpigmented' cells. The distribution of the different cells in the skin gives this chimera a stripy coat.

egg. There is no evidence that any signals from other regions of the embryo are necessary for their specification, except in the case of some endoderm that forms in the marginal zone. The mesoderm, however, is different.

The formation of mesoderm in amphibians is totally dependent on inducing signals from the vegetal region of the blastula, which convert a band of adjacent animal cells from an ectodermal fate to a mesodermal one (Fig. 3.22). The standard experiment for studying mesoderm induction is to take a small piece of tissue from the animal half of a blastula (the **animal cap**), which would normally produce only ectoderm, and place it in contact with tissue from the vegetal region. This combination is cultured for 3 days and examined for the presence of mesoderm. Mesoderm can be distinguished by its histology as, after 3 days culture, it may contain muscle, notochord, blood, and loose mesenchyme (connective tissue). It can also be identified by the typical proteins that cells of mesodermal origin may produce, such as muscle-specific actin, which can be detected with antibodies.

Using these criteria, one finds that the animal portion of the combined explant not only forms epidermis, but also a substantial amount of mesoderm, which is restricted to the region in immediate contact with the vegetal tissue (see Fig. 3.22). One can confirm that it is indeed the animal cap cells that are forming mesoderm, and not the vegetal cells, by pre-labeling the animal region of the blastula with a cell lineage marker such as diI and showing that the labeled cells form the mesoderm. Clearly, the vegetal region is producing a signal or signals that can induce mesoderm. Some endoderm is also induced in this region, probably by similar signals.

Fig. 3.22 Induction of mesoderm by the vegetal region in the *Xenopus* blastula. Top panels: explants of animal cap cells or vegetal cells on their own from a late blastula form only ectoderm or endoderm, respectively. Explants from the equatorial region, where animal and vegetal regions are adjacent, form mesodermal tissues (mesenchyme, blood cells like erythrocytes, notochord, and muscle), showing that mesoderm induction has taken place. The reason for differences between the mesodermal tissues formed by ventral and dorsal explants at this stage are explained later. Bottom panels: when pieces of animal and vegetal regions from an early blastula are combined and cultured for a few days, mesoderm is induced from the animal cap tissue. This mesoderm contains notochord, muscle, blood, and loose mesenchyme.

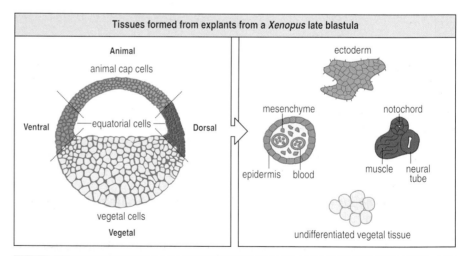

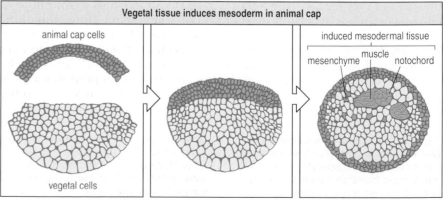

3-13   The mesoderm is induced by a diffusible signal during a limited period of competence.

The explant system in which pieces of tissue are placed in contact with one another (described in the previous section) is well suited for experimental investigation of the mesoderm-inducing signal and the animal cap cells' response. If the explanted animal and vegetal fragments are separated by a filter with pores too small to allow cell contacts to develop, induction still takes place. This suggests that the inducing signal is in the form of secreted molecules that diffuse across the extracellular space, and does not pass directly from cell to cell via cell junctions (see Fig. 1.20).

The distance over which the signal acts to induce muscle is small, about 80 μm, or four cell diameters in the blastula. This can be shown by blocking both cell movement and cell division in the animal cap explant with the drug cytochalasin; the boundary between the inducing vegetal tissue and the induced mesodermal tissue can then be clearly distinguished. Of course, the distance of 80 μm reflects only the response of the induced cells, and the signal may well be present further away, but at a concentration below that necessary for mesoderm induction to occur.

The animal cap is **competent** to respond to the inducing signal only for a limited time. Using explanted tissues from embryos of different ages, it has been shown that mesoderm induction starts at around the 32-cell stage and is almost complete by the time gastrulation starts. Only a short period of contact is required between the inducing vegetal region and the responding animal cap cells: 2 hours is sufficient to give some induction of muscle, and 5 hours' contact leads to complete induction of mesoderm tissues. The animal cap loses its competence to respond about 11 hours after fertilization.

Induction of a mesodermal tissue, muscle, appears to depend on a **community effect** in the responding cells. A few animal cap cells placed on vegetal tissue will not be induced to express muscle-specific genes. Even when a small number of individual cells are placed between two groups of vegetal cells, induction does not occur. By contrast, larger aggregates of animal cap cells respond by strongly expressing muscle-specific genes (Fig. 3.23). The explanation for this is that the induced cells produce a factor that has to reach a sufficiently high concentration for muscle differentiation to occur. This concentration is reached only when there are sufficient cells within a confined volume.

What then determines the extent of mesoderm induction during normal development? One possibility is that the inducing signal from the vegetal region forms a gradient with a threshold below which mesoderm induction does not occur. We look next at the mechanism controlling the temporal sequence of events following induction.

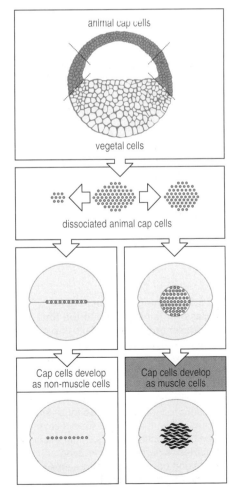

Fig. 3.23 The community effect. One or a small number of animal cap cells in contact with vegetal tissue are not induced to become mesodermal cells and do not begin to express mesodermal markers such as muscle-specific proteins. A sufficiently large number of animal cap cells must be present for induction of muscle differentiation to occur.

| 3-14 | **An intrinsic timing mechanism controls the time of expression of mesoderm-specific genes.** |

Developmental events need to be coordinated both in space and time. Relatively little attention has been given to the timing of developmental processes, which is nevertheless of great importance. For example, following mesoderm induction a cascade of events leads eventually to gastrulation, and we need to understand how these are timed so that they occur in the correct sequence.

As a result of mesoderm induction, muscle-specific genes begin to be expressed in the mesoderm at the mid-gastrula stage. One might expect the timing of this gene expression to be closely coupled to the time at which the mesoderm is induced—but it is not. There is a period of about 7 hours at the blastula stage during which animal cells are competent to respond to a mesoderm-inducing signal. Mesoderm-specific gene expression always starts about 5 hours after the end of this time. For some induction to occur, the animal cap cells require exposure to inducing signal only for a period of about 2 hours. Irrespective of when during the 7-hour period of competence the cells are exposed to the 2-hour induction, the time at which muscle-specific gene expression starts remains the same (Fig. 3.24). Muscle gene expression can occur as early as 5 hours after induction, if induction occurs late in the period of competence, or as late as 9 hours after induction, if induction occurs early in the competent period. These results suggest that there is an independent timing mechanism by which the cells monitor the time elapsed since fertilization and then, provided they have been induced, express muscle-specific genes.

The time at which animal cells lose the ability to respond to the mesoderm-inducing signal also seems to be fixed by an intrinsic timing mechanism. This timing is unaffected by blocking cleavage or by the time of the onset of zygotic gene expression at the mid-blastula transition (see Section 3-19). It also persists even if animal cap tissue is dissociated into single cells several hours before the normal time of transition and cultured so that the cells cannot communicate with each other. A timing mechanism in which the concentration of some protein increases or decreases to a threshold level, is ruled out as, surprisingly, no new protein synthesis is required. Possibly the timing mechanism is based on the breakdown of some protein or the synthesis of some other class of molecule.

An extended period of competence gives the embryo a certain latitude as to when mesoderm induction actually takes place. This means that the timing of the inductive signal does not have to be rigorously linked to the time when the animal region is competent.

Fig. 3.24 Timing of muscle gene expression is not linked to the time of mesoderm induction. Animal cap cells isolated from an early *Xenopus* blastula are competent to respond to mesoderm-inducing signals only for a period of about 7 hours, between 4 and 11 hours after fertilization. For expression to occur, exposure to inducer must be for at least 2 hours within this period. Irrespective of when the induction occurs within this competence period, muscle gene expression occurs at the same time —16 hours after fertilization.

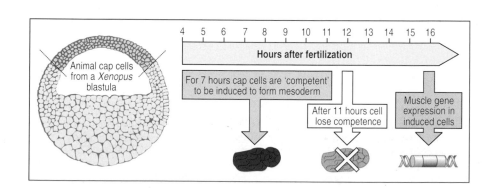

3-15 **Several signals induce and pattern the mesoderm in the *Xenopus* blastula.**

One can see from the blastula fate map (Fig. 3.25) that the mesoderm is divided into a number of regions along the dorso-ventral axis, with the notochord originating in the most dorsal region, and the blood-forming tissue most ventrally. But we can see from the specification map that, at the same blastula stage, only a small region on the dorsal side is specified as somites that give rise to muscle, whereas the fate map shows that a great deal of muscle will come from somites arising from more lateral regions. Thus, explants from the dorsal marginal zone of a blastula, taken after mesoderm induction has begun but before it is completed, behave much in line with their normal fate; they develop into notochord and muscle, and the explants even mimic gastrulation movements by converging and extending. By contrast, ventral and lateral marginal zone explants develop into mesenchyme and blood-forming tissue only (see Fig. 3.22). They do not give rise to any muscle, although their normal fate in the embryo is to form considerable amounts.

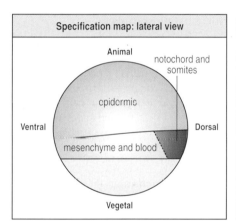 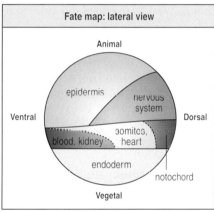

Fig. 3.25 The difference between the fate map and specification map of a *Xenopus* blastula. The fate of a region when isolated and placed in culture is shown on the specification map (left panel), while the normal fate of the blastula regions is shown in the fate map (right panel). There is a clear difference in specification of dorsal and ventral mesoderm. While the notochord's fate and specification correspond at this stage, that of the rest of the mesoderm is much more labile, and the specification of most of the somites and other mesodermal tissues has yet to occur. That involves signals from the region of the Spemann organizer, which acts just before and during gastrulation, as well as signals from the ventral region.

These results, together with other evidence we discuss later, allow us to construct a picture of mesoderm induction. Induction by the vegetal region involves at least two sets of signals: one is a general mesoderm inducer, broadly specifying a ventral-type mesoderm, which can be considered the ground or default state; the second signal, acting simultaneously or a little later, specifies the dorsal-most mesoderm that will contain the Spemann organizer and form notochord. There are two further sets of signals, patterning the ventral mesoderm along the dorso-ventral axis, which subdivide it into prospective muscle, kidney, and blood. The third set of signals comes from the ventral region, while the fourth set comes from the organizer region and modifies the ventralizing action of the third set of signals. We are therefore proposing the requirement for a four-signal model for mesoderm induction and patterning (Fig. 3.26).

Fig. 3.26 Four signals involved in mesoderm induction. Two signals originate in the vegetal region, one on the dorsal side from the region of the Nieuwkoop center (1), and the other from the ventral region (2). The dorsal signal specifies the Spemann organizer (O) and dorsal mesoderm, and the second specifies ventral mesoderm. The third signal (3), from the ventral region, ventralizes the mesoderm and the fourth signal (4) dorsalizes it by inhibiting the action of the third signal.

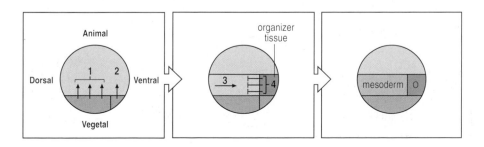

The model in no way implies that only four distinct signaling molecules are required, or indeed that all four signals are qualitatively different. It is quite possible that each 'signal' represents the actions of more than one molecule, or that different signals represent the same molecule acting at different concentrations.

3-16   Sources of the mesoderm-inducing signals.

Direct evidence for at least two signals coming from the vegetal region is provided by comparing the inducing effects of dorsal and ventral vegetal regions (Fig. 3.27). Dorsal vegetal tissue containing the Nieuwkoop center induces notochord and muscle from animal cap cells, while ventral vegetal tissue induces mainly blood-forming tissue, and little muscle. A minimum of two signals from the vegetal region can thus specify the broad differences between dorsal and ventral mesoderm. These signals are, however, insufficient to explain all of the patterning. In normal development, the ventral mesoderm makes a major contribution to the somites and hence to muscle, yet isolated explants of presumptive ventral mesoderm from an early blastula do not make muscle. Another signal or set of signals that pattern the ventral mesoderm further must be required.

The third signal, which emanates from the ventral region of the embryo, ventralizes the mesoderm and interacts with the dorsalizing signal which limits its influence. The fourth dorsalizing signal originates in the Spemann organizer itself. Evidence for this signal comes from combining a fragment of dorsal marginal zone from a late blastula with a fragment of the ventral presumptive mesoderm. The ventral fragment will form substantial amounts of muscle, whereas isolated ventral mesoderm after vegetal induction will form mainly blood-forming tissue and mesenchyme.

A dramatic demonstration of the action of the fourth signal is to graft the Spemann organizer into the ventral marginal zone of an early gastrula (Fig. 3.28). The graft induces a complete new dorsal side, and the result is a twinned embryo. This is the famous experiment, carried out by Hans Spemann and Hilde Mangold in the 1920s (see Fig. 1.10), that first identified this key signaling region. The Spemann organizer and the dorsal mesoderm are specified by the Nieuwkoop center in the presumptive endoderm,

Fig. 3.27 Differences in mesoderm induction by dorsal and ventral vegetal regions. The dorsal vegetal region of the *Xenopus* blastula, which contains the Nieuwkoop center, induces notochord and muscle from animal cap tissues, while ventral vegetal cells induce blood and associated tissues. This is good evidence for different inducing signals coming from the dorsal and ventral vegetal regions.

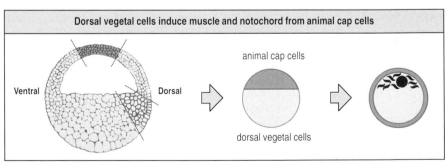

Dorsal vegetal cells induce muscle and notochord from animal cap cells

animal cap cells

Ventral   Dorsal

dorsal vegetal cells

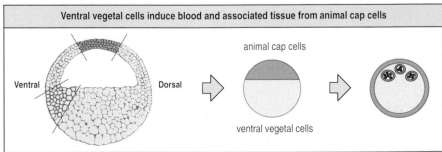

Ventral vegetal cells induce blood and associated tissue from animal cap cells

animal cap cells

Ventral   Dorsal

ventral vegetal cells

which lies just vegetal to it. We can now see that the dorsalizing effect of the Nieuwkoop center, observed in the experiments described in Section 3-2, is due to its induction of the Spemann organizer. We next consider some of the molecules that seem to be involved in mesoderm induction and patterning.

3-17 Candidate mesoderm inducers have been identified in *Xenopus*.

From the explant experiments outlined above, the mesoderm-inducing signals produced by the vegetal region seem most likely to be secreted. Two main approaches are used to test for mesoderm-inducing and patterning factors. One is to apply the suspected factor directly to isolated animal caps in culture. The other is to inject the mRNA encoding the suspected inducer into the animal pole cells of the early blastula.

The most obvious candidates for the vegetal mesoderm-inducing signals are the maternal factors localized in the vegetal region of the egg. A prime candidate is Vg-1, a maternally expressed member of the TGF-β family (see Box 3A, page 64), whose mRNA is localized in the vegetal region. Like all TGF-β family proteins, newly synthesized Vg-1 has to be proteolytically processed before it becomes active. Although the precursor protein is abundant in the vegetal region, the injection of neither its mRNA nor the precursor protein itself into the animal cap have any significant effect. This suggests that Vg-1 activity is regulated at the post-translational level and that the animal cap cells are unable to process the Vg-1 precursor efficiently. The cortical rotation described in Section 3-2 may be required for the proper processing and activation of Vg-1 in the dorsal vegetal region.

Mature, properly processed Vg-1 protein does indeed have a pronounced mesoderm-inducing effect on animal cap cells. Production of mature Vg-1 protein in animal cap cells was achieved by constructing a hybrid mRNA containing the coding sequence of the mature protein fused with that of a related protein to ensure correct post-translational modification. When this RNA construct is injected into animal cap explants, the expression of mature active Vg-1 induces dorsal mesoderm. Vg-1 expression also rescues embryos ventralized by UV irradiation (see Section 3-3). Treatment of isolated animal caps with purified active Vg-1 protein induces embryoids with clear axial organization and head structures.

Vg-1 is therefore a very likely candidate for a mesoderm inducer. At high concentrations Vg-1 induces dorsal mesoderm, while at lower concentrations it induces ventral-type mesoderm. Thus, it is even possible that Vg-1 at different concentrations could provide both the first and second signals. A high concentration could provide the dorsal vegetal signal (signal 2) while a lower concentration would provide the ventral vegetal signal (signal 1).

Another member of the TGF-β family, activin, also has mesoderm-inducing activity. Activin was isolated from the culture fluid of a *Xenopus* cell line by its powerful inducing activity. The response of animal caps to purified activin is also concentration dependent: at higher concentrations notochord develops, together with muscle, whereas at lower concentrations only muscle is induced. Although activin-like activity can be detected in extracts from oocytes and early embryos, there is no evidence so far for maternal activin mRNA in the egg. Activin itself may therefore not be the primary inducing signal *in vivo*. Different members of the TGF-β family may bind and act through the same receptors, so the activin applied to the animal cap cells could be acting through the same pathway as, say, Vg-1.

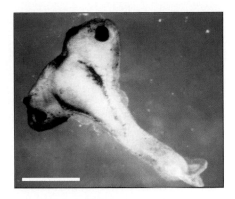

Fig. 3.28 Transplantation of the Spemann organizer can induce a new axis in *Xenopus*. The third set of signals required for mesoderm induction and patterning come from the Spemann organizer region. Their effect can be seen by transplanting the Spemann organizer into the ventral region of another gastrula. The resulting embryo has two distinct heads; one of which was induced by the Spemann organizer. The organizer therefore produces signals that not only pattern the mesoderm dorso-ventrally, but induce neural tissue and anterior structures. Scale bar = 1 mm. Photograph courtesy of J.Smith.

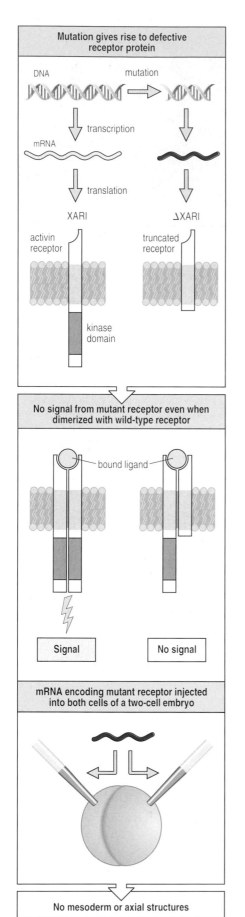

Mutation gives rise to defective receptor protein

DNA   mutation

transcription

mRNA

translation

XARI   ΔXARI

activin   truncated
receptor   receptor

kinase
domain

No signal from mutant receptor even when dimerized with wild-type receptor

bound ligand

Signal   No signal

mRNA encoding mutant receptor injected into both cells of a two-cell embryo

No mesoderm or axial structures

Fig. 3.29 A mutant activin receptor blocks mesoderm induction.
Receptors for factors of the TGF-β family function as dimers. Ligand binding causes dimerization, which activates a serine-threonine kinase in the cytoplasmic region of the receptor. Receptor function can be blocked by introducing mRNA encoding a mutant receptor subunit. The introduced construct encodes a receptor that lacks most of the cytoplasmic domain and so cannot function. It can bind ligand and form heterodimers with normal receptor subunits but cannot signal. It thus acts as a dominant-negative mutation of receptor function. When mRNA encoding the mutant receptor subunit is injected into cells of the two-cell *Xenopus* embryo, subsequent mesoderm formation is blocked. No mesoderm or axial structures are formed except for the cement gland, the most anterior structure of the embryo. XARI = *Xenopus* activin receptor.

By itself, the ability simply to induce mesoderm in culture does not prove that a particular protein is a natural inducer in the embryo. Rigorous criteria must be met before such a conclusion can be reached. These criteria include the presence of the protein in the right concentration and place, and at the right time in the embryo; the demonstration that the appropriate cells can respond to the factor; and the demonstration that blocking the response prevents induction taking place. On all these criteria, however, the evidence for a key role for a member or members of the TGF-β family in mesoderm induction is rather good.

Of particular importance in proving this are experiments that block the response of cells to factors such as activin by preventing their receptors from being activated. If mesoderm induction is also prevented by this treatment, it is a clear sign that protein factors that can bind to the receptor in question, are causally involved. A receptor for several TGF-β family growth factors, including activin and Vg-1, is the activin type II receptor, which is expressed and uniformly distributed throughout the early *Xenopus* blastula. TGF-β family receptors only function as dimers and their function can be blocked by the presence of a mutant subunit, which associates with a normal subunit to produce an inactive receptor (Fig. 3.29). If mRNA for a mutant subunit of the activin receptor is injected into the early *Xenopus* embryo, mesoderm formation is prevented. The presence of a mutant activin receptor subunit has the same effect as a **dominant-negative mutation** in the gene coding for the receptor, that is it inhibits receptor function. This direct biochemical intervention is particularly useful in *Xenopus* where there are no means of producing suitable genetic mutations.

These experiments show that a protein of the TGF-β family is involved in mesoderm induction, and could provide a key element of the first two signals, but can give no indication as to which one it is. However, since Vg-1 is present at the right place and at the right time it is clearly a likely candidate. Another factor that is involved in mesoderm induction is the *Xenopus* equivalent of fibroblast growth factor (FGF). This is present in the blastula, mainly in the animal region, and may be necessary to potentiate the response of animal cap cells to TGF-β-like molecules.

3-18   Mesoderm patterning factors are produced within the mesoderm.

A number of proteins seem to be involved in patterning the mesoderm along the dorso-ventral axis once it has been induced (Fig. 3.30). The gene *noggin*, which was identified during a screen for factors that could rescue UV-irradiated embryos, is expressed in the region of the Spemann organizer (Fig. 3.31). The noggin protein is a secreted protein unrelated to any of the known growth factor families. *noggin* expression does not induce mesoderm in animal pole explants but can dorsalize explants of ventral marginal zone tissue, making it a good candidate for one of the fourth class of signals that patterns the mesoderm along the dorso-ventral axis. The chordin protein, another protein secreted by the organizer, can also induce dorsal tissues. Yet another secreted protein in this region is frizbee.

The third set of signals promote ventralization of the embryo. A member of the TGF-β family, bone morphogenetic protein-4 (BMP-4), is expressed throughout the late blastula, and Xwnt-8 is expressed in the future mesoderm. As gastrulation proceeds, BMP-4 is no longer expressed in dorsal regions. When the action of BMP-4 is blocked by introducing a dominant-negative receptor, the embryo is dorsalized, with ventral cells now differentiating as both muscle and notochord. The secreted protein Xwnt-8, which is expressed in the future mesoderm, can also ventralize the embryo. How do the dorsalizing and ventralizing factors interact? The action of the dorsalizing agents is not on the presumptive mesodermal cells, but on the ventralizing factors. Thus, noggin acts by interacting with BMP-4, and prevents it binding to its receptor. Chordin acts in a similar manner, and frizbee acts by binding to Wnt proteins. Fig. 3.32 summarizes the main signals so far identified as mesoderm inducers and patterning factors in *Xenopus*.

Although we are largely ignorant of the mechanism by which mesoderm is specified in the mouse, chick, and zebrafish, TGF-β family members have also been implicated as mesoderm-inducing signals in chick and mouse embryos. In the chick, mesoderm specification occurs during primitive streak formation. Chick epiblast isolated before streak formation will form some mesoderm containing blood vessels, blood cells, and muscle. Treatment of the epiblast with activin results in the additional appearance of dorsal axial structures such as notochord and more muscle. In early chick embryos, activin-secreting cells implanted into the marginal zone induce only a transitory primitive streak-like axis, but when combined with cells secreting Wnt-family growth factors, a complete new axis can be induced. This suggests that while activin can induce dorsal mesoderm, full development of the axis, including anterior structures, requires the action of Wnt proteins. The chick homolog of Vg-1 can, however, induce a whole new axis when cells secreting it are grafted to the margin of an early chick blastoderm.

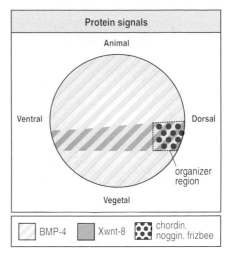

Fig. 3.30 Distribution of protein signals in the *Xenopus* blastula. The signals from the organizer block the action of BMP-4 and Xwnt-8.

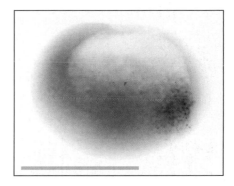

Fig. 3.31 Expression of *noggin* in the *Xenopus* blastula. *noggin* expression is shown as the dark staining area in the region of the Spemann organizer. Scale bar = 1 mm. Photograph courtesy of R. Harland, from Smith, W.C., *et al.*: 1992.

Fig. 3.32 Signals in early *Xenopus* development.

Signals in early *Xenopus* development		
Factor	**Protein family**	**Effects**
Vg-1	TGF-β family	mesoderm induction
activin	TGF-β family	mesoderm induction
bone morphogenetic factor (e.g. BMP-4)	TGF-β family	ventral mesoderm patterning
XWnt-8	Wnt family	ventralizes mesoderm
fibroblast growth factor (FGF)	FGF	ventral mesoderm induction
noggin		dorsalizes—binds BMP-4
chordin		dorsalizes—binds BMP-4
frizbee		dorsalizes—binds Wnt proteins

In the mouse, the gene *Nodal*, which encodes another member of the TGF-β family, is expressed in the primitive streak at the time of mesoderm formation. This protein is likely to be involved in mesoderm induction because, in homozygous mutants of *Nodal*, mesoderm does not form during gastrulation. However, mice lacking activin or a type II activin receptor still develop mesoderm, which suggests that neither activin nor the type II receptor are necessary in mesoderm induction in mammals.

A variety of protein growth factors seem to be involved in mesoderm induction and patterning. Members of the TGF-β family are very likely to play a part, and other factors may interact with them to modify tissue response. We now look at the target genes for these factors, and the further patterning of the mesoderm. The mesoderm inducers act on the embryo's own genes, the **zygotic genes**. Zygotic gene expression begins in earnest at the mid-blastula transition (in fact quite late in the blastula stage), and we first look at this transition.

3-19   Zygotic gene expression begins at the mid-blastula transition in *Xenopus*.

The *Xenopus* egg contains quite large amounts of maternal mRNA, which is laid down during oogenesis. In addition, there are large amounts of stored proteins; there is, for example, sufficient histone protein for the assembly of more than 10,000 nuclei. On fertilization, the rate of protein synthesis increases 1½-fold and during cleavage a large number of new proteins begin to be synthesized, as shown by two-dimensional electrophoretic analysis of extracts of whole embryos. All these proteins are synthesized by translation of preformed maternal mRNA. There is, however, very little new mRNA synthesis until 12 cleavages have taken place and the embryo contains 4096 cells. This point is the so-called **mid-blastula transition** (although it actually happens in the late blastula, just before gastrulation starts). Transcription of the embryo's own genes begins at the mid-blastula transition, and paternal genes are transcribed for the first time in the life of the embryo.

The start of transcription coincides, more or less, with several other changes in the blastula. The first cleavages take place at regular 35 minute intervals, but at the 12th cleavage they become asynchronous as cells take different amounts of time to complete the next cell cycle (Fig. 3.33). At the same time, cells become more motile and can be seen to form small outpushings. It is the coincidence of all these events, which may not be causally related, that leads to the stage being referred to as the mid-blastula transition.

How is the mid-blastula transition triggered? Suppression of cleavage, but not DNA synthesis, by cytochalasin B does not alter the timing of transcriptional activation, and so it is not linked directly to cell division. Neither are cell–cell interactions involved, as dissociated blastomeres undergo the transition at the same time as intact embryos. The key factor in triggering the mid-blastula transition seems to be the ratio of DNA to cytoplasm—the quantity of DNA present per unit mass of cytoplasm.

Direct evidence for this comes from increasing the amount of DNA artificially by allowing more than one sperm to enter the egg or by injecting extra DNA into the egg. In both cases transcriptional activation occurs prematurely, suggesting that there may be some fixed amount of a general repressor of transcription present initially in the egg cytoplasm. As the egg cleaves, the amount of cytoplasm does not increase, but the amount of DNA does. The amount of repressor in relation to DNA gets smaller and smaller until there is insufficient to bind to all the

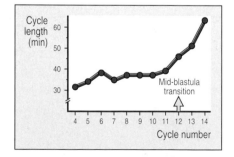

Fig. 3.33 Timing of the cell cycle during cleavage in *Xenopus*. While the cell cycles of early cleavages in *Xenopus* are short and synchronous, later cleavages are longer and asynchronous. The mid-blastula transition in *Xenopus* occurs at the 12th cleavage.

available sites on the DNA and the repression is lifted. Timing of the mid-blastula transition thus seems to fit with a timing model of the hour-glass egg-timer type (Fig. 3.34): something has to accumulate, probably DNA, until a threshold is reached, this threshold being determined by the initial concentration of some cytoplasmic factor.

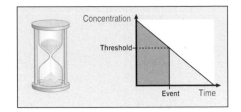

Fig. 3.34 Timing mechanism that could operate in development. A mechanism based on an analogy to an egg-timer could measure time to mid-blastula transition. The decrease in concentration of some molecule, such as a repressor, could occur with time, and the transition could occur when the repressor reaches a critically low threshold concentration. This would be equivalent to all the sand running into the bottom of the egg-timer.

| 3-20 | **Mesoderm induction activates genes that pattern the mesoderm.** |

The signals described in Sections 3-17 and 3-18 pattern the mesoderm by turning on groups of genes that control mesoderm differentiation. An early response gene characterizing mesoderm is *Brachyury*, which encodes a transcription factor. This gene was first identified in mice, where it is required for mesoderm formation (see Section 2-3). In all vertebrates, *Brachyury* is initially expressed throughout the presumptive mesoderm (Fig. 3.35), later becoming confined to the notochord and to posterior mesoderm. The *Xenopus* homolog of *Brachyury* is switched on in presumptive ectoderm treated with mesoderm-inducing factors such as activin. The maintenance of *Brachyury* depends on the expression of a fibroblast growth factor (FGF), and *Brachyury* activates expression of FGF.

Overexpression of *Brachyury* in presumptive ectoderm after injection of its mRNA causes formation of ventral mesoderm. These results strongly suggest a key role for *Brachyury* in mesoderm patterning. This notion is further strengthened by the finding that the *no-tail* mutant in zebrafish, which results in absence of posterior mesoderm, is due to a mutation in the zebrafish homolog of *Brachyury*.

On the dorsal side of the *Xenopus* embryo, the Nieuwkoop center induces the Spemann organizer in the dorsal mesoderm. The organizer is not only involved in patterning the dorso-ventral axis of the mesoderm but, as we shall see in Chapter 4, is involved in patterning the antero-posterior axis of both the mesoderm and the nervous system. One of the first zygotic genes to be expressed in the organizer region is *goosecoid*, which was identified by screening a cDNA library taken from the *Xenopus* dorsal mesoderm region. *goosecoid* is a homeobox-containing gene (see Box 4A, page 104), encoding a transcription factor with a homeodomain somewhat similar to that of both the gooseberry and bicoid proteins of *Drosophila*—hence the name. *goosecoid* is a zygotic gene that is expressed in the mesoderm after the mid-blastula transition. It is expressed in the dorsal marginal zone, which includes the Spemann organizer.

In line with its presence in the organizer region, microinjection of *goosecoid* mRNA into the ventral region of the blastula mimics to a great extent the transplantation of the Spemann organizer (see Fig. 3.28), resulting in the formation of a secondary axis which may have complete head structures. Genes for other transcription factors are also expressed in the organizer region (Fig. 3.36). These include *Pintallavis* and *HNF-3β*, both of which code for proteins with so called forkhead domains, and Xnot and *Xlim-1*, which code for proteins with a homeodomain. While their function is as yet unknown, they may be involved in a cascade of gene interactions that give the organizer its special properties and could activate genes for secreted proteins such as noggin and chordin.

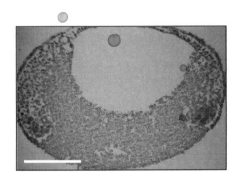

Fig. 3.35 Expression of *Brachyury* in the *Xenopus* blastula. A cross-section through the embryo along the animal-vegetal axis shows that *Brachyury* (red) is expressed in the future mesoderm. Scale bar = 0.5 mm. Photograph courtesy of M. Sargent and L. Essex.

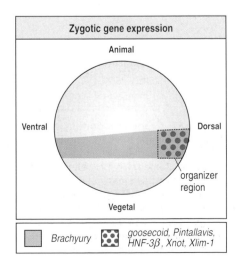

Fig. 3.36 Zygotic gene expression in a late *Xenopus* blastula. Expression domains of a number of zygotic genes that code for transcription factors correspond quite well to demarcations on the specification map. The gene

Brachyury is expressed in a ring around the embryo corresponding quite closely to the future mesoderm. Several transcription factors are expressed in the region of the dorsal mesoderm that corresponds to the Spemann organizer.

There is also some apparent redundancy, as mice in which the *goosecoid* gene has been deleted by gene knock-out still develop through embryogenesis.

A key gene essential for notochord formation has been identified in zebrafish. Mutation in the gene *floating head*, which is expressed in the presumptive notochord region, results in complete absence of the notochord and some increase in muscle. *floating head* codes for a homeodomain transcription factor. Its homolog in *Xenopus* is *Xnot*, expressed in the organizer region, which is also the region of the presumptive notochord. Its expression, like that of *Brachyury*, is induced by mesoderm inducers such as activin. Overexpression of *Xnot* results in a notochord that is larger than normal, further evidence that it plays a central role in notochord specification.

3-21 Gradients in protein signaling factors and threshold responses could pattern the mesoderm.

While possible signaling factors have been identified, it is still not clear how they turn on genes like *goosecoid* and *Brachyury* in the right place. One model for patterning the mesoderm, and also other tissues, proposes that positional information is provided by a dorso-ventral gradient of a morphogen. Several of the factors identified as possible mesoderm-patterning agents in *Xenopus* are indeed expressed in a graded fashion.

Activin provides an interesting example of how a diffusible growth factor could pattern a tissue by turning on specific genes at specific threshold concentrations. Animal cap cells from a *Xenopus* blastula respond to increasing doses of activin by different genes being activated at a threshold concentration. Increasing activin concentration by as little as 1½-fold results in a dramatic alteration in molecular marker expression and tissue differentiation, causing, for example, a change from homogeneous muscle formation to notochord. Increasing concentrations of activin can specify several different cell states that correspond to the different regions along the dorso-ventral axis. At the lowest concentrations of activin, only epidermis develops. Then, as the concentration increases, *Brachyury* is expressed together with muscle genes such as those coding for actin. With a further increase in activity *goosecoid* is expressed, and this corresponds to the most dorsal region of the mesoderm—the organizer (Fig. 3.37). Similar results can be obtained by injecting increasing quantities of activin mRNA. One can thus see how graded signals could activate transcription factors in particular regions and so pattern tissues.

Further support for the idea that a diffusion gradient and thresholds could pattern the mesoderm was provided by experiments in which vegetal tissues were injected with increasing amounts of activin mRNA. These were then placed in contact with an animal cap. The results showed that activin diffused into the animal cap and while *Brachyury* was turned on at some distance from the source, *goosecoid* was expressed nearest to the source, which fits with the idea of a concentration gradient turning on genes at a specific threshold concentration.

goosecoid expression is graded from dorsal to ventral zones in the mesoderm, being highest in dorsal and lowest in ventral cells. Ventral marginal zone tissue responds to increasing levels of *goosecoid* expression by developing into more dorsal mesoderm. *goosecoid* is thus a good candidate for controlling dorso-ventral patterning in the mesoderm, with different levels of expression specifying differences along the mesoderm. We do not yet know how the graded pattern of expression of *goosecoid* is established. The *goosecoid* protein is a transcription

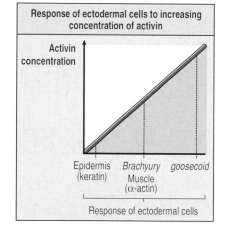

Fig. 3.37 Graded responses of early *Xenopus* tissue to increasing concentrations of activin. When animal cap cells are treated with increasing concentrations of activin, particular genes are activated at specific concentrations. At intermediate concentrations, *Brachyury* is induced, whereas *goosecoid*, which is typical of the organizer region, is only induced at high concentrations.

factor restricted to the nucleus of the cells in which it is synthesized, so its graded expression along the mesoderm must be a response to a gradient in some other protein factor, possibly one of the mesoderm-inducing growth factors. It is likely that other members of the TGF-β family like BMP-4, only act over short distances, and a relay mechanism involving a sequence of short-range signals may be involved.

We are now in a position to consider the final emergence of the typical vertebrate body plan. Further patterning of the germ layers occurs during gastrulation, along both the antero-posterior and dorso-ventral axes, and this is discussed in the next chapter.

Summary.

Once the antero-posterior and dorso-ventral axes are established, one can begin to construct a fate map for the germ layers. There are strong similarities in the fate maps of the amphibian, zebrafish, chick, and mouse at later stages. Even though there is good evidence for the maternal specification of some regions such as the future ectoderm and endoderm in amphibians, the embryo can still undergo considerable regulation at the blastula stage. This implies that interactions between cells, rather than intrinsic factors, have a central role even in early amphibian development. This strategy is particularly pronounced in the mouse, where there is no evidence for any maternal specification, and it is position that determines cell fate.

In *Xenopus*, the mesoderm and some endoderm are induced from animal cap tissue by the vegetal region, which contains the Nieuwkoop center. Early patterning of the mesoderm can be accounted for by a four-signal model. The first signal is a general mesoderm inducer, specifying a ventral-type mesoderm. The second specifies dorsal mesoderm, including the organizer, while the third signal comes from the ventral side and ventralizes the mesoderm. The fourth signal originates from the organizer and establishes further pattern within the mesoderm by interacting with the third signal.

Protein growth factors such as members of the TGF-β family are excellent candidates for the natural mesoderm-inducing factors. Other signaling factors such as the noggin protein and members of the Wnt family are involved in specifying the dorsal and ventral mesoderm, respectively. *Brachyury* and *goosecoid* are early mesodermally expressed genes coding for transcription factors and their pattern of expression may be specified by gradients in signaling proteins, the genes being turned on at particular threshold concentrations.

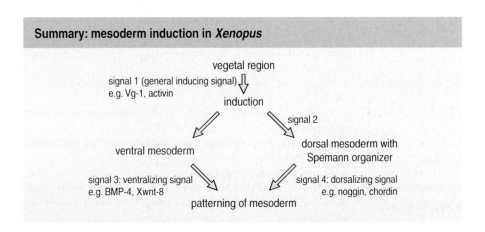

Summary: mesoderm induction in *Xenopus*

vegetal region

signal 1 (general inducing signal)
e.g. Vg-1, activin

induction

signal 2

ventral mesoderm

dorsal mesoderm with Spemann organizer

signal 3: ventralizing signal e.g. BMP-4, Xwnt-8

signal 4: dorsalizing signal e.g. noggin, chordin

patterning of mesoderm

Summary to Chapter 3.

All vertebrates have the same basic body plan. During early development, the antero-posterior and dorso-ventral axes of this body plan are set up. The mechanism is different in frog, chick, zebrafish, and mouse but can involve localized maternal determinants, external signals, and cell–cell interactions. This early patterning also establishes bilateral asymmetry. Once the axes are established, it is possible to construct a fate map for the three germ layers—mesoderm, endoderm, and ectoderm. The fate maps of the different vertebrates have strong similarities. At this early stage the embryos are still capable of considerable regulation and this emphasizes the essential role of cell–cell interactions in development. In *Xenopus*, at least four separate signals are involved in mesoderm induction and early patterning. Good candidates for these signals have been identified and include members of the TGF-β family. These signals activate mesoderm-specific genes such as *Brachyury* at particular concentrations and so their gradients could pattern the mesoderm. The summary table below lists all genes considered in this chapter in relation to *Xenopus*.

Summary: genes involved in patterning of axes and germ layers

Gene	Maternal/Zygotic	Type of protein	Where expressed	Effects
Xwnt-8	Z	secreted	mesoderm	ventralizes mesoderm
Xwnt-11	M	secreted	vegetal egg	mesoderm induction
Vg-1	M	TGF-β	vegetal egg	mesoderm induction
GSK-3	M	protein kinase	?	suppresses dorsalizing signals
Brachyury	Z	transcription factor	early mesoderm	–
goosecoid	Z	transcription factor	organizer	–
activin	Z	TGF-β	?	mesoderm induction
FGF	Z	growth factor	blastula	mesoderm induction
noggin	M/Z	secreted	organizer	dorsalizes mesoderm
chordin	Z	secreted	organizer	dorsalizes mesoderm
BMP-4	Z	TGF-β	gastrula	ventralizes mesoderm
Xnot	Z	transcription factor	notochord	–
Xlim-1	Z	?	–	–
HNF-3β	–	transcription factor	organizer	–
Pintallavis	–	–	–	–

General references.

Slack, J.M.W.: *From Egg to Embryo*, 2nd edn. Cambridge: Cambridge University Press, 1991.

Section references.

3-1   The animal-vegetal axis of *Xenopus* is maternally determined.

Chen, Y., Struhl, G.: **Dual roles for patched in sequestering and transducing hedgehog.** *Cell* 1996, **87**:553–563.

Foristall, C., Pondel, M., Chen, L., King, M.L.: **Patterns of localization and cytoskeletal association of two vegetally localized RNAs, Vg-1 and Xcat-2.** *Development* 1995, **121**: 201–208.

Hogan, B.L.M.: **Bone morphogenetic proteins in development.** *Curr. Opin. Genet. & Devel.* 1996, 6:432–438

Lardelli, M., Williams, R., Lendahl, U.: ***Notch*-related genes in animal development.** *Int. J. Dev. Biol.* 1995, **39**:769–780.

Massagué, J.: **TGFβ signaling: receptors, transducers, and Mad proteins.** *Cell* 1996, **85**:947–950.

Nusse, R., Varmus, H.E.: ***Wnt* genes.** *Cell* 1992, 69:1073–1087.

Weeks, D.L., Melton, D.A.: **A maternal mRNA localized to the vegetal hemisphere in *Xenopus* eggs codes for a growth factor related to TGF-β.** *Cell* 1987, **51**: 861–867.

Wilkie, A.O.M., Morriss-Kay, G.M., Jones, E.Y., Heath, J.K.: **Functions of fibroblast growth factors and their receptors.** *Curr. Biol.* 1995, **5**:500–507.

3-2 The dorso-ventral axis of amphibian embryos is determined by the site of sperm entry.

Gerhart, J., Danilchik, M., Doniach, T., Roberts, S., Browning, B., Stewart, R.: **Cortical rotation of the *Xenopus* egg: consequences for the antero-posterior pattern of embryonic dorsal development.** *Dev. Suppl.* 1989, 37–51.

3-4 Maternal proteins with dorsalizing and ventralizing effects have been identified.

He, X., Saint-Jennet, J-P., Woodgett, J.R., Varmus, H.E., Dawid, T.B.: **Glycogen synthase kinase-3 and dorsoventral patterning in *Xenopus* embryos.** *Nature* 1995, **374**:617–622.

Kessler, D.S., Melton, D.A.: **Induction of dorsal mesoderm by soluble, mature Vg1 protein.** *Development* 1995, **121**:2155–2164.

Smith, W.C., Harland, R.M.: **Injected Xwnt-8 RNA acts early in *Xenopus* embryos to promote formation of a vegetal dorsalizing center.** *Cell* 1991, **67**: 753–765

Sokol, S., Christian, J.L., Moon, R.T., Melton, D.A.: **Injected Wnt RNA induces a complete body axis in *Xenopus* embryos.** *Cell* 1991, **67**:741–752.

3-5 The dorso-ventral axis of the chick blastoderm is specified in relation to the yolk and the antero-posterior axis is set by gravity.

Khaner, O., Fyal-Giladi, H.: **The chick's marginal zone and primitive streak formation. I. Coordinative effect of induction and inhibition.** *Dev. Biol.* 1989, **134**:206–214.

Kochav, S., and Eyal-Giladi, H.: **Bilateral symmetry in chick embryo determination by gravity.** *Science* 1971, **171**:1027–1029.

Seleiro, E.A.P., Connolly, D.J., Cooke, J.: **Early developmental expression and experimental axis determination by the chicken Vg-1 gene.** *Curr. Biol.* 1996, **11**:1476–1486.

3-6 The axes of the mouse embryo are specified by cell–cell interactions.

Hillman, N., Sherman, M.I., Graham, C.: **The effect of spatial arrangement on cell determination during mouse development.** *J. Emb. Exp. Morph.* 1972, **28**: 263–278.

Lewis, N.E., Rossant, J.: **Mechanism of size regulation in mouse embryo aggregates.** *J. Emb. Exp. Morph.* 1982, **72**:169–181.

3-7 Specification of left–right handedness of internal organs requires special mechanisms.

Brown, N.A., Wolpert, L.: **The development of handedness in left/right asymmetry.** *Development* 1990, **109**:1–9.

3-8 Organ handedness in vertebrates is under genetic control.

Hyatt, B.A., Lohr, J.L., Yost, H.J.: **Initiation of vertebrate left-right axis formation by maternal Vg-1.** *Nature* 1996, **384**:62–65.

King, T., Brown, N.A.: **Embryonic asymmetry: Left TGF-β at the right time?** *Curr. Biol.* 1997, **7**:212–215.

Levin, M.: **Left-right asymmetry in vertebrate embryogenesis.** *Bio Essays* 1997, **19**:287–296.

Yokoyama, T., Copeland, N.G., Jenkins, N.A., Montgomery, C.A., Elder, F.F., Overbeek, P.A.: **Reversal of left-right symmetry: a situs inversus mutation.** *Science* 1993, **260**:679–682.

3-9 A fate map of the amphibian blastula is constructed by following the fate of labeled cells.

Dale, L., Slack, J.M.W.: **Fate map for the 32 cell stage of *Xenopus laevis*.** *Development* 1987, **99**:527–551.

3-10 The fate maps of vertebrates are variations on a basic plan.

Beddington, R.S.P., Morgenstern, J., Land, H., Hogan, A.: **An *in situ* transgenic enzyme marker for the midgestation mouse embryo and the visualization of inner cell mass clones during early organogenesis.** *Development* 1989, **106**: 37–46.

Gardner, R.L., Rossant, J.: **Investigation of the fate of 4–5 day post-coitum mouse inner cell mass cells by blastocyst injection.** *J. Emb. Exp. Morph.* 1979, **52**: 141–152.

Helde, K.A., Wilson, E.T., Cretehos, C.J., Grunwald, D.J.: **Contribution of early cells to the fate map of the zebrafish gastrula.** *Science* 1994, **265**: 517–520.

Kimmel, C.B., Warga, R.M., Schilling, T.F.: **Origin and organization of the zebrafish fate map.** *Development* 1990, **108**:581–594.

Lawson, K.A., Meneses, J.J., Pedersen, R.A.: **Clonal analysis of epiblast fate during germ layer formation in the mouse embryo.** *Development* 1991, **113**: 891–911.

Stern, C.D.: **The marginal zone and its contribution to the hypoblast and primitive streak of the chick embryo.** *Development* 1990, **109**:667–682.

Stern, C.D., Canning, D.R.: **Origin of cells giving rise to mesoderm and endoderm in chick embryo.** *Nature* 1990, **343**:273–275.

3-11 Cells of early vertebrate embryos do not yet have their fates determined.

Snape, A., Wylie, C.C., Smith, J.C., Heasman, J.: **Changes in states of commitment of single animal pole blastomeres of *Xenopus laevis*.** *Dev. Biol.* 1987, **119**:503–510.

Wylie, C.C., Snape, A., Heasman, J., Smith, J.C.: **Vegetal pole cells and commitment to form endoderm in *Xenopus laevis*.** *Dev. Biol.* 1987, **119**: 496–502.

3-13 The mesoderm is induced by a diffusible signal during a limited period of competence.

Gurdon, J.B., Lemaire, P., Kato, K.: **Community effects and related phenomena in development.** *Cell* 1993, **75**:831–834.

3-14 An intrinsic timing mechanism controls the time of expression of mesoderm-specific genes.

Cooke, J., Smith, J.C.: **Measurement of developmental time by cells of early embryos.** *Cell* 1990, **60**:891–894.

Ffrench-Constant, C.: **How do embryonic cells measure time?** *Curr. Biol.* 1994, **4**: 415–419.

3-15 Several signals induce and pattern the mesoderm in the *Xenopus* blastula.

Klein, P.S., Melton, D.A.: **Hormonal regulation of embryogenesis: the formation of mesoderm in *Xenopus laevis*.** *Endocrine Reviews* 1994, **15**:326–341.

Slack, J.M.W.: **Inducing factors in *Xenopus* early embryos.** *Curr. Biol.* 1994, **4**:116–126.

3-17   Candidate mesoderm inducers have been identified in *Xenopus*.

Amaya, E., Musci, T.J., Kirschner, M.W.: **Expression of a dominant negative mutant of the FGF receptor disrupts mesoderm formation in *Xenopus* embryos.** *Cell* 1991, **66**:257–270.

Kemmati-Brivanlou, A., Melton, D.A.: **A truncated activin receptor inhibits mesoderm induction and formation of axial structures in *Xenopus* embryos.** *Nature* 1992, **359**:609–614.

3-18   Mesoderm patterning factors are produced within the mesoderm.

Cooke, J., Takado, S., McMahon, A.: **Experimental control of axial pattern in the chick blastoderm by local expression of *Wnt* and activin; the role of HNK-1 positive cells.** *Dev. Biol.* 1994, **164**:513–527.

Leyns, L., Bouwmeester, T., Kim, S-H., Piccolo, S., De Robertis, E.M.: **Frzb-1 is a secreted antagonist of Wnt signaling expressed in the Spemann organizer.** *Cell* 1997, **88**:747–756.

Moon, R.T., Brown, J.D., Yang-Snyder, J.A., Miller, J.R.: **Structurally related receptors and antagonists compete for secreted Wnt ligands.** *Cell* 1997, **88**:725–728.

Piccolo, S., Sasai, Y., Lu, B., De Robertis, E.M.: **Dorsoventral patterning in *Xenopus*: inhibition of ventral signals by direct binding of chordin to BMP-4.** *Cell* 1996, **86**:589–598.

Smith, J.: **Angles on activin's absence.** *Nature* 1995, **374**:311–312.

Smith, J.C.: **Mesoderm-inducing factors and mesodermal patterning.** *Curr. Opin. Cell Biol.* 1995, **7**:856–861.

Thomsen, GH., Melton, D.A.: **Processed Vg1 protein is an axial mesoderm inducer in *Xenopus*.** *Cell* 1993, **74**:433–441.

Zhou, X., Sasaki, H., Lowe, L., Hogan, BL., Kuehn, M.R.: **Nodal is a novel TGF-β-like gene expressed in the mouse node during gastrulation.** *Nature* 1993, **361**:543–547.

Zimmerman, L.B., De Jesús-Escobar, J.M., Harland, R.M.: **The Spemann organizer signal noggin binds and inactivates bone morphogenetic protein 4.** *Cell* 1996, **86**:599–606.

3-19   Zygotic gene expression begins at the mid-blastula transition in *Xenopus*.

Davidson, E.: *Gene Activity In Early Development*. New York: Academic Press, 1986.

Yasuda, G.K., Schübiger, G.: **Temporal regulation in the early embryo: is MBT too good to be true?** Trends Genet. 1992, **8**:124–127.

3-20   Mesoderm induction activates genes that pattern the mesoderm.

Ang, S-L., Rossant, J.: **HNF-3β is essential for node and notochord formation in mouse development.** *Cell* 1994, **78**:561–574.

Isaacs, H.V., Pownall, M.E., Slack, J.M.W.: **eFGF regulates *Xbra* expression during *Xenopus* gastrulation.** *EMBO* 1994, **13**:4469–4481.

Schulte-Merker, S., Smith, J.C.: **Mesoderm formation in response to *Brachyury* requires FGF signalling.** *Curr. Biol.* 1995, **5**:62–67.

Taira, M., Jamrich, M., Good, P.J., Dawid, L.B.: **The LIM domain-containing homeobox gene Xlim-1 is expressed specifically in the organizer region of *Xenopus* gastrula embryos.** *Genes Dev.* 1992, **6**:356–366.

3-21   Gradients in protein signaling factors and threshold responses could pattern the mesoderm.

Gurdon, J.B., Harger, P., Mitchell, A., Lemaire, P.: **Activin signalling and response to a morphogen gradient.** *Nature* 1994, **371**:487–492.

Green, J.B.A., New, H.V., Smith, J.C.: **Responses of embryonic *Xenopus* cells to activin and FGF are separated by multiple dose thresholds and correspond to distinct axes of the mesoderm.** *Cell* 1992, **71**:731–739.

Jones, C.M., Armes, N., Smith, J.C.: **Signaling by TGF-β family members: short-range effects of Xnr-2 and BMP-4 contrast with the long-range effects of activin.** *Curr. Biol.* 1996, **6**:1468–1475.

Niehrs, C., Steinbeisser, H., De Robertis, E.M.: **Mesodermal patterning by a gradient of the vertebrate homeobox gene *goosecoid*.** *Science* 1994, **263**:817–820.

Reilly, K.M., Melton, D.A.: **Short-range signaling by candidate morphogens of the TGF-β family and evidence for a relay mechanism of induction.** *Cell* 1996, **86**:743–754.

Patterning the Vertebrate Body Plan II: The Mesoderm and Early Nervous System

4

Somite formation and patterning.

The role of the organizer region and neural induction.

"Early on, we behaved as a single unit, as there were no barriers between us. Once our positions were known, we became separate and joined different groups. We only spoke to those in our own group."

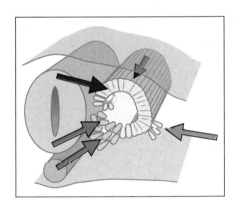

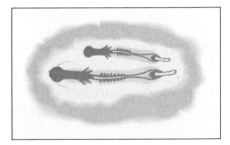

In the previous chapter we have seen how the body axes are set up and how the three germ layers are initially specified in various vertebrate embryos. Although amphibian, chick, and mouse embryos share some features in common at these stages, there are many significant differences. As we approach the phylotypic stage—the embryonic stage common to all vertebrates (see Fig. 2.2)—the similarity between vertebrate embryos becomes greater, and so we can consider the patterning of the vertebrate body plan in an integrated way. By the phylotypic stage, the embryo has undergone gastrulation, and the main axial structures characteristic of vertebrate embryos—somites, notochord, and neural tube—are well developed and already show signs of regional organization, especially along the antero-posterior axis. In this chapter we shall look at how this patterning is achieved.

During gastrulation, the germ layers—mesoderm, endoderm, and ectoderm—move to the positions in which they will develop into the structures of the larval or adult body. The antero-posterior body axis of the vertebrate embryo emerges clearly, with the head at one end and the future tail at the other (Fig. 4.1). In this chapter, we focus mainly on the patterning of the somitic mesoderm that forms the skeleton and muscles of the trunk, and of the ectoderm that will develop into the future nervous system. The phenomenon of gastrulation and the action of the organizer region are crucial to establishing the vertebrate body plan (see Section 1-6), and will be discussed in this chapter in relation to their role in the patterning processes. A detailed discussion of the behavior of cells and tissues during gastrulation will, however, be deferred to Chapter 8.

After gastrulation, the part of the mesoderm that comes to lie along the dorsal side of the embryo, under the ectoderm, gives rise to the notochord and somites, and to a small amount of head mesoderm anterior to the notochord. During gastrulation, cells of the dorsal-most mesoderm

Fig. 4.1 Rearrangement of the presumptive germ layers during gastrulation and neurulation in *Xenopus*. The mesoderm (pink and red), which is in an equatorial band at the blastula stage, moves inside to give rise to the notochord, somites, and lateral mesoderm (not shown). The endoderm (orange) moves inside to line the gut. The neural tube (dark blue) forms and the ectoderm (light blue) covers the whole embryo. The antero-posterior axis emerges, with the head at the anterior end.

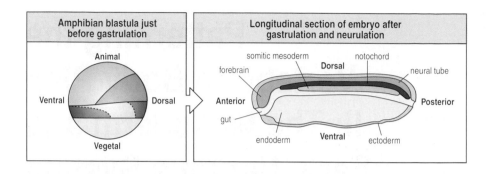

(the organizer region) are internalized, and eventually form a rigid rod-like notochord along the dorsal midline, flanked on each side by blocks of somites, which are derived from cells lying on either side of the organizer region in the marginal zone mesoderm of the blastula (see Fig. 3.15). In vertebrates, the notochord is a transient structure, and its cells eventually become incorporated into the vertebral column. During neurulation, the neural tube is formed from the ectoderm overlying the notochord, and develops into the brain and spinal cord. The somites, now positioned on either side of the neural tube, give rise to the vertebrae and ribs, to the muscles of the trunk and limbs, and also contribute to the dermis of the skin. Neural crest cells migrating away from the neural tube develop into a variety of tissues that include skeletal elements of the head, the sensory and autonomic nervous systems, and pigment cells.

Both the mesodermally derived structures along the antero-posterior axis of the vertebrate trunk and the ectodermally derived nervous system have a distinct antero-posterior organization. The vertebrae, for example, have characteristic shapes in each of the four anatomical regions: cervical, thoracic, lumbar, and sacral. In this chapter, we first examine the development of the somites and how they are patterned. We then deal with how their positional identity along the antero-posterior axis is specified. In later sections, we consider the function of the vertebrate organizer and the movements of gastrulation in establishing the antero-posterior organization of the embryo and its co-ordination with the dorso-ventral organization that we discussed in the previous chapter. Finally, the induction and early patterning of the nervous system will be discussed.

Somite formation and patterning.

In Chapter 3, we discussed the early specification of the mesoderm and its patterning along the dorso-ventral axis. The fate maps of the various vertebrates (see Fig. 3.20) show that the notochord develops from the most dorsal region of the mesoderm, and somites from a more ventral region on either side. During gastrulation, the mesoderm moves inside the embryo. The notochord then develops as a rod in the dorsal midline, neurulation begins, and the prospective somitic mesoderm segments into blocks, which will eventually flank the neural tube. The somites give rise to the body and limb muscles, the cartilage that forms the vertebrae and ribs, and the dermis. Their patterning thus provides much of the antero-posterior organization of the body. In this section, we will look at the initial formation of the somites after gastrulation and how they are patterned.

| 4-1 | **Somites are formed in a well-defined order along the antero-posterior axis.** |

In the chick embryo, somite formation occurs in the mesodermal region anterior to the regressing Hensen's node (see Fig. 2.15). Between the node and the most recently formed somite, there is an unsegmented region—the pre-somitic mesoderm—which will segment into four or five somites. Changes in cell shape and intercellular contacts in the pre-somitic mesoderm result in the formation of distinct blocks of cells—the somites. Somites are formed in pairs, one on either side of the notochord, with each pair of somites forming simultaneously. In all vertebrate embryos, somite formation begins at the anterior 'head' end and proceeds in a posterior direction. This sequence seems to reflect an even earlier patterning process.

The sequence of somite formation in the unsegmented region is unaffected by transverse cuts in the plate of pre-somitic mesoderm, suggesting that somite formation is an autonomous process and that, at this time, no signal specifying antero-posterior position is involved. Even if a small piece of the unsegmented mesoderm is rotated through 180°, each somite still forms at the normal time, but with the sequence of formation running in the opposite direction to normal in the inverted tissue (Fig. 4.2). So, before somite formation begins, a molecular pattern that specifies the time of formation of each somite has already been laid down in the mesoderm. The order of somite formation is most likely due to some graded property in the unsegmented mesoderm that was set up earlier and that may also be involved in patterning the antero-posterior axes. Somite formation starting at the top of the gradient would thus establish the temporal sequence.

Somites differentiate into particular axial structures depending on their position along the antero-posterior axis. Anterior somites, for example, will form cervical vertebrae while more posterior ones will develop as thoracic vertebrae with ribs. Specification by position has occurred before somite formation begins during gastrulation; if unsegmented somitic mesoderm from, for example, the presumptive thoracic region is grafted to replace the presumptive mesoderm of the neck region, it will still form thoracic vertebrae with ribs (Fig. 4.3). How then is the pre-somitic mesoderm patterned so that somites acquire their identity

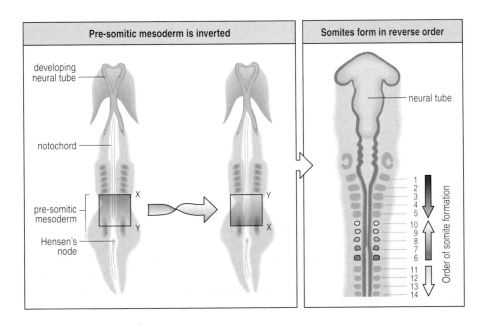

Fig. 4.2 The temporal order of somite formation is specified early in embryonic development. Somite formation in the chick proceeds in an antero-posterior direction. Somites form sequentially in the pre-somitic region between the last-formed somite and Hensen's node, which moves posteriorly. If the antero-posterior axis of the pre-somitic mesoderm is inverted through 180°, as shown by the arrow, the temporal order of somite formation is not altered—somite 6 still develops before somite 10.

Fig. 4.3 The pre-somitic mesoderm has a positional identity before somite formation. Somitic mesoderm that will give rise to thoracic vertebrae is grafted to an anterior region of a younger embryo that will develop into cervical vertebrae. The grafted mesoderm develops according to its original position and forms ribs in the cervical region.

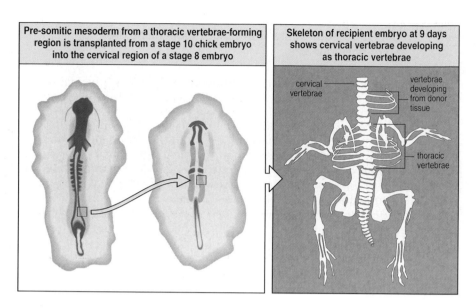

Pre-somitic mesoderm from a thoracic vertebrae-forming region is transplanted from a stage 10 chick embryo into the cervical region of a stage 8 embryo

Skeleton of recipient embryo at 9 days shows cervical vertebrae developing as thoracic vertebrae

cervical vertebrae

vertebrae developing from donor tissue

thoracic vertebrae

and form particular vertebrae? We consider this below, but first we deal with the patterning of the individual somite with respect to the different tissues it gives rise to.

4-2   The fate of somite cells is determined by signals from the adjacent tissues.

The somites of the vertebrate embryo give rise to major axial structures: the cartilage cells of the vertebrae and ribs; all the skeletal muscles, including those of the limbs; and the dermis. The fate map of a somite has been made by grafting somites from a quail into a corresponding position in a chick embryo at a similar stage of development and following the fate of the quail cells. These can be distinguished from chick cells by their distinctive nuclei, which can be detected in histological sections.

Cells located in the dorsal and lateral regions of a newly formed somite make up the **dermamyotome**, which expresses the *Pax3* homeobox gene (see Box 4A, page 104). This forms the **myotome**, which gives rise to muscle cells, and the **dermatome**, an epithelial-like sheet over the myotome which gives rise to the dermis. Cells from the medial region of the somite form mainly axial and back muscles, and express the muscle-specific transcription factor MyoD and related proteins. Lateral somite cells migrate to give rise to abdominal and limb muscles. The ventral part of the medial somite contains **sclerotome** cells that express the *Pax1* homeobox gene and migrate ventrally to surround the notochord and develop into vertebrae (Fig. 4.4). The lateral and medial parts of chick somites are of different origins, and are brought together during gastrulation: the medial portion comes from cells in the primitive streak close to Hensen's node, while the lateral portion comes from more posterior cells.

Which cells will form cartilage, muscle, or dermis is not yet determined at the time of somite formation. Specification of these fates requires signals from tissues adjacent to the somite. This is clearly shown by experiments in which the dorso-ventral orientation of newly formed somites is altered; they still develop normally. Both the neural tube and notochord produce signals that pattern the somite and are required for its future development. If the notochord and neural tube are removed, the somites undergo necrosis; neither vertebrae nor axial muscles develop although, surprisingly, limb musculature still does.

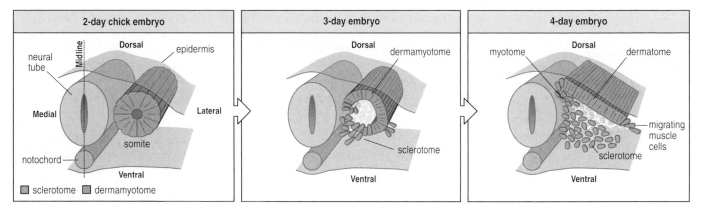

Fig. 4.4 The fate map of a somite in the chick embryo. The ventral medial quadrant (blue) gives rise to the sclerotome cells, which migrate to form the cartilage of the vertebrae. The rest of the somite, the dermamyotome, forms the dermatome and myotome, which gives rise to the dermis, and all the trunk muscles. It also gives rise to muscle cells that migrate into the limb bud.

The role of the notochord in specifying somitic cells has been shown by experiments in the chick, in which an extra notochord is implanted to one side of the neural tube, adjacent to the somite. This has a dramatic effect on somite differentiation, provided the operation is carried out on unsegmented pre-somitic mesoderm: when the somite develops, there is an almost complete conversion to cartilage precursors (Fig. 4.5), suggesting that the notochord is an inducer of cartilage. The neural tube also has a cartilage-inducing effect on somites, which is mediated by the most ventral region of the tube, the floor plate (which is itself induced by the notochord—see Section 11-6). There is also evidence for a signal from the lateral plate mesoderm, which is involved in specifying the lateral part of the dermamyotome, and a signal from the overlying ectoderm (Fig. 4.6).

Possible signals that pattern the somite have been identified. In the chick, both the notochord and the ventral neural tube express the gene *Sonic hedgehog*, which encodes a secreted protein that seems to be a key molecule for positional signaling in a number of developmental situations (see Section 10-4). (We also met *Sonic hedgehog* in Chapter 4 as a gene involved in asymmetry of structures about the midline. Here, this gene is being expressed at a quite different stage of development and in different tissues.) One model proposes that the signal generated by *Sonic hedgehog* specifies the ventral region of the somite. Signals from the dorsal neural tube and from the overlying non-neural ectoderm would specify the dorsal region. The protein factor BMP-4 and secreted signaling proteins of the Wnt family are good candidates for lateral and dorsal signals, respectively.

Regulation of the Pax homeobox genes in the somite by signals from the notochord and neural tube seems to be important in determining cell fate. *Pax3* is initially expressed in all cells that will form somites. Its expression is then modulated by the BMP-4 and Wnt family proteins so that it becomes confined to muscle precursors. It is then further down-regulated in cells that differentiate as the muscles of the back, but remains switched on in the migrating presumptive muscle cells that populate the limbs. Mice that lack a functional *Pax3* gene—*Splotch* mutants—lack limb muscles.

Fig. 4.5 A signal from the notochord induces sclerotome formation. A graft of an additional notochord to the dorsal region of a somite in a 10-somite embryo suppresses the formation of the dermamyotome from the dorsal portion of the somite, and induces the formation of sclerotome, which develops into cartilage. The graft also affects the shape of the neural tube.

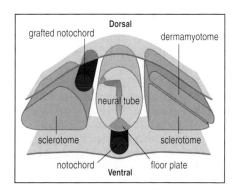

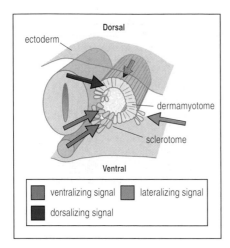

Fig. 4.6 A model for patterning of somite differentiation. The sclerotome is thought to be specified by a diffusible signal, probably the Sonic hedgehog protein, from the noto-chord and the floor plate of the neural tube (blue arrows). Signals from the dorsal neural tube and ectoderm (pink arrows) would specify the dermamyotome, together with lateral signals (green arrows) from the lateral plate mesoderm. After Johnson, R.L.: 1994.

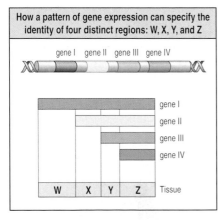

Fig. 4.7 Gene activity can provide positional values. The model shows how the pattern of gene expression along a tissue can specify the distinct regions W,X,Y, and Z. For example, only gene I is expressed in region W but all four genes are expressed in region Z.

Having seen how individual somites are formed and are patterned after gastrulation, we now discuss the patterning of the pre-somitic meso-derm along the antero-posterior axis that gives each somite its individual character.

<table>
<tr><td>4-3</td></tr>
</table>

Positional identity of somites along the antero-posterior axis is specified by Hox gene expression.

The antero-posterior patterning of the mesoderm is most clearly seen in the differences in the vertebrae, each vertebra having well-defined anatomical characteristics depending on its location along the axis. The most anterior vertebrae are specialized for attachment and articulation of the skull, while the cervical vertebrae of the neck are followed by the rib-bearing thoracic vertebrae and then those of the lumbar region which do not bear ribs, and finally, those of the sacral and caudal regions. This antero-posterior patterning is a process distinct from the patterning of individual somites in respect of the cells that will produce muscle, cartilage, and dermis, and occurs earlier, while the pre-somitic mesoderm is still unsegmented. It is based on the mesodermal cells acquiring a positional value that reflects their position along the axis and so determines their subsequent development. Mesodermal cells that will form thoracic vertebrae, for example, have different positional values from those that will form cervical vertebrae.

Patterning along the antero-posterior axis in all vertebrates involves the expression of a set of genes that specify positional identity along the axis. These are the **Hox genes**, members of the large family of **homeobox genes** that are involved in many aspects of development (Box 4A). The concept of positional identity, or positional value, has important implications for developmental strategy; it implies that a cell or a group of cells in the embryo acquires a unique state related to its position at a given time, and that this determines its later development (see Section 1-13). Unfortunately, we do not yet know how positional information is set up during gastrulation.

Homeobox genes specifying positional identity along the antero-posterior axis were originally identified in the fruit fly *Drosophila* and, to developmental biologists' delight, it turned out that related genes are also involved in patterning the vertebrate axis. As we see in the final part of this chapter, patterning along the antero-posterior axis by Hox genes and other homeobox genes is not confined to mesodermal structures; the hindbrain, for example, is also divided into distinct regions.

All the homeobox genes whose functions are known encode DNA-binding proteins that act as transcription factors. The subset known as the Hox genes are the vertebrate counterparts of a cluster of homeobox genes in *Drosophila* that is involved in specifying the identities of the different segments of the insect body. Vertebrates have four separate clusters of Hox genes that are thought to have arisen by duplications of the genes within a cluster, and of the clusters themselves (see Box 4A). A particular feature of Hox gene expression in both insects and vertebrates is that the genes in each cluster are expressed in a temporal and spatial order that reflects their order on the chromosome.

A simple idealized model illustrates the key features by which a Hox gene cluster records positional identity. Consider four genes I, II, III, and IV, arranged along a chromosome in that order (Fig. 4.7). The genes are expressed in a corresponding order along the antero-posterior axis of a tissue. Thus, gene I is expressed throughout the tissue with its ante-rior boundary at the anterior end. Gene II has its anterior boundary in a more posterior position and expression continues posteriorly. The same principles apply to the two other genes. This pattern of expression

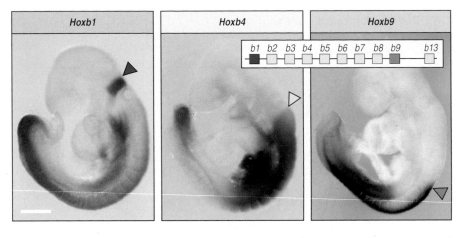

| b1 | b2 | b3 | b4 | b5 | b6 | b7 | b8 | b9 | b13 |

Fig. 4.8 Hox gene expression in the mouse embryo (after neurulation). The three panels show lateral views of 9½ days post-coitum embryos immunostained with antibodies specific for the protein products of the *Hoxb1*, *Hoxb4*, and *Hoxb9* genes. The arrow-heads indicate the anterior boundary of expression of each gene within the neural tube. The position of the three genes within the Hoxb gene complex is indicated (inset). Scale bar = 0.5 mm. Photographs courtesy of A. Gould.

defines four distinct regions. If the amount of gene product is varied within each expression domain, for example by interactions between the genes, many more regions can be specified.

The role of the Hox genes in vertebrate axial patterning has been best studied in the mouse, which has four Hox clusters. As in all vertebrates, the Hox genes begin to be expressed, with the 'anterior' genes expressed first, in mesoderm cells at an early stage of gastrulation when the meso-derm cells begin to leave the primitive streak. Since the posterior pattern develops later, clearly defined patterns of Hox gene expression are most easily seen in the mesoderm and neural tube, after somite formation and neurulation, respectively (Fig. 4.8). Hox genes are expressed as gastrulation proceeds. Typically, the pattern of expression is charac-terized by a relatively sharp anterior border and usually a much less well defined posterior border. Although there is considerable overlap in expression, almost every region in the anterior part of the antero-posterior axis is characterized by a particular set of expressed Hox genes (Fig. 4.9). For example, the most anterior somites are characterized

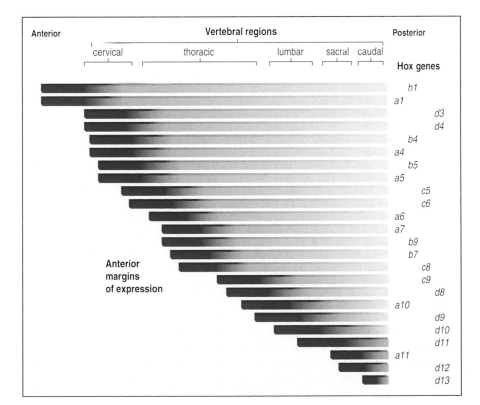

Fig. 4.9 Hox gene expression along the antero-posterior axis of the mouse mesoderm. The anterior border of each gene is shown by the dark red blocks. Expression usually extends backward some distance but the posterior margin of expression is usually poorly defined. The pattern of Hox gene expression could specify the identity of the tissues at different positions. For example, the pattern of expression is quite different in anterior and posterior regions of the body axis.

Box 4A Homeobox genes.

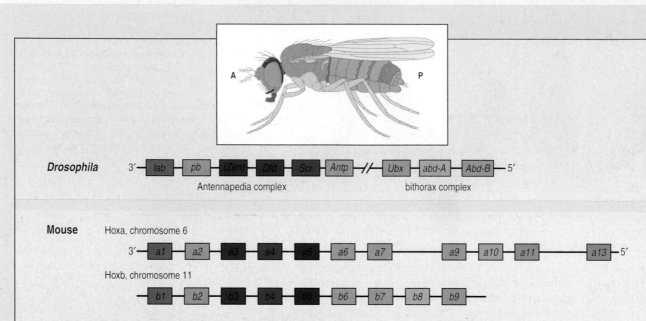

The homeobox gene family encodes a large group of transcription factors which all contain a similar DNA-binding region of around 60 amino acids called the **homeodomain**. The homeodomain contains a helix-turn-helix DNA-binding motif which is characteristic of many DNA-binding proteins. This domain is encoded by a DNA sequence of 180 base pairs termed the **homeobox**. Many homeobox genes are involved in development, and the homeobox was originally identified in genes that control patterning in *Drosophila* development.

The name 'homeobox' comes from the fact that mutations in some of these genes result in what is known as a **homeotic** transformation, in which one structure replaces another. For example, in one homeotic mutation in *Drosophila,* a segment in the fly's body that does not normally bear wings is transformed into an adjacent segment that does bear wings, resulting in a fly with four wings.

Clusters of homeotic genes involved in specifying segment identity were first discovered in the fruit fly *Drosophila*. Similar complexes of homeotic genes have been identified in many animals. In vertebrates, the related clusters are known as the Hox complexes, and the homeoboxes of the genes are related to the Antennapedia homeobox of *Drosophila*. In the mouse, there are four unlinked Hox complexes, designated Hoxa, Hoxb, Hoxc, and Hoxd, (originally called Hox1, Hox2, Hox3, and Hox4) located on chromosomes 6, 11, 15, and 2, respectively (see figure).

The four vertebrate clusters have arisen by gross duplication of an ancestral cluster, possibly related to the single Hox cluster in the lancelet amphioxus, a simple chordate. Thus, corresponding genes within the four clusters resemble each other closely. The original cluster is thought to have formed by gene duplication and divergence and all Hox genes thus resemble each other to some extent; the homology is most marked within the homeobox and less marked in sequences outside it. Genes that have arisen by duplication and divergence within a species are known as **paralogs**, and the corresponding genes in the different clusters (e.g. *Hoxa4*, *Hoxb4*, *Hoxc4*, *Hoxd4*) are usually known as a **paralogous subgroup**. In the mouse there are 13 paralogous groups.

The Hox gene clusters and their role in development are of ancient origin. The mouse and frog genes are similar to each other and to those of the fruit fly *Drosophila* both in their coding sequences and in their order on the chromosome. In both *Drosophila* and vertebrates, these homeotic genes are involved in specifying regional identity along the antero-posterior axis. The Hox clusters in mice and in *Drosophila* (where they are called HOM genes) almost certainly arose by gene duplication in some common ancestor of vertebrates and insects.

Most genes that contain a homeobox do not, however, belong to a homeotic complex, nor are they involved in homeotic transformations.

Other subfamilies of homeobox genes in vertebrates include the **Pax genes**, which contain a homeobox and another conserved domain known as *paired*. All these genes encode transcription factors with various functions in development and cell differentiation.

The homeobox genes are the most striking example of a widespread conservation of developmental genes in animals. It is widely believed that there are common mechanisms underlying the development of all animals. This implies that if a gene is identified as playing a central role in the development of one animal it is worth looking to see where it is present in another animal and whether it has a similar function. This strategy of comparing genes by sequence homology has proved extremely successful in identifying genes involved in development in vertebrates. Numerous genes first identified in *Drosophila*, in which the genetic basis for development is far better understood than in any other animal, have proved to have counterparts involved in development in vertebrates. Illustration after Coletta, P., *et al.*: 1994.

by expression of genes *Hoxa1* and *Hoxb1*, and no other Hox genes are expressed in this region. By contrast, all the Hox genes are expressed in the most posterior regions. (The most anterior regions of the vertebrate body—the anterior head, forebrain, and midbrain—are characterized by expression of the homeobox genes *etx* and *otx*, and not by Hox genes.)

If we focus on just one set of Hox genes, those of the Hoxa complex, we find that the most anterior border of expression in the mesoderm is that of *Hoxa1* in the posterior head mesoderm, while *Hoxa11*, the most posterior gene in the Hoxa cluster, has its anterior border of expression in the sacral region (see Fig. 4.9). This exceptional correspondence, or co-linearity between the order of the genes on the chromosome and their order of spatial and temporal expression along the antero-posterior axis, is typical of all the Hox complexes. The genes of each Hox complex are expressed in an orderly sequence, with the gene lying most 3' in the cluster being expressed the earliest and in the most anterior position. The correct expression of the Hox genes is dependent on their position in the cluster and anterior genes must be expressed before more posterior genes.

Evidence that the Hox genes are involved in controlling regional identity comes from comparing their patterns of expression in mouse and chick with the well-defined regions—cervical, thoracic, and so on (Fig. 4.10). Hox gene expression corresponds nicely with the different regions. For example, even though the number of cervical vertebrae in birds (14) is twice that of mammals, the anterior boundaries of *Hoxc5* and *Hoxc6* gene expression in both chick and mouse lie on either side of the cervical/thoracic boundary. A correspondence between Hox gene expression and region is also similarly conserved among vertebrates at other anatomical boundaries.

It must be emphasized that the summary picture of Hox gene expression given in Fig. 4.9 does not represent a 'snapshot' of expression at a particular time but rather the total overall pattern of expression. Some genes are switched on earlier than others and are then downregulated, while others are expressed considerably later than others; the most

Fig. 4.10 Patterns of Hox gene expression in the mesoderm of chick and mouse embryos, and their relation to regionalization. The posterior margins of expression of Hox genes in the mesoderm vary along the axes. The vertebrae are derived from somites, 40 of which are shown. The vertebrae have characteristic shapes in each of the five regions: cervical (C), thoracic (T), lumbar (L), sacral (S) and caudal (Ca). Which somites form which vertebrae differs in chick and mouse. For example, thoracic vertebrae start at somite 20 in the chick, but at somite 12 in the mouse. The transition from one region to another corresponds with the pattern of Hox gene expression, so *Hoxc5* and *Hoxc6* are on either side of the cervical and thoracic vertebral transition in both chick and mouse. Similarly, *Hoxd9* and *Hoxd10* are at the transition between lumbar and sacral regions. After Burke, A.C.: 1995.

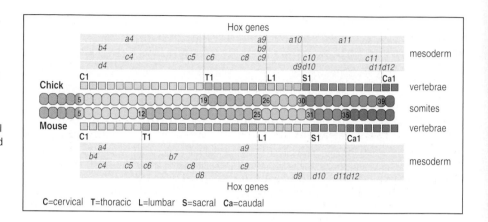

posterior Hox genes such as *d12* and *13*, for example, are expressed in the post-anal tail, which develops later. Moreover, this summary picture reflects the general expression of the genes in embryonic regions; not all Hox genes expressed in a region are expressed in all the cells of that region. Nevertheless, the overall pattern suggests that the combination of Hox genes provides positional identity. In the cervical region, for example, each somite, and thus each vertebra, could be specified by a unique pattern of Hox gene expression.

If the Hox genes do provide positional values that determine a region's subsequent development then, if their pattern of expression is altered, one would expect morphological changes. This is indeed the case, as we see next.

4-4 Deletion or overexpression of Hox genes causes changes in axial patterning.

In order to see how Hox genes control patterning, either their expression can be prevented by mutation, or they can be expressed in abnormal positions. Hox gene expression can be eliminated from the developing mouse embryo by gene knock-out techniques (Box 4B). Experiments along these lines have shown that absence of a Hox gene affects patterning in a way that accords with the idea that Hox gene activity provides the cells with positional identity. For example, mice in which the gene *Hoxa3* has been deleted show complex structural defects in the region of the head and thorax, where this gene is normally strongly expressed, and tissues derived from both ectoderm and mesoderm are affected. But the Hox genes seem to specify positional identity in rather complex ways. There is undoubtedly some **redundancy** between the effect of some of the genes, and when one gene is removed, another may serve in its place, which can make it difficult to interpret the results of experiments in which a particular Hox gene has been inactivated. There is also interaction between the individual genes, and this can further complicate results. For example, with the mutated *Hoxa3* gene described above, more posterior axial structures, where the inactivated gene is also normally expressed, show no evident defects.

This illustrates a general principle of Hox gene expression, which is that more posteriorly expressed Hox genes tend to inhibit the action of the Hox genes normally expressed anterior to them; this phenomenon is known as **posterior dominance** or **posterior prevalence**. This means that a change in Hox gene expression usually affects the most anterior regions in which the gene is expressed, leaving posterior structures relatively unaffected. The effects of a Hox gene knock-out can also be tissue specific, so that certain tissues in which a Hox gene is normally expressed appear normal, while other tissues at the same position along the antero-posterior axis are affected. The apparent absence of an effect may be due to redundancy, with **paralogous genes** from another complex being able to compensate. For example, *Hoxb1* is expressed in the same region as *Hoxa1* (see Fig. 4.9), and so may be largely able to fulfill the function of an absent *Hoxa1* gene.

Loss of Hox gene function often results in **homeotic transformation** —the conversion of one body part into another—as in the case of a knock-out mutation of *Hoxc8*. In normal embryos, *Hoxc8* is expressed in the thoracic and more posterior regions of the embryo from late gastrulation onward. Mice homozygous for mutant *Hoxc8* die within a few days of birth, and have abnormalities in patterning between the seventh thoracic vertebra and the first lumbar vertebra. The most obvious homeotic transformations are the attachment of an eighth pair of ribs to the sternum and the development of a 14th pair of ribs on the first lumbar vertebra (Fig. 4.11). Thus, the absence of *Hoxc8* modifies the development of some of the cells that would normally express it. Its absence gives them a more anterior positional value, and they develop accordingly. In mice in which *Hoxd11* is mutated, anterior sacral vertebrae are transformed into lumbar vertebrae. Another example of the homeotic transformation of a structure into one normally anterior to it can be seen in knock-out mutations of *Hoxb4*. In normal mice, *Hoxb4* is expressed in the mesoderm that will give rise to the axis (the second cervical vertebra), but not in that giving rise to the atlas (the first cervical vertebra). In *Hoxb4* knock-out mice, the axis is transformed into another atlas.

By contrast, abnormal expression of Hox genes in anterior regions that normally do not express them can result in transformations of anterior structures into structures that are normally more posterior. For example, when *Hoxa7*, whose normal anterior border of expression is in the thoracic region, is expressed throughout the whole antero-posterior axis, the basal occipital bone of the skull is transformed into a pro-atlas structure, normally the next most posterior skeletal structure.

There are synergistic interactions between Hox genes of the same paralogous group. Thus, knock-outs of *Hoxa3* do not affect the first cervical vertebra—the atlas—or the basal occipital bone of the skull to which it connects, even though *Hoxa3* is expressed in the mesoderm that gives rise to these bones. However, knock-outs of *Hoxd3* (which is also expressed in this region) cause a homeotic transformation of the atlas into the adjacent basal occipital bone. A double knock-out of *Hoxa3* and *Hoxd3* results in complete deletion of the atlas. The complete absence of this bone in the absence of Hox gene expression suggests that one target of Hox gene action is the cell proliferation required to build such a structure from the somite cells. Since we do not know how the pattern of Hox gene expression is specified, it is of interest that the pattern can be altered by retinoic acid, a diffusible molecule.

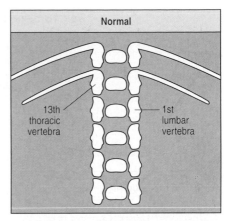

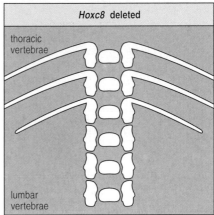

Fig. 4.11 Homeotic transformation of vertebrae due to deletion of *Hoxc8* in the mouse. In loss-of-function homozygous mutants of *Hoxc8*, the first lumbar vertebra is transformed into a rib-bearing thoracic vertebra. The mutation has resulted in the transformation of the lumbar vertebra into a more anterior structure.

| 4-5 | Retinoic acid can alter positional values. |

Retinoic acid is a small hydrophobic molecule—a derivative of vitamin A—which has an important role in local signaling in vertebrate development. In a similar way to steroid and thyroid hormones, it diffuses across the plasma membrane unaided and binds to intracellular receptors; the complex of receptor and retinoic acid then functions as a transcription factor. A variety of experiments have shown that retinoic acid can alter cells' positional values in limb development (see Section 10-5), and it can also have effects on the antero-posterior axis.

Box 4B Gene targeting: insertional mutagenesis and gene knock-out.

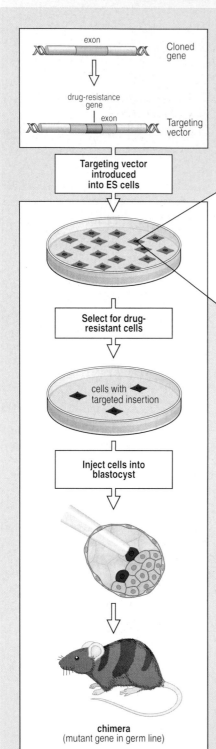

In order to study the function of a gene controlling development, it is highly desirable to be able to introduce an altered gene into the animal to see what effect this has. Mice into which an additional or altered gene has been introduced are known as **transgenic** mice. Two main techniques for generating transgenic mice are currently in use. One is to inject DNA containing the required gene directly into the nuclei of fertilized eggs; the other is to alter or add a gene to the genome of **embryonic stem cells** (ES cells) in culture, and then to inject the genetically altered cells into the blastocyst, where they become part of the inner cell mass.

ES cells can be genetically altered by techniques that can be used to create a mutation in a particular gene. A vector DNA molecule that is introduced into an ES cell by **transfection** will usually insert randomly in the genome. However, it is possible to tailor the vector DNA in such a way that only those DNA molecules that insert at a specific predetermined site by **homologous recombination**, and thus mutate and inactivate a particular gene, will be selected. The DNA to be introduced must contain enough sequence homology with the target gene that it will insert within the target gene in at least a few cells in the culture, even though most insertions will be random. These mutated ES cells can then be introduced into the blastocyst, producing a transgenic mouse carrying a mutation in a known gene. The use of homologous recombination to inactivate a gene is known as **gene knock-out** when the animal is homozygous for the inactivated gene. Many mutations produced by this technique do not result in a knock-out, but the mutation alters gene function.

The enormous advantage of using ES cells over microinjection methods for generating transgenic mice is that it is possible to design a selection procedure to isolate just those rare cells in which the DNA has incorporated at the desired site. These can then be used to generate the chimeric embryo. The selection procedure is based on including certain genes for drug-resistance and drug-sensitivity in the DNA construct such that when the DNA inserts at the correct site, only the cells with the correctly targeted DNA can be selected.

The mutated ES cells are then introduced into the cavity of an early blastocyst, which is then returned to the uterus. They become incorporated into the inner cell mass and thus into the embryo, where they can give rise to germ cells and gametes. Once the mutant gene has entered the germ line, strains of mice heterozygous and homozygous for the altered gene can be bred (see the figure on the opposite page) and the effect of completely inactivating and so knocking out the gene can be examined.

A significant number of experiments in which a gene is mutated and effectively knocked out result in mice developing without any obvious abnormality or with fewer and less severe abnormalities than might be expected from the normal pattern of gene activity. A striking

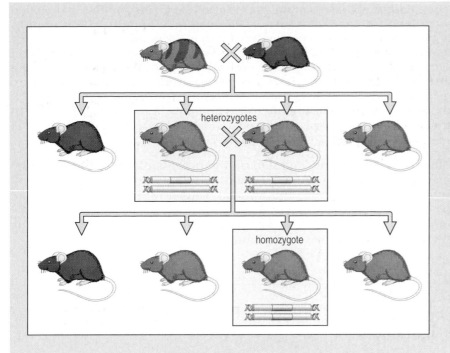

example is that in the absence of *myoD*, a key gene involved in muscle differentiation, the mice are anatomically normal, although they do have a reduced survival rate. This could mean that *myoD* is genuinely redundant, in that other genes can substitute for its functions.

However, it is extremely unlikely that any gene is without any value at all to an animal. It is much more likely that there is an altered phenotype in these apparently normal animals, which is too subtle to be detected under the artificial conditions of life in a laboratory. Redundancy is thus probably apparent rather than real. A further complication is provided· by the possibility that related genes with similar functions may increase their activity to compensate for the mutated gene.

The developmental abnormalities induced by retinoic acid are apparently the result of its interference with the normal establishment of the Hox gene expression pattern. For example, treatment of early mouse embryos with retinoic acid results in homeotic transformation of vertebrae, both anterior and posterior transformations being induced, depending on the time of treatment. It is likely that this effect of retinoic acid is mediated by its action on Hox gene expression, and studies of cells in culture show that the Hox genes can be induced by retinoic acid in a concentration-dependent manner. In *Xenopus*, the gene *Xlhbox6* (related to mouse *Hoxb9*) is normally expressed in posterior regions, but in embryos treated with retinoic acid, the expression of *Xlhbox6* extends anteriorly, and dorsal mesoderm and anterior structures are defective. As there is evidence for a gradient in retinoic acid along the antero-posterior axis in the mouse embryo, it could be important in activating Hox genes in normal antero-posterior patterning.

Summary.

Somites are blocks of mesodermal tissue that are formed after gastrulation from mesoderm that was originally immediately lateral to the dorsal-most mesoderm (the organizer region) of the marginal zone. They form sequentially in pairs on either side of the notochord, starting at the anterior end of the embryo. The somites give rise to the vertebrae, to the muscles of the trunk and limbs, and to the dermis of the skin. The pre-somitic mesoderm is patterned in an antero-posterior dimension before somite formation, and the first manifestation of this pattern is the compartmentalized expression of the Hox genes in the mesoderm. The somites are also patterned by signals from the notochord, neural tube, and ectoderm, which induce particular regions of each somite to give rise to muscle, cartilage, or dermis.

The regional character of the mesoderm that forms somites is specified even before the somites form. The positional identity of the somites appears to be specified by combinatorial expression of genes of the Hox complexes along the antero-posterior axis, with the order of expression of these genes along the axis corresponding to their order along the chromosome. Mutation or overexpression of a Hox gene results, in general, in localized defects in the anterior parts of the regions in which the gene is expressed, and can cause homeotic transformations. We can think of Hox genes as providing positional information that specifies the identity of a region and its later development.

Although we have concentrated on the expression of Hox genes in the mesoderm, they are also expressed in a patterned way in the neural tube after its induction, and we shall return to this aspect of antero-posterior regionalization later. But next, we look at the role of the crucially important organizer region both in neural induction and in organizing the antero-posterior axis in vertebrate embryos.

The role of the organizer region and neural induction.

The Spemann organizer of amphibians, Hensen's node in the chick, and the equivalent node region in the mouse all have a similar global organizing function in vertebrate development. They can all induce a complete body axis if transplanted to another embryo at an appropriate stage, and so are able to organize and coordinate both dorso-ventral and antero-posterior aspects of the body plan, as well as to induce neural tissue from ectoderm.

During gastrulation, the ectoderm lying along the dorsal midline of the embryo becomes specified as neural plate; during the subsequent stage of neurulation, this folds to form the neural tube, and then differentiates into the brain and spinal cord (see Section 2-1). The brain and spinal cord must develop in the correct relationship with other body structures, particularly the mesodermally derived structures that give rise to the skeleto-muscular system. Thus, patterning of the nervous system must be linked to that of the mesoderm. In this section, we consider the patterning of the neural tube up to shortly after its closure, although consideration of the hindbrain will take us to the stage when that region becomes segmented and the neural crest cells have migrated.

The function of the organizer has been best studied in amphibians, and we have already emphasized its role in the dorso-ventral patterning of the mesoderm of *Xenopus* (see Section 3-15). We now discuss its profound effects on the antero-posterior axis.

4-6   The organizer can specify a new antero-posterior axis.

In amphibians, the action of the organizer is dramatically demonstrated in what is classically known as primary embryonic induction. Grafts of the Spemann organizer (which corresponds to the dorsal lip of the amphibian blastopore) to the ventral side of the marginal zone of another embryo result in a twinned embryo. This second embryo can have a well defined head and trunk region and even a tail, but will be joined to the main embryo along the axis. A variety of other treatments, such as grafting dorsal vegetal blastomeres to the ventral side, produce a similar result (see Fig. 3.5). What all these treatments have in common is that, directly or indirectly, they result in the formation of a new Spemann organizer. In contrast, if the organizer is removed or its formation prevented, the embryo still gastrulates and internalizes the remaining mesoderm and endoderm, but develops as a cylindrically symmetrical, ventralized embryo, lacking dorsal and anterior structures such as the head, the neural tube, notochord and somites (see Fig. 3.4).

As we saw earlier in the book, and will consider in more detail in Chapter 8, the cells of the organizer region give rise during gastrulation to a small amount of prospective head mesoderm (the pre-chordal mesoderm), and to the notochord, which clearly provide signals for induction of head structures and the neural plate, respectively. However, understanding how the organizer mesoderm organizes the global pattern of the antero-posterior axis is not so straightforward. In amphibians, the inductive properties of the Spemann organizer region appear to change throughout gastrulation. In transplantation experiments of the type described above, a dorsal lip taken from an early gastrula induces a complete additional embryo, a dorsal lip from a mid-gastrula induces a trunk and tail but no head, while a dorsal lip from a late gastrula induces only a tail (Fig. 4.12). As gastrulation proceeds, the antero-posterior axis becomes specified, and so at later stages cells in the blastopore induce only posterior structures.

The avian equivalent to the Spemann organizer is Hensen's node, the region at the anterior end of the primitive streak in the chick blastoderm. It contributes to both notochord and somites, and can induce an additional axis, complete with somites, if grafted beneath a chick epiblast at the head process stage of development (Fig. 4.13). This stage is reached when elongation of the primitive streak is complete, the notochord (head process) has started to form anteriorly, but Hensen's node has not yet started to regress (see Fig. 2.14). Induction only occurs if the graft is placed quite close to the streak; it can then induce normally non-axial mesoderm to form somites and other axial structures. The node at the anterior region of the primitive streak in the mouse can induce a similar duplication of the mouse axis on transplantation.

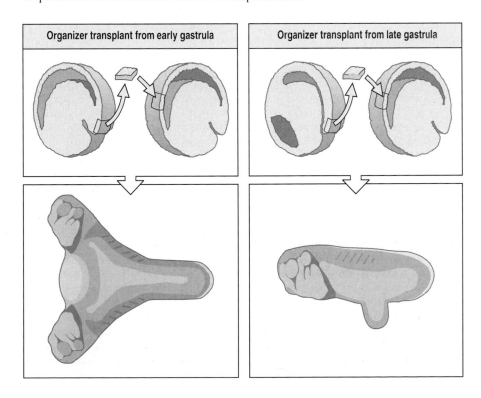

| Organizer transplant from early gastrula | Organizer transplant from late gastrula |

Fig. 4.12 The inductive properties of the organizer change during gastrulation. A graft of the organizer region, from the dorsal lip of the blastopore of an early frog gastrula to the ventral side of another gastrula, results in the development of an additional anterior axis at the site of the graft (left panels). A graft from the dorsal lip region of a late gastrula only induces formation of tail structures (right panels).

Fig. 4.13 Hensen's node can induce a new axis in avian embryos. When Hensen's node from a quail embryo is grafted to a position lateral to the primitive streak of a chick embryo at the same stage of development, a new axis forms at the site of transplantation. Histological examination shows that although some of the somites of this new axis are formed from the graft itself (quail tissue can easily be distinguished from chick tissue), others have been induced from host tissue that does not normally form somites.

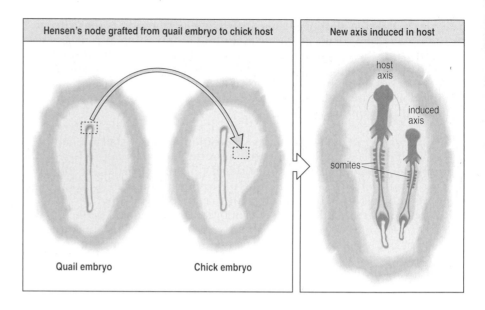

A number of genes are specifically expressed in the organizer of both *Xenopus* and the mouse (Fig. 4.14) and, as we have already seen in Section 3-18, some of these genes have a role in patterning the mesoderm along the future antero-posterior axis. For example, mice lacking the *Brachyury* gene have no posterior mesoderm-derived structures and the absence of the related gene *no-tail* in zebrafish gives a similar phenotype. FGF, or fibroblast growth factor, seems to be required to maintain *Brachyury* expression in *Xenopus* development. Inhibition of the FGF receptor in zebrafish leads to complete loss of the trunk and tail. In the mouse, the homeobox gene *Lim-1* is crucially involved in specifying head structures. Mouse embryos in which this gene has been inactivated by mutation lack all anterior head structures, but have normal trunk and tail development, even though *Lim-1* is normally expressed in both the head and the rest of the axial mesoderm. We have already noted that the mutation in the *floating head* gene in zebrafish, a homolog of *Xnot1*, results in the absence of notochord and anterior structures (see Section 3-20).

We still do not know what switches on the Hox genes in the mesoderm during gastrulation. In all vertebrates, the Hox genes begin to be expressed (the anterior-most genes being expressed first) at an early stage of gastrulation, when the mesodermal cells begin their gastrulation movements. If the position of a Hoxd gene is relocated to the 5' end of the Hoxd complex, its expression resembles that of the neighboring *Hoxd13*. This shows that the structure of the complex is crucial in determining Hox gene expression. One way the antero-posterior pattern of Hox gene expression might be established in the somitic mesoderm is through linking Hox gene expression to movement of the mesoderm cells through the blastopore at gastrulation. Such a mechanism relies on the striking correspondence in the order of genes on the chromosome with their spatial and temporal expression (see Section 4-3). One could propose a mechanism that activates expression of a Hox gene complex at its 3' end in pre-somitic mesoderm cells as they start to move inside the embryo during gastrulation. If activation of the Hox complex then spreads along the chromosome in the 5' direction but the spread ceases when cells move over the dorsal lip of the blastopore, then more of the complex will be expressed in posterior cells, which enter last, than in anterior cells, which enter first. At present, such a mechanism is entirely speculative.

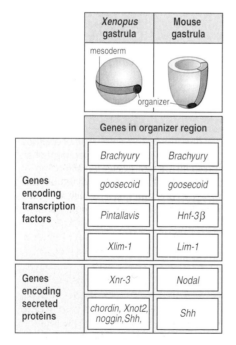

Xenopus gastrula	Mouse gastrula
mesoderm ... organizer	organizer
Genes in organizer region	

	Xenopus	Mouse
Genes encoding transcription factors	*Brachyury*	*Brachyury*
	goosecoid	*goosecoid*
	Pintallavis	*Hnf-3β*
	Xlim-1	*Lim-1*
Genes encoding secreted proteins	*Xnr-3*	*Nodal*
	chordin, Xnot2, noggin, Shh,	*Shh*

Fig. 4.14 Genes expressed in the Spemann organizer region of the *Xenopus* gastrula, and in Hensen's node in the mouse gastrula. There is a similar pattern of gene activity in the two animals with homologous genes being expressed. *Shh = Sonic hedgehog.*

| 4-7 | **The neural plate is induced by mesoderm.** |

The induction of neural tissue from ectoderm was first indicated by the organizer transplant experiment in frogs, which is described in Fig. 4.12; in the secondary embryo that forms at the site of transplantation, a nervous system develops from the host ectoderm that would normally have formed ventral epidermis. This suggests that neural tissue can be induced from as yet unspecified ectoderm by signals emanating from the mesoderm of the organizer region. The requirement for induction is confirmed by experiments that exchange prospective neural plate ectoderm for prospective epidermis before gastrulation; the transplanted prospective epidermis develops into neural tissue (Fig. 4.15). This shows that the formation of the nervous system is dependent on an inductive signal.

An enormous amount of effort in the 1930s and 1940s was devoted to trying to identify the signals involved in neural induction in amphibians. Researchers were encouraged by the finding that a dead organizer region could still induce neural tissue. It seemed to be merely a matter of hard work to isolate the chemicals responsible. Alas, the search was fruitless, for it appeared that an enormous variety of substances were capable of varying degrees of neural induction. As it turned out, this was because newt ectoderm, the main experimental material used, seems to have a high propensity to develop into neural tissue on its own. This is not the case with *Xenopus* ectoderm, although prolonged culture of disaggregated ectodermal cells can result in their differentiation as neural cells. Whatever the nature of the signal, it seems that it is a molecule that can diffuse through a Nuclepore filter (which prevents cell contact but allows the passage of quite large molecules, such as proteins), and that contact lasting about 2 hours is required for induction to occur. The molecules responsible for neural induction have still not been definitively identified, although there are some intriguing candidates. However, contrary to what was previously thought, the inducing molecules do not act directly on the cells that will form neural tissue, but act instead on molecules that inhibit the cells from forming neural tissue.

The secreted growth factor BMP-4 plays a pivotal role in neural induction, since it inhibits cells from forming neural tissue. If BMP signaling is disinhibited, then neural tissue develops. One such inhibitor of BMP signaling is the protein encoded by the gene *noggin*. As discussed in

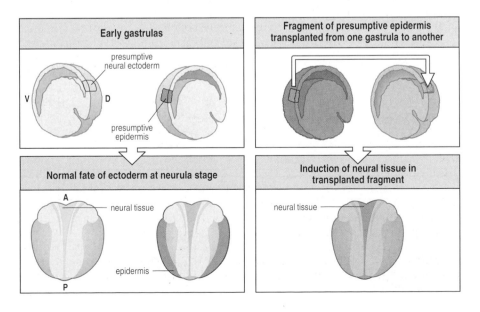

Fig. 4.15 The nervous system of *Xenopus* is induced during gastrulation. The left panels show the normal development fate of ectoderm at two different positions in the early gastrula. The right panels show the transplantation of a piece of ventral ectoderm, whose normal fate is to form epidermis, from the ventral side of an early gastrula to the dorsal side of another, where it replaces a piece of dorsal ectoderm whose normal fate is to form neural tissue. In its new location, the transplanted prospective epidermis develops not as epidermis but as neural tissue, and forms part of a normal nervous system. This shows that the ventral tissue has not yet been determined at the time of transplantation, and that neural tissue is induced during gastrulation.

Section 3-18, noggin protein is secreted by the *Xenopus* organizer region and is probably one of the signals dorsalizing the presumptive mesoderm. The noggin protein may also be a neural induction factor: if it is added to isolated blastula animal caps at a high concentration, neural markers are induced. The protein noggin has the expression pattern and activity expected of a neural inducer. Another secreted protein, chordin, is expressed in the organizer-derived mesoderm underlying the future neural plate, and it too has neuralizing activity and acts by binding to BMP-4, thus neutralizing its action. The antagonistic actions of BMP-4 and chordin in neural induction could be similar to that seen in the dorso-ventral patterning of the mesoderm itself (see Section 3-18). Chordin and BMP-4 are homologs of proteins that also have antagonistic functions in patterning the dorso-ventral axis of *Drosophila*, as we see in Chapter 5.

4-8   The nervous system can be patterned by signals from the mesoderm.

Pieces of mesoderm taken from different positions along the antero-posterior axis of a newt neurula and placed in the blastocoel of an early newt embryo induce neural structures at the site of transplantation. Positional specificity in this induction is indicated by the fact that the structures formed correspond more or less to the original position of the transplanted mesoderm. Pieces of anterior mesoderm induce a head with a brain, whereas posterior pieces induce a trunk with a spinal cord (Fig. 4.16). Another indication of positional specificity in induction comes from the observation that pieces of the neural plate themselves induce similar regional neural structures in adjacent ectoderm when transplanted beneath the ectoderm of a gastrula. Indications that gene expression in the mesoderm may be influencing gene expression in the ectoderm come from the observation of coincident expression of several Hox genes in the notochord and in the presomitic mesoderm and ectoderm at the same position along the antero-posterior axis: *Xlhbox1* in *Xenopus* and *Hoxb1* in the mouse are examples of genes that have coincident expression of this type.

In the anterior-most mesoderm of the mouse, Hox genes are not expressed, but the gene *otx-2* (related to the *Drosophila* gene *orthodenticle*) is expressed in the endoderm and mesoderm cells in this region. Mutations in this gene result in the absence of forebrain and midbrain regions, suggesting that the endodermal and mesodermal cells are necessary to specify these brain regions.

Fig. 4.16 Induction of the nervous system by the mesoderm is region specific. Mesoderm from different positions along the dorsal antero-posterior axis of early newt neurulas induces structures specific to its region of origin when transplanted to ventral regions of early gastrulas. Anterior mesoderm induces a head with a brain (top panels), whereas posterior mesoderm induces a posterior trunk with a spinal cord ending in a tail (bottom panels). After Mangold, O.: 1933.

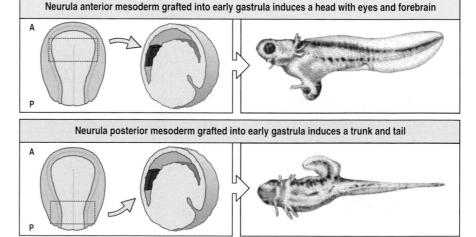

Neurula anterior mesoderm grafted into early gastrula induces a head with eyes and forebrain

Neurula posterior mesoderm grafted into early gastrula induces a trunk and tail

One model for neural ectoderm patterning suggests that qualitatively different mesodermal inducers are present at different positions along the antero-posterior axis. However, many experiments fit quite well with a simpler two-signal model of neural patterning (Fig. 4.17), in which differences are due to quantitative rather than qualitative differences in the inducing signal. In this model, the first signal is produced by the whole mesoderm and induces the ectoderm to become anterior neural tissue. Chordin and noggin are good candidates for this signal. The second signal transforms part of this tissue so that it acquires a more posterior identity. This latter signal would be graded in the mesoderm with the highest concentration at the posterior end. The model thus proposes a progressively posterior specification; it starts by inducing anterior tissue, after which posterior tissue is specified. In line with this model, *Xenopus* ectoderm that has already been specified to make the cement gland, which occupies a position anterior to the neural tube, can be induced to acquire a more posterior fate by posterior mesoderm. Fibroblast growth factor, Wnt proteins, and retinoic acid are good candidates for the posteriorizing signal.

Mesoderm also induces neural tissue in chick and mouse embryos. Neural tissue can be induced in the chick epiblast—in both the area pellucida and the area opaca—by grafts from the primitive streak. Inductive activity is initially located in the anterior primitive streak and later, during regression of Hensen's node, becomes confined to the region just anterior to the node. By the four-somite stage inductive activity has disappeared, but the competence of the ectoderm to respond only disappears later, at the head process stage.

Hensen's node in the chick embryo can also induce neural gene expression in amphibian (*Xenopus*) ectoderm (Fig. 4.18), which is of great interest as it suggests that there has been an evolutionary conservation of inducing signals. Moreover, early nodes induce gene expression characteristic of anterior amphibian neural structures, while older nodes induce gene expression typical of posterior structures. These results are in line with the theory that the node is able to specify different antero-posterior positional values, and they also confirm the essential similarity of Hensen's node to the Spemann organizer.

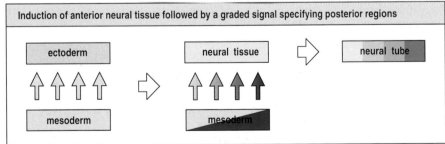

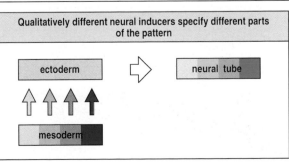

Fig. 4.17 Models of neural patterning by induction. Top panel: in the two-signal model, one signal from the mesoderm first induces anterior tissue throughout the corresponding ectoderm. A second, graded signal from the mesoderm then specifies more posterior regions. Bottom panel: in an alternative model, qualitatively different inducers are localized in the mesoderm. After Kelly, O.G., *et al.*: 1995.

Fig. 4.18 Hensen's node from a chick embryo can induce gene expression characteristic of neural tissue in *Xenopus* ectoderm. Tissues from different parts of the primitive streak stage of a chick epiblast are placed between two fragments of animal cap tissue (prospective ectoderm) from a *Xenopus* blastula. The induction in the *Xenopus* ectoderm of genes that characterize the nervous system is detected by looking for expression of mRNAs for neural cell adhesion molecule (N-CAM) and neurogenic factor-3 (NF-3), which are expressed specifically in neural tissue in stage 30 *Xenopus* embryos. Only transplants from Hensen's node induce the expression of these neural markers in the *Xenopus* ectoderm. (EF-1A is a common transcription factor expressed in all cells.) After Kintner, C.R., *et al.*: 1991.

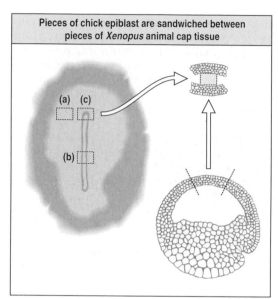

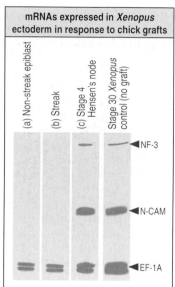

4-9   **Signals that pattern the neural plate may travel within the neural plate itself.**

Development of neural tissue was originally thought only to occur if the mesoderm came to lie immediately beneath the ectoderm and in contact with it. This seemed to be clearly shown in what are known as **exogastrulas**, which can be formed by the embryos of newts and other urodeles. Exogastrulas can be induced by treating embryos at the start of gastrulation with hypertonic salt solutions; in these embryos the mesoderm does not enter the embryo but comes to lie outside it. In such exogastrulas the ectodermal region does not appear to form a nervous system (Fig. 4.19). However, closer examination of *Xenopus* exogastrulas shows that some genes characteristic of neural tissue, such as that for the neural cell adhesion molecule N-CAM, are expressed in the ectoderm. Moreover, these neural genes are expressed in the correct antero-posterior order in posterior neural tissue; anterior neural markers are, however, absent. Thus, it now appears that there are two routes by which the mesoderm can induce the nervous system. One is by the traditional vertical or transverse route from the mesoderm to the overlying ectoderm, while the other is planar, the signal being generated within the neural plate itself and traveling within the ectodermal sheet.

Evidence that signals within the plate may also be important comes from explants of early *Xenopus* gastrulas that consist of dorsal mesoderm, including the Spemann organizer, and ectoderm that will normally give

Fig. 4.19 Exogastrulation results in the separation of mesoderm from ectoderm. In exogastrulation, instead of the mesoderm moving inward as in normal gastrulation, it moves outward (left panel) and becomes only tenuously joined to the ectoderm (right panel). The ectoderm (blue) does not form neural tissue but does express some neural markers; the mesoderm forms an axis complete with typical mesodermal structures such as notochord and somites. After Holtfreter, J., *et al.*: 1955.

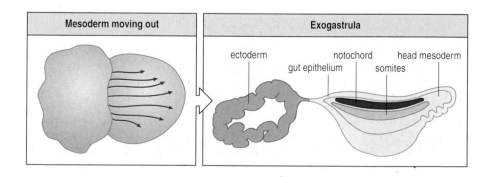

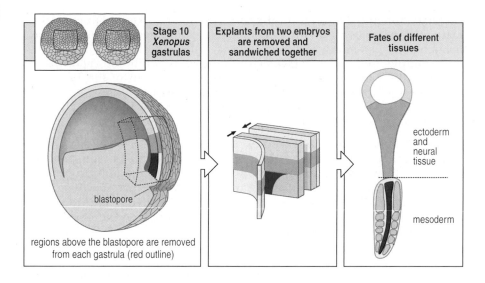

| Stage 10 *Xenopus* gastrulas | Explants from two embryos are removed and sandwiched together | Fates of different tissues |

regions above the blastopore are removed from each gastrula (red outline)

blastopore

ectoderm and neural tissue

mesoderm

Fig. 4.20 Induction and patterning of the nervous system involves planar signals originating within the ectoderm. Explants containing both mesoderm (pink and red) and prospective neural ectoderm (blue) are taken from the marginal zone of early gastrulas at the time when the mesoderm begins to invaginate. Two such explants are sandwiched together (to prevent curling of the strips), and cultured. In these explants, mesoderm lies posterior and in the same plane as the ectoderm, with only a small point of contact, rather than underneath the ectoderm and in contact with it along its whole length as it is in the embryo. Nevertheless, the explants differentiate into both neural tissue and mesoderm. The ectoderm and mesoderm converge and extend in the explants, and genes such as *engrailed-2* and *Krox-20*, which are normally expressed in region-specific patterns within neural tissue, are expressed. After Doniach, T., *et al*.: 1992.

rise to neural tissue after induction. Two of these explants are cultured as a sandwich in such a way as to keep the strip flat and prevent it rounding up. Both ectoderm and mesoderm in such explants converge and extend as they do during normal gastrulation (see Section 8-9), and the ectoderm differentiates into neural tissue, as shown by the expression of the neural-specific cell adhesion molecule N-CAM (Fig. 4.20).

The neural ectoderm of such explants shows clear spatial patterning, as indicated by the correct order of expression of genes that are normally localized to specific regions along the antero-posterior axis of the neural plate, namely *engrailed-2*, *Krox-20*, *Xlhbox1*, and *Xlhbox6*. Moreover, two of the genes (*engrailed-2* and *Xlhbox1*) are also expressed in the mesoderm of the explants, as they are in the mesoderm of intact embryos. Thus, the expression of these two genes occurs independently in mesoderm and ectoderm without the close apposition of these two tissues that normally occurs in embryos *in vivo*. The most likely explanation of these results is that signals confined to the plane of either the ectoderm or mesoderm are both involved in patterning the ectoderm, perhaps in the form of a gradient with its high point at the site of the organizer. Within the developing nervous system, the early patterning of the hindbrain region has been best studied and will be considered next.

4-10 The hindbrain is segmented into rhombomeres by boundaries of cell lineage restriction.

Patterning of the posterior region of the head and the hindbrain involves segmentation of the neural tube along the antero-posterior axis. This does not occur elsewhere along the spinal cord, where the segmental pattern, which consists of dorsal root ganglia and ventral motor nerves at regular intervals—one pair per somite—is imposed by the somites. In the chick embryo, three segmented systems can be seen in the posterior head region by 3 days of development: the mesoderm on either side of the notochord is subdivided into somites, the hindbrain (the rhombocephalon) is divided into eight **rhombomeres**, and the lateral mesoderm has formed a series of branchial arches (Fig. 4.21).

Development of the head in the hindbrain region involves several interacting components. The neural tube gives rise both to the segmentally arranged cranial nerves that innervate the face and neck, and to neural crest cells, which in turn give rise both to peripheral nerves and to

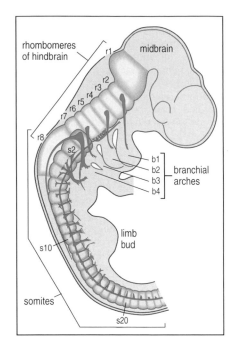

Fig. 4.21 The nervous system in a 3-day chick embryo. The hindbrain is divided into eight rhombomeres (r1 to r8). The positions of the cranial nerves III to XII are shown in green. b1 to b4 are the four branchial arches. b1 gives rise to the jaws. s = somites. Adapted from Lumsden, A.: 1991.

skeletal elements. In addition, the otic vesicle gives rise to the ear. The main skeletal elements of the head in this region develop from the three branchial arches into which neural crest cells migrate (see Section 2-2). For example, the first arch gives rise to the jaws, while the second arch develops into the bony parts of the ear. This region of the head is a particularly valuable model for studying patterning along the antero-posterior axis because of the presence of the numerous different structures ordered along it.

Immediately after the neural tube of the chick embryo closes in this region, the future hindbrain becomes constricted at evenly spaced positions to define eight rhombomeres (see Fig. 4.21). The cellular basis for these constrictions is not understood but may involve differential cell division or changes in cell shape. Whatever the underlying cause, it seems that the boundaries between rhombomeres are barriers of **lineage restriction**; that is, once the boundaries form, cells and their descendants are confined within a rhombomere and do not cross from one side of a boundary to the other. Marking of individual cells shows that before the constrictions become visible, the descendants of a given labeled cell can populate two adjacent rhombomeres. After the constrictions appear, however, descendants of cells then within a rhombomere never cross the boundaries and are thus confined to a single rhombomere (Fig. 4.22). It seems that the cells of a rhombomere share some adhesive property that prevents them mixing with those of adjacent rhombomeres. This implies that cells in each rhombomere may be under control of the same genes, and that the rhombomere is a developmental unit. Rhombomeres are thus behaving like a compartment, which is a common feature of insect development as we see in Chapter 5, but seems to be rare in vertebrates.

The idea that each rhombomere is a developmental unit is supported by the observation that when an odd-numbered and an even-numbered rhombomere from different positions along the antero-posterior axis are placed next to each other after a boundary between them has been surgically removed, a new boundary forms. Moreover, no boundaries form when different odd-numbered rhombomeres are placed next to each other, suggesting that their cells have similar surface properties.

Fig. 4.22 Lineage restriction in rhombomeres of the embryonic chick hindbrain. Single cells are injected with a label (rhodamine dextran) at an early stage (left panel) or a later stage (right panel) of neurulation, and their descendants are mapped 2 days later. Cells injected before rhombomere boundaries form give rise to some clones that span two rhombomeres (dark red) as well as those that do not cross boundaries (red). Clones marked after rhombomere formation never cross the boundary of the rhombomere that they originate in (blue). Adapted from Lumsden, A.: 1991.

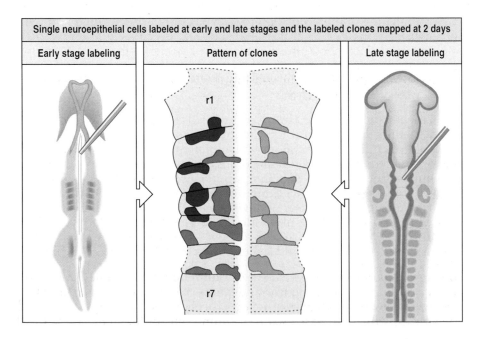

Single neuroepithelial cells labeled at early and late stages and the labeled clones mapped at 2 days

| Early stage labeling | Pattern of clones | Late stage labeling |

The division of the hindbrain into rhombomeres has functional significance in that each has a unique identity and this determines how each will develop. As we see below, their development is under the control of Hox gene expression. We first consider the behavior of the neural crest, which comes from the neural tube and initially migrates over the rhombomeres and gives rise to structures such as those that come from branchial arches, including the lower jaw.

4-11   Neural crest cells have positional values.

The cranial neural crest that migrates out from the rhombomeres of the dorsal region of the hindbrain also has a segmental arrangement. This has been revealed by labeling chick neural crest cells *in vivo* and following their migration pathways. The patterns of crest emergence and migration correlate closely with the rhombomere from which the crest cells come. Thus, branchial arches 1, 2, and 3 become populated by crest cells from rhombomeres 2, 4, and 6, respectively. In the chick, crest cells from rhombomeres 3 and 5 are apparently largely eliminated from further development by **programmed cell death**.

The crest cells have already acquired a positional value before they begin to migrate. When crest cells of rhombomere 4 are replaced by cells from rhombomere 2 taken from another embryo, these cells enter the second branchial arch but develop into structures characteristic of the first arch, to which they would normally have migrated. This can result in the development of an additional lower jaw in the chick embryo.

4-12   Hox genes provide positional identity in the hindbrain region.

Hox gene expression provides a possible molecular basis for the positional identity of both the rhombomeres and the neural crest. Hox genes are expressed in the mouse embryo hindbrain in a well defined pattern, which closely correlates with the segmental pattern (Fig. 4.23). For example, *Hoxb3* has its most anterior region of expression at the border of rhombomeres 4 and 5, while *Hoxb2* has its anterior border at the border of rhombomeres 2 and 3 (Fig. 4.24). In general, the paralogous genes of the different Hox complexes have similar patterns of expression. It is clear that the three paralogous groups involved have different anterior margins of expression, paralog 1 (i.e. *Hoxa1*, *Hoxb1*, etc.) being most anterior, followed by paralogs 2 and 3. The pattern of Hox gene expression in the ectoderm and branchial arches at a particular position along the antero-posterior axis is similar to that in the neural tube and neural crest, and it may be that the crest cells induce these positional values in the overlying ectoderm during their migration.

Transplantation of rhombomeres from an anterior to a more posterior position alters the pattern of Hox gene expression to that normally expressed at the new location. The signals responsible for this originate from the neural tube itself and not the surrounding tissues.

Studies of the control of Hox gene expression at the molecular level have provided some indication as to how their pattern of expression is controlled. For example, although the *Hoxb2* gene is expressed in the three contiguous rhombomeres 3, 4, and 5, its expression in rhombomeres 3 and 5 is controlled quite independently from its expression in rhombomere 4. The regulatory regions of the *Hoxb2* gene carry two separate enhancer elements which regulate its expression in these three rhombomeres. Expression in rhombomeres 3 and 5 is controlled

Fig. 4.23 Expression of Hox genes in the branchial region of the head. The expression of genes of three paralogous Hox complexes in the hindbrain (rhombomeres r1 to r8), neural crest, branchial arches (b1 to b4), and surface ectoderm is shown. *Hoxa1* and *Hoxd1* are not expressed at this stage. The arrows indicate the migration of neural crest cells into the arches. Note the absence of neural crest migration from r3 and r5. After Krumlauf, R.: 1993.

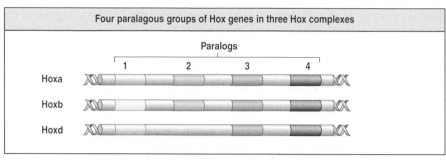

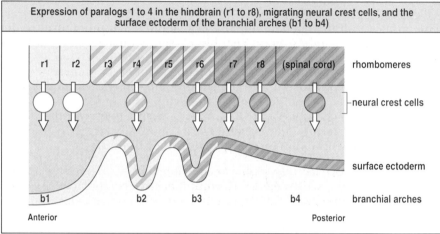

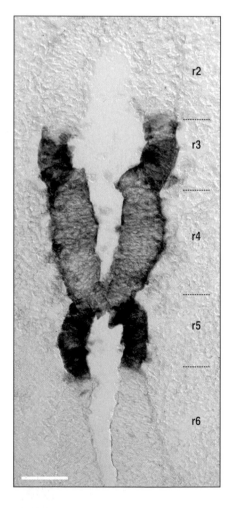

through one of these enhancers, while expression in rhombomere 4 is controlled through the other (see Fig. 4.24). In rhombomeres 3 and 5, *Hoxb2* is activated in part by the zinc finger transcription factor encoded by the gene *Krox-20*, which is expressed in these rhombomeres but not in rhombomere 4. There are binding sites for the Krox-20 protein in the enhancer element that activates expression of *Hoxb2* in rhombomeres 3 and 5. How the spatially organized expression of transcription factors such as Krox-20 is achieved is not yet known.

Gene knock-out in mice has also shown that the Hox genes are involved in patterning of the hindbrain region, though the results are not always easy to interpret; knock-out of a particular Hox gene can affect different populations of neural crest cells in the same animal, such as those that will form neurons and those that will form skeletal structures. Knock-out of the *Hoxa2* gene, for example, results in skeletal defects in that region of the head corresponding to the normal domain of expression of the gene which extends from rhombomere 3 backward. Segmentation itself is not affected, but the skeletal elements in the second branchial arch, all of which come from neural crest cells derived from rhombomere 4, are abnormal. The usual elements, such as the stapes of the inner ear, are absent, but instead some of the skeletal elements normally

Fig. 4.24 Gene expression in the hindbrain. The photograph shows a coronal section through the hindbrain of a 9½ days post-coitum mouse embryo, which is transgenic for two reporter constructs. The first construct contains the *lacZ* gene under the control of an enhancer from *Hoxb2*, which directs expression in rhombomeres 3 and 5 (revealed as blue staining). The second construct contains an alkaline phosphatase gene under the control of an enhancer from *Hoxb1*, which directs expression in rhombomere 4 (revealed as brown staining). A similar enhancer directing expression in rhombomere 4 exists for *Hoxb2*. Anterior is uppermost, and the positions of five of the rhombomeres are indicated (r2 to r6). Scale bar = 0.1 mm. Photograph courtesy of J. Sharpe, from Krumlauf, R., *et al.*: 1996.

formed by the first arch develop, such as Meckel's cartilage, which is a precursor element in the lower jaw. Thus, suppression of *Hoxa2* causes a partial homeotic transformation of one segment into another.

These observations, together with those described earlier in this chapter, show that during gastrulation the cells of vertebrates acquire positional value along the antero-posterior axis that is encoded by the genes of the Hox complexes. Many of the anatomical differences between vertebrates are probably simply due to differences in the subsequent targets of Hox gene actions, which result in the emergence of different but homologous skeletal structures—the mammalian jaw or the bird's beak for example. These principles will be elaborated further in Chapter 5, which deals with the development of the fruit fly *Drosophila*, the animal in which Hox-like genes were first identified, and the concept of their role in regional specification first formulated.

| 4-13 | **The embryo is patterned by the neurula stage into organ-forming regions that can still regulate.** |

At the neurula stage, the body plan has been established and the regions of the embryo that will form limbs, eyes, heart, and other organs have become determined (Fig. 4.25).

This contrasts sharply with the blastula stage, at which such determination has not yet occurred. The basic vertebrate phylotypic body plan is thus established during gastrulation. But although the positions of various organs are fixed, there is no overt sign yet of differentiation. Numerous grafting experiments have shown that the potential to form a given organ is now confined to specific regions. Each of these regions has, however, considerable capacity for regulation, so that if part of the region is removed a normal structure can still form. For example, the region of the neurula that will form a forelimb will, when transplanted to a different region, still develop into a limb. If part of a future limb region is removed, the remaining part can still regulate to develop a normal limb.

| | **Summary.** |

Patterning along both the antero-posterior and dorso-ventral axes is closely related to the action of the Spemann organizer and its morphogenesis during gastrulation. When grafted to the ventral side of an early gastrula, the Spemann organizer induces both a new dorso-ventral axis and a new antero-posterior axis, with the development of a second twinned embryo. In chick development, Hensen's node serves a function similar to that of the Spemann organizer, and it too can specify a new antero-posterior axis.

The vertebrate nervous system, which forms from the neural plate, is induced by cells that give rise to the mesoderm and come to lie beneath prospective neural plate ectoderm during gastrulation. While molecules that can induce neural tissue, such as the noggin protein, have been identified, induction is due to inhibition of BMP activity. Patterning of the neural plate, which includes the movement of signals within the plate itself, can partly be accounted for by a two-signal model: the ectoderm is first specified as anterior neural tissue and then a second signal, possibly graded, specifies more posterior structures.

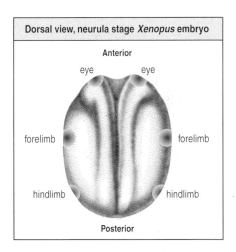

Dorsal view, neurula stage *Xenopus* embryo

Anterior

eye   eye

forelimb   forelimb

hindlimb   hindlimb

Posterior

Fig. 4.25 The *Xenopus* embryo has become regionalized by the neurula stage. Various organs such as limbs, heart, and eyes will develop from specific regions (red) of the neurula after gastrulation is complete. Some of these regions, like the limb buds, are already determined at this stage and will not form any other structure. The boundaries of the regions are not sharply defined, however, and within each region or 'field' considerable regulation is still possible.

The hindbrain is segmented into rhombomeres, with the cells of each rhombomere respecting their boundaries. Neural crest cells from the hindbrain populate specific mesodermal regions such as the branchial arches in a position-dependent fashion. A Hox gene code provides positional values for the rhombomeres and neural crest cells of the hindbrain region. By the neurula stage, after gastrulation, the body plan has been established.

Summary to Chapter 4.

The germ layers specified during blastula formation become patterned along the antero-posterior and dorso-ventral axes during gastrulation. The Spemann organizer in amphibians and its counterpart, Hensen's node, in chick and mouse embryos, are involved in the initial patterning that underlies the regionalization of the antero-posterior axis. Positional identity of cells along the antero-posterior axis is encoded by the combinatorial expression of genes of the four Hox complexes. There is both spatial and temporal co-linearity between the order of Hox genes on the chromosomes and the order in which they are expressed along the antero-posterior axis of the embryo. Inactivation or overexpression of Hox genes can lead both to localized abnormalities and to homeotic transformations of one 'segment' of the axis into another, indicating that these genes are crucial in specifying regional identity. At the end of gastrulation, the basic body plan has been laid down and the nervous system induced. Specific regions of each somite give rise to cartilage, muscle, and dermis, and these regions are specified by signals from the notochord, neural tube, and epidermis. Induction and patterning of the nervous system involves both signals from the underlying mesoderm, and planar signals arising within the neural plate itself. In the hindbrain, Hox gene expression provides positional values for both neural tissue and neural crest cells.

Summary: patterning of the vertebrate axial body plan

gastrulation and Spemann organizer activity

⇩

the four Hox gene complexes are expressed along the antero-posterior axis

⇩

Hox gene expression establishes positional identity for mesoderm, endoderm, and ectoderm

↙   ↘

mesoderm develops into notochord,   mesoderm induces neural
somites, and lateral plate mesoderm   plate from ectoderm

⇩   ⇩

somites receive signals from notochord,   mesoderm and planar ectodermal signals
neural tube, and ectoderm   give regional identity to neural tube

⇩   ⇩

somite develops into sclerotome   rhombomeres and neural crest in the
and dermamyotome   hindbrain are characterized by regional
                                    patterns of Hox gene expression

References.

| 4-1 | Somites are formed in a well-defined order along the antero-posterior axis. |

Kieny, M., Mauger, A., Sengel P.: **Early regionalization of somitic mesoderm as studied by the development of the axial skeleton of the chick embryo.** *Dev. Biol.* 1972, **28**:142–161.

| 4-2 | The fate of somite cells is determined by signals from the adjacent tissues. |

Cossu, G., Tajbakhsh, S., Buckingham, M.: **How is myogenesis initiated in the embryo?** *Trends Genet.* 1996, **12**:218–223.

Duboule, D. (Ed.).: *Guidebook to the Homeobox Genes.* Oxford: Oxford University Press, 1994.

Fan, C.M., Porter, J.A., Chiang, C., Chang, D.T., Beachy, P.A., Tessier-Lavigne, M.: **Long range sclerotome induction by Sonic hedgehog: direct role of the amino-terminal cleavage product and modulation by the cyclic AMP signaling pathway.** *Cell* 1995, **81**:457–465.

Fan, C., Tessier-Lavigne, M.: **Patterning of mammalian somites by surface ectoderm and notochord: evidence for sclerotome induction by a hedgehog homolog.** *Cell* 1994, **79**:1175–1186.

Pourquié, O., Coltey, M., Feillet, M.A., Ordahl, C., Le Douarin, N.M.: **Control of dorso-ventral patterning of somite derivatives by notochord and floor plate.** *Proc. Nat. Acad. Sci.* 1993, **90**:5242–5246.

Pourquié, O., Fan, C-M., Coltey, M., Hirsinger, E., Watanabe, Y., Bréant, C., Francis-West, P., Brickell, P., Tessier-Lavigne, M., Le Douarin, N.M.: **Lateral and axial signals involved in avian somite patterning: a role for BMP-4.** *Cell* 1996, **84**:461–471.

Selleck, M., Stern, C.D.: **Fate mapping and cell lineage analysis of Hensen's node in the chick embryo.** *Development* 1991, **112**:615–626.

Williams, B.A., Ordahl, C.P.: **Pax-3 expression in segmental mesoderm marks early stages in myogenic cell specification.** *Development* 1994, **120**:785–796.

| 4-3 | Positional identity of somites along the antero-posterior axis is specified by Hox gene expression. |

Burke, A.C., Nelson, C.E., Morgan, B.A., Tabin, C.: **Hox genes and the evolution of vertebrate axial morphology.** *Development* 1995, **121**:333–346.

Godsave, S., Dekker, E.J., Holling, T., Pannese, M., Boncinelli, E., Durston, A.: **Expression patterns of Hoxb in the *Xenopus* embryo suggest roles in antero-posterior specification of the hindbrain and in dorso-ventral patterning of the mesoderm.** *Dev. Biol.* 1994, **166**:465–476.

Hunt, P., Krumlauf, R.: **Hox codes and positional specification in vertebrate embryonic axes.** *Ann. Rev. Cell Biol.* 1992, **8**:227–256.

Kessel. M., Gruss, P.: **Murine developmental control genes.** *Science* 1990, **249**:374–379.

Krumlauf, R.: **Hox genes in vertebrate development.** *Cell* 1994, **78**:191–201.

McGinnis, W., Krumlauf, R.: **Homeobox genes and axial patterning.** *Cell* 1992, **68**:283–302.

| 4-4 | Deletion or overexpression of Hox genes causes changes in axial patterning. |

Condie, B.G., Capecchi, M.R.: **Mice with targeted disruptions in the paralogous genes *hoxa3* and *hoxd3* reveal synergistic interactions.** *Nature* 1994, **370**:304–307.

Duboule, D.: **Vertebrate Hox genes and proliferation: an alternative pathway to homeosis?** *Curr. Opin. Genet. Devel.* 1995, **5**:525–528.

Favier, B., Le Meur, M., Chambon, P., Dollé, P.: **Axial skeleton homeosis and forelimb malformations in *Hoxd11* mutant mice.** *Proc. Nat. Acad. Sci.* 1995, **92**:310–314.

| 4-5 | Retinoic acid can alter positional values. |

Condie, B.G., Capecchi, M.R.: **Mice with targeted disruptions in the paralogous genes *Hoxa3* and *Hoxd3* reveal synergistic interactions.** *Nature* 1994, **370**:304–307.

Conlon, R.A.: **Retinoic acid and pattern formation in vertebrates.** *Trends Genet.* 1995, **11**:314–319.

Duboule, D., Morata, G.: **Colinearity and functional hierarchy among genes of the homeotic complexes.** *Trends Genet.* 1994, **10**:358–364.

Jegalian, B.G., De Robertis, E.M.: **Homeotic transformations in the mouse induced by overexpression of a human Hox 3.3 transgene.** *Cell* 1992 **71**:901–910.

Kessel, M., Gruss, P.: **Homeotic transformations of moving vertebrae and concomitant alteration of the codes induced by retinoic acid.** *Cell* 1991, **67**:89–104.

Le Mouellic, H., Lallemand, Y., Brulet, P.: **Homeosis in the mouse induced by a null mutation in the *Hox 3.1* gene.** *Cell* 1992, **69**:251–264.

Ruiz-i-Altaba, A., Jessell, T.: **Retinoic acid modifies mesodermal patterning in early *Xenopus* embryos.** *Genes Dev.* 1991, **5**:175–187.

Ruiz-i-Altaba, A., Melton, D.: **Involvement of the *Xenopus* homeobox gene *XHox3* in pattern formation along the anterior-posterior axis.** *Cell* 1989, **57**:317–326.

Sive, H.L., Cheng, P.F.: **Retinoic acid perturbs the expression of *Xhox.lab* genes and alters mesodermal determination in *Xenopus laevis*.** *Genes Dev.* 1991, **5**:1321–1332.

| 4-6 | The organizer can specify a new antero-posterior axis. |

Griffin, K., Patient, R., Holder, N.: **Analysis of FGF function in normal and no tail zebrafish embryos reveals separate mechanisms for formation of the trunk and tail.** *Development* 1995, **121**:2983–2994.

Shawlot, W., Bohringer, R.R.: **Requirement for *Lim-1* in head organizer function.** *Nature* 1995, **374**:425–430.

Slack, J.M.W., Tannahill, D.: **Mechanism of antero-posterior axis specification in vertebrates. Lessons from amphibians.** *Development* 1992, **114**:285–302.

| 4-7 | The neural plate is induced by mesoderm. |

Hawley, S.H.B., Wünnerberg-Stapleton, K., Hashimoto, C., Laurent, M.N., Watabe, T., Blumberg, B.W., Cho, K.W.Y.: **Disruption of BMP signals in embryonic *Xenopus* ectoderm leads to direct neural induction.** *Genes & Devel.* 1995, **9**:2923–2935.

Kemmati-Brivanlou, A., Melton, D.: **Vertebrate embryonic cells will become nerve cells unless told otherwise.** *Cell* 1997, **88**:13–17.

Sasai, Y., Lu, B., Steinbesser, H., De Robertis, E.M.: **Regulation of neural induction by the Chd and BMP-4 antagonistic patterning signals in *Xenopus*.** *Nature* 1995, **376**:333–336.

Wilson, P., Kemmati-Brivanlou, A.: **Induction of epidermis and inhibition of neural fate by BMP-4.** *Nature* 1995, **376**:331–333.

4-8 The nervous system can be patterned by signals from the mesoderm.

Acampora, D., Mazan, S., Lallemand, Y., Avantaggiato, V., Maury, M., Simeone, A., Brulet, P.: **Forebrain and midbrain regions are deleted in Otx2-/- mutants due to a defective anterior neuroectoderm specification during gastrulation.** *Development* 1995, **121**:3279–3290.

Ang, S.L.,, Rossant, J.: **HNF-3β is essential for node and notochord formation in mouse development.** *Cell* 1994, **78**:561–574

Blitz, I.L., Cho, K.W.Y.: **Anterior neurectoderm is progressively induced during gastrulation: the role of the *Xenopus* homeobox gene *orthodenticle*.** *Development* 1995, **121**:993–1004.

Doniach, T.: **Basic FGF as an inducer of antero-posterior neural pattern.** *Cell* 1995, **85**:1067–1070.

Kelly, O.G., Melton, D.A.: **Induction and patterning of the vertebrate nervous system.** *Trends Genet.* 1995, **11**:273–278.

Kintner, C.R., Dodd, J.: **Hensen's node induces neural tissue in *Xenopus* ectoderm. Implications for the action of the organizer in neural induction.** *Development* 1991, **113**:1495–1505.

Mason, I.: **Neural induction: do fibroblast growth factors strike a cord?** *Curr. Biol.* 1996, **6**:672–675.

Ruiz-i-Altaba, A., Jessell, T.M.: ***Pintallavis*, a gene expressed in the organizer and midline cells of frog embryos: involvement in the development of the neural axis.** *Development* 1992, **116**:81–93.

Sasai, Y., De Robertis, E.M.: **Ectodermal patterning in vertebrate embryos.** *Dev. Biol.* 1997, **182**:5–20.

Storey, K., Crossley, J.M., De Robertis, E.M., Norris, W.E., Stern, C.D.: **Neural induction and regionalization in the chick embryo.** *Development* 1992, **114**:729–741.

4-9 Signals that pattern the neural plate may travel within the neural plate itself.

Doniach, T., Phillips, C.R., Gerhart, J.C.: **Planar induction of antero-posterior pattern in the developing central nervous system of *Xenopus laevis*.** *Science* 1992, **257**:542–545.

Ruiz-i-Altaba, A., Melton, D.: **Interaction between peptide growth factors and homeobox genes in the establishment of antero-posterior polarity in frog embryos.** *Nature* 1989, **341**:33–38.

Sive, H.L., Hattori, K., Weintraub, H.: **Progressive determination during formation of antero-posterior axis in *Xenopus laevis*.** *Cell* 1989, **58**:171–180.

4-10 The hindbrain is segmented into rhombomeres by boundaries of cell lineage restriction.

Lumsden, A.: **Cell lineage restrictions in the chick embryo hindbrain.** *Phil. Trans. Roy. Soc. Lond. B* 1991, **331**:281–286.

4-11 Neural crest cells have positional values.

Keynes, R., Lumsden, A.: **Segmentation and the origin of regional diversity in the vertebrate central nervous system.** *Neuron* 1990, **4**:1–9.

4-12 Hox genes provide positional identity in the hindbrain region.

Grapin-Botton, A., Bonnin, M-A., McNaughton, L.A., Krumlauf, R., Le Douarin, N.M: **Plasticity of transposed rhombomeres: Hox gene induction is correlated with phenotypic modifications.** *Development* 1995, **121**:2707–2721.

Hunt, P., Krumlauf, R.: **Hox codes and positional specification in vertebrate embryonic axes.** *Ann. Rev. Cell Biol.* 1992, **8**:227–256.

Krumlauf, R.: **Hox genes and pattern formation in the branchial region of the vertebrate head.** *Trends Genet.* 1993, **9**:106–112.

Nonchev, S., Maconochie, M., Vesque, C., Aparicio, S., Ariza-McNaughton, L., Manzanares, M., Maruthainar, K., Kuroiwa, A., Brenner, S., Charnay, P., Krumlauf, R.: **The conserved role of *Krox-20* in directing *Hox* gene expression during vertebrate hindbrain segmentation.** *Proc. Natl. Acad. Sci.* 1996, **93**:9339–9345.

Rijli, F.M., Mark, M., Lakkaraju, S., Dierich, A., Dolle, P., Chambon, P.: **A homeotic transformation is generated in the rostral branchial region of the head by disruption of *Hoxa2*, which acts as a selector gene.** *Cell* 1993, **75**:1333–1349.

4-13 The embryo is patterned by the neurula stage into organ-forming regions that can still regulate.

De Robertis, E.M., Morita, E.A., Cho, K.W.Y.: **Gradient fields and homeobox genes.** *Development* 1991, **112**:669–678.

Development of the *Drosophila* Body Plan

5

"From the beginning we knew where we were even though there were no boundaries between us. Later, we divided into separate groups and were given a proper address."

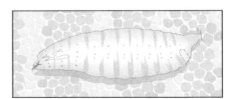

We are much more like flies in our development than you might think. Astonishing discoveries in developmental biology over the past ten years have revealed that many of the genes that control the development of the fruit fly *Drosophila* are similar to those controlling development in vertebrates, and indeed in many other animals. It seems that once evolution had found a satisfactory way of patterning animal bodies, it tended to use the same mechanisms and molecules over and over again with, of course, some important modifications.

Drosophila is the best understood of all developmental systems, especially at the genetic level, and although it is an invertebrate it has had an enormous impact on our understanding of the genetic basis of vertebrate development. We have already seen this with the Hox genes (see Box 4A, page 104), which were first discovered in *Drosophila*. The pre-eminent place of *Drosophila* in modern developmental biology was recognized by the award of the 1995 Nobel Prize for Physiology or Medicine for work that led to a fundamental understanding of how genes control development in the fly embryo. This was only the second time that the Nobel Prize had been awarded for work in developmental biology. Many questions that still remain unanswered in vertebrates have been solved at the molecular level for *Drosophila*; these include the mechanisms of axis determination in the egg, and the identification and mechanism of action of key signaling circuits and transcriptional regulators in pattern formation. While insect and vertebrate development may seem to be very different, much has been learnt that can be applied to vertebrate development; indeed, many of the key genes in vertebrate development were originally identified as developmental genes in *Drosophila*.

Drosophila, like many other insects, hatches from the egg as a larva, which grows and subsequently undergoes metamorphosis into the adult (see Fig. 2.29). In this chapter, we look at how the basic body plan of the *Drosophila* larva is established. We see how the antero-posterior and dorso-ventral axes are determined, how the embryo becomes divided into a series of segments each with its unique identity, and how the mesoderm and ectoderm become specified. The first half of the chapter concentrates on the development of the embryo up to the stage at which it becomes segmented. In the second half of the chapter, we consider how the segments are patterned and acquire their unique identities. We leave until Chapter 10 the development of the imaginal discs—groups of cells set aside in the embryo that eventually give rise to adult structures, such as wings and legs, at metamorphosis. The imaginal discs provide continuity between the pattern of the larval body and that of the adult, even though the processes of metamorphosis intervene.

Like all animals with bilateral symmetry, the *Drosophila* larva is patterned along two distinct and largely independent axes: the antero-posterior and dorso-ventral axes, which are at right angles to each other. Along the antero-posterior axis, the larva appears regularly segmented, and is divided into several broad anatomical regions. At the anterior end is the head, behind which are three thoracic segments followed by eight abdominal segments (Fig. 5.1). Each segment has its own unique character, as revealed by both its external cuticular structure and its internal organization. At each end of the larva are specialized structures —the acron at the head end and the telson at the tail end.

Fig. 5.1 Patterning of the *Drosophila* embryo. The body plan is patterned along two distinct axes. The antero-posterior and dorso-ventral axes are at right angles to each other and are laid down in the egg. In the early embryo, the dorso-ventral axis is divided into four regions: mesoderm (red), ventral ectoderm (yellow), dorsal ectoderm (orange), and amnioserosa (an extra-embryonic membrane, green). The ventral ectoderm gives rise to both ventral epidermis and neural tissue, the dorsal ectoderm to epidermis. The antero-posterior axis becomes divided into different regions that later give rise to the head, thorax, and abdomen. After the initial division into broad body regions, segmentation begins. The future segments can be visualized as transverse stripes by staining for specific gene activity; these stripes demarcate 14 parasegments, 10 of which are marked. The embryo develops into a segmented larva. By the time the larva hatches, the 14 para-segments have been converted into thoracic (T1–T3) and abdominal (A1–A8) segments, which are offset from the parasegments by one half segment. Different segments are distinguished by the patterns of bristles and denticles on the cuticle. Specialized structures, the acron and telson, develop at the head and tail ends, respectively.

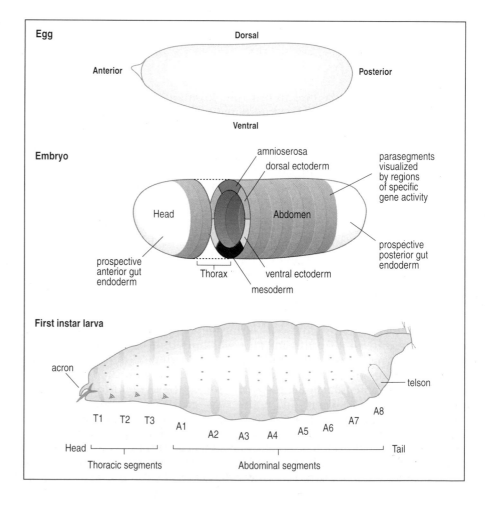

Early in embryogenesis, the dorso-ventral axis becomes divided up into four regions, which give rise to the dorso-ventral organization of the larval body. Organization along the antero-posterior and dorso-ventral axes of the early embryo develops more or less simultaneously but is specified by independent mechanisms and by different sets of genes in each axis.

Early *Drosophila* development is peculiar to insects, as early patterning occurs within a multinucleate **syncytial blastoderm** formed by repeated rounds of nuclear division without any corresponding cytoplasmic division (see Fig. 2.30). Only after the beginning of segmentation does the embryo become truly multicellular. The absence of cells in the early *Drosophila* embryo represents an important difference from other organisms. At the syncytial stage, the whole embryo can be considered as a multinucleate single cell, and many proteins, including those that are not normally secreted from cells, such as transcription factors, can diffuse through the blastoderm and enter the nuclei. Concentration gradients, which can provide positional information for the nuclei, can thus be set up (see Section 1-13).

Early development is essentially two-dimensional; patterning occurs mainly in the blastoderm, the superficial layer of the embryo that consists first of nuclei and later of cells. But the larva is a three-dimensional object, with internal structures. This third dimension develops later, at gastrulation, when parts of the surface layer move into the interior to form the gut, the mesodermal structures that will give rise to muscle, and the ectodermally derived nervous system.

We start this chapter by looking at the appearance of the first level of antero-posterior and dorso-ventral organization in the syncytial embryo, before returning to the formation of the *Drosophila* egg to show how the positional information that organizes the early embryo is originally laid down in the developing oocyte by the mother.

Maternal genes set up the body axes.

The earliest stage of *Drosophila* development is guided by preformed mRNAs and proteins that are synthesized and laid down in the egg by the mother fly. Several of these become localized at the ends of the egg while it is being formed in the ovary. The genes responsible for this maternal contribution are known as **maternal genes** since they must be expressed by the mother and not by the embryo; they are expressed in the tissues of the ovary during oogenesis. By contrast, **zygotic genes** are those required during the development of the embryo; they are expressed in the nuclei of the embryo itself.

About 50 maternal genes are involved in setting up the two axes and a basic framework of positional information, which is then interpreted by the embryo's own genetic program. All later patterning, which involves expression of the zygotic genes, is built on this framework (Fig. 5.2). Maternal gene products establish the axes and set up regional differences along each axis in the form of spatial distributions of RNA and proteins. These proteins then activate zygotic genes in the nuclei at particular positions along both axes for the next round of patterning. The sequential activities of the maternal and zygotic genes pattern the embryo in a series of steps: broad regional differences are established first, and these are then refined to produce a larger number of smaller developmental

Fig. 5.2 The sequential expression of different sets of genes establishes the body plan along the antero-posterior axis. After fertilization, maternal gene products laid down in the egg, such as *bicoid* mRNA, are translated. They provide positional information which activates the zygotic genes. The four main classes of zygotic genes acting along the antero-posterior axis are the gap genes, the pair-rule genes, the segment polarity genes, and the selector, or homeotic, genes. The gap genes define regional differences that result in the expression of a periodic pattern of gene activity by the pair-rule genes, which define the parasegments and foreshadow segmentation. The segment polarity genes elaborate the pattern in the segments, and segment identity is determined by the selector genes. The functions of each of these classes of genes are discussed in this chapter.

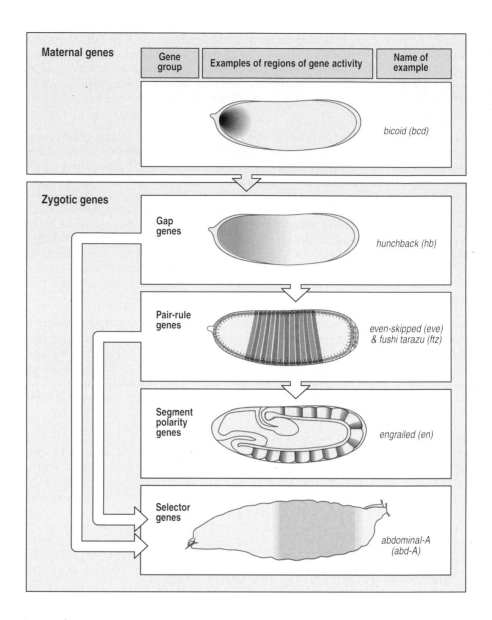

domains, each characterized by a unique profile of zygotic gene activity. Developmental genes act in a strict temporal sequence. They form a hierarchy of gene activity in which the action of one set of genes is essential for another set of genes to be activated, and thus for the next stage of development to occur. We first look at how maternal gene products specify the antero-posterior axis.

5-1   Three classes of maternal genes specify the antero-posterior axis.

Maternal gene expression creates differences in the egg along the antero-posterior axis even before the egg is fertilized. These differences distinguish the future head and posterior ends of the adult. Maternal genes are identified by mutations that, when present in the mother, do not damage her but have effects on the development of her progeny and cannot be rescued by genes from wild-type sperm. The roles of the maternal genes can be deduced from the effects of these **maternal-effect mutations** on the larva. The mutations fall into three classes: those that affect anterior regions; those that affect posterior regions; and those that affect

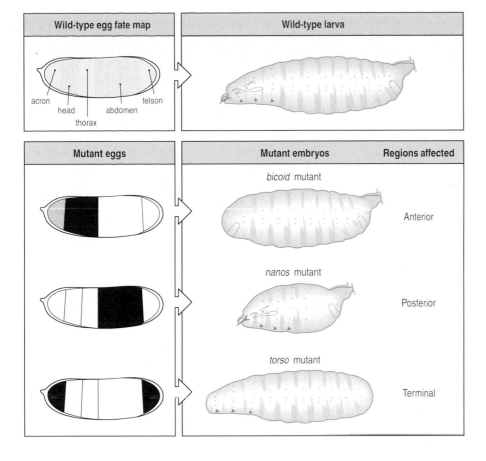

Fig. 5.3 The effects of mutations in the maternal gene system. Mutations in maternal genes lead to deletions and abnormalities in anterior, posterior, or terminal structures. The wild-type fate map shows which regions of the egg give rise to particular regions and structures in the larva. Regions that are affected in mutant eggs and which lead to lost or altered structures in the larva are shaded in red. In *bicoid* mutants there is a partial loss of anterior structures and the appearance of a posterior structure—the telson—at the anterior end. *nanos* mutants lack a large part of the posterior region. *torso* mutants lack both acron and telson.

both terminal regions (Fig. 5.3). Mutations of genes in the anterior class, such as *bicoid*, lead to a reduction or loss of head and thoracic structures, and in some cases, their replacement with posterior structures. Posterior class mutations, such as *nanos*, (the Spanish word for dwarf) cause the loss of abdominal regions, leading to a smaller than normal embryo; and those of the terminal class, such as *torso*, affect the specialized structures—the acron and telson—at the head and tail ends of the embryo. Each class of genes acts more or less independently of the others. The apparently idiosyncratic naming of genes in *Drosophila* usually reflects the attempts by the discoverer to describe the mutant phenotype. In this chapter, we meet quite a number of gene names; all these are listed, together with their functions, where known, in the table at the end of this chapter.

Of the 50 or so maternal genes, the products of four in particular —*bicoid, hunchback, nanos,* and *caudal*—become distributed along the antero-posterior axis and are crucial in establishing it.

5-2 The *bicoid* gene provides an antero-posterior morphogen gradient.

In the unfertilized egg, *bicoid* mRNA is localized in the anterior end. After fertilization it is translated, and the bicoid protein diffuses from the anterior end and forms a concentration gradient along the antero-posterior axis. This provides the positional information required for further patterning along this axis. Historically, the bicoid protein gradient provided the first reliable evidence for the existence of the morphogen gradients that had been postulated to control pattern formation (see Section 1-13).

The role of the *bicoid* gene was first elucidated by a combination of genetic and physical experiments on the *Drosophila* embryo. Female flies lacking *bicoid* gene expression give rise to embryos that have disrupted anterior segments and thus have no proper head or thorax (see Fig. 5.3). They also have a telson instead of an acron at the head end. In a separate line of investigation into the role of localized cytoplasmic factors in anterior development, normal eggs were pricked at their anterior ends and some cytoplasm allowed to leak out. The embryos that developed bore a striking resemblance to *bicoid* mutant embryos. This suggested that normal eggs have some factor(s) in the cytoplasm at their anterior end which is absent in *bicoid* mutant eggs. And indeed, *bicoid* mutant embryos can be partially rescued, in the sense that they will develop more normally if anterior cytoplasm of wild-type embryos is injected into their anterior regions (Fig. 5.4). Moreover, if normal anterior cytoplasm is injected into the middle of a fertilized *bicoid* mutant egg, head structures develop at the site of injection and the adjacent segments become thoracic segments, setting up a mirror-image body pattern at the site of injection. The simplest interpretation of these experiments is that the *bicoid* gene is necessary for the establishment of the anterior structures because it establishes a gradient in some substance whose source and highest level are at the anterior end; this substance is the bicoid protein.

Using *in situ* hybridization (see Box 3B, page 65), *bicoid* mRNA has been shown to be present in the anterior region of the unfertilized egg, where it is attached to the cytoskeleton. This mRNA is not translated until after fertilization. Staining with an antibody against the bicoid protein

Fig. 5.4 The *bicoid* gene is necessary for the development of anterior structures. Embryos whose mothers lack the *bicoid* gene lack anterior regions (upper middle panels). Transfer of anterior cytoplasm from wild-type embryos to *bicoid* mutant embryos causes some anterior structure to develop at the site of injection (lower middle panels). If wild-type anterior cytoplasm is transplanted to the middle of a *bicoid* mutant egg or early embryo, head structures develop at the site of injection, flanked on both sides by thoracic-type segments (bottom panels). These results can be interpreted in terms of the anterior cytoplasm setting up a gradient of bicoid protein with the high point at the site of injection (see graphs, bottom left panel).

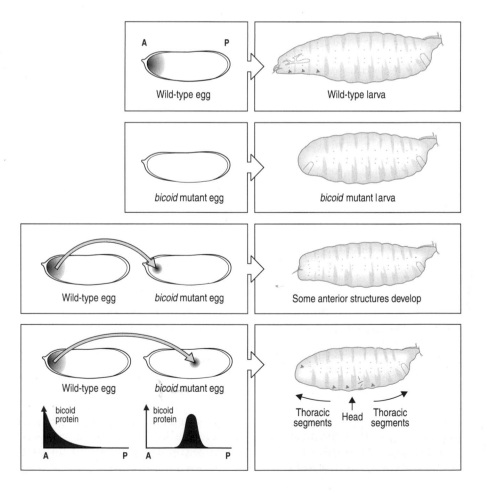

shows that the protein is absent from the unfertilized egg, but that after fertilization the mRNA is translated into protein, which forms a gradient with the high point at the anterior end of the egg, at the site of its synthesis (Fig. 5.5). As the bicoid protein diffuses through the embryo it also breaks down—it has a half-life of about 30 minutes—and this breakdown is important in establishing the antero-posterior concentration gradient.

The bicoid protein is a transcription factor that acts as a morphogen, as described in more detail in Section 5-11. It switches on certain zygotic genes at different threshold concentrations, so initiating a new pattern of gene expression along the axis. Thus, *bicoid* is a key maternal gene in early *Drosophila* development. The other maternal genes of the anterior group are mainly involved in the localization of *bicoid* mRNA to the anterior end of the egg during oogenesis, and in the control of its translation.

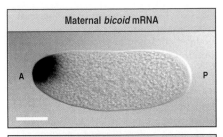

Maternal *bicoid* mRNA

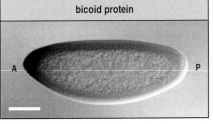

bicoid protein

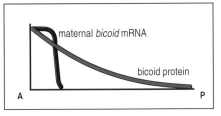

maternal *bicoid* mRNA

bicoid protein

Fig. 5.5 The distribution of the maternal mRNA for *bicoid* in the egg and the gradient of bicoid protein after fertilization. Top panel: the mRNA is visualized by *in situ* hybridization. Middle panel: the bicoid protein is stained with a labeled antibody. Bottom panel: translation of *bicoid* mRNA and diffusion of bicoid protein from its site of synthesis produces an antero-posterior gradient of bicoid protein in the embryo. Scale bars = 0.1 mm. Photographs courtesy of R. Lehmann, from Suzuki, D.T., *et al.*: 1996.

| **5-3** | **The posterior pattern is controlled by the gradients of nanos and caudal proteins.** |

For proper patterning along an axis, both ends need to be specified, and bicoid protein defines only the anterior end of the antero-posterior axis. The posterior end is specified by the actions of nine maternal genes—the posterior group genes—which are involved in posteriorly localizing a specific maternal factor in the oocyte. Just as mutations in the *bicoid* gene result in larvae in which head and thoracic regions do not develop normally, mutations in the posterior group genes result in larvae in which abdominal development is abnormal. These mutant embryos are shorter than normal because they have no abdomen (see Fig. 5.3). One of the actions of the maternal posterior group genes, such as *oskar*, is to localize *nanos* mRNA at the extreme posterior pole of the unfertilized egg. Like *bicoid* mRNA, *nanos* mRNA is translated after fertilization to give a concentration gradient of nanos protein, in this case with the highest level at the posterior end of the embryo.

However, unlike the bicoid protein, nanos protein does not act directly as a morphogen to specify the abdominal pattern. It has a quite different role. Its function is to suppress, in a graded way, the translation of the mRNA of another maternal gene, *hunchback*. The reason that translation of maternal *hunchback* RNA has to be suppressed is that the *hunchback* gene is also expressed zygotically in the embryo. In the early embryo, zygotic *hunchback* is activated at the anterior end of the embryo by high concentrations of bicoid protein, resulting in an antero-posterior gradient of hunchback protein, which acts as a morphogen for the next stage of patterning. But maternal *hunchback* mRNA is also uniformly distributed (at a low level) throughout the egg and its translation would result in too high a concentration of hunchback protein in the posterior region. To establish a clear antero-posterior gradient of zygotic hunchback protein in the posterior of the embryo, translation of the posterior maternal *hunchback* mRNA has to be inhibited, and this is what the nanos protein does (Fig. 5.6), by specifically binding to a complex of *hunchback* mRNA and the protein encoded by the gene *pumilio*.

The fourth crucial maternal product, *caudal* mRNA is, like *hunchback*, uniformly distributed throughout the egg initially. A posterior to anterior gradient of the caudal protein is established by inhibition of caudal protein synthesis by the bicoid protein. Because the concentration of

Fig. 5.6 Establishment of a maternal gradient in hunchback protein.
Left panel: maternal *hunchback* mRNA (turquoise) is present at a relatively low level throughout the egg, whereas *nanos* mRNA (yellow) is located posteriorly. The photograph is an *in situ* hybridization showing nanos mRNA. Right panel: after fertilization, *nanos* mRNA is translated and nanos protein blocks translation of *hunchback* mRNA in the posterior regions, giving rise to a shallow antero-posterior gradient in maternal hunchback protein. The photograph shows the distribution of nanos protein, detected with a labeled antibody. Photographs courtesy of R. Lehmann, from Suzuki, D.T., *et al.*: 1996.

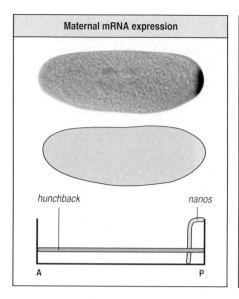

Maternal mRNA expression

hunchback   nanos

A   P

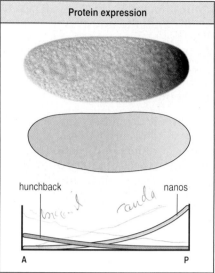

Protein expression

hunchback   nanos

A   P

bicoid protein is low at the posterior end of the embryo, caudal protein concentration is highest there. Mutations in the *caudal* gene result in abnormal development of abdominal segments.

Soon after fertilization, therefore, several gradients of maternal proteins have been established along the antero-posterior axis. Two gradients—bicoid and hunchback proteins—run in an anterior to posterior direction, while caudal protein is graded posterior to anterior. We next look at the different mechanism that specifies the two termini of the embryo.

5-4   The anterior and posterior extremities of the embryo are specified by cell-surface receptor activation.

A third group of maternal genes specifies the structures at the extreme ends of the antero-posterior axis—the acron and the head region at the anterior end, and the telson and the most posterior abdominal segments at the posterior end. A key gene in this group is *torso*; mutations in *torso* can result in embryos developing neither acron nor telson (see Fig. 5.3). This indicates that the two terminal regions, despite their topographical separation, are not specified independently, but use the same pathway.

The terminal regions are specified by an interesting mechanism, which also involves a maternal gene product that has been localized to particular regions in the egg. Terminal specification is due to activation, at the two poles only, of a receptor protein encoded by the maternal gene *torso*, which then transmits a signal to the adjacent cytoplasm. The torso receptor is uniformly distributed throughout the egg plasma membrane, but is only activated at the ends of the fertilized egg because the protein ligand for the receptor is present only there. During oogenesis the ligand is laid down at these two sites in the vitelline envelope outside the egg plasma membrane.

Before fertilization, the ligand is immobilized within the vitelline envelope and cannot come into contact with the receptor. Only when development begins, after fertilization, is the ligand released into the perivitelline space where it can bind to the torso receptor. The ligand is only present

Fig. 5.7 The gene *torso* is involved in specifying the terminal regions of the embryo. The receptor protein encoded by the gene *torso* is present throughout the egg plasma membrane. Its ligand is laid down in the vitelline membrane at each end of the egg during oogenesis. After fertilization the ligand is released and diffuses across the perivitelline space to activate the torso receptor protein at the ends of the embryo only.

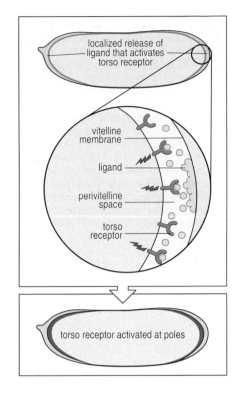

in small quantities and so most becomes bound to torso receptors at the poles, with little left to diffuse further away. In this way, a localized area of receptors activation is set up at each pole (Fig. 5.7). Stimulation of the torso receptor produces a signal that is transduced across the plasma membrane to the interior of the developing embryo. This signal directs the activation of zygotic genes at both poles, thus defining the two extremities of the embryo. The torso protein is one of a large group of transmembrane receptors known as receptor tyrosine kinases; they possess an intrinsic protein tyrosine kinase on the cytoplasmic face of the receptor that is activated by ligand binding and transmits the signal onward by phosphorylating cytoplasmic proteins.

This ingenious mechanism for setting up a localized area of receptor activation is not confined to determination of the terminal regions of the embryo, but is also used in setting up the dorso-ventral axis, which we consider next.

5-5 The dorso-ventral polarity of the egg is specified by localization of maternal proteins in the vitelline envelope.

The dorso-ventral axis is specified by a set of maternal genes separate from those that specify the antero-posterior axis. But, like the antero-posterior axis, it is initially established in the unfertilized egg in the ovary. The ventral end of the axis is set by the localized deposition of a maternal protein in the extra-embryonic vitelline membrane on one side of the egg only. This will become the future ventral region of the embryo.

This localized protein lies dormant in the vitelline envelope until after fertilization. It then initiates a set of reactions locally within the envelope, involving the products of several other maternal genes. This series of reactions results in the processing of the spätzle protein, which has been secreted by the egg uniformly into the perivitelline space, to produce a spätzle fragment. Like the ligand for the torso receptor, the spätzle fragment is the ligand for a receptor protein that is distributed throughout the egg plasma membrane. In this case the receptor is the product of the maternal gene *Toll*.

Because of the localized nature of the processing reactions, the spätzle protein fragment is only produced in the ventral perivitelline space. Thus, it locally activates the Toll receptor protein only in the future ventral region of the embryo. This sends a signal to the adjacent cytoplasm of the embryo. Toll receptor activation is greatest where the concentration of its ligand is highest, but falls off rapidly, probably due to the limited amount of ligand being mopped up by the receptors. At this stage, the embryo is still a syncytial blastoderm and stimulation of the Toll receptor produces a signal that causes a maternal gene product in the adjoining cytoplasm—the dorsal protein—to enter nearby nuclei (Fig. 5.8). This protein, encoded by the *dorsal* gene, is a transcription factor with a vital role in organizing the dorso-ventral axis.

Fig. 5.8 Toll protein activation results in a gradient of intranuclear dorsal protein along the dorso-ventral axis. Before receptor activation, the dorsal protein (red) is distributed throughout the peripheral band of cytoplasm. The Toll protein is a receptor that is only activated in the ventral region, by a maternally-derived ligand (the spätzle fragment), which is processed in the perivitelline space after fertilization. The localized activation of the Toll receptor results in the entry of dorsal protein into nearby nuclei. The intranuclear concentration of dorsal protein is greatest in ventral nuclei, resulting in a ventral to dorsal gradient.

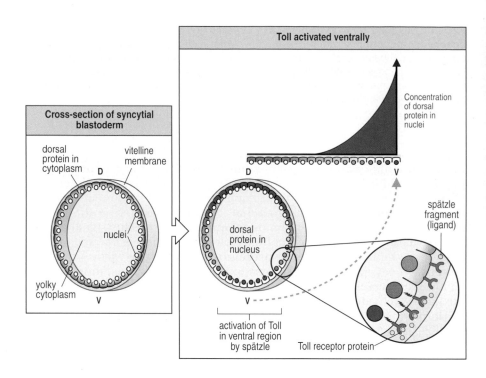

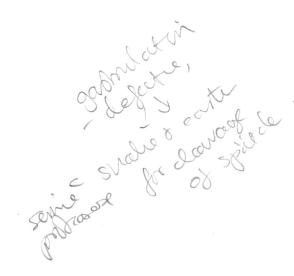

5-6 Positional information along the dorso-ventral axis is provided by the dorsal protein.

The initial dorso-ventral organization of the embryo is established at right angles to the antero-posterior axis at about the same time that this axis is being divided into terminal, anterior, and posterior regions. The embryo initially becomes divided into four regions along the dorso-ventral axis (see Fig. 5.1). Patterning along the dorso-ventral axis is controlled by the distribution of the maternal protein, dorsal.

The dorsal protein is uniformly distributed along the dorso-ventral axis of the embryo. Initially it is restricted to the cytoplasm, but under the influence of signals from the ventrally activated Toll receptor it enters nuclei in a graded fashion, with the highest concentration in ventral nuclei and the concentration progressively decreasing in a ventral to dorsal direction, as the Toll signal becomes weaker (see Fig. 5.8). The greater the number of Toll receptors activated by the spätzle fragment, the more dorsal protein enters the nuclei. There is little or no dorsal protein in the nuclei in the dorsal regions of the embryo. The role of the Toll receptor was first established by the observation that embryos lacking it are strongly 'dorsalized'—that is, no ventral structures develop. In these embryos, the dorsal protein does not enter the nuclei but remains uniformly distributed in the cytoplasm. Transfer of wild-type cytoplasm into *Toll* mutant embryos results in the specification of a new dorso-ventral axis, the ventral region always corresponding to the site of injection. The Toll receptor present in the wild-type cytoplasm enters the membrane at the site of the injection of cytoplasm. The spätzle fragments, now diffusing throughout the perivitelline space, activate these localized Toll receptors and set in motion the chain of events that leads to dorsal protein entering nearby nuclei, thus defining the ventral region at the site of injection.

In the absence of a signal from the Toll receptor, dorsal protein is prevented from entering nuclei by being bound in the cytoplasm to another maternal gene product, the cactus protein. As a result of Toll receptor activation, cactus protein is degraded and no longer binds the dorsal protein, which is then free to enter the nuclei (Fig. 5.9). In embryos lacking cactus

Fig. 5.9 The mechanism of localization of dorsal protein to the nucleus. In unfertilized eggs, dorsal protein is present in the cytoplasm bound to the cactus protein, which prevents it entering nuclei. The signal delivered by Toll receptor activation is transmitted along an intracellular signaling pathway involving other maternal gene products (e.g. those of *tube* and *pelle*), with the end result that cactus protein is degraded and so no longer binds to dorsal protein, which can then enter the nucleus.

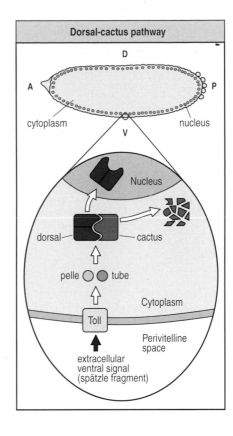

protein, almost all of the dorsal protein is found in the nuclei; there is a very poor concentration gradient and the embryos are 'ventralized' —that is, no dorsal structures develop.

The interaction between the dorsal and cactus proteins is of more than local interest: dorsal protein is a transcription factor with considerable homology to the vertebrate transcription factor NF-κB, which is involved in regulation of gene expression in the B cells of the immune system. NF-κB is also found in B-cell cytoplasm, bound to another protein, I-κB, which prevents it from entering the nucleus before the cell has received the appropriate signal that dissociates the complex. I-κB has homology with the *Drosophila* cactus protein. So what might seem at first sight a rather specialized mechanism for confining transcription factors to the cytoplasm until it is time for them to enter the nucleus, is likely to be widely used for controlling cell differentiation.

Having considered the importance of localized maternal gene products in the egg in setting the basic framework for future development, we now look at how they come to be localized so precisely.

Summary.

Maternal genes act in the ovary of the mother fly to set up differences in the egg in the form of localized deposits of mRNAs and proteins. After fertilization, maternal mRNAs are translated and provide the embryonic nuclei with positional information in the form of protein gradients or nuclear localization. Along the antero-posterior axis there is an anterior to posterior gradient of maternal bicoid protein, which controls patterning of the anterior region. For normal development it is essential that maternal hunchback protein is absent from the posterior region and that its suppression is the function of the posterior to anterior gradient of nanos protein. The extremities of the embryo are specified by localized activation of the torso receptor at the poles. The dorso-ventral axis is established by intranuclear localization of the dorsal protein in a graded manner (ventral to dorsal), as a result of ventrally localized activation of the Toll membrane receptor by a fragment of the protein spätzle.

Summary: maternal gene action in the fertilized egg of *Drosophila*	
Antero-posterior	**Dorso-ventral**
mRNAs: *bicoid* forms anterior to posterior gradient; *hunchback* uniform; *nanos* and *caudal* posterior to anterior	spätzle protein activates Toll receptor on ventral side
⬇	⬇
bicoid protein gradient formed: *hunchback* mRNA translation suppressed in posterior region by nanos	dorsal protein enters ventral nuclei, giving ventral to dorsal gradient
Termini: torso receptor activated at ends of egg	

Polarization of the body axes during oogenesis.

When the *Drosophila* egg is released from the ovary it already has a well-defined organization: *bicoid* mRNA is located at the anterior end and *nanos* and *caudal* mRNAs at the opposite end. The ligand for the torso protein is present in the vitelline envelope at both poles, and other maternal proteins are localized in the ventral vitelline envelope. Numerous other mRNAs and proteins, such as Toll, torso, dorsal, and cactus proteins, are distributed uniformly. How do these maternal mRNAs and proteins get into the egg during its period of development in the ovary—oogenesis—and how are they localized in the correct places?

The development of an egg in the *Drosophila* ovary is shown in Fig. 5.10. A stem cell undergoes four mitotic divisions to give 16 cells with cytoplasmic bridges between each other. One of these 16 cells will become the **oocyte**, the other 15 will develop into **nurse cells**, which produce large quantities of proteins and RNAs that are exported into the egg through the cytoplasmic bridges. Somatic ovarian cells form a sheath of **follicle cells** around the nurse cells and oocyte to form the egg chamber, and they play a key role in patterning the egg's axes. There are various types of follicle cell, which express different genes and have different effects on the oocyte (Fig. 5.11). Follicle cells also secrete the materials of the vitelline envelope and eggshell that surround the mature egg.

| 5-7 | **Antero-posterior and dorso-ventral axes of the oocyte are specified by interactions with follicle cells.** |

The first visible sign of antero-posterior polarization during oogenesis is the movement of the oocyte toward one end of the egg chamber, where it comes into contact with the follicle cells (Fig. 5.12). It induces these follicle cells to adopt a posterior fate, while the follicle cells at the other end of the egg chamber, which are not in contact with the oocyte, remain unaffected and become the anterior follicle cells. The inductive signal from the oocyte is transmitted by the gurken protein, which belongs to the transforming growth factor-α (TGF-α) family. The gurken protein is synthesized and secreted by the oocyte at the posterior end,

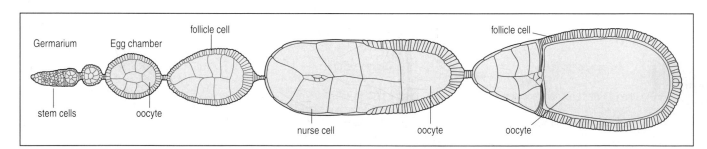

Fig. 5.10 Egg development in *Drosophila*. Oocyte development begins in a germarium, with stem cells at one end. One stem cell will divide four times to give 16 cells with cytoplasmic connections between each other. One of the cells that is connected to four others will become the oocyte, the others will become nurse cells. The nurse cells and oocyte become surrounded by follicle cells and the resulting structure buds off from the germarium as an egg chamber. Successively produced egg chambers are still attached to each other at the poles. The oocyte grows as the nurse cells provide material through the cytoplasmic bridges. The follicle cells play a key role in patterning the oocyte.

where the oocyte nucleus is located at the time. It binds to a receptor protein in the follicle cell plasma membrane encoded by the *torpedo* gene. The torpedo protein is a transmembrane receptor tyrosine kinase similar to the epidermal growth factor receptor.

The posterior follicle cells send a signal back to the oocyte that results in a reorganization of the oocyte's microtubule cytoskeleton into an array of microtubules stretching from the anterior end toward the posterior end. This reorganization is essential for the localization of *bicoid* mRNA at the anterior end of the egg. *bicoid* mRNA is made by nurse cells located next to the anterior end of the developing oocyte, and is transferred from them to the egg. The *bicoid* mRNA interacts with the microtubule array in such a way that it is moved toward, and retained at, the anterior end. Similarly, *oskar* mRNA, which specifies the egg posterior germ plasm that gives rise to the germ cells, is delivered into the oocyte by nurse cells and moved to the posterior end through its interaction with the microtubule array. *nanos* mRNA is also localized at the posterior end. Several maternal genes are necessary for the localization of *bicoid* mRNA. Mother flies lacking the gene *exuperantia*, for example, have eggs in which the *bicoid* mRNA is distributed throughout the egg and not restricted to the anterior. Therefore, *exuperantia* must be involved in the anterior localization process.

The setting-up of the egg's dorso-ventral axis involves a later set of oocyte–follicle cell interactions, which occur after the posterior end of the oocyte has been specified, and depend on the previous reorganization of the microtubule array. The oocyte nucleus moves along the microtubules

Fig. 5.11 *Drosophila* oocyte development. A developing *Drosophila* oocyte (right) is shown attached to its 15 nurse cells (left) and surrounded by a monolayer of 700 follicle cells. The oocyte and follicle layer are cooperating at this time to define the future dorso-ventral axis of the egg and embryo, as indicated by the expression of a gene only in the follicle cells overlying the dorsal anterior region of the oocyte (blue staining). Photograph courtesy of A. Spradling.

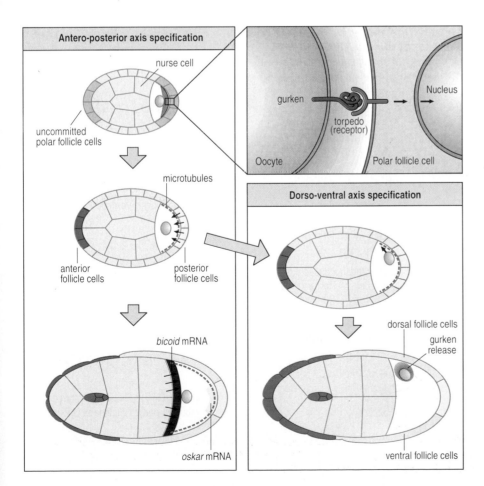

Fig. 5.12 Specification of the antero-posterior and dorso-ventral axes during *Drosophila* oogenesis. The oocyte moves to the posterior end of the egg chamber and comes into contact with the polar follicle cells. It is separated from follicle cells at the anterior end (blue) by the nurse cells. *gurken* mRNA is synthesized and the gurken protein secreted locally. The binding of this protein to the torpedo receptor protein on the adjacent follicle cell initiates their specification as posterior polar follicle cells (yellow). They send a signal back to the oocyte that reorganizes the oocyte cytoskeleton (green microtubules). This directs the localization of the bicoid and oskar proteins to the anterior and posterior ends of the oocyte, respectively, thus defining the antero-posterior axis. Subsequent movement of the nucleus toward the future dorsal side and the local release of gurken protein then specifies the adjacent follicle cells as dorsal follicle cells and that side of the oocyte as the future dorsal side. After Gonzáles-Reyes, A., *et al.*: 1995.

from the posterior of the oocyte to a site on the anterior margin. In this new position, the *gurken* gene is expressed in the oocyte nucleus again. This time, the locally secreted gurken protein acts as a signal to adjacent follicle cells on one side of the oocyte, specifying them as dorsal follicle cells; the side away from the nucleus thus becomes the ventral region by default. The ventral follicle cells produce proteins that are only deposited in the ventral vitelline envelope.

The gurken protein can polarize both axes by its interactions with different sets of follicle cells, which implies that an earlier mechanism has already made some follicle cells, such as the polar follicle cells at each end of the egg chamber, different from the others. Such a difference would ensure that only polar follicle cells could respond to the gurken protein signal to become posterior cells.

The ligand for the torso protein, which distinguishes the termini, is synthesized and secreted by both posterior and anterior follicle cells, but not by the other follicle cells. It is thus only deposited in the vitelline envelope at both ends of the egg during oogenesis.

Summary.

Nurse cells surrounding the *Drosophila* oocyte in the ovarian follicle provide it with large amounts of mRNA and proteins, some of which become localized in particular sites. The oocyte produces a local signal, which induces follicle cells at one end to become posterior follicle cells. The posterior follicle cells cause a reorganization of the oocyte cytoskeleton that localizes *bicoid* mRNA to the anterior end and other mRNAs to the posterior end of the oocyte. The dorso-ventral axis of the oocyte is also initiated by a local signal from the oocyte to certain follicle cells, which then become dorsal follicle cells. Follicle cells on the opposite side of the oocyte specify the ventral side of the oocyte by deposition of maternal proteins in the ventral vitelline envelope. Follicle cells at either end of the oocyte specify the termini.

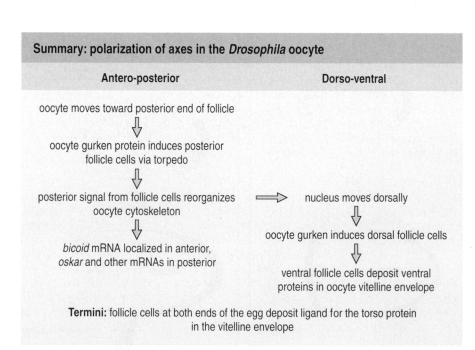

Summary: polarization of axes in the *Drosophila* oocyte

Antero-posterior	Dorso-ventral
oocyte moves toward posterior end of follicle	
⇩	
oocyte gurken protein induces posterior follicle cells via torpedo	
⇩	
posterior signal from follicle cells reorganizes oocyte cytoskeleton ⟹	nucleus moves dorsally
⇩	⇩
bicoid mRNA localized in anterior, *oskar* and other mRNAs in posterior	oocyte gurken induces dorsal follicle cells
	⇩
	ventral follicle cells deposit ventral proteins in oocyte vitelline envelope

Termini: follicle cells at both ends of the egg deposit ligand for the torso protein in the vitelline envelope

Zygotic genes pattern the early embryo.

Understanding in such detail how the main body axes of *Drosophila* are specified is a major achievement, and those who work on other animals such as the frog and chick, where very little is known about their developmental genetics, are justifiably somewhat envious. We have seen how gradients of bicoid, hunchback, and caudal proteins are established along the antero-posterior axis, and how intranuclear dorsal protein is graded along the dorso-ventral axis. This maternally derived framework of positional information is interpreted and elaborated on by zygotic genes to give each region of the embryo an identity. Most of the zygotic genes first activated along the antero-posterior and dorso-ventral axes encode transcription factors, which are thus localized along the axes and activate yet more zygotic genes. We first consider the patterning along the dorso-ventral axis, which is somewhat simpler than that along the antero-posterior axis.

5-8   The expression of zygotic genes along the dorso-ventral axis is controlled by dorsal protein.

After dorsal protein has entered the nuclei, its effects on gene expression divide the dorso-ventral axis into well-defined regions and specify the ventral-most cells as prospective mesoderm. Going from ventral to dorsal, the main regions are mesoderm, ventral ectoderm (prospective neurectoderm), dorsal ectoderm (prospective dorsal epidermis), and prospective amnioserosa (an extra-embryonic membrane on the dorsal side of the embryo that is sloughed off as embryonic development is completed). The mesoderm gives rise to internal soft tissues such as muscle and connective tissue; the ventral ectoderm gives rise to epidermis as well as all nervous tissues; the dorsal ectoderm gives rise only to epidermis. The third germ layer, the endoderm, which is located at either end of the embryo, and which we do not consider here, gives rise to the midgut.

Patterning along the axes poses a problem like that of the French flag (see Section 1-13). Expression of zygotic genes in localized regions along the dorso-ventral axis is initially controlled by the graded concentration of the intranuclear dorsal protein, which falls off rapidly in the dorsal half of the embryo; little dorsal protein is found in nuclei above the equator. In the ventral region, dorsal protein has two main functions —it activates certain genes at specific positions in the ventral region and it represses the activity of other genes, which are therefore only expressed in the dorsal region (Fig. 5.13).

In the ventral-most region, where concentrations of intranuclear dorsal protein are highest, the zygotic genes *twist* and *snail* are activated by dorsal protein in a strip of nuclei along the ventral side of the embryo; soon after this the blastoderm becomes cellular. This ventral strip of cells will form the mesoderm. The expression of *twist* and *snail* is required both for development of the cells as mesoderm and for gastrulation, during which the ventral band of cells moves into the interior of the embryo (see Section 8-8). In the future neurectoderm, which will give rise both to the nervous system and to larval ventral epidermis, the gene *rhomboid* is activated at low levels of dorsal protein, but is not expressed in more ventral regions because it is repressed by the snail protein.

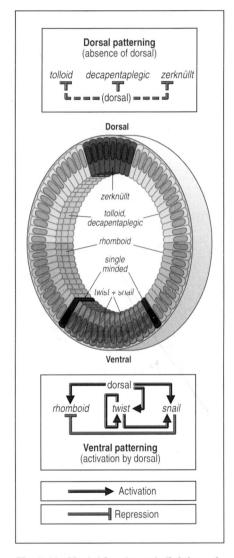

Fig. 5.13 Model for the subdivision of the dorso-ventral axis into different regions by the gradient in nuclear dorsal protein. In the dorsal region, where nuclear dorsal protein is absent, *tolloid*, *zerknüllt*, and *decapentaplegic* are not repressed. In the ventral region, the dorsal protein activates the genes *twist*, *snail*, and *rhomboid*. *twist* is autoregulatory, maintaining its own expression, and also activates *snail*; snail protein inhibits *rhomboid* expression. The gene *single-minded* is activated by the dorsal protein but its localization is dependent on the action of other genes.

The genes *decapentaplegic, tolloid,* and *zerknüllt* are repressed by dorsal protein and so their activity is confined to the more dorsal regions of the embryo, where there is virtually no dorsal protein in the nuclei. *zerknüllt* is expressed most dorsally and appears to specify the amnioserosa. *decapentaplegic* is a key gene in the specification of pattern in the dorsal part of the dorso-ventral axis and its role is considered in detail in the next section.

Mutations in the maternal dorso-ventral genes can cause dorsalization or ventralization of the embryo. In dorsalized embryos, dorsal protein is excluded uniformly from the nuclei. This has a number of effects, one of which is that the *decapentaplegic* gene is expressed everywhere, in line with the observation that it is normally repressed in the ventral region by high intranuclear concentrations of the dorsal protein. By contrast, *twist* and *snail* are not expressed at all in dorsalized embryos, as they are activated by high intranuclear concentrations of dorsal protein. Just the opposite result is obtained in ventralized embryos, where the dorsal protein is present at high concentration in all the nuclei; *twist* and *snail* are expressed throughout and *decapentaplegic* is not expressed at all (Fig. 5.14).

Genes whose expression is regulated by the dorsal protein, such as *twist, snail,* and *decapentaplegic,* contain binding sites for dorsal protein in their regulatory regions that activate or repress gene expression at particular concentrations of the protein. This threshold effect on gene expression is the result of the integrating function of these regulatory binding sites. The ability of genes to respond in a threshold-like manner to varying concentrations of dorsal protein is due to the presence of both high-affinity and low-affinity binding sites for dorsal protein in their regulatory regions. In the most ventral regions (a strip 12–14 cells wide), where the concentration of dorsal protein is high, low-affinity sites delimit gene expression, whereas high-affinity sites control expression in slightly more dorsal regions (up to 20 cells from the ventral

Fig. 5.14 The nuclear gradient in dorsal protein is interpreted by the activation of other genes, such as *twist* and *decapentaplegic*. Left panels: in normal larvae, the *twist* gene is activated above a certain threshold concentration (green line) of dorsal protein, whereas above a lower threshold (yellow line), the *decapentaplegic* gene is repressed. Right panels: in ventralized larvae, the dorsal protein is present in all nuclei; *twist* is now also expressed everywhere, whereas *decapentaplegic* is not expressed at all, because dorsal protein is above the threshold level required for its repression everywhere.

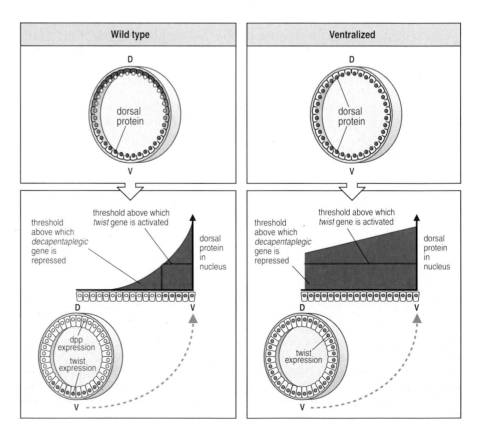

midline). It is very likely that the threshold response involves cooperativity between the different binding sites; that is, binding at one site makes binding at a nearby site easier and so facilitates further binding. In addition, inhibitory interactions with other gene products are involved. For example, the snail protein represses the expression of certain genes in the ventral regions, and thus helps to confine the expression of genes, such as *rhomboid*, to the neurectoderm.

The gradient of dorsal protein is therefore effectively acting as a morphogen gradient along the dorso-ventral axis, activating specific genes at different threshold concentrations, and so defining the dorso-ventral pattern. The regulatory sequences in these genes can be thought of as developmental switches, which when thrown by the binding of transcription factors, activate genes and set cells off along new developmental pathways. The dorsal protein gradient provides one solution to the French flag problem, but it is not the whole story—yet another gradient is also involved.

5-9 The decapentaplegic protein acts as a morphogen to pattern the dorsal region.

As with the antero-posterior axis, each end of the dorso-ventral axis is specified by different proteins. The gradient of dorsal protein, with its high point in the ventral-most region, specifies the initial pattern of zygotic gene activity, and patterns the ventral mesoderm. But the dorsal region is not similarly specified by a low-level gradient in dorsal protein. Indeed, there is little or no dorsal protein in the nuclei of the dorsal half of the embryo. The more dorsal part of the dorso-ventral pattern is thought to be determined by a gradient in the activity of the decapentaplegic protein.

Soon after the gradient of intranuclear dorsal protein has become established, membranes form between the nuclei, the embryo becomes cellular, and transcription factors can no longer diffuse between nuclei. Secreted or transmembrane proteins and their corresponding receptors must now be used to transmit signals between cells. The decapentaplegic protein is one such secreted signaling protein. It is a member of the transforming growth factor-β (TGF-β) family of vertebrate growth factors (see Box 3A, page 64) and, as we shall see, is involved in a variety of signaling processes throughout *Drosophila* development.

The *decapentaplegic* gene is expressed throughout the dorsal region, where dorsal protein is not present in the nuclei. Evidence for a gradient in decapentaplegic protein activity comes from experiments in which *decapentaplegic* mRNA is introduced into an early wild-type embryo. As more mRNA is introduced and the concentration of decapentaplegic protein increases above the normal level, the cells along the dorso-ventral axis adopt a more dorsal fate than they would normally. Ventral ectoderm becomes dorsal ectoderm and, at very high concentrations of *decapentaplegic* mRNA, all the ectoderm develops as the dorsal-most region—the amnioserosa. The graded activity of decapentaplegic protein along the dorso-ventral axis at the cellular blastoderm stage is due to its interaction with a secreted protein, such as short gastrulation, which is expressed in the ventral region and diffuses into the dorsal region (Fig. 5.15).

Thus, the patterning of the dorso-ventral axis involves two gradients, one of dorsal protein and one of decapentaplegic protein, with high points at opposite ends. Together they result in the dorso-ventral axis becoming divided up into several regions of unique gene activity and developmental fate. We now return to the antero-posterior axis at an earlier stage, while the embryo is still acellular.

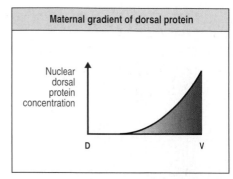

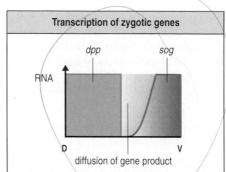

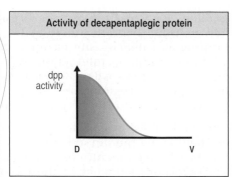

Fig. 5.15 A dorso-ventral gradient in decapentaplegic protein activity is produced by the antagonistic activity of the short gastrulation protein. The maternal gradient of dorsal protein in the nuclei represses *decapentaplegic* (*dpp*) gene transcription ventrally but not dorsally, where dorsal protein is not present in the nuclei. The gene *short gastrulation* (*sog*) is expressed in the ventral region of the embryo. The short gastrulation protein diffuses into the dorsal region and antagonizes the activity of the decapentaplegic protein, giving a dorso-ventral gradient in the activity of decapentaplegic, thus providing positional information in the dorsal region.

5-10   **The antero-posterior axis is divided up into broad regions by gap gene expression.**

The **gap genes** are the first zygotic genes to be expressed along the antero-posterior axis, and they all code for transcription factors. Their expression is initiated by the antero-posterior gradient of bicoid protein while the embryo is still a syncytial blastoderm. bicoid protein primarily activates anterior expression of the gap gene *hunchback*, which in turn is instrumental in switching on the expression of the other gap genes, including *giant*, *Krüppel*, and *knirps*, which are expressed in this order along the antero-posterior axis (Fig. 5.16). (*giant* is in fact expressed in two bands, one anterior and one posterior, but its posterior expression does not concern us here.)

Gap genes were initially recognized by their mutant phenotypes, in which quite large sections of the body pattern along the antero-posterior axis are missing. Although the mutant phenotype of a gap gene usually shows a gap in the antero-posterior pattern in more or less the region in which the gene is normally expressed, there are also more wide-ranging effects. This is because gap gene expression is also essential for later development along the axis.

As the blastoderm is still acellular at the stage at which the gap genes are expressed, the gap gene proteins can diffuse away from their sites of synthesis. They are short-lived proteins with a half-life of minutes. Their distribution therefore extends only slightly beyond the region in which the gene is expressed, and this typically gives a bell-shaped protein concentration profile. The hunchback protein is exceptional in this respect as its gene is expressed over a broad anterior region and it has a steep antero-posterior protein gradient. The control of zygotic *hunchback* expression by bicoid protein is well understood and will be considered next.

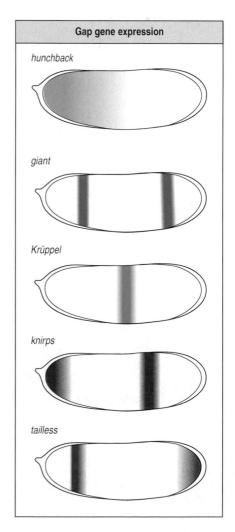

Fig. 5.16 The expression of the gap genes *hunchback*, *Krüppel*, *giant*, *knirps*, and *tailless* in the early *Drosophila* embryo. Gap gene expression at different points along the antero-posterior axis is controlled by the concentration of bicoid and hunchback proteins, together with interactions between the gap genes themselves. The expression pattern of the gap genes provides an aperiodic pattern of transcription factors along the antero-posterior axis, which delimits broad body regions.

5-11 bicoid protein provides a positional signal for the anterior expression of *hunchback*.

Zygotic expression of the *hunchback* gene in normal embryos occurs over most of the anterior half of the embryo. This zygotic expression is superimposed on a low level of maternal *hunchback* mRNA, whose translation is suppressed posteriorly by nanos protein (see Section 5-3). This results in a gradient of hunchback protein in the posterior half of the embryo, running anterior to posterior.

The localized anterior expression of *hunchback* is an interpretation of the positional information provided by the bicoid protein gradient. The *hunchback* gene is switched on only when bicoid protein, a transcription factor, is present at a certain threshold concentration. This level is attained only in the anterior third of the embryo, close to the site of bicoid protein synthesis, which restricts *hunchback* expression to this region. There is some evidence that maternal hunchback protein is also necessary for spatial control of zygotic *hunchback* expression.

The relationship between bicoid protein concentration and *hunchback* gene expression can be illustrated by looking at how *hunchback* expression changes when the bicoid protein concentration gradient is changed by increasing the maternal dosage of the *bicoid* gene (Fig. 5.17). The result is that expression of *hunchback* extends more posteriorly, because the region in which the concentration of the bicoid protein is above the threshold for *hunchback* gene activation is also extended posteriorly. By calibrating the extension of *hunchback* expression with the maternal dosage of the *bicoid* gene, it can be calculated that a twofold increase in the concentration of bicoid protein can switch *hunchback* gene expression from off to on.

The bicoid protein is a member of the homeodomain family of transcriptional activators and activates the *hunchback* gene by binding to regulatory sites within the promoter region. Direct evidence of *hunchback* activation by bicoid protein has been obtained by gene transfer experiments (Box 5A) using a fusion gene constructed from *hunchback* promoter regions and a bacterial reporter gene, *lacZ*, and introduced into the fly genome. *lacZ* codes for the enzyme β-galactosidase, which is visualized by histochemical staining. The extent of *lacZ* expression in embryos carrying this transgene parallels normal *hunchback* expression exactly if the promoter region of the transgene is complete, but not if a large part of it is deleted (Fig. 5.18). The large promoter region required for completely normal gene expression can be cut down to an essential sequence of 263 base pairs that will still give almost normal activity in this situation. This sequence has several sites at which bicoid protein can bind, and it seems that cooperative binding of bicoid proteins is involved in establishing the threshold response.

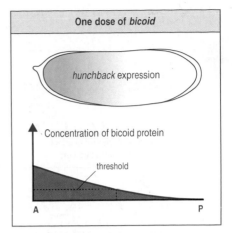

One dose of *bicoid*

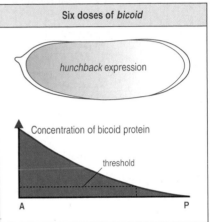

Six doses of *bicoid*

Fig. 5.17 Maternal bicoid protein controls zygotic *hunchback* expression. If the dose of maternal *bicoid* is increased sixfold, the extent of the bicoid protein gradient also increases. The activity of the *hunchback* gene is determined by the threshold concentration of bicoid protein, so at the higher dose, its region of expression is extended toward the posterior end because the region in which bicoid protein concentration exceeds the threshold level also extends more posteriorly (see graph, bottom panel).

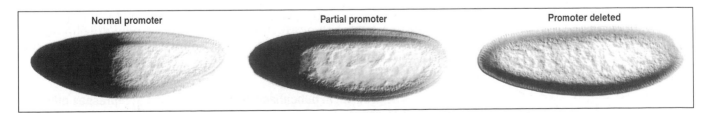

Normal promoter | Partial promoter | Promoter deleted

Fig. 5.18 Zygotic *hunchback* expression is controlled by bicoid protein. *hunchback* expression is visualized by joining the bacterial *lacZ* reporter gene to the control region of the *Drosophila hunchback* gene and inserting this construct into the fly's genome. With the normal *hunchback* control region, *lacZ* is expressed in the anterior half of the embryo (left); with only a partial control region its expression is more restricted (center); and when the construct lacks a bicoid-binding site, *lacZ* expression is absent (right). *lacZ* expression is visualized by histochemical staining for the *lacZ* product, the enzyme β-galactosidase. Photographs courtesy of D. Tautz.

Box 5A Transgenic flies.

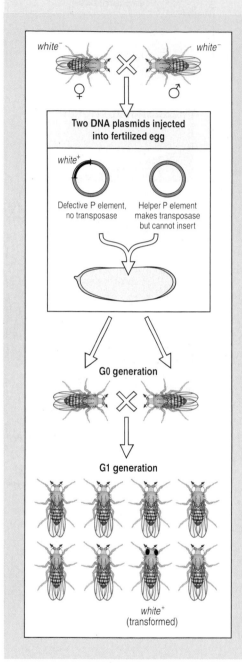

white⁻   white⁻

Two DNA plasmids injected
into fertilized egg

white⁺

Defective P element,
no transposase

Helper P element
makes transposase
but cannot insert

G0 generation

G1 generation

white⁺
(transformed)

As discussed in Boxes 3C (page 81) and 4B (page 108), transgenic mice can be made in which the function of developmental genes can be studied. Transgenic fruit flies can also be created, and have contributed greatly to *Drosophila* developmental genetics. They are made by inserting a known sequence of DNA into the *Drosophila's* chromosomal DNA, using as a carrier, a **transposon** that occurs naturally in some strains of *Drosophila*. This transposons known as a **P element** and the technique as P element mediated transformation.

P elements can insert at almost any site on a chromosome, and can also hop from one site on a chromosome to another within the germ cells, an action which requires an enzyme called transposase. As hopping can cause genomic instability, carrier P elements have had their own transposase gene removed. The transposase required to insert the P element initially is instead provided by a helper P element, which cannot itself insert into the host chromosomes and is thus quickly lost from cells. The two elements are injected together into the posterior end of the egg where the germ cells are made.

Usually a marker gene, such as the wild-type *white⁺* gene, is added to the P element. When *white⁺* is the marker, the P element is inserted into host flies homozygous for the mutant *white⁻* gene (which have white eyes rather than the red eyes of the wild-type *Drosophila*). Red eyes are dominant over white and so flies in which the P element has become integrated into the chromosome, and is being expressed, can be detected by their red eyes.

In the first generation, all flies have white eyes, since any P element that has integrated is still restricted to the germ cells. But in the second generation, a few flies will have wild-type red eyes, showing that they carry the inserted P element in their somatic cells.

This technique can be used to increase the number of copies of a particular gene, to introduce a mutated gene that, for example, has its control or coding regions altered in a known way, or to introduce new genes. It is also possible to introduce genes that carry a marker coding sequence such as *lacZ* (encoding the bacterial enzyme β-galactosidase), whose expression is detectable by histochemical staining. Genes can be given a heat-shock promoter, which is switched on by a sudden rise in the surrounding temperature. By adjusting the temperature, the timing of expression of genes attached to this promoter can be controlled; the effects of expressing a gene at different stages of development can be studied in this way.

The regulatory sequences of genes such as these are yet further examples of developmental switches that direct cells along a new developmental pathway. We will encounter many more examples of these transcriptional switches in *Drosophila* early development.

5-12   **The gradient in hunchback protein activates and represses other gap genes.**

Expression of the other gap genes is localized in bands across the antero-posterior axis (see Fig. 5.16). The hunchback protein is itself a transcription factor and acts as a morphogen to which the other gap genes respond. The bands of gap gene expression are delimited by mechanisms that depend on the gene control regions being sensitive to different concentrations of

Fig. 5.19 *Krüppel* gene activity is specified by hunchback protein. Top panel: above a threshold concentration of hunchback protein, the *Krüppel* gene is repressed; at a lower concentration, above another threshold value, it is activated. Bottom panel: in mutants lacking the *bicoid* gene, and thus also lacking zygotic *hunchback* gene expression, only maternal hunchback protein is present, which is located at the anterior end of the embryo at a relatively low level. In these mutants, *Krüppel* is activated at the anterior end of the embryo, giving an abnormal pattern.

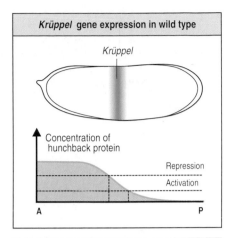

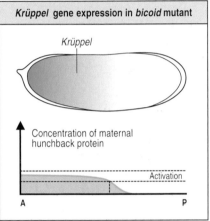

hunchback protein, and also to other proteins, including bicoid protein. Expression of the *Krüppel* gene, for example, is activated by a combination of bicoid protein and low levels of hunchback, but is repressed at high concentrations of hunchback protein. Within this concentration 'window' *Krüppel* remains activated (Fig. 5.19, top panel). But below the lower threshold concentration of hunchback it is not activated. In this way, the gradient in hunchback protein precisely locates a band of *Krüppel* gene activity near the center of the embryo. Refined spatial localization is brought about by repression of *Krüppel* by other gap gene proteins.

Such relationships were worked out by altering the concentration profile of hunchback protein systematically, while all other known influences were eliminated or held constant. Increasing the dose of hunchback protein, for example, results in a posterior shift in its concentration profile, and this results in a posterior shift in the posterior boundary of *Krüppel* expression. In another set of experiments on embryos lacking bicoid protein so that only the maternal hunchback protein gradient is present, the level of hunchback protein is such that *Krüppel* is even activated at the anterior end of the embryo (see Fig. 5.19, bottom panel)

The hunchback protein is also involved in specifying the anterior borders of the bands of expression of the gap genes *knirps* and *giant*, again by a mechanism involving thresholds for repression and activation of these genes. At high concentrations of hunchback protein, *knirps* is repressed, and this specifies its anterior margin of expression. The posterior margin of the *knirps* band is specified by a similar type of interaction with the product of another gap gene, *tailless*. Where the regions of expression of the gap genes overlap, there is extensive cross-inhibition between them, their proteins all being transcription factors. These interactions are essential to sharpen and stabilize the pattern of gap gene expression.

The antero-posterior axis becomes divided into a number of unique regions on the basis of the overlapping and graded distributions of different transcription factors. This beautifully elegant method of delimiting regions can, however, only work in an embryo such as the acellular blastoderm of *Drosophila*, where the transcription factors are able to diffuse throughout the embryo. This regional distribution of the gap gene products provides the starting point for the next stage in development—the activation of the pair-rule genes and the beginning of segmentation.

Summary.

Gradients of maternally derived transcription factors along the dorso-ventral and antero-posterior axes provide positional information that activates zygotic genes at specific locations along these axes. The dorso-ventral axis becomes divided into four regions: ventral mesoderm, ventral ectoderm (neurectoderm), dorsal ectoderm (dorsal epidermis), and amnioserosa. A ventral to dorsal gradient of maternal dorsal protein both specifies the ventral mesoderm and defines the dorsal region; a second gradient, of the decapentaplegic protein, specifies the dorsal ectoderm. Along the antero-posterior axis the zygotic gap genes are activated by the bicoid protein gradient to specify general body regions. Interactions between the gap genes, all of which code for transcription factors, help to define their borders of expression. Patterning along the dorso-ventral and antero-posterior axes divides the embryo into a number of discrete regions, each characterized by a unique pattern of zygotic gene activity.

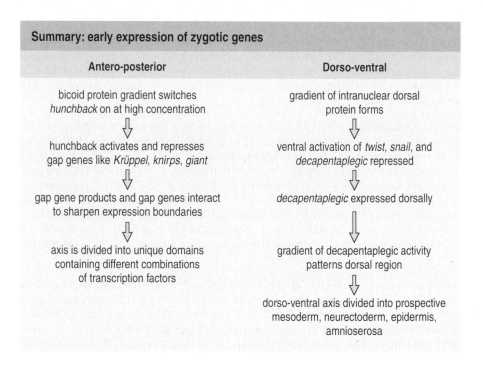

Summary: early expression of zygotic genes

Antero-posterior	Dorso-ventral
bicoid protein gradient switches *hunchback* on at high concentration	gradient of intranuclear dorsal protein forms
⇩	⇩
hunchback activates and represses gap genes like *Krüppel, knirps, giant*	ventral activation of *twist, snail*, and *decapentaplegic* repressed
⇩	⇩
gap gene products and gap genes interact to sharpen expression boundaries	*decapentaplegic* expressed dorsally
⇩	⇩
axis is divided into unique domains containing different combinations of transcription factors	gradient of decapentaplegic activity patterns dorsal region
	⇩
	dorso-ventral axis divided into prospective mesoderm, neurectoderm, epidermis, amnioserosa

Segmentation: activation of the pair-rule genes.

The most obvious feature of a *Drosophila* larva is the regular segmentation of the larval cuticle along its antero-posterior axis, each segment carrying cuticular structures that define it as, for example, thorax or abdomen. This pattern of segmentation is carried over into the adult, in which each segment has its own identity. Adult appendages such as wings, halteres, and legs are attached to particular segments (Fig. 5.20). But the discernible segments of the mature larva are not in fact the first units of segmentation along this axis. The basic developmental modules, whose definition we will follow in some detail, are the **parasegments**, which are specified first, and from which the segments derive.

5-13   **Parasegments are delimited by expression of pair-rule genes in a periodic pattern.**

The first visible signs of segmentation in the embryo are transient grooves that appear on the surface of the embryo after gastrulation. These grooves define the parasegments. There are 14 parasegments,

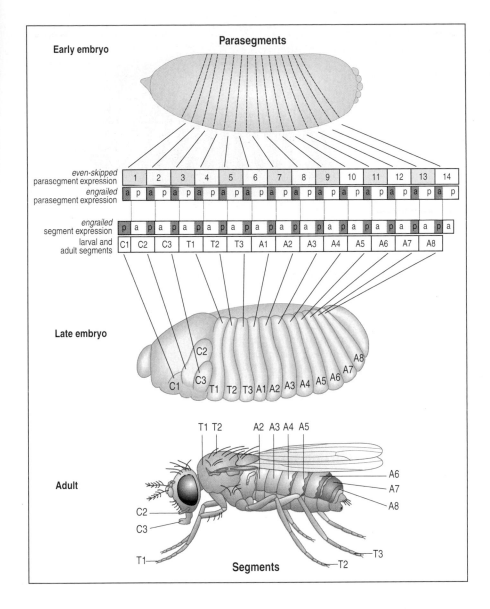

Fig. 5.20 The relationship between parasegments and segments in the early embryo, late embryo, and adult fly. Initially, pair-rule genes are expressed in the embryo as stripes in every second parasegment. *even-skipped* (yellow) is expressed in odd-numbered parasegments. The segment polarity selector gene *engrailed* (blue) is expressed in the anterior region of every parasegment, and delimits the anterior margin of each parasegment. Each larval segment is composed of the posterior region of one parasegment and the anterior region of the next. The anterior region of a parasegment becomes the posterior portion of a segment. Segments are thus offset from the original parasegments by about half a segment, and *engrailed* is expressed in the posterior region of each segment. In this figure, a and p refer to the anterior and posterior compartments of segments or parasegments. The segment specification is carried over into the adult and results in particular appendages, such as legs and wings, developing on specific segments only. C1, C2, and C3 represent segments which become fused to form the head region. T= thoracic segments; A = abdominal segments. After Lawrence, P.: 1992.

which are the fundamental units in the segmentation of the *Drosophila* embryo. Once each parasegment is delimited, it behaves as an independent developmental unit, under the control of a particular set of genes, and in this sense at least, the embryo can be thought of as being built up piecemeal. The parasegments are initially similar but each will eventually acquire its own unique identity. They are also out of register with the final segments by half a segment; each segment is made up of the posterior region of one parasegment and the anterior region of the next (see Fig. 5.20). In the head region the segmental arrangement is lost when the anterior parasegments fuse.

The parasegments are delimited by the action of the **pair-rule genes**, each of which is expressed in a series of seven transverse stripes along the embryo, each stripe corresponding to every second parasegment. When pair-rule gene expression is visualized by staining for the pair-rule proteins, a striking zebra-striped embryo is revealed (Fig. 5.21).

The positions of the stripes of pair-rule gene expression are determined by the pattern of gap gene expression—a non-repeating pattern of gap gene activity is converted into repeating stripes of pair-rule gene expression. We now consider how this is achieved.

Fig. 5.21 The striped patterns of activity of pair-rule genes in the *Drosophila* embryo just before cellularization. Parasegments are delimited by pair-rule gene expression, each pair-rule gene being expressed in alternate parasegments. Expression of the pair-rule genes *even-skipped* (blue) and *fushi tarazu* (brown) is visualized by staining with antibody for their proteins products. *even-skipped* is expressed in odd-numbered parasegments, *fushi tarazu* in even-numbered parasegments. Scale bar = 0.1 mm. Photograph from Lawrence, P.: 1992.

Anterior   Posterior

5-14   Gap gene activity positions stripes of pair-rule gene expression.

Pair-rule genes are expressed in stripes, with a periodicity corresponding to alternate parasegments. Mutations in these genes thus affect alternate segments. Some pair-rule genes (e.g. *even-skipped*) define odd-numbered parasegments, whereas others (e.g. *fushi tarazu*) define even-numbered parasegments. The striped pattern of expression of pair-rule genes is present even before cells are formed, while the embryo is still a syncytium, although cellularization occurs soon after expression begins. Each pair-rule gene is expressed in seven stripes, each of which is only a few cells wide. With some genes, such as *even-skipped*, the anterior margin of the stripe corresponds to the anterior boundary of a parasegment; the domains of expression of other pair-rule genes, however, cross parasegment boundaries.

The striped expression pattern appears gradually; the *even-skipped* gene, for example, is initially expressed at a low level in all nuclei. A single broad stripe of gene expression then appears anteriorly and narrows as the other stripes develop. Each stripe is initially fuzzy, but eventually acquires a sharp anterior margin. At first sight this type of patterning would seem to require some underlying periodic process, such as the setting up of a wave-like concentration of a morphogen, with each stripe forming at the crest of a wave. It was surprising, therefore, to discover that each stripe is specified independently.

As an example of how the pair-rule stripes are generated, we will look in detail at the expression of the second *even-skipped* stripe (Fig. 5.22). The appearance of this stripe depends on the normal expression of *bicoid* and of the three gap genes *hunchback*, *Krüppel*, and *giant* (only the anterior band of *giant* expression is involved in specifying the second *even-skipped* stripe, see Fig. 5.16). bicoid and hunchback proteins are required to activate the *even-skipped* gene, but they do not define the boundaries of the stripe. These are defined by the Krüppel and giant proteins, by a mechanism based on repression of *even-skipped*. When concentrations of Krüppel and giant proteins are above certain threshold levels, *even-skipped* is repressed, even if bicoid and hunchback proteins are present. The anterior edge of the stripe is localized at the point of threshold concentration of giant protein, whereas the posterior border is similarly specified by Krüppel protein.

The independent localization of each of the stripes by the gap gene transcription factors requires that, in each stripe, pair-rule genes respond to different concentrations and combinations of the gap gene transcription factors. The pair-rule genes thus require complex control regions with multiple binding sites for each of the different factors. Examination

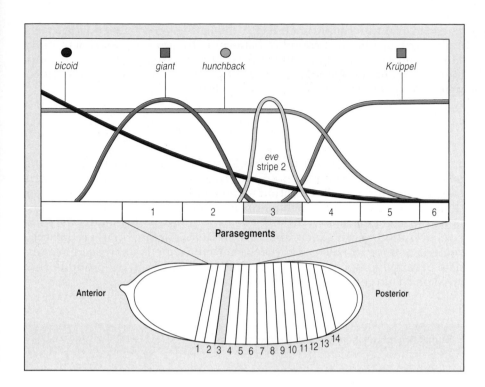

Fig. 5.22 The specification of the second *even-skipped* (*eve*) stripe by gap gene proteins. The different concentrations of transcription factors encoded by the gap genes *hunchback*, *giant*, and *Krüppel* localizes *even-skipped* —expressed in a narrow stripe at a particular point along their gradients— in parasegment 3. bicoid and hunchback proteins activate the gene in a broad domain, and the anterior and posterior borders are formed through repression by giant and Krüppel proteins, respectively.

of the regulatory regions of the *even-skipped* gene reveals a number of separate regions, each controlling the localization of a different stripe. Using the *lacZ* reporter gene technique described in Box 5A (page 144), regulatory regions of around 500 base pairs have been isolated, and each determines the expression of a single stripe (Fig. 5.23).

The presence of control regions which, when activated, can lead to gene expression in a specific position in the embryo is an important principle controlling gene action in development. Other examples are provided by the localized expression of the gap genes and of genes along the dorso-ventral axis.

Each of the regulatory regions on such genes contains binding sites for different transcription factors, some of which activate the gene, while others repress it. In this way, the gap genes regulate pair-rule gene expression in each parasegment. Some pair-rule genes, such as *fushi tarazu*, may not be regulated by the gap genes directly, but may depend on the prior expression of primary pair-rule genes, such as *even-skipped* and *hairy*. With the initiation of pair-rule gene expression the embryo becomes segmented; it is now divided into a number of unique regions, which are characterized mainly by the combinations of transcription factors being expressed in each. These include proteins encoded by the gap genes, the pair-rule genes and the genes expressed along the dorso-ventral axis.

Fig. 5.23 Sites of action of activating and repressing transcription factors in the region of the *even-skipped* promoter involved in expression of the second *even-skipped* stripe. A promoter region of around 500 base pairs, located between 1070 and 1550 base pairs upstream of the transcription start site, directs formation of the second *even-skipped* stripe. Gene expression occurs when the bicoid and hunchback transcription factors are present above a threshold concentration, with the giant and Krüppel proteins acting as repressors where they are above threshold levels. The repressors may act by preventing binding of activators.

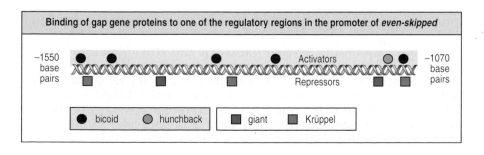

The transcription factors encoded by the pair-rule genes set up the spatial framework for the next round of patterning by transcriptional activation. This involves the further patterning of the parasegments, the development of the final segmentation and the acquisition of segment identity, which is considered in the following sections.

Summary.

The activation of the pair-rule genes by the gap genes results in the transformation of the embryonic pattern along the antero-posterior axis from an aperiodic regionalization to a periodic one. The pair-rule genes define 14 parasegments. Each parasegment is defined by narrow stripes of pair-rule gene activity. These stripes are uniquely defined by the local concentration of gap gene transcription factors acting on the regulatory regions of the pair-rule genes. Each pair-rule gene is expressed in alternate parasegments—some in odd-numbered, others in even-numbered. Most pair-rule genes code for transcription factors.

Summary: pair-rule genes and segmentation

production of local combinations of gap gene transcription factors

activation of each pair-rule gene in seven transverse stripes along the antero-posterior axis

pair-rule gene expression defines 14 parasegments, each pair-rule gene being expressed in alternate parasegments

---

Segment polarity genes and compartments.

The expression of the pair-rule genes defines the anterior boundaries of all 14 parasegments but, like the gap genes, their activity is only temporary. Moreover, at this time the blastoderm becomes cellularized. How, therefore, are the positions of parasegment boundaries fixed, and how do the final segment boundaries of the larval epidermis become established? This is the role of the **segment polarity genes**. Unlike the gap genes and pair-rule genes, which encode transcription factors, the segment polarity genes are a diverse group of genes that bear no obvious relation to each other in their protein products or mechanism of operation. They are known as segment polarity genes because the effect of their mutant forms is generally to upset the antero-posterior polarity of the segments, producing mirror-image or tandem duplications of either anterior or posterior parts.

Segment polarity genes are activated in response to pair-rule gene expression. They are each expressed in 14 transverse stripes, one stripe corresponding to each parasegment. During pair-rule gene expression, the blastoderm becomes cellularized, so the segment polarity genes are acting in a cellular rather than a syncytial environment. One of the segment polarity genes to be activated by the pair-rule genes is the transcription factor *engrailed*, which is expressed in the anterior region of every parasegment. This gene is of particular interest as its expression delimits a boundary of cell lineage restriction. Moreover, *engrailed* is also a **selector gene**, a gene that confers a particular identity on a region, or regions, by controlling the activity of other genes, and which continues to act for an extended period.

| 5-15 | **Expression of the *engrailed* gene delimits a cell lineage boundary and defines a compartment.** |

The *engrailed* gene plays a key role in segmentation and, unlike the pair-rule and gap genes, whose activity is transitory, it is expressed throughout the life of the fly. *engrailed* activity first appears at the time of cellularization as a series of 14 transverse stripes. Fig. 5.24 shows expression at the later extended germ-band stage, when part of the ventral blastoderm (the germ band) has extended over the dorsal side of the embryo. *engrailed* is initially expressed in a single line of cells at the anterior margin of each parasegment, which is itself only about three cells wide (Fig. 5.25). It is likely that this periodic pattern of *engrailed* activity is the result of the combinatorial action of transcription factors encoded by pair-rule genes, including *fushi tarazu* and *even-skipped*. Evidence that the pair-rule genes do control *engrailed* expression is provided, for example, by embryos carrying mutations in *fushi tarazu*, in which *engrailed* expression is absent only in even-numbered parasegments (in which *fushi tarazu* is normally expressed).

The anterior margin of the parasegment has a very important property —it is a boundary of **cell lineage restriction** (such as the boundaries between the rhombomeres of the vertebrate hindbrain discussed in Section 4-10). Cells and their descendants from one parasegment never move into adjacent ones. This implies that the cells in a parasegment are under some common genetic control, which both prevents them from mixing with their neighbors, and controls their later development. Such domains of lineage restriction are known as **compartments**.

The existence of compartments can be detected by cell lineage studies. Cell lineage can be followed by marking single cells in the early embryo so that all their descendants (**clones**) can be identified at later stages of development. One technique is to inject the egg with a harmless fluorescent compound which is incorporated into all the cells of the embryo. In the early embryo, a fine beam of ultraviolet light directed onto a single cell activates the fluorescent compound. All the descendants of that cell will be fluorescent and can therefore be identified. Examination of these clones together with an analysis of *engrailed* expression shows that cells at the anterior margin of the parasegment have no descendants that lie on the other side of that margin. The anterior margin is therefore a boundary of lineage restriction; that is, cells and their descendants that are on one or other side of the boundary, when it is formed, never cross it.

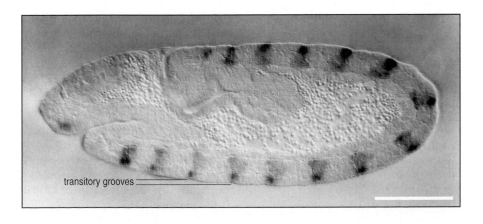

transitory grooves

Fig. 5.24 The expression of the *engrailed* gene in a late (stage 11) *Drosophila* embryo. The gene is expressed in the anterior region of each parasegment and the transitory grooves between parasegments can be seen. At this stage of development, the germ band has temporarily extended and is curved over the back of the embryo. Scale bar = 0.1 mm. Photograph from Lawrence, P.:1992.

Fig. 5.25 The expression of the pair-rule genes *fushi tarazu* (blue), *even-skipped* (pink), and *engrailed* (purple dots) in parasegments. *engrailed* is expressed at the anterior margin of each stripe and delimits the anterior border of each parasegment. The boundaries of the parasegments become sharper and straighter later on. After Lawrence, P.: 1992.

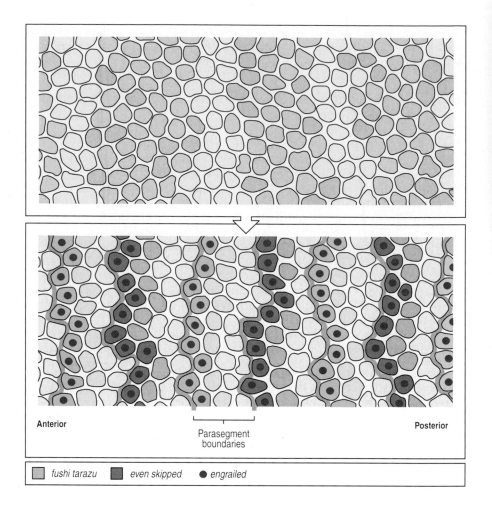

Anterior   Parasegment boundaries   Posterior

☐ *fushi tarazu*   ■ *even skipped*   ● *engrailed*

The lineage restriction of the anterior margin of the parasegment is carried over into the segments of the larva and the adult. But as the anterior part of a parasegment becomes the posterior part of a segment, the lineage restriction within the segment is between anterior and posterior regions. The segment is thus divided into anterior and posterior compartments, with *engrailed* expression defining the posterior compartment. A compartment can be defined as a region in the embryo that contains all the descendants of the cells present when the compartment is set up, and no others.

In a compartment, the cells may also all be under common genetic control. Imagine a group of cells becoming divided into two separate regional populations by means of signals that turn on different genes in each region. The regions will be compartments if there is never any mixing of the descendants of the original cells across the boundary between them. This is illustrated by the behavior of cells in the wing of the adult fly. The adult wing is chosen as it is easier to distinguish the lineage restriction in the adult structures, in which there has been considerable cell division, compared with embryonic and early larval structures where there has been little cell division since the initial determining events.

Compartments are made visible by making mosaic flies composed of two distinguishable kinds of cells (Box 5B). Single cells in the embryonic blastoderm, or in the epidermis of the larva, can be given a distinctive phenotype by X-ray induced mitotic recombination. The fate of all the descendants of this marked cell can then be followed. Their behavior

Box 5B Genetic mosaics and mitotic recombination.

Genetic mosaics are embryos derived from a single genome but in which there is a mixture of cells with re-arranged or inactivated genes. In flies, genetic mosaics can be generated by inducing rare mitotic recombination events in the embryo or larva. Chromosome breaks induced with X rays result in the exchange of material between homologous chromosomes, just after the chromosomes have replicated into chromatids. Such an event can generate a single cell with a unique genetic construction that will be inherited by all the cell's descendants, which in flies usually form a coherent patch of tissue.

Easily distinguishable mutations like the recessive *multiple wing hairs* can be used to identify the marked clone; if a cell homozygous for this mutation is generated by mitotic recombination in a heterozygous larva, all of its descendant cells will have multiple hairs.

Marked epidermal clones made by this method are usually small as there is little cell proliferation after the recombination event. Larger clones can be made using the *Minute* technique. The cells in flies carrying a mutation in the *Minute* gene grow more slowly than those in the wild type. By using flies heterozygous for the *Minute* mutation, clones can be made in which the mitotic recombination event has generated a marked cell that is normal because it has lost the *Minute* mutation, and so is wild type. This normal cell proliferates faster than the slower-growing background *Minute* heterozygous cells and thus large clones of the marked cells are produced. The mitotic recombination technique has many applications. If clones of marked cells are generated at different stages of development, one can trace the fate of the altered cells and thus see what structures they are able to contribute to. This can provide information on their state of determination or specification at different developmental stages. The technique can also be used to study the localized effects of homozygous recessive mutations that are lethal in present in the homozygous state throughout the whole animal. Illustration after Lawrence, P.:1992.

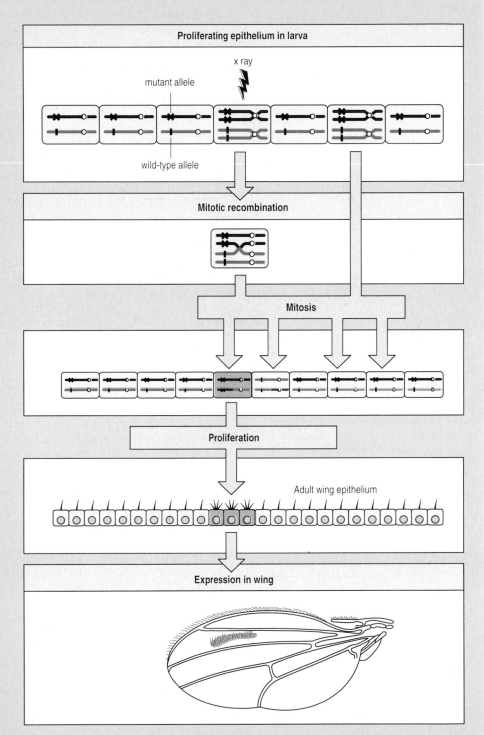

depends on the stage of development at which the founder nucleus was marked. Descendants of nuclei marked at early stages during cleavage become part of many tissues and organs, but those of nuclei marked at the blastoderm stage or later have a more restricted fate. They are found only in the anterior or posterior part of each segment (or of an appendage like a wing), but never throughout the whole segment.

Because the cells of imaginal discs divide only about ten times after the cellular blastoderm stage, the clones of marked cells are small even in the adult, and it is not easy to detect a boundary of lineage restriction (Fig. 5.26, top panel). Clone size can be increased by the *Minute* technique, which results in the marked cell dividing many more times than the other cells (see Box 5B). A single clone of such cells can almost fill either the anterior or posterior part of the wing, meaning that the boundary that the cells never transgress becomes much more evident (see Fig. 5.26, middle panel). This boundary separates the anterior and posterior compartments. In a normal wing, each compartment is constructed by all the descendants of a set of founding cells. The compartment boundary is remarkably sharp and straight and does not correspond to any structural features in the wing. These experiments also show that the pattern of the wing is in no way dependent on cell lineage. A single marked embryonic cell can give rise to about one twentieth of the cells of the adult wing or, using the *Minute* technique to increase clone size, about half the wing. The lineage of the wing cells in each case is quite different, yet the wing's pattern is quite normal.

The compartment pattern in the adult *Drosophila* wing is carried over from its initial specification into the imaginal discs. When epidermal cells are set aside in the embryo to form the imaginal discs, each disc carries over the compartment pattern of the parasegments from which it arises. A wing disc, for example, develops at the boundary between the two parasegments that contribute to the second thoracic segment. Thus, a wing is divided into an anterior compartment and a posterior compartment, the compartment boundary (the old parasegment boundary) running in a straight line more or less down the middle of the wing.

The specification of cells as the posterior compartment of a segment (the anterior of a parasegment) initially occurs when the parasegments are set up, and is due to the *engrailed* gene. Expression of *engrailed* is required both to confer a 'posterior segment' identity on the cells and to change their surface properties so that they cannot mix with the cells adjacent to them, hence setting up the parasegment (compartment) boundary.

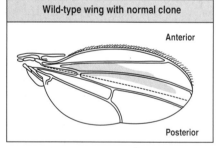

Wild-type wing with normal clone

Anterior

Posterior

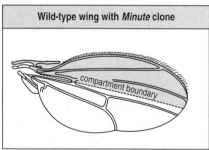

Wild-type wing with *Minute* clone

compartment boundary

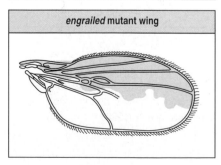

***engrailed* mutant wing**

Fig. 5.26 The boundary between anterior and posterior compartments in the wing can be demonstrated by marked cell clones. Top panel: in the wild-type wing, clones marked by mitotic recombination in the embryo, which gives marked cells a phenotype different from other cells of the wing, are too small to demonstrate the compartment boundary. Middle panel: the use of the *Minute* technique (see Box 5B) produces an increased rate of cell division in the marked cell and gives clones large enough to visualize that cells from one compartment do not cross the boundary into an adjoining compartment. Bottom panel: in the wing of an *engrailed* mutant, in which cells do not express the *engrailed* gene, there is no posterior compartment or boundary. Clones in the anterior part of the wing cross over into the posterior region and the posterior region is transformed into a more anterior-like structure, bearing anterior-type hairs on its margin. The *engrailed* gene is required for the maintenance of the character of the posterior compartment, and for the formation of the boundary.

Direct evidence for this comes from examining the behavior of clones of wing cells in an *engrailed* mutant (see Fig. 5.26, bottom panel). In the absence of normal *engrailed* expression, clones are not confined to anterior or posterior parts of the segment and there is no compartment boundary. Moreover, in *engrailed* mutants, the posterior compartment is partly transformed so that it comes to resemble the pattern of the anterior part of the wing. For example, bristles normally found only at the anterior margin of the wing are also found at the posterior margin.

The *engrailed* gene needs to be expressed continuously throughout larval and pupal stages and into the adult to maintain the character of the posterior compartment of the segment. Thus, *engrailed* is also an example of a selector gene—a gene whose activity is sufficient to cause cells to adopt a particular fate. Selector genes can control the development of a region such as a compartment and, by controlling the activity of other genes, give the region a particular identity.

<table>
<tr><td>5-16</td><td>**Segment polarity genes pattern the segments and stabilize parasegment and segment boundaries.**</td></tr>
</table>

Each larval segment has a well defined antero-posterior pattern, which is easily seen on the ventral epidermis of the abdomen: the anterior region of each segment bears denticles (outgrowths of the chitinous cuticle), and the posterior region is naked (Fig. 5.27). The rows of denticles make a distinct pattern, and this is thought to reflect an antero-posterior polarity gradient in each segment. Mutations in segment polarity genes often alter the denticle pattern, and this is how they were first discovered. For example, a mutation in the segment polarity gene *wingless* results in the whole of the ventral abdomen being covered in denticles, but in the posterior half of each segment the denticle pattern is reversed. In this mutant, the anterior region of each segment is duplicated, but with mirror-image polarity, and the posterior pattern is lost. Mutations in the segment polarity gene *hedgehog* give a similar phenotype. The genes *hedgehog* and *wingless* encode secreted signaling proteins and are related to the vertebrate *Sonic hedgehog* and *Wnt* genes, respectively, which are key players in signaling during pattern formation in vertebrates.

The larval denticle patterns depend on the correct establishment and maintenance of parasegment boundaries. Segment polarity genes are expressed in restricted domains within each parasegment (Fig. 5.28).

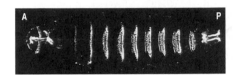

Fig. 5.27 Each segment of the larva has a characteristic pattern of denticles on its ventral surface. Denticles are confined to the anterior regions of segments, with each segment having its own pattern.

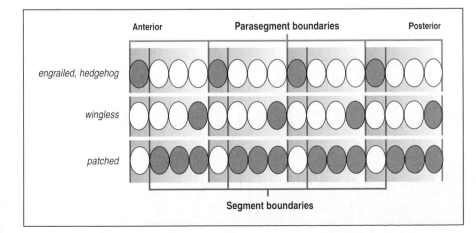

Fig. 5.28 The domains of expression of the segment polarity genes. At the cellular blastoderm stage, the *engrailed* gene is expressed at the anterior margin of each parasegment together with *hedgehog*, while *wingless* is expressed at the posterior margin. After gastrulation, *patched* is expressed in all of the cells expressing neither *engrailed* nor *hedgehog*. About two cell divisions occur between the time that the parasegments are delimited and when the embryo hatches.

Maintenance of a parasegment boundary depends on an intercellular signaling circuit being set up between adjacent cells on either side of the boundary, and involves interactions between *engrailed*, *wingless*, *hedgehog*, and other segment polarity genes. On one side of the boundary, cells expressing *engrailed* also express *hedgehog*, which codes for a secreted protein. This hedgehog protein maintains *wingless* expression in the adjacent cell on the other side of the boundary. The *wingless* gene encodes a secreted glycoprotein, which in turn provides feedback to maintain *hedgehog* and *engrailed* expression, thus maintaining the compartment boundary (Fig. 5.29).

We shall meet the wingless and hedgehog proteins many times again, as they, and their counterparts in other animals, are highly conserved developmental signaling molecules, which have been recruited into a variety of positional signaling systems.

Other segment polarity genes provide receptors for these signaling molecules and components of the signaling pathways involved in maintaining parasegment and segment boundaries: the segment polarity gene *patched* encodes receptors for the hedgehog protein. This complex intercellular circuitry consolidates the compartment boundaries; gradients of signals that pattern the segment, of which wingless protein may be one, are then thought to be set up within these boundaries.

Fig. 5.29 Interactions between *hedgehog*, *wingless*, and *engrailed* genes and proteins at the compartment boundary control denticle pattern. Top left panel: the *engrailed* gene, which encodes a homeodomain transcription factor, is expressed in cells along the anterior margin of the parasegment; these cells also express the segment polarity gene *hedgehog* and secrete the hedgehog protein. hedgehog protein activates and maintains expression of the segment polarity gene *wingless* in adjacent cells across the compartment boundary. wingless protein feeds back on *engrailed*-expressing cells to maintain expression of the *engrailed* gene and *hedgehog*. These interactions stabilize and maintain the compartment boundary. Top right panel: in mutants where the *wingless* gene is inactivated and the wingless protein is absent, neither *hedgehog* nor *engrailed* genes are expressed. This leads to loss of the compartment boundary and the normally well-defined pattern of denticles within each abdominal segment. Bottom left panel: in the wild-type larva the denticle bands in the ventral cuticle are confined to the anterior part of the segment and are dependent on the activity of *hedgehog* and *wingless* genes. Bottom right panel: in the *wingless* mutant, denticles are present across the whole ventral surface of the segment in what looks like a mirror-image repeat of the anterior segment pattern. Photographs from Lawrence, P.:1992.

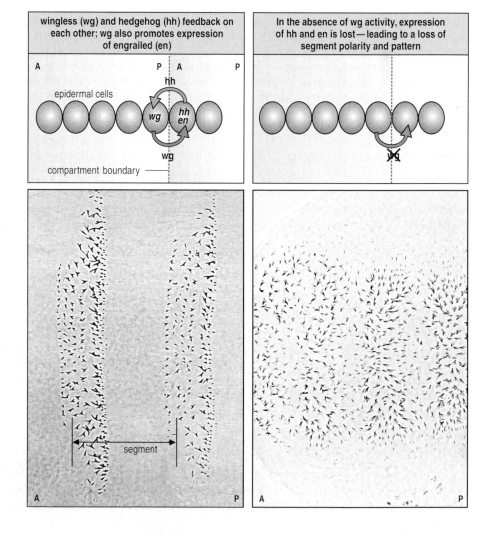

5-17 | **Compartment boundaries are involved in patterning and polarizing segments.**

Evidence for the role of a compartment boundary, in this case a segment boundary, in controlling pattern and polarization of the segment comes from structures on the epidermis of two other insects—the plant-feeding bug *Oncopeltus*, and the wax moth *Galleria*.

Oncopeltus adults have a large number of hairs covering each segment, and in most individuals all the hairs point in an anterior to posterior direction, just as if they were little arrows (Fig. 5.30, left panel). Some individuals have a gap in the segment boundary and this results in a rather precise pattern of hairs with altered orientation: many of the hairs near the gap point in the reverse direction (see Fig. 5.30, center and right panels). One can explain this by assuming that there is a gradient—possibly of a morphogen—running from the anterior boundary to the posterior boundary of each segment. If the slope of the gradient gives the hairs their polarity, they will always point down the gradient. A gap in a segment boundary would result in a local change in the gradient, the sharp change in concentration at the normal boundary being smoothed out and forming a local gradient running in the opposite direction. This would result in the hairs in this region pointing in the opposite direction.

Evidence for such a gradient, providing not only polarity but positional information, is suggested by grafting experiments on the larval cuticle of *Galleria* (Fig. 5.31). In the adult moth there are seven different types of cuticle in each segment. Grafting a piece of larval cuticle to a more anterior position results in a striking repatterning in the region of the graft, which can best be explained in terms of a gradient in positional information determining the character and polarity of the cuticle in each region. While both the above observations are open to other interpretations, even these involve gradients.

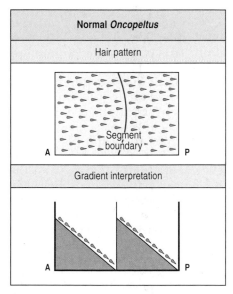
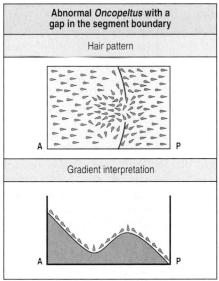
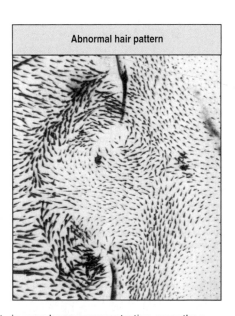

Fig. 5.30 Gradients could specify polarity in the segments of *Oncopeltus*. The hairs on the cuticle point in a posterior direction and this may reflect an underlying gradient in a morphogen that is partly maintained by the segment boundary (left panels). When there is a gap in the boundary (center panels) the sharp discontinuity in morphogen concentration smoothes out, with a resulting local reversal of the gradient and of the direction in which the hairs point. This is illustrated in the photograph (right panel). Photograph from Lawrence, P.:1992. Illustration after Lawrence, P.:1992.

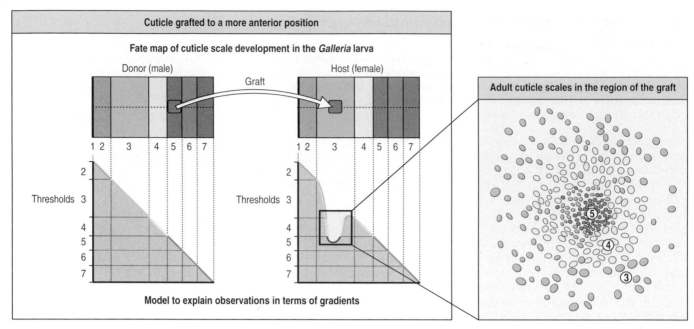

Fig. 5.31 Positional information could specify the cuticle pattern in *Galleria*. There are seven types of cuticle scale in each segment of the adult moth, arranged in consecutive bands. If a small piece of cuticle from a male larva is grafted to a more anterior position in the segment of a female, there is a local alteration in both the pattern and orientation of the cuticle scales in the adult. Grafts are carried out between males and females as female cells have distinctive nuclei, so the transplanted cells and their descendants can be identified. The new pattern can be understood in terms of changes in a gradient in the segment that provides positional information for scale development.

5-18   **Some insects use different mechanisms for patterning the body plan.**

Drosophila belongs to an evolutionarily advanced group of insects, one of its main characteristics being that all segments are specified more or less at the same time, as shown both by the striped patterns of pair-rule gene expression in the acellular blastoderm and by the appearance of all the segments shortly after gastrulation. This type of development is known as **long-germ development**, since the blastoderm corresponds to the whole of the future embryo. All of the segments form at about the same time. Many other insects, such as the flour beetle *Tribolium*, have a **short-germ development**. In short-germ development, the blastoderm is short and forms only anterior segments. The posterior segments are formed after completion of the blastoderm stage and gastrulation. Thus, most segments are formed from a cellular blastoderm and posterior segments are added by growth in the posterior region (Fig. 5.32). In spite of early differences between them, the mature germ band stages of both long-germ and short-germ insect embryos still look similar. This, therefore, is a common stage —the **phylotypic stage**—through which all insect embryos develop (see Fig. 2.2).

The question clearly arises as to what processes are common to the specification of the body plan in long-germ and short-germ insects. One clear difference is that while the patterning of the body plan in the long-germ band of *Drosophila* takes place before cell boundaries form, much of the body plan in short-germ insects is laid down at a later stage, when the posterior segments are generated, during growth. At this stage the embryo is multicellular, so are the same genes involved?

Fig. 5.32 Differences in the development of long-germ and short-germ insects. Top panel: the general fate map of long-germ insects such as *Drosophila* shows that the whole of the body plan —head (H), thorax (Th), and abdomen (Ab)—is present at the time of initial germ-band formation. Bottom panel: in short-germ insects, only the anterior regions of the body plan are present at this embryonic stage. Most of the abdominal segments develop later, after gastrulation, from a posterior growth zone (Gz).

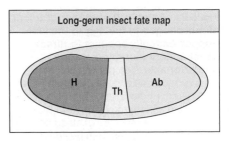

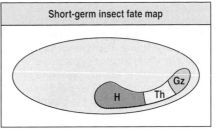

There is very good evidence that the same genes and developmental processes are involved in the patterning of *Tribolium* and *Drosophila*. For example, the gap gene *Krüppel* is expressed at the posterior end of the *Tribolium* embryo at the blastoderm stage and not in the middle region, as in *Drosophila* (Fig. 5.33). It therefore seems to be specifying the same part of the body in both insects. Only two repeats of the pair-rule stripes are present at the blastoderm stage, together with a posterior cap of pair-rule gene expression, in contrast to the seven in *Drosophila*. The genes *wingless* and *engrailed* are also expressed in a similar relation to that in *Drosophila*.

Although not many genes have been studied in detail in other insects, at least one, the segment polarity gene *engrailed*, is known to be expressed in the posterior region of segments in a variety of insects. The pair-rule gene *even-skipped* (see Section 5-14), while present in the grasshopper (a short-germ insect) may not have a similar role in segmentation. However, it is involved later in the development of the nervous system in the grasshopper, and is also expressed at the posterior end of the growing germ band.

Experiments with another insect, the leaf-hopper *Euscelis*, show that it has a mechanism of antero-posterior axis determination with a strong resemblance to the mechanism involving the bicoid gradient (see Section 5-2). *Euscelis* has a mode of development intermediate between long-germ and short-germ insects. It has a rather long egg which contains a ball of symbiotic bacteria at its posterior end. This ball can be moved with a microneedle to more anterior regions. Some posterior cytoplasm is moved along with it, thus providing a serendipitous cytoplasmic marker, which is useful in transplantation experiments.

Two experiments provide evidence for a morphogen gradient in the *Euscelis* egg with the high point at the posterior end. First, if a ligature is tied around the fertilized egg to prevent communication between the anterior and posterior regions, the result is a gap in the body plan, with some regions failing to develop. Second, if the ball of bacteria with posterior cytoplasm is moved anteriorly, and then a ligature is tied behind

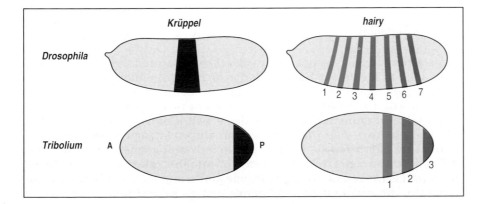

Fig. 5.33 Gap and pair-rule gene expression in long-germ and short-germ insects at the time of germ band formation. *Krüppel* (red) is a gap gene and *hairy* (green) is a pair-rule gene. The position of the *Krüppel* stripe in the short-germ embryo of *Tribolium* indicates that posterior regions of the body are not yet present at this stage. Similarly, only three *hairy* stripes, corresponding to the first three *hairy* stripes in a long-germ embryo, are present in *Tribolium*.

Fig. 5.34 An antero-posterior gradient of morphogen in the egg of the leaf-hopper *Euscelis* appears to direct development along the antero-posterior axis. Top panels: *Euscelis* eggs carry a ball of symbiotic bacteria at the posterior pole, and some posterior cytoplasm can be moved toward the middle of the embryo by pushing the ball forward. The egg can then be ligatured so that postulated diffusible factors emanating from the transplanted cytoplasm and the posterior pole cannot cross the ligature. This manipulation results in alteration of the pattern of segments in line with the presence of a morphogen gradient in the normal egg, with a high point in the posterior cytoplasm. Bottom panel: isolated germ-band stage embryo of normal *Euscelis*. Capital letters indicate body regions identified in the experiments illustrated in the top panels.

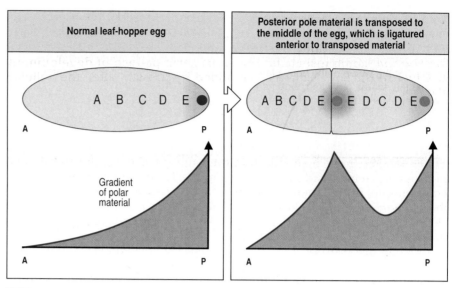

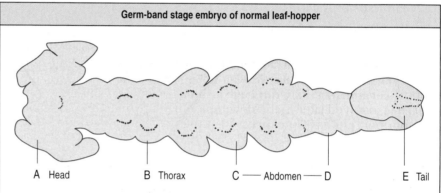

it (Fig. 5.34), a typical result is that a complete set of structures forms anterior to the ligature, while a mirror-image set of incomplete structure forms posteriorly. Both these results can be accounted for in terms of a diffusible morphogen gradient, with its high point and source at the posterior end.

Looking at some other insects, differences in early development are much more dramatic. In certain parasitic wasps, the egg is small and undergoes cleavage to form a ball of cells, which then falls apart. Each of the resulting small clusters of cells—there can be as many as 400—can develop into a separate embryo. This wasp's development apparently does not depend on maternal information to specify the body axes, and in this respect resembles that of the early mammalian embryo.

Summary.

The segment polarity genes are involved in patterning the parasegments. One of the first genes to be activated is *engrailed*, which is expressed at the anterior margin of each parasegment, delineating a group of cells which becomes the posterior compartment of a segment. *engrailed* is also a selector gene in that it provides a group of cells with a long-term regional identity. *engrailed* expression shows lineage restriction: cells expressing *engrailed* define the posterior compartment of a segment, and cells never cross from the posterior into the anterior compartment. *engrailed* is turned on by the pair-rule genes and its expression is maintained by the segment polarity genes, *wingless* and *hedgehog*, which stabilize the compartment boundary. Studies in other insects suggest that a discrete gradient of positional information is set up in each segment, delimited by the boundaries which pattern the segment. In contrast to *Drosophila*, some insects have a short-germ pattern of development in which posterior segments are added by growth, after the cellular blastoderm stage.

Summary: segment polarity gene expression defines segment compartments

pair-rule gene expression

⇩

segment polarity and selector gene *engrailed* activated in anterior of each parasegment, defining anterior compartment of parasegment and posterior compartment of segment

⇩

engrailed-expressing cells also express segment polarity gene *hedgehog*

⇩

cells on other side of compartment boundary express segment polarity gene *wingless*

⇩

engrailed expression maintained and compartment boundary stabilized by wingless and hedgehog proteins

⇩

compartment boundary provides signaling center from which segment is patterned

---

Segmentation: selector and homeotic genes.

Each segment has a unique identity, most easily seen in the characteristic pattern of denticles on the ventral surface of the larva (see Fig. 5.27). Since the same segment polarity genes are turned on in each segment, what makes the segments different from each other? Specification of segment identity is carried out by a class of master regulatory genes, known as **homeotic selector genes**, which set the future developmental pathway of each segment. A selector gene controls the activity of other genes and is required throughout development to maintain this pattern of gene expression. The *Drosophila* selector genes that control segment identity are organized into two gene complexes (Fig. 5.35), which together are

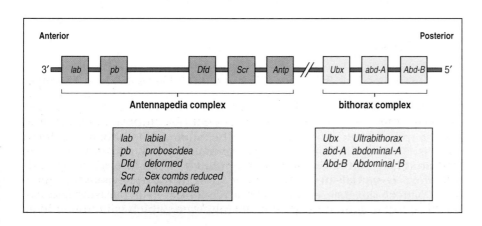

Fig. 5.35 The Antennapedia and bithorax homeotic selector gene complexes. The order of the genes from 3′ to 5′ in each complex reflects both the order of their spatial expression (anterior to posterior) and the timing of expression (3′ earliest).

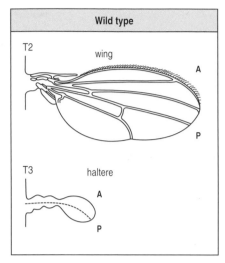

Wild type

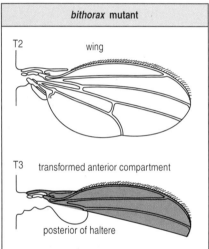

bithorax mutant

Fig. 5.36 Homeotic transformation of the wing and haltere by mutations in the bithorax complex. Top panel: in the normal adult, both wing and haltere are divided into an anterior (A) and a posterior (P) compartment. Middle panel: the *bithorax* mutation transforms the anterior compartment of the haltere into an anterior wing region. The mutation *postbithorax* acts similarly on the posterior compartment, converting it into posterior wing (not illustrated). Bottom panel: if both mutations are present together, the effect is additive and the haltere is transformed into a complete wing, producing a four-winged fly. Photograph courtesy of E. Lewis, from Bender, W., *et al.*: 1983. Illustration after Lawrence, P.:1992.

broadly homologous to a single Hox gene complex in vertebrates (see Box 4A, page 104). They are collectively known as HOM genes, since each codes for a transcription factor containing a homeobox. As we saw in Chapter 4, such genes also control patterning in vertebrates; however, they were first identified in *Drosophila*, and it is in *Drosophila* that their actions are best understood.

The two homeotic complexes in *Drosophila*, the bithorax and Antennapedia gene complexes, are named after the unusual and striking mutations that first revealed their existence. Flies with the *bithorax* mutation have part of the haltere (the balancing organ on the third thoracic segment) transformed into part of a wing (Fig. 5.36), whereas flies with the dominant *Antennapedia* mutation (that first allowed identification of this gene complex) have their antennae transformed into legs. Genes identified by such mutations are called **homeotic genes**, because when mutated they result in **homeosis**—the transformation of a whole segment or structure into another related one, as in the transformation of antenna to leg. These bizarre transformations arise out of the homeotic selector genes' key role as positional identity specifiers. They control the activity of other genes in the segments, thus determining, for example, that a particular imaginal disc will develop as wing or haltere. The bithorax complex controls the development of parasegments 5–14 while the Antennapedia complex controls the identity of the more anterior parasegments. The action of the bithorax complex of selector genes is the best understood and will therefore be discussed first.

5-19   Homeotic selector genes of the bithorax complex are responsible for diversification of the posterior segments.

The bithorax complex of *Drosophila* comprises three homeobox genes: *Ultrabithorax*, *abdominal-A*, and *Abdominal-B*. These genes are expressed in the parasegments in a combinatorial manner (Fig. 5.37, top panel). *Ultrabithorax* is expressed in all parasegments from 5 onward, *abdominal-A* is expressed more posteriorly from parasegment 7 onward, and *Abdominal-B* more posteriorly still from parasegment 10 onward. Because the genes are also active to varying extents in different parasegments, their combined activities define the character of each parasegment. *Abdominal-B* also suppresses *Ultrabithorax* such that *Ultrabithorax* expression is very low by parasegment 14, as *Abdominal-B* expression increases. The pattern of activity of the bithorax complex genes is determined by gap and pair-rule genes.

The role of the bithorax complex genes was first indicated by classical genetic experiments. In larvae lacking the whole of the bithorax complex (see Fig. 5.37, second panel), every parasegment from 5–13 develops in the same way and resembles parasegment 4. The bithorax complex is therefore essential for the diversification of these segments, whose basic pattern is represented by parasegment 4. This parasegment can be considered as being a type of 'default' state, which is modified in all

Fig. 5.37 The spatial pattern of expression of genes of the bithorax complex characterizes each para-segment. In the wild-type embryo (top panel) the expression of the genes *Ultrabithorax*, *abdominal-A*, and *Abdominal-B* are required to confer an identity on each parasegment. Mutations in the bithorax complex result in homeotic transformations in the character of the parasegments and the segments derived from them. When the bithorax complex is completely absent (second panel), parasegments 5–13 are converted into 9 parasegments 4 (corresponding to segment T2 in the larva), as shown by the denticle and bristle patterns on the cuticle. The three lower panels show the transformations caused by the absence of different combinations of genes. When the *Ultrabithorax* gene alone is absent (bottom panel), parasegments 5 and 6 are converted into 4. In each case the spatial extent of gene expression is detected by *in situ* hybridization (see Box 3B, page 65). Note that the specification of parasegment 14 is relatively unaffected by the bithorax complex.

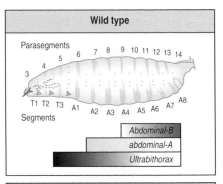

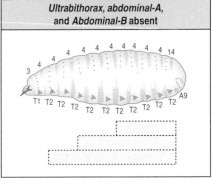

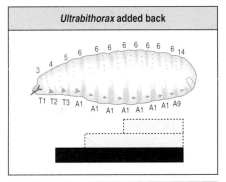

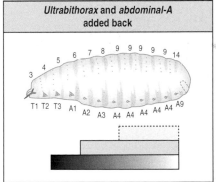

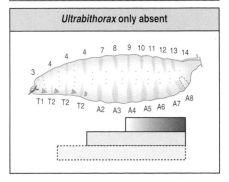

the parasegments posterior to it by the proteins encoded by the bithorax complex. It is because the genes of the bithorax complex are able to superimpose a new identity on the default state that they are called selector genes.

An indication of the role of each gene of the bithorax complex can be deduced by looking at embryos constructed so that the genes are put one at a time into embryos lacking the whole complex (see Fig. 5.37, bottom three panels). If only the *Ultrabithorax* gene is present, the resulting larva has one parasegment 4, one parasegment 5, and eight parasegments 6. Clearly, *Ultrabithorax* has some effect on all parasegments from 5 backward and can specify parasegments 5 and 6. If *abdominal-A* and *Ultrabithorax* are put into the embryo, then the larva has parasegments 4, 5, 6, 7, and 8, followed by five parasegments 9. So *abdominal-A* affects parasegments from 7 backward, and in combination *Ultrabithorax* and *abdominal-A* can specify the character of parasegments 7, 8, and 9. Similar principles apply to *Abdominal-B*, whose domain of influence extends from parasegment 10 backward, and which is expressed most strongly in parasegment 14. Differences between segments may reflect differences in the spatial and temporal pattern of HOM gene expression.

These results illustrate an important principle, namely that the character of the parasegments is specified by the genes of the bithorax complex acting in a combinatorial manner. Their combinatorial effect can also be seen by taking the genes away, one at a time, from the wild type. In the absence of *Ultrabithorax*, for example, parasegments 5 and 6 become converted to parasegment 4 (see Fig. 5.37, bottom panel). There is a further effect on the cuticle pattern in parasegments 7–14 in that structures characteristic of the thorax are now present in the abdomen, showing that *Ultrabithorax* exerts an effect in all these segments. Such abnormalities may result from expression of 'nonsense' combinations of the bithorax genes. For example, in such a mutant, abdominal-A protein is present in parasegments 7–9 without Ultrabithorax protein, and this is a combination never found normally.

While the gap and pair-rule proteins control the pattern of HOM gene expression, these proteins disappear after about 4 hours. The continued correct expression of the homeotic genes involves two groups of genes—the *polycomb* and *trithorax* groups. The proteins of the *polycomb* group maintain transcriptional repression of homeotic genes where they are initially off, while the *trithorax* group maintain expression in those cells where it has been turned on. These genes thus maintain expression of the HOM complex.

| 5-20 | **The Antennapedia complex controls specification of anterior regions.** |

The Antennapedia complex comprises five homeobox genes (see Fig. 5.35), which control the behavior of the parasegments anterior to parasegment 5 in a manner similar to the bithorax complex. Since there are no novel principles involved we just mention its role briefly. Several genes within the complex are critically involved in specifying particular parasegments. Mutations in the gene *deformed* affect the ectodermally derived structures of parasegments 0 and 1, those in *Sex combs reduced* affect parasegments 2 and 3, and those in the *Antennapedia* gene affect parasegments 4 and 5.

| 5-21 | **The order of HOM gene expression corresponds to the order of genes along the chromosome.** |

The bithorax and Antennapedia complexes possess some striking features of gene organization. In both, the order of the genes in the complex is the same as the spatial and temporal order in which they are expressed along the antero-posterior axis during development. *Ultrabithorax*, for example, is on the 3' side of *abdominal-A* on the chromosome, is more anterior in its pattern of expression, and is activated earlier. As we have already seen, the related Hox gene complexes of vertebrates (see Box 4A, page 104), whose ancestors diverged from those of arthropods hundreds of millions of years ago, show the same correspondence between gene order and order of expression. This highly conserved temporal and spatial co-linearity of gene order and expression must be related to the mechanisms that control the expression of these genes.

The complex yet subtle control to which the genes of the bithorax complex are subject is seen in experiments in which Ultrabithorax protein is forced to be produced in all segments. This is achieved by linking the Ultrabithorax protein-coding sequence to a heat-shock promoter—a promoter that is activated at 29°C—and introducing the novel DNA construct into the fly genome using a P element (see Box 5A, page 144). When this transgenic embryo is given a heat shock for a few minutes, the extra *Ultrabithorax* gene is transcribed and the protein is made at high levels in all cells. This has no effect on posterior parasegments (in which Ultrabithorax protein is normally present), with the exception of parasegment 5 which, for unknown reasons, is transformed into parasegment 6 (this may reflect a quantitative effect of the Ultrabithorax protein). However, all parasegments anterior to 5 are also transformed into parasegment 6. While this seems a reasonably simple and expected result, consider what happens to parasegment 13.

Transcription of *Ultrabithorax* is normally suppressed in parasegment 13 in wild-type embryos, yet when the protein itself is produced as a result of heat shock, it has no effect. By some unknown mechanism the Ultrabithorax protein is rendered inactive in this parasegment. This phenomenon is quite common in the specification of the parasegments and is known as phenotypic suppression or posterior prevalence (see Section 4-4): HOM gene products normally expressed in anterior regions are suppressed by more posterior products.

While the role of the bithorax and Antennapedia complexes in controlling segment identity is well established, we do not yet know very much about their interactions with the downstream target genes which specify the structures that give the segments their unique identities. What, for example, is the pathway that leads to a thoracic rather than an abdominal segment structure? We shall return to this topic in Chapter 10, when we look at appendage development in *Drosophila*. The role of the Hox genes in the evolution of different body patterns will be considered in Chapter 15. We finish this chapter by considering one downstream pathway which has been partly worked out—the influence of HOM gene expression in the mesoderm on the segment-specific development of the endoderm.

5-22	**HOM gene expression in visceral mesoderm controls the structure of the adjacent gut.**

Discussion of the actions of the gene *engrailed* and of the bithorax and Antennapedia complexes has so far referred only to their effects on ecto-dermal structures, especially the epidermis or cuticle. These genes are also expressed in internal tissues, such as the somatic and visceral mesoderm of the embryo. The somatic mesoderm gives rise to the main body muscles, while the visceral mesoderm generates the smooth muscle surrounding the gut. In the somatic mesoderm, the pattern of express-ion of the bithorax complex is simpler than in the ectoderm, but in general corresponds with it. The visceral mesoderm, however, deserves special attention as HOM gene expression there appears to induce pattern in the gut endoderm.

The developing midgut has three constrictions, the second of which occurs in parasegment 7. Most of the HOM selector genes are not expressed in the endoderm, and its segment-specific character is conferred by induction as a result of HOM gene expression in the visceral mesoderm surround-ing it. The detailed pattern of expression of the bithorax complex in the visceral mesoderm is somewhat different from that in both ectoderm and somatic mesoderm. The *Ultrabithorax* gene, for example, is expre-ssed only in the parasegment of the visceral mesoderm adjacent to the second of the three gut constrictions (Fig. 5.38, top panel). If *Ultrabithorax* expression is absent, this constriction does not develop and the gut is abnormal.

Ultrabithorax itself is not expressed in the endoderm. Rather, it appears to exert its effect on the gut by its action on two other genes—*decapenta-plegic* (see Section 5-9), and *labial*, which is part of the Antennapedia complex. In normal embryos, both these genes are expressed in the region of the gut constriction, *decapentaplegic* in the visceral meso-derm and *labial* in the endoderm, but in the absence of *Ultrabithorax*, both are hardly expressed at all. Thus, the Ultrabithorax protein is necessary for the activation of the *decapentaplegic* gene in the visceral mesoderm. The decapentaplegic protein then diffuses from the visceral mesoderm to the adjacent endoderm where it stimulates a signaling pathway that results in the activation of the *labial* gene, which is involved in gut morphogenesis (see Fig. 5.38, bottom panel). Thus, the visceral mesoderm may pattern the endoderm by transfer of positional information from one germ layer to another by means of extracellular signals, a process reminiscent of the induction of the nervous system in vertebrates (see Chapter 4).

	Summary.

Segment identity is conferred by action of selector or homeotic genes, which direct the development of different parasegments so that each acquires a unique identity. Two clusters of selector genes are involved in specification of segment identity in *Drosophila*: the Antennapedia complex, which controls parasegment identity in the head and first tho-racic segment, and the bithorax complex, which acts on the remaining

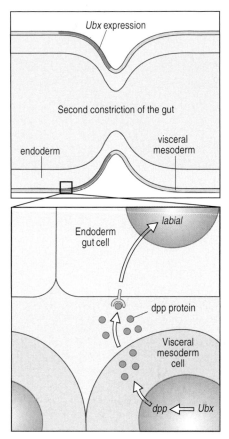

Fig. 5.38 Gene expression in the visceral mesoderm patterns the underlying gut endoderm. Top panel: *Ultrabithorax* (*Ubx*) is expressed in the visceral mesoderm near the second constriction of the gut. Bottom panel: it activates the gene *decapentaplegic* (*dpp*) in visceral mesoderm cells. This results in secretion of decapentaplegic protein from the visceral mesoderm, which induces expression of the gene *labial* in the adjacent cells of the gut endoderm. Patterning laid down in the visceral mesoderm is thus transferred to the gut endoderm.

parasegments. Segment identity seems to be determined by the combination of genes active in a particular region. Selector genes need to be switched on continuously throughout development to maintain the required phenotype. The spatial pattern of expression of selector genes along the embryo body is largely determined by the preceding gap gene activity. Mutations in genes of the Antennapedia and bithorax complexes can bring about homeotic transformations, in which one segment or structure is changed into a related one, for example an antenna into a leg. The Antennapedia and bithorax complexes are remarkable in that gene order on the chromosome corresponds to the spatial and temporal order of gene expression in the body. Selector genes of the two complexes are expressed in the ectoderm and in the underlying somatic and visceral mesoderm. They are not expressed in the endoderm from which the gut develops, but induction of the endoderm by visceral mesoderm seems to transfer to the endoderm some of the segment-specific pattern conferred on the mesoderm by selector gene activity.

Summary to Chapter 5.

How genes control early development is better understood in the fruit fly, *Drosophila*, than in any other organism. The earliest stages of development in *Drosophila* take place when the embryo is a multinucleate syncytium. Maternal gene products deposited in the egg in a particular spatial pattern during oogenesis determine the main body axes and lay down a framework of positional information; which activates a cascade of zygotic gene activity that further patterns the body. The first zygotic genes to be activated along the antero-posterior axis are the gap genes, all encoding transcription factors, whose pattern of expression divides the embryo into a number of regions. The transition to a segmented body organization then starts with the activity of the pair-rule genes, whose sites of expression are specified by the gap gene proteins and which divide the axis into 14 parasegments. Along the dorso-ventral axis, zygotic gene expression also defines several regions, including the future mesoderm and future neural tissue. At the time of pair-rule gene expression, the embryo becomes cellularized and is no longer a syncytium. Segment polarity genes pattern the parasegment. The identity of the segments is determined by two homeotic gene complexes which contain selector genes. The spatial pattern of expression of these genes is largely determined by gap gene activity. The order of genes in the complexes corresponds with their spatial order of expression. Patterning of the gut endoderm involves transfer of patterning from the visceral endoderm.

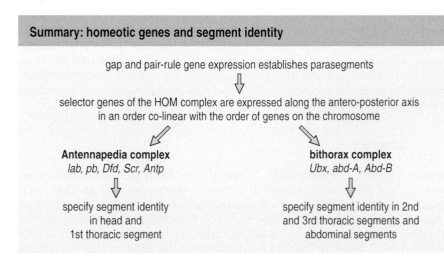

Summary: homeotic genes and segment identity

gap and pair-rule gene expression establishes parasegments

selector genes of the HOM complex are expressed along the antero-posterior axis in an order co-linear with the order of genes on the chromosome

| **Antennapedia complex** | **bithorax complex** |
| *lab, pb, Dfd, Scr, Antp* | *Ubx, abd-A, Abd-B* |

specify segment identity in head and 1st thoracic segment

specify segment identity in 2nd and 3rd thoracic segments and abdominal segments

The main genes involved in specifying pattern in the early *Drosophila* embryo

	Gene	Maternal/zygotic	Nature of protein	Transcription factor (T), receptor (R), or signal protein (S)	Function (where known)
Antero-posterior System	*bicoid*	M	Homeodomain	T	Morphogen, activates zygotic *hunchback* and other gap genes
	hunchback	M	Zinc fingers	T	Morphogen, activates gap genes
	nanos	M			Inactivates maternal *hunchback* mRNA/protein; involved in making gradient of maternal hunchback protein
	gurken	M	Secreted protein of TGF-α family	S	Specifies oocyte axis
	exuperantia	M			Localization of maternal RNAs (e.g. *bicoid* mRNA)
	oskar	M			Specification of germ plasm
Terminal system	*torso*	M	Receptor tyrosine kinase	R	Activation specifies termini
	torso-like	M		S	Ligand for torso protein
Gap genes	*hunchback*	Z	Zinc fingers	T	
	Krüppel	Z	Zinc fingers	T	
	knirps	Z	Zinc fingers	T	Localize pair-rule gene expression
	giant	Z	Leucine zipper	T	
	tailless	Z	Zinc fingers	T	
Pair-rule genes	*even-skipped*	Z	Homeodomain	T	Delimits odd numbered parasegments
	fushi tarazu	Z	Homeodomain	T	Delimits even numbered parasegments
	hairy	Z	Helix-loop-helix	T	
Segment polarity genes	*engrailed*	Z	Homeodomain	T	Defines anterior region of parasegment and posterior region of segment
	hedgehog	Z	Membrane or secreted	S	
	wingless	Z	Secreted	S	
	gooseberry	Z	Homeodomain	T	
	patched	Z	Membrane	R	
	smoothened	Z	Membrane	R	
Selector genes bithorax complex	*Ultrabithorax*	Z	Homeodomain	T	
	abdominal-A	Z	Homeodomain	T	Combinatorial activity confers identity on parasegments 5–13
	Abdominal-B	Z	Homeodomain	T	
Antennapedia complex	*Deformed*	Z	Homeodomain	T	
	Sex combs reduced	Z	Homeodomain	T	
	Antennapedia	Z	Homeodomain	T	Combinatorial activity confers identity on parasegments anterior to 5
	labial	Z	Homeodomain	T	
Maintenance genes	*polycomb*	Z		T	Maintain state of homeotic gene expression
	bithorax	Z		T	

Dorso-ventral System					
Maternal genes	*Toll*	M	Membrane	R	Activation results in dorsal protein entering nucleus
	spätzle	M		S	Ligand for Toll protein
	dorsal	M		T	Morphogen, sets dorso-ventral polarity
	cactus	M			Binds dorsal protein and prevents it entering nucleus
	pelle	M/Z			
	tube	M			
	gurken	M	Secreted protein of TGF-α family	S	Specifies oocyte axis
Zygotic genes	*twist*	Z	Helix-loop-helix	T	Define mesoderm
	snail	Z	Zinc finger	T	
	rhomboid	Z	Membrane protein		
	single-minded	Z			
	zerknüllt	Z	Homeodomain	T	
	decapentaplegic	Z	Secreted protein of TGF-β family	S	Confer regional identity on dorso-ventral axis
	tolloid	Z	BMP-2 family	S	
	short gastrulation	Z			

General references.

Lawrence, P.A.: *The Making of a Fly*. Oxford: Blackwell Scientific Publications, 1992.

Section references.

5-1 Three classes of maternal genes specify the antero-posterior axis.

St Johnston, D., Nüsslein-Volhard, C.: **The origin of pattern and polarity in the** Drosophila embryo. *Cell* 1992, **68**:201–219.

5-2 The *bicoid* gene provides an antero-posterior morphogen gradient.

Driever, W., Nüsslein-Volhard, C.: **The bicoid protein determines position in the** *Drosophila* **embryo in a concentration dependent manner**. *Cell* 1988, **54**:95–104.

5-3 The posterior pattern is controlled by the gradients of nanos and caudal proteins.

Irish, V., Lehmann, R., Akam, M.: **The** *Drosophila* **posterior-group gene** *nanos* **functions by repressing** *hunchback* **activity**. *Nature* 1989, **338**:646–648.
Murafta, Y., Wharton, R.P.: **Binding of pumilio to maternal** *hunchback* **mRNA is required for posterior patterning in** *Drosophila* **embryos**. *Cell* 1995, **80**:747–756.
Rivera-Pomar, R., Lu, X., Perrimon, N., Taubert, H., Jackle, H.: **Activation of posterior gap gene expression in the** *Drosophila* **blastoderm**. *Nature* 1995, **376**:253–256.
Struhl, G.: **Differing strategies for organizing anterior and posterior body pattern in** *Drosophila* **embryos**. *Nature* 1989, **338**:741–744.

5-4 The anterior and posterior extremities of the embryo are specified by cell-surface receptor activation.

Casanova, J., Struhl, G.: **Localized surface activity of** *torso*, **a receptor tyrosine kinase, specifies terminal body pattern in** *Drosophila*. *Genes Dev.* 1989, **3**:2025–2038.

5-5 The dorso-ventral polarity of the egg is specified by localization of maternal proteins in the vitelline envelope.

Morisato, D., Anderson, K.V.: **The** *spätzle* **gene encodes a component of the extracellular signaling pathway establishing the dorsal-ventral pattern of the** *Drosophila* **embryo**. *Cell* 1994, **76**:677–688.

5-6 Positional information along the dorso-ventral axis is provided by the dorsal protein.

Belvin, M.P., Anderson, K.V.: **A conserved signaling pathway: the** *Drosophila* **Toll-dorsal pathway**. *Ann. Rev. Cell Dev. Biol.* 1996, **12**:393–416.
Roth, S., Stein, D., Nüsslein-Volhard, C.: **A gradient of nuclear localization of the dorsal protein determines dorso-ventral pattern in the** *Drosophila* **embryo**. *Cell* 1989, **59**:1189–1202.
Steward, R., Govind, R.: **Dorsal-ventral polarity in the** *Drosophila* **embryo**. *Curr. Opin. Genet. Dev.* 1993, **3**:556–561.

5-7 Antero-posterior and dorso-ventral axes of the oocyte are specified by interactions with follicle cells.

Gavis, E.R.: **Pattern formation** *Gurken* **meets** *torpedo* **for the first time**. *Curr. Biol.* 1995, **5**:1252–1254.
Gonzalez-Reyes, A., Elliott, H., St Johnston, D.: **Polarization of both major body axes in** *Drosophila* **by gurken-torpedo signaling**. *Nature* 1995, **375**: 654–658.
Roth, S., Neuman-Silberberg, F.S., Barcelo, G., Schupbach, T.: **Cornichon and the EGF receptor signaling process are necessary for both anterior-posterior and dorsal-ventral pattern formation in** *Drosophila*. *Cell* 1995, **81**:967–978.
St Johnston, D.: **The intracellular localization of messenger RNAs**. *Cell* 1995, **81**:161–170.

5-8 The expression of zygotic genes along the dorso-ventral axis is controlled by dorsal protein.

Jiang, G., Levine, M.: **Binding affinities and cooperative interactions with HLH activators delimit threshold responses to the dorsal gradient morphogen**. *Cell* 1993, **72**:741–752.

5-9 The decapentaplegic protein acts as a morphogen to pattern the dorsal region.

Rusch, J., Levine, M.: **Threshold responses to the dorsal regulatory gradient and the subdivision of primary tissue territories in the** *Drosophila* **embryo**. *Curr. Opin. Genet. Devel.* 1996, **6**:416–423.
Wharton, K.A., Ray, R.P., Gelbart, W.M.: **An activity gradient of decapentaplegic is necessary for the specification of dorsal pattern elements in the** *Drosophila* **embryo**. *Development* 1993, **117**:807–822.

5-10 The antero-posterior axis is divided up into broad regions by gap gene expression.

Hülskamp, M., Tautz, D.: **Gap genes and gradients—the logic behind the gaps**. *BioEssays* 1991, **13**:261–268.

5-11 bicoid protein provides a positional signal for the anterior expression of *hunchback*.

Simpson-Brose, M., Treisman, J., Desplan, C.: **Synergy between the hunchback and bicoid morphogens is required for anterior patterning in** *Drosophila*. *Cell* 1994, **78**:855–865.
Struhl, G., Struhl, K., Macdonald, P.M.: **The gradient morphogen bicoid is a concentration-dependent transcriptional activator**. *Cell* 1989, **57**:1259–1273.

5-12 The gradient in hunchback protein activates and represses other gap genes.

Rivera-Pomar, R., Jäckle, H.: **From gradients to stripes in** *Drosophila* **embryogenesis: filling in the gaps**. *Trends Genet.* 1996, **12**:478–483.
Struhl, G., Johnston, P., Lawrence, P.A.: **Control of** *Drosophila* **body pattern by the hunchback morphogen gradient**. *Cell* 1992, **69**:237–249.

5-14 Gap gene activity positions stripes of pair-rule gene expression.

Small, S., Levine, M.: **The initiation of pair-rule stripes in the** *Drosophila* **blastoderm**. *Curr. Opin. Genet. Dev.* 1991, **1**:255–260.

5-15 Expression of the *engrailed* gene delimits a cell lineage boundary and defines a compartment.

Gray, S., Cai, H., Barolo, S., and Levine, M.: **Transcriptional repression in the** ***Drosophila*** **embryo**. *Phil. Trans. R. Soc. Lond.* 1995, **349**:257–262

Harrison, D.A., and Perrimon, N.: **Simple and efficient generation of marked clones in** ***Drosophila***. *Curr. Biol.* 1993, **3**:424–433.

Lawrence, P.A.: **The present status of the parasegment**. *Development (Suppl.)* 1988, **104**:61–69.

Vincent, J. P, O'Farrell, P.H.: **The state of engrailed expression is not clonally transmitted during early** ***Drosophila*** **development**. *Cell* 1992, **68**:923–931.

5-16 Segment polarity genes pattern the segments and stabilize parasegment and segment boundaries.

Dinardo, S., Heemskerk, J., Dougan, S., O'Farrell, P.H.: **The making of a maggot: patterning the** ***Drosophila*** **embryonic epidermis**. *Curr. Opin. Genet. Dev.* 1994, **4**:529–534.

Bejsovec, A., Martinez-Arias, A.: **Roles of wingless in patterning the larval epidermis of** ***Drosophila***. *Development* 1991, **113**:471–485

Kornberg, T.B., Tabata, T.: **Segmentation of the** ***Drosophila*** **embryo**. *Curr. Opin. Genet. Dev.* 1993, **3**:585–594.

Sampedro, J., Lawrence, P.A.: ***Drosophila*** **development after the first three hours**. *Development* 1993. **119**:971–976.

Heemskerk, J., Diardo, S.: ***Drosophila*** **hedgehog acts as a morphogen in cellular patterning**. *Cell* 1994, **76**:449–460.

5-18 Some insects use different mechanisms for patterning the body plan.

French, V.: **Segmentation (and** *eve***) in very odd insect embryos**. *BioEssays* 1996, **18**:435–438.

Nagy, L.M.: **A glance posterior**. *Curr. Biol.* 1994, **4**:811–814.

Sander, K.: **Pattern formation in the insect embryo**. In *Cell Patterning, Ciba Found. Symp. 29*. London: Ciba Foundation, 1975: 241–263.

Akam, M., Dawes, R.: **More than one way to slice an egg**. *Curr. Biol.* 1992, **8**:395–398.

Tautz, D., Sommer, R.J.: **Evolution of segmentation genes in insects**. *Trends Genet.* 1995, **11**:23–27.

5-19 Homeotic selector genes of the bithorax complex are responsible for diversification of the posterior segments.

Castelli-Gair, J., Akam, M.: **How the Hox gene** *Ultrabithorax* **specifies two different segments: the significance of spatial and temporal regulation within metameres**. *Development* 1995, **121**:2973–2982.

Duncan, I.: **How do single homeotic genes control multiple segment identities?** *BioEssays* 1996, **18**:91–94.

Lawrence, P.A., Morata, G.: **Homeobox genes: their function in** ***Drosophila*** **segmentation and pattern formation**. *Cell* 1994, **78**:181–189.

Simon, J.: **Locking in stable states of gene expression: transcriptional control during** ***Drosophila*** **development**. *Curr. Opin. Cell Biol.* 1995, **7**:376–385.

5-21 The order of HOM gene expression corresponds to the order of genes along the chromosome.

Morata, G.: **Homeotic genes of** ***Drosophila***. *Curr. Opin. Genet. Dev.* 1993, **3**:606–614.

5-22 HOM gene expression in visceral mesoderm controls the structure of the adjacent gut.

Immergluck, K., Lawrence, P.A., Bienz, M.: **Induction across germ layers in** ***Drosophila*** **mediated by a genetic cascade**. *Cell* 1990, **62**:261–268

Development of Invertebrates, Ascidians, and Slime Molds

6

Nematodes.

Molluscs.

Annelids.

Echinoderms.

Ascidians.

Cellular slime molds.

"Sometimes we did things in a different way, were given our identity one by one, and spoke only to our neighbors."

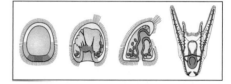

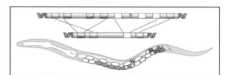

This chapter considers aspects of body plan development in a variety of invertebrate organisms—nematodes, molluscs, annelids, echinoderms, and ascidians—finishing with a short discussion on the cellular slime molds, which represent a very simple developmental system. These organisms will highlight the similarities and differences in developmental mechanisms compared with those organisms we have already considered. The evolutionary relationships between the organisms discussed in this chapter are shown in Fig. 6.1. All, with the exception of the cellular slime mold, conform to the general plan of animal development: cleavage leads to a blastula, which undergoes gastrulation with the emergence of a body plan.

There is an old, and now less fashionable, distinction sometimes made between so-called regulative and mosaic development—the former involving mainly cell–cell interactions, while the latter is based on localized cytoplasmic factors and their distribution through asymmetric cell divisions (see Section 1-10). The organisms discussed in earlier chapters

Fig. 6.1 Phylogenetic tree showing relationships between the organisms considered in this book. Those organisms discussed in this chapter are highlighted in blue.

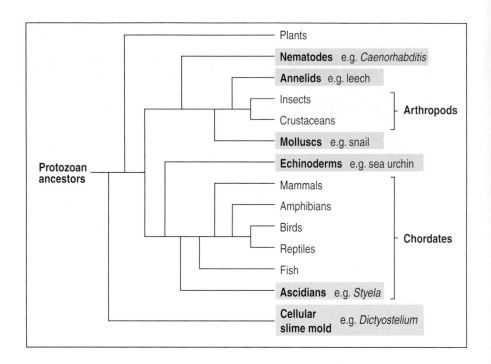

are examples of mainly regulative development, whereas some of those discussed in this chapter have a mosaic-like development, although there is an element of both processes in most organisms.

An important feature of the development of some invertebrates, such as nematodes, molluscs, annelids, and the chordate ascidians, is that cell fate is often specified on a cell-by-cell basis rather than in groups of cells as in flies and vertebrates, and in general does not rely on positional information established by gradients of morphogens. The early embryos of many invertebrates contain far fewer cells than those of vertebrates or flies, with each cell acquiring a unique identity at an early stage of development. For example, in the nematode there are only 26 cells when gastrulation starts, compared to thousands in vertebrates. Specification on a cell-by-cell basis often seems to make use of a developmental mechanism that is less commonly found in insects and vertebrates, namely **asymmetric cell division** and the unequal distribution of cytoplasmic factors as a means of determining cell fate (see Section 1-15). Daughter cells resulting from asymmetric cell division often adopt different fates, due not to extracellular signals but autonomously, as a result of the unequal distribution of some factor between them. However, asymmetric cell division in the early stages of development does not mean that cell–cell interactions are absent or unimportant in these organisms.

We begin our discussion with the nematode worm *Caenorhabditis elegans*, which has been studied intensively and in which many key developmental genes have been identified. In this animal, specification is largely on a cell-by-cell basis. We then look at the early embryonic development of molluscs and annelids, each of which displays a different way of laying down the body plan and in which cell fate specification is again mainly cell-by-cell. We then consider echinoderms, which develop much more like vertebrates, in that their embryos rely heavily on intercellular interactions, are highly regulative, and patterning involves groups of cells. Then we consider ascidians, with particular emphasis on the role of cytoplasmic localization in their early development. Finally, we look at patterning in the cellular slime mold, which represents a primitive and very different developmental system.

Nematodes.

It is a triumph of direct observation that, with the aid of Nomarski interference microscopy, the complete lineage of every cell in the nematode *Caenorhabditis elegans* has been worked out (see Fig. 2.37). The pattern of cell division is invariant—it is the same in every embryo. The larva, when is hatches, is made up of 558 cells, and after four further molts this number has increased to 959, excluding the germ cells which vary in number. This is not the total number of cells derived from the egg, as 113 cells die during development. As the fate of every cell at each stage is known, a fate map can be accurately drawn at any stage and thus has a precision not found in any vertebrate. However, as with any fate map, even where there is an invariant cell lineage this precision in no way implies that the lineage determines the fate or that the fate of the cells cannot be altered. As we shall see, cell interactions play a major role in determining cell fate in the nematode.

6-1 The developmental axes are determined by asymmetric cell division and cell–cell interactions.

The first cleavage of the nematode egg is unequal, dividing the egg into a large anterior AB cell and a smaller posterior P_1 cell. This asymmetry defines the antero-posterior axis. The P_1 cell behaves rather like a **stem cell**; at each further division it produces one P-type cell, and one daughter cell that will embark on another developmental pathway. For the first three divisions, the P cell daughters give rise to body cells but after the fourth cleavage they only give rise to germ cells (Fig. 6.2). Division of

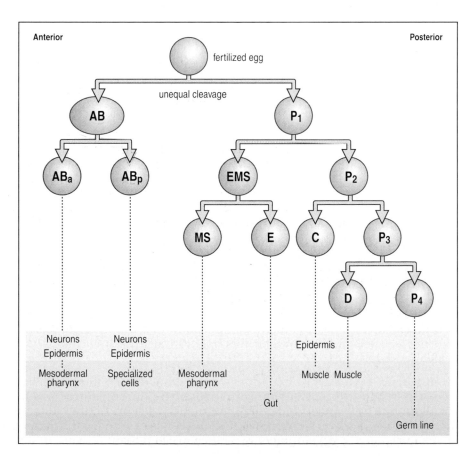

Fig. 6.2 Early cell lineage in the nematode *Caenorhabditis elegans*. The cleavage is invariant. The first cleavage divides the egg into a large AB cell and a smaller P_1 cell. The descendants of these cells have constant lineages and fates. For example, germ cells all come from the P_4 cell, a descendant of P_1, and the gut from the E cell. After Sulston, J.E., *et al.*: 1983.

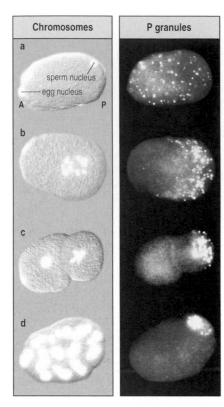

Fig. 6.3 Localization of P granules after fertilization. (a) Fertilized egg with egg nucleus at anterior end and sperm nucleus at posterior end (left panel). Right panel: the P granules are distributed throughout the egg. (b) Nuclear fusion and P granules are confined to the posterior end. (c) Two-cell stage. (d) 26-cell stage. All the P granules are in the P$_4$ cell. Photographs courtesy of W. Wood, from Strome, S., et al.: 1983.

the AB cell gives rise to anterior and posterior AB daughter cells. The anterior cell AB$_a$ gives rise to typically ectodermal tissues, such as the epidermis (hypodermis) and nervous system, but also to a portion of the mesoderm of the pharynx. The posterior daughter AB$_p$ also makes neurons and epidermis as well as some specialized cells. At second cleavage, P$_1$ divides asymmetrically, to give P$_2$ and EMS, which will subsequently divide into MS and E. MS gives rise to mesodermal pharynx, while E is the sole precursor of the gut. The C daughter cell, which is derived from the P$_2$ cell at the third cleavage, forms epidermis and muscle; D from the fourth cleavage of the P cell produces only muscle. All the cells undergo further invariant divisions, and by about 100 minutes after fertilization, gastrulation begins.

Before fertilization, there is no evidence of any asymmetry in the nematode egg. The first cleavage, which is both unequal and asymmetrical, is related to the point of sperm entry and defines the future antero-posterior axis, with the large AB cell marking the anterior end and the smaller P$_1$ cell, the posterior. Preceding this cleavage, a cap of actin microfilaments forms at the anterior end, and a set of granules, so-called P granules, become localized at the posterior end (Fig. 6.3). The importance of the microfilaments can be shown by destroying their integrity with a short pulse of cytochalasin D during the first cleavage. This results in aberrant early divisions and development.

Clearly, some sort of structural inhomogeneity in the egg after fertilization controls the early manifestation of antero-posterior polarity. It cannot, however, be the P granules, whose distribution is a reflection rather than the cause of the polarity. Experiments that remove egg cytoplasm suggest the presence of some component in the posterior cytoplasm that determines the polarity. If more than 25% of the egg cytoplasm is extruded from the posterior end after fertilization, the unequal pattern of early cleavage is lost, but up to 40% of cytoplasm can be extruded from the anterior end without altering the early pattern.

A very early marker of the antero-posterior axis is the protein Par-1, encoded by the maternal *par-1* gene, which is localized in the future posterior region of the egg after fertilization. Maternal mutations in *par-1* disrupt the asymmetry of the first cleavage. In such mutants, P granules fail to localize, and further cleavage and development is abnormal. Par-1 probably acts by interacting with the egg cell's cytoskeleton. After first cleavage the Par-1 protein is localized in the P$_1$ cell.

In spite of the highly determinate cell lineage in the nematode, cell–cell interactions are involved in specifying the dorso-ventral axis. At the time of the second cleavage, if the future anterior AB$_a$ cell is pushed and rotated with a glass needle, not only is the antero-posterior order of the daughter AB cells reversed, but the cleavage of P$_1$ is also affected, so that the position of the P$_1$ daughter cell, EMS, relative to the AB cells is inverted (Fig. 6.4). The manipulated embryo develops completely normally, but the dorso-ventral axis is reversed, showing firstly that the axis is not yet determined at this stage, for by inverting the position of the EMS cell the dorso-ventral axis is also inverted, and secondly that cell interactions are involved in specifying the cells of the early embryo. By inverting the position of the EMS, the dorso-ventral behavior of P$_1$ is also inverted. This has the further implication that the left-right axis is also not yet determined; indeed this axis can also be inverted as we shall now see.

Adult worms have a well-defined left-right asymmetry in their internal structures, and this axis is determined after the dorso-ventral axis is established. The pattern of development on the left and right sides of the nematode embryo shows striking differences; indeed, the differences during early development are more marked than in many other

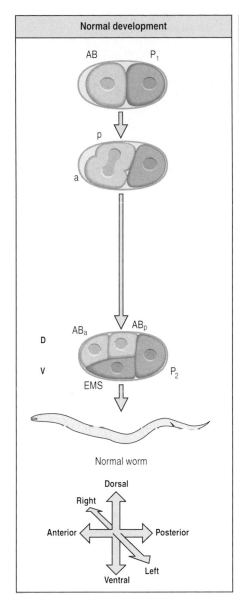

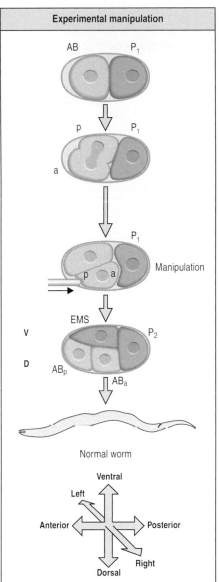

Fig. 6.4 Reversal of dorso-ventral polarity at the four-cell stage of the nematode. In normal embryos, the AB cell rotates at cleavage so that AB_a is anterior. If at second cleavage the AB cell is mechanically rotated in the opposite direction, the AB_p cell is now anterior. This manipulation also displaces the P_1 cell so that when it divides, the position of the EMS daughter cell is reversed with respect to the AB cells. However, development is still normal, but the dorso-ventral axis is inverted and so too is left-right asymmetry. After Sulston, J.E., *et al.*: 1983.

animals. Not only do cell lineages on left and right differ, but some cells even cross from one side to the other. The concept of left and right, as discussed previously (see Section 3-7), only has meaning once the antero-posterior and dorso-ventral axes are defined. Since the dorso-ventral axis in the nematode can still be reversed at the two-cell to four-cell stage, left and right are also not yet specified.

Specification of left and right occurs at the third cleavage, and handedness can be reversed by experimental manipulation at this stage. Following the division of the AB cell into anterior and posterior blastomeres AB_a and AB_p, each of these divides at the third cleavage to produce laterally disposed right and left daughter cells. The plane of cleavage is, however, slightly asymmetric so that the left daughter cell lies just a little anterior to its right-hand sister. If the cells are manipulated with a glass rod during this cleavage, their positioning can be reversed so that the right-hand cell lies slightly anterior (Fig. 6.5). This small manipulation is sufficient to reverse the handedness of the animal.

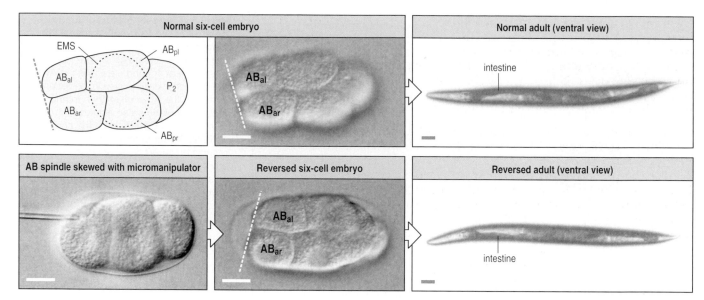

Fig. 6.5 Reversal of handedness in *C. elegans*. At the six-cell stage in a normal embryo, the ABa cell on the left side (Abal) is slightly anterior to that on the right (top left panels, scale bar = 10 μm). Manipulation that makes the ABa cell on the right side more anterior results in an animal with reversed handedness (bottom left panels, scale bars = 10 μm). The right panels show the resulting normal and reversed adults, scale bars = 50 μm. Photographs courtesy of W. Wood, from Wood, W.B.: 1991.

6-2   Cell–cell interactions specify cell fate in the early nematode embryo.

Although cell lineage in the nematode is invariant, experimental evidence, such as that described above, shows that cell–cell interactions are of crucial importance in specifying cell fate in the early embryo. Otherwise, the reversal of ABa and ABp (which normally have different cell fates) by micromanipulation, just as they are being formed (see Fig. 6.4), would not give rise to a normal worm. ABa and ABp must initially be equivalent and their fate must be specified by interactions with adjacent cells. Evidence for such an interaction comes from removing P1 at the first cleavage, since pharyngeal cells, a normal product of ABa, are then not made.

What then are the interactions that specify the non-equivalence of the two AB descendants? The P2 blastomere is responsible for specifying ABp since if ABp is prevented from contacting P2 it develops as an ABa cell. The induction of ABp by P2 involves proteins encoded by the maternal genes *glp-1* and *apx-1*, which are respectively similar to the *Notch* and *Delta* genes that are involved in many interactions between adjacent cells (see Box 3A, page 64).

The protein Glp-1 is a transmembrane receptor and is one of the earliest proteins to be spatially localized during embryogenesis. Although *glp-1* mRNA is uniformly present throughout the embryo, its translation is repressed in the posterior P cell; at the two-cell stage the Glp-1 protein is thus only expressed in the anterior AB cell. This strategy for protein localization is highly reminiscent of the localization of maternal hunchback protein in *Drosophila* (see Section 5-3).

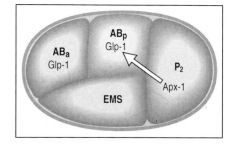

Fig. 6.6 An early inductive event specifies ABp. ABp becomes different from ABa as a result of a signal emanating from the adjacent P2 cell, thought to be the protein Apx-1. This signal is received by the receptor Glp-1 on the ABp cell. After Mello, C.C., *et al.*: 1994.

After second cleavage the ABa and ABp cells both contain the Glp-1 receptor protein. The two cells are then directed to become different by a local inductive signal, sent to the ABp cell from the P2 cell at the four-cell stage (Fig. 6.6). This signal is thought to be the protein Apx-1 (produced by the P2 cell) acting as an activating ligand for the Glp-1 receptor protein. As a result of this induction, the descendants of ABa and ABp respond differently to later signals from the adjacent MS cell (a daughter of EMS).

Other maternal-effect genes involved in these early cell–cell interactions have been identified. Specification of the EMS cell involves the expression of the *skin excess* (*skn*) genes. The two cells which EMS gives rise to, E and MS, produce intestinal and muscle cell types, respectively. Mutations in *skn-1* result in E making mainly muscle instead of intestinal cells. mRNA is uniform at the two-cell stage but the protein is at much higher levels in the nucleus of P_1 than in AB.

Gut development is also dependent on induction. The nematode gut develops from a single blastomere, the E cell, which is formed at the eight-cell stage as one of the daughters of EMS, which divides at third cleavage to give an anterior MS cell and a posterior E cell (see Fig. 6.2). If isolated from the influence of its neighbors at the four-cell stage, an EMS cell can develop gut structures, whereas an isolated P_2 cell never does. However, this property of EMS depends crucially on just when it is isolated. If isolated at the beginning of the four-cell stage, EMS cannot develop gut structures, suggesting that its ability to form gut requires an interaction with other cells at the beginning of this stage. The cell necessary to induce gut formation from EMS is the P_2 cell, since removal of P_2 at the early four-cell stage results in no gut being formed. Recombining isolated EMS and P_2 cells restores gut development, while recombining EMS with other cells from the four-cell stage has no effect.

We thus begin to see how cell–cell interactions and localized cytoplasmic determinants together specify cell fate in the early nematode embryo. The results of killing individual cells by laser ablation at the 32-cell stage, when gastrulation begins, suggest that many of the cell lineages are now determined, as when a cell is destroyed at this stage there is no regulation and its normal descendants are absent. Cell–cell interactions are, however, required later to effect their final differentiation.

6-3   A small cluster of homeobox genes specify cell fate along the antero-posterior axis.

Although the nematode differs greatly in body plan from vertebrates and *Drosophila*, including the absence of segmental pattern along the antero-posterior axis, homeobox-containing genes (see Box 4A, page 104) are involved in specifying cell fate along this axis, as they also are in those organisms. The nematode contains a large number of homeobox genes, of which only four are similar to the Antennapedia class of *Drosophila* Hox genes and the Hox genes of vertebrates (Fig. 6.7). These four genes are all arranged in a cluster, the Hox cluster, and in a similar order on the chromosome to their *Drosophila* homologs. A fifth homeobox gene of the cluster, *ceh-23*, is less related to the Antennapedia class. The Hox genes—*lin-39*, *mab-5*, and *egl-5*—are expressed in different

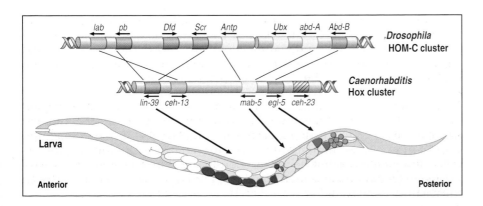

Fig. 6.7 The *C. elegans* Hox gene cluster and its relation to the HOM-C gene cluster of *Drosophila*. The nematode contains a cluster of five Hox genes with homologies to four of those of the Antennapedia complex of the fly. The pattern of expression of three of the genes in the larva is shown. After Bürglin, T.R., *et al.*: 1993.

positions along the antero-posterior axis during embryonic development, which are co-linear with their order along the chromosome. Although they are expressed during embryonic stages, it seems that their primary function is in post-embryonic larval development, since mutations in these Hox genes only affect larval development. Mutations in these genes can, however, cause cells in one part of the body region to adopt fates characteristic of other body regions. For example, a mutation in *lin-39* can result in mid-body cells expressing fates characteristic of more anterior or posterior body regions.

Despite the fact that nematode Hox gene expression occurs in a regional pattern, the pattern is itself not position dependent. For example, in the larva, cells expressing the Hox gene *mab-5* all occur in the same region (see Fig. 6.7) but are quite unrelated by lineage. Yet, despite appearances, the expression of *mab-5* is not due to extracellular positional signals but is determined autonomously and independently in each of the lineages. For example, some of the cells that will express *mab-5* migrate to their final positions during development; if the migration is blocked they will still express *mab-5*, but at an abnormal site. Another feature of Hox gene expression in *C. elegans* is that in a cell lineage expressing a Hox gene, the gene is turned on and off in different cells in the lineage and so does not reflect sustained positional identity.

6-4   **Genes control graded temporal information in nematode development.**

Because each cell in the developing nematode can be identified by its lineage and position, genes that control the fates of individual cells at specific times in development can also be identified. This enables the genetic control of timing during development to be studied. The order of developmental processes is of central importance, as well as the time at which they occur. Genes must be expressed at both the right place and time. We have already discussed this in relation to early *Xenopus* development and mesoderm induction (see Chapter 3). One well-studied example of timing in nematode development is the generation of different patterns of cell division and differentiation in the four larval stages of *C. elegans*, which are easily distinguished in the developing cuticle.

Mutations in two genes, *lin-4* and *lin-14*, change the timing of cell divisions in many tissues and cell types. Mutations that alter the timing of developmental events are called **heterochronic**. Mutations in *lin-4* and *lin-14* can result in both 'retarded' and 'precocious' development so that, for example, some stage-specific events such as molting and larval cuticle synthesis are repeated at abnormally late stages, leading to retardation of normal events such as adult cuticle synthesis.

Examples of changes in developmental timing that result from mutations in *lin-14* are provided by the lateral hypodermal T-cell lineage called T.ap (Fig. 6.8). In wild-type embryos, the T cell generates a lineage that gives rise to epidermal cells, neurons, and their support cells in both the first (L1) and second (L2) larval stages. During the later larval stages, L3 and L4, some of the T cell descendants divide to give rise to other structures. Gain-of-function (gf) mutations in *lin-14* result in retarded development. Post-embryonic development begins normally, but the developmental patterns of the first or second larval stages are repeated. Loss-of-function (lf) mutations in *lin-14* result in a precocious phenotype—the pattern of cell divisions seen in early larval stages is lost and post-embryonic development starts with cell divisions normally seen in the second larval stage.

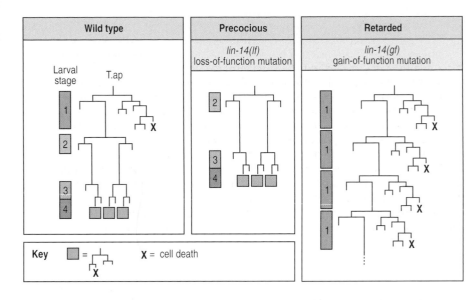

Fig. 6.8 Cell lineage patterns in wild-type and heterochronic mutants of *C. elegans*. The lineage of the T blast cell (T.ap) continues through four larval stages (left panel). Mutants in the gene *lin-14* show disturbance in the timing of cell division, resulting in changes in the patterns of cell lineage. Loss-of-function mutations result in a precocious lineage pattern, with the pattern of development of early stages being lost (center panel). Gain-of-function mutations result in retarded lineage patterns, with the patterns of the early larval stage being repeated (right panel).

It has been suggested that the genes that control timing of developmental events may do so by controlling the concentration of some substance, causing it to decrease with time (Fig. 6.9). This temporal gradient could control development in much the same way that a spatial gradient can control patterning. This type of timing mechanism seems to operate in *C. elegans* since the concentration of lin-14 protein drops tenfold between the first and later larval stages. Differences in the concentration of lin-14 protein at different stages of development may specify the fates of cells, with high concentrations specifying early fates and low concentrations later fates. Thus, the decrease of lin-14 protein during development could provide the basis of a precisely ordered temporal sequence of cell activities. Dominant gain-of-function mutations of *lin-14* keep lin-14 protein levels high, and so in these mutants the cells continue to behave as if they are at an earlier larval stage. In contrast, loss-of-function mutations result in abnormally low concentrations of lin-14 protein, and the larvae therefore behave as if they were at a later larval stage.

The concentration of lin-14 protein is post-transcriptionally regulated in an interesting and unusual way. The translation of *lin-14* mRNA can be repressed by *lin-4* RNA, which complexes with the *lin-14* mRNA. Increasing synthesis of *lin-4* RNA during the later larval stages could thus generate the temporal gradient in lin-14 protein. Support for this idea comes from mutations in *lin-4*. Loss-of-function mutations in *lin-4* have the same effect as gain-of-function mutations in *lin-14*.

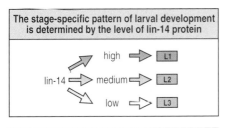

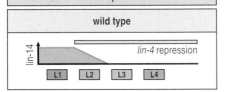

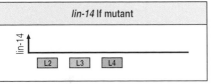

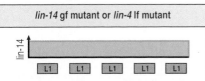

Fig. 6.9 A model for the control of the temporal pattern of *C. elegans* larval development. Top panel: the stage-specific pattern of larval development is determined by a temporal gradient of lin-14 protein, which decreases during larval development. A high concentration in the early stages specifies the pattern of early-stage development as shown in Fig. 6.8, left panel. The reduction in lin-14 protein at later stages is due to the inhibition of *lin-14* mRNA translation by *lin-4* mRNA. Bottom panels: loss-of-function (lf) mutations in *lin-14* result in the absence of the first larval stage (L1) lineage, whereas gain-of-function (gf) mutations, which maintain a high level of lin-14 protein throughout development, keep the lineage in an L1 phase. Loss-of-function mutations in *lin-4* result in a lifting of repression of *lin-14*, a continued high activity of lin-14 protein, and a repetition of the L1 lineage.

Summary.

Specification of cell fate in the nematode embryo is intimately linked to the pattern of cleavage, and provides an excellent example of the subtle relationships between maternally specified cytoplasmic differences and very local and immediate cell–cell interactions. The antero-posterior axis is specified at the first cleavage and specification of the dorso-ventral and left-right axes involves local cell–cell interactions. Gene products are asymmetrically distributed during the early cleavage stages but specification of cell fates in the early embryo is crucially dependent on local cell–cell interactions. The development of the gut, which is derived from a single cell, requires an inductive signal by an adjacent cell. A small cluster of homeobox genes provides positional identity along the antero-posterior axis of the larva. The timing of developmental events in the larva may be based on the concentration of a substance that decreases with time.

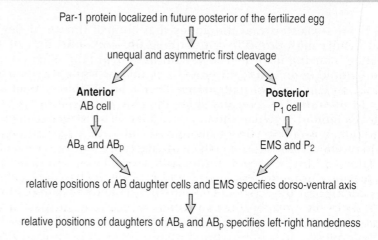

Summary: specification of axes in early nematode development

Par-1 protein localized in future posterior of the fertilized egg

unequal and asymmetric first cleavage

Anterior
AB cell

Posterior
P₁ cell

AB_a and AB_p

EMS and P₂

relative positions of AB daughter cells and EMS specifies dorso-ventral axis

relative positions of daughters of AB_a and AB_p specifies left-right handedness

Molluscs.

In contrast to the radial pattern of cleavage that occurs in vertebrate and echinoderm eggs (see Fig. 6.19), many invertebrates have a spiral pattern of cleavage. Spiral cleavage can be highly stereotyped and invariant, as in the nematode, and the fate of cells tends to be specified on a cell-by-cell basis. The spiral pattern is most easily seen in some molluscs, such as the snail *Lymnaea*, at the eight-cell stage; after the third cleavage, which divides the embryo into animal and vegetal tiers of four cells each, the animal cells do not sit directly on top of the vegetal cells but are displaced sideways, usually in a clockwise fashion when viewed from the animal pole (Fig. 6.10). This clockwise spiral arrangement is termed dextral, and displacement in the opposite direction, sinistral. Dextral cleavage patterns are by far the most common. The spiral arrangement reflects the oblique orientation of the mitotic spindles in relation to the egg axis.

Although there is considerable variation in spiral cleavage patterns between mollusc species, two features are quite common. First, in many molluscs the first and second cleavages are unequal; this results

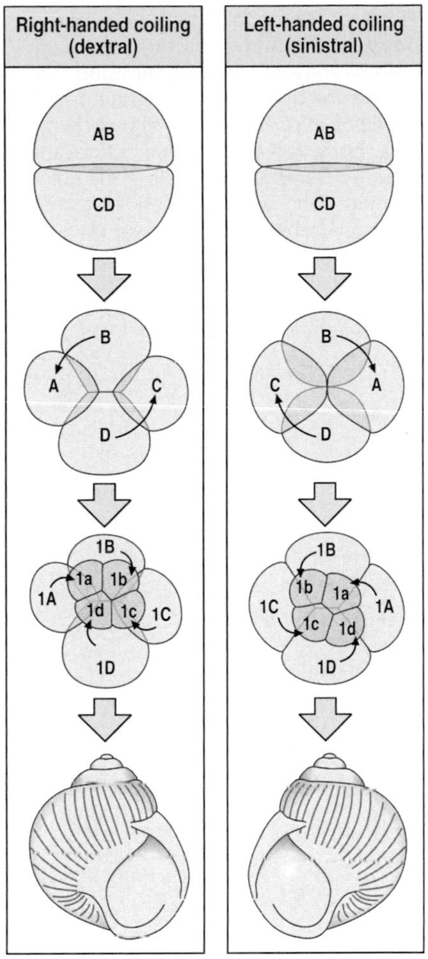

Fig. 6.10 Spiral cleavage in the snail *Lymnaea*. Cleavage results in a clockwise (dextral) displacement of the four animal blastomeres (1a to 1d) with respect to the larger vegetal blastomeres (1A to 1D) by the eight-cell stage. In sinistral cleavage the displacement is in the opposite direction. The handedness of the spiral cleavage corresponds with the later direction of coiling of the shell. After Morgan, T.H.: 1927.

in one of the cells at the four-cell stage being larger than the others. This so-called D blastomere (see Fig. 6.10) marks the posterior-dorsal region of the embryo, and so early cleavage is related to setting up the body axes. Second, by the 64-cell to 128-cell stage, the fate of single cells can be mapped. Of particular interest is that most of the mesoderm comes from one of the smaller **micromeres** that arise from unequal cleavage of the D cell—in this case the 4d cell formed at the sixth cleavage. Another micromere derived from the D cell, the 2d micromere formed at the fourth cleavage, gives rise to the gland that secretes the shell in the larva. Gastrulation and further development results in the free-living trochophore larva, which has a characteristic ciliated band around its middle (Fig. 6.11).

6-5 | The handedness of spiral cleavage is specified maternally.

Many molluscs, such as snails, have spirally coiled shells and these spirals have a consistent handedness, usually being dextral. The direction of coiling of the shell corresponds to the handedness of cleavage in the embryo (see Fig. 6.10), although there is as yet no explanation as to why they should be related. Both the direction of cleavage and shell coiling are maternally determined and are controlled by a recessive gene called *sinistral*, although other genes may be involved. Right-handedness is dominant over left-handedness so that races of molluscs with left-handed shells are rare. Injection of cytoplasm from wild-type dextral cleaving eggs into sinistral eggs causes them to adopt the dextral pattern, illustrating the dominance of the dextral pattern.

6-6 | Body axes in molluscs are related to early cleavages.

The adult bodies of molluscs such as snails and oysters have a morphology that does not quite fit with the orthogonal axes we are used to, with antero-posterior and dorso-ventral axes at right angles to each other. Nevertheless, most mollusc eggs do have a clear animal-vegetal axis, which often corresponds with the orientation of the oocyte in the ovary, implying that this axis is maternally determined. However, in some species it is possible to alter the original axis. In the snail *Lymnaea*, for example, the orientation of the animal-vegetal axis is determined by the orientation of the spindle of the second meiotic division that gives rise to the second polar body. Normally, this polar body forms at the animal pole and so defines the animal-vegetal axis. However, if the spindle is experimentally rotated through 90°, the animal-vegetal axis is similarly rotated (Fig. 6.12).

Although it is not easy to identify an antero-posterior axis in an early mollusc embryo, the dorsal and posterior region is more clearly recognizable and in most species is associated with the large D blastomere that is evident at the four-cell stage. This arises as a result of unequal first and second cleavages. In some species, the initial asymmetric cleavage is related to the point of sperm entry, which biases the direction

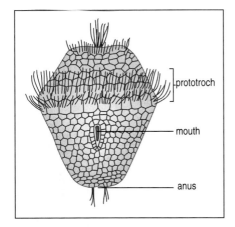

Fig. 6.11 The trochophore larva of molluscs. The trochophore is a free-living, feeding larval stage with a characteristic ciliated band, the prototroch, around the equator. After Wilmer, P.: 1990.

Fig. 6.12 The site of second polar body formation determines the animal-vegetal axis in a mollusc. Top panels: the animal-vegetal axis normally develops in relation to the site of second polar body formation, which defines the animal pole. Bottom panels: if the spindle forming in the second meiotic division (during which the second polar body is produced) is moved, then the polar body, which has increased in size, forms at a new site and the animal-vegetal axis is also shifted.

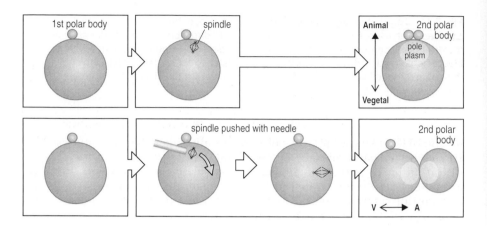

of the movement of the spindle so as to cause the asymmetric division. But it is the asymmetric division itself and not the point of sperm entry that is the key determining factor in future development.

In some molluscs the D blastomere is specified by asymmetric cleavages that segregate special cytoplasm, which is present at the vegetal pole. In the mud snail *Ilyanassa*, a polar lobe forms at the vegetal pole during the first cleavage (Fig. 6.13). The polar lobe is attached to just one of the blastomeres by a thin stalk. Following cleavage, the lobe is reabsorbed and becomes part of the cell, making it about twice the size of the other. By convention this blastomere is called CD and the smaller one is called AB. At the next cleavage, blastomeres A, B, C, and D are formed; the polar lobe again protrudes and is then withdrawn into the D blastomere, which is therefore larger than the others.

The importance of the polar lobe cytoplasm can be demonstrated by manipulating the *Ilyanassa* embryo at the first cleavage so that cytoplasm from the polar lobe enters both CD and AB blastomeres. This results in many structures in the larva being duplicated. If the lobe is removed at the first cleavage, the resulting embryos are also abnormal. Descendants of all blastomeres are affected, with the descendants of the D blastomere being most affected. This suggests that the D blastomere, by virtue of the polar lobe cytoplasm it contains, has an organizing role involving intercellular interactions. In some mollusc species (e.g. *Bythynaea*), a cytoplasmic inclusion is visible in the vegetal region where the polar lobe forms. If this inclusion is moved out of the polar lobe into the future D cell by centrifugation, removal of the polar lobe now has little effect on development, suggesting that the cytoplasmic inclusion contains key determinants affecting the fate of the D blastomere. In molluscs with no polar lobe, cleavage is equal, and more complex interactions are involved in specifying the dorso-ventral axis.

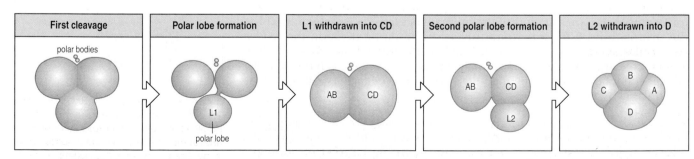

Fig. 6.13 Early cleavage and the polar lobe in molluscs. During the first cleavage division, a polar lobe (L1) is extruded from the vegetal region and is then withdrawn into the CD cell. A second polar lobe (L2) is extruded from CD during the second cleavage and is withdrawn into the D cell, which is larger than the other cells. After van den Biggelaar, J.A.M.: 1991.

Summary.

As in the nematode, specification of cell fate in mollusc embryos is on a cell-by-cell basis. Many molluscs have a spiral pattern of cleavage, which is usually dextral (right-handed) and whose handedness is the same as that seen in the coiling of their shells. The body axes are defined by early cleavages, and the localization of cytoplasmic determinants and asymmetric cleavage are thought to play an important part in specifying the fate of early blastomeres in some molluscs.

Annelids.

Annelids, a phylum which includes earthworms and leeches, are segmented animals with body plans quite unlike those of molluscs. However, their embryos undergo spiral cleavage and may be quite similar to those of molluscs in the very early stages. Here, we consider the later process of segmentation in the leech, and compare the mechanism with that in *Drosophila* and vertebrates.

6-7   The teloblasts are specified by localization of cytoplasmic factors.

In the leech, both mesodermal and ectodermal segmental structures are derived from cells known as **teloblasts**, which arise from the D blastomere. The specification of the D blastomere is associated with special cytoplasm the **teloplasm**—that segregates at the animal and vegetal poles before cleavage (Fig. 6.14). Just before first cleavage occurs, the teloplasm is distributed with respect to the animal-vegetal axis such that at the end of the third cleavage the D macromere (a large blastomere) contains most of it. If the distribution of the teloplasm is shifted by gentle centrifugation before the second cleavage so that macromere C acquires as much teloplasm as D, then both give rise to teloblasts. If centrifugation results in the C cell acquiring most of the teloplasm, it develops as the D cell, despite being smaller than the normal D cell. Thus, it seems likely that only cells that inherit teloplasm can generate teloblasts. The blastomeres that give rise to the teloblasts are known as DM and DNOPQ and their origin is shown in Fig. 6.14. The first three cleavages of the fertilized egg are both spiral and unequal, resulting in four micromeres (small blastomeres) in the animal region overlying four macromeres, of which the D macromere is the largest. At the next cleavage the division of the D macromere is obliquely equatorial and gives rise to an animal cell, DNOPQ, which is the ectodermal precursor, and DM. The DNOPQ cell divides several times to eventually give rise to four teloblasts (N, O, P, and Q), which produce ectoderm, while the DM cell gives rise to two M teloblasts (Fig. 6.15), which produce the mesoderm.

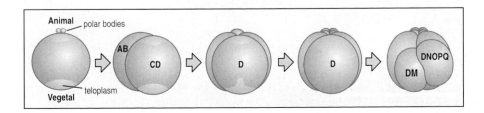

Fig. 6.14 Cleavage in the leech embryo and the origin of the teloblasts. After fertilization, teloplasm (yellow) forms at the animal and vegetal poles. Successive cleavages result in a large D macromere containing most of the teloplasm. This divides into DM and DNOPQ, which give rise to the precursors of the teloblasts.

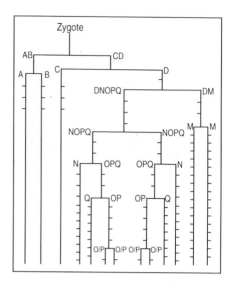

Fig. 6.15 Cell lineage in the leech embryo. The D blastomere gives rise to all five teloblasts M, N, O, P, and Q. The short lines indicate micromeres. After Bissen, S.T., *et al*.: 1996.

| 6-8 | **Antero-posterior patterning and segmentation in the leech is linked to a lineage mechanism.** |

Segmentation in the leech, *Helobdella triserialis*, which has 32 segments along the antero-posterior axis (Fig. 6.16), involves a mechanism quite different from that operating in a long germ insect such as *Drosophila*. In the fly, segmentation results from a positional type of mechanism in which genes delimiting the segments are activated in stripes by varying concentrations of transcription factors positioned along the antero-posterior axis (see Fig. 5.21). Segmentation in the leech, in contrast, seems to depend on a lineage mechanism: that is, the periodic pattern of segmentation is related to when the cells along the antero-posterior axis were born. A cell's character in this system is determined by the order in which it was generated from a stem cell and not by positional information. There is a fixed correlation between a cell's birth rank and its final location, with the first-born cells contributing to the most anterior regions. Thus in the leech, spatial periodicity and segmentation along the axis is determined by the temporal sequence of events.

Segmentation of both the ectoderm and the mesoderm can be traced back to the two sets of five teloblasts (N, O, P, Q, and M), one set on each side of the embryo. Each teloblast undergoes repeated asymmetric cell divisions in which one of the daughter cells retains the character of the parent cell while the other, the smaller blast cell, proceeds along a different developmental pathway; the teloblasts thus behave like stem cells. The repeated divisions generate a long bandlet of daughter blast cells (which stay joined together) from each teloblast (Fig. 6.17). The bandlets from the teloblasts on one side, although arising from cells located in quite separate parts of the embryo, come together to form a germinal band which joins with the germinal band from the other side to form the germinal plate. The bandlets in each germinal band are arranged in a precise order (see Fig. 6.17). Blast cells from the M teloblast, which will develop into the mesoderm, form a bandlet (m) underneath the other four bandlets, which will form the ectoderm and neural tissue. These are derived from the N, O, P, and Q teloblasts, and have the order, from the midline outward, of n, o, p, q. As they come together in the germinal plate, the n and q bandlets slide forward with respect to m, o, and p. The individual blast cells now start to divide, each blast cell generating a clone of about 100 descendants in a stereotyped lineage. There is no transverse mingling of blast cell clones up to this point (the beginning of blast cell division). These blast cell clones have a distinct but complex relationship to the embryonic segments.

Individual blast cell clones are restricted to particular segments, with those formed first contributing to the anterior-most segments. Individual

Fig. 6.16 Dorsal view of an adult leech, *Helobdella triserialis*. Segmentation along its antero-posterior axis can be clearly seen with segment boundaries visible as rows of pale dots. Scale bar = 1 mm. Photograph courtesy of D. Weisblat.

Fig. 6.17 Generation of blast cells in the leech embryo. The segmented tissue arises from two sets of five teloblasts (M, N, O, P, and Q) located on each side of the midline (for simplicity only one set is illustrated here). M forms mesoderm, and N, O, P, and Q form ectoderm and neural tissue. Each teloblast acts as a stem cell, undergoing repeated unequal divisions to bud off blast cells, which stay together to form long bandlets that come together in a germinal band on each side of the embryo (the m bandlet forms underneath the other four and so cannot be seen on this diagram). The germinal bands from each side come together in the germinal plate and the blast cells start to divide, each undergoing a stereotyped pattern of cell division. After Wedeen, C.J., *et al.*: 1991.

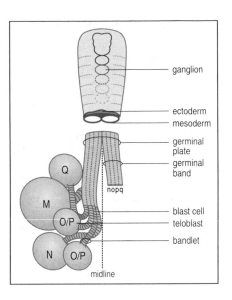

blast cell clones derived from the m, o, and p bands each contribute to two adjacent segments, an example of which is shown in Fig. 6.18 for the o and n bands, while individual clones derived from the n and q bands contribute to just one segment, each segment containing two blast cell clones from each of these bandlets.

Since the linear order of the blast cells along the axis is determined entirely by the 'time' or 'division number' at which the blast cells were born, this timing could provide a unique identity defining the blast cell's experimental manipulations that alter the relationships between the blast cells in the bandlets; a blast cell whose position is altered (for example by killing other blast cells) result in a blast cell at a particular position giving rise to cell types that relate to its time of birth rather than its final position. Thus, segment identity is normally determined by cell lineage, and not by the cells' final position.

The effect of altering the position of blast cells is particularly clear in studies on the expression of the leech homeobox gene *Lox2*, which is related to genes of the bithorax complex of *Drosophila*. The gene is expressed in all five teloblast lineages, with a sharp anterior border of expression at segment 6. Displacing the blast cells of one teloblast lineage by killing some cells still results in expression of *Lox2* in the remaining cells that would normally have expressed it, even though they are now out of register with the other lineages. This strongly suggests that this gene expression is autonomous and is not dependent on positional signals.

The number of segments, by contrast, seems to be determined by a positional mechanism. More blast cells are produced than required and the excess cells die, but just how segment number is specified is not yet known. Death of ectodermal cells is known to be determined by their position, because forcing cells that would normally give rise to segments into the region of supernumerary cells causes them to die. A positional mechanism for determining segment number is, however, not easy to reconcile with a lineage mechanism specifying segment identity.

While leech segments seem to be generated by a mechanism different from that in *Drosophila*, there may be similarities with segmentation mechanisms in short germ insects where, unlike *Drosophila*, segment formation requires cell division and growth (see Section 5-18).

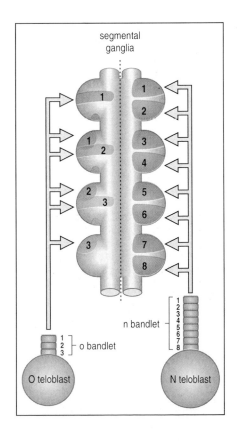

Fig. 6.18 Distribution of blast cells in the nervous system. Each blast cell undergoes a sterotyped series of divisions to give rise to a clone of cells. The first formed blast cells give rise to the anterior-most segments. Each M, O, and P blast cell clone contributes to two segments—as shown here for the O lineage in the nervous system. Each N and Q blast cell clone is confined to a single segment and the N and Q lineages contribute two blast cell clones per segment. To obtain the correct segmental registration the N and Q lineages must slide anteriorly past M, O, and P. After Shankland, M.: 1994.

Summary.

In annelids such as the leech, the segmental structures are derived from a set of cells called teloblasts, which are specified early in embryonic development by cytoplasmic determinants. The teloblasts give rise to blast cells from which the segments derive. Patterning and segmentation along the antero-posterior axis is based on both cell lineage and on temporal mechanisms, the first-formed blast cells always giving rise to the most anterior segments. In some cases cell fate may also be specified by position.

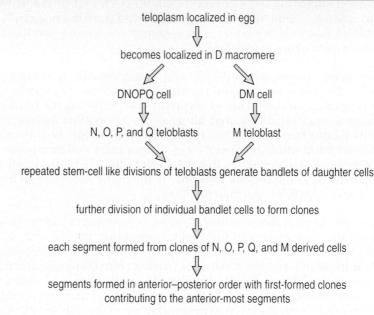

Summary: segmentation in the leech

teloplasm localized in egg

⇩

becomes localized in D macromere

↙   ↘

DNOPQ cell   DM cell

⇩   ⇩

N, O, P, and Q teloblasts   M teloblast

↘   ↙

repeated stem-cell like divisions of teloblasts generate bandlets of daughter cells

⇩

further division of individual bandlet cells to form clones

⇩

each segment formed from clones of N, O, P, Q, and M derived cells

⇩

segments formed in anterior–posterior order with first-formed clones contributing to the anterior-most segments

Echinoderms.

Echinoderms include the sea urchins and starfish. Because of their transparency and ease of handling, sea urchin embryos have long been used as a model developmental system. Developmental studies are confined to the development of the larva, as metamorphosis into the adult is a complex and poorly understood process. The sea urchin embryo is classically regarded as a model of regulative development. It formed the basis for Driesch's ideas at the beginning of the century, that the position of the cells in an embryo determined their fate (see Section 1-3).

The sea urchin egg divides by radial cleavage. The first three cleavages are symmetric but the fourth cleavage is asymmetric, producing four small micromeres at one pole of the egg, the vegetal pole (Fig. 6.19), thus defining the animal-vegetal axis of the egg. The first two cleavages divide the egg along the animal-vegetal axis, while the third cleavage is equatorial and divides the embryo into animal and vegetal halves. At the next cleavage the animal cells divide in a plane parallel with the animal-vegetal axis but the vegetal cells divide asymmetrically to produce four macromeres and four micromeres. Continued cleavage results in a hollow spherical blastula composed of about 1000 ciliated cells that form an epithelial sheet enclosing the blastocoel.

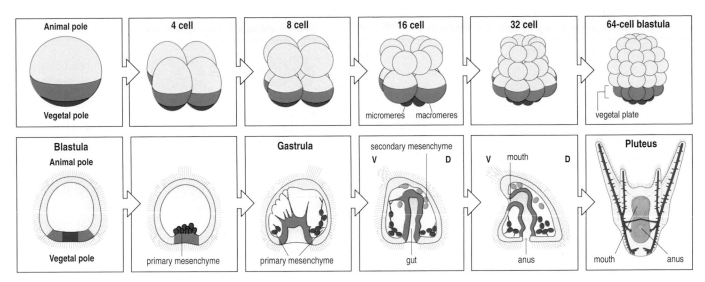

Fig. 6.19 Development of the sea urchin embryo. Top panels: external view of cleavage to the 64-cell stage. The first two cleavages divide the egg along the animal-vegetal axis. The third cleavage divides the embryo into animal and vegetal halves. At the fourth cleavage, which is unequal, four small micromeres (blue) are formed at the vegetal pole, only two of which can be seen here. Further cleavage results in a hollow blastula. Bottom panels: gastrulation and development of the pluteus larva. Sections of the developing embryo through the plane of the animal-vegetal axis are shown, starting from a hollow blastula. Gastrulation begins at the vegetal pole, with the entry of about 40 primary mesenchyme cells into the interior of the blastula. The gut invaginates from this site and fuses with the mouth, which invaginates from the opposite side of the embryo. During further development, growth of skeletal rods laid down by the primary mesenchyme results in the extension of the four 'arms' of the pluteus larva. The pluteus is depicted from its outside with the mouth uppermost.

Gastrulation in sea urchin embryos, whose mechanisms we consider in detail in Chapter 8, starts about 10 hours after fertilization, with the mesoderm and endoderm moving inside from the vegetal region. The first event is the entry into the blastocoel of about 40 primary mesenchyme (mesoderm) cells at the vegetal pole. They migrate along the inner face of the blastula wall to form a ring in the vegetal region and lay down calcareous skeletal rods. The endoderm, together with the secondary mesenchyme, then starts to invaginate at the vegetal pole, the invagination eventually stretching right across the blastocoel where it fuses with a small invagination in the future mouth region on the ventral side. Thus the mouth, gut, and anus are formed. Before the invaginating gut fuses with the mouth, secondary mesenchyme cells at the tip of the invagination migrate out as single cells and give rise to mesoderm, such as muscle and pigment cells. The embryo is then a feeding **pluteus** larva (see Fig. 6.19).

6-9   The sea urchin egg is polarized along the animal-vegetal axis.

The sea urchin egg has a well-defined animal-vegetal polarity that appears to be related to the site of attachment of the egg in the ovary. Polarity is marked in some species by a fine canal at the animal pole, whereas in other species there is a band of pigment granules in the vegetal region. Early development is intimately linked to this egg axis, and it can be considered as corresponding to the antero-posterior axis of the larva. The first two planes of cleavage are always parallel to the animal-vegetal axis and at the fourth cleavage, which is unequal (see Fig. 6.19), the micromeres form at the vegetal pole. The micromeres give rise to the primary mesenchyme, and both the micromeres and the primary mesenchyme may initially be specified by cytoplasmic factors localized at the vegetal pole of the egg.

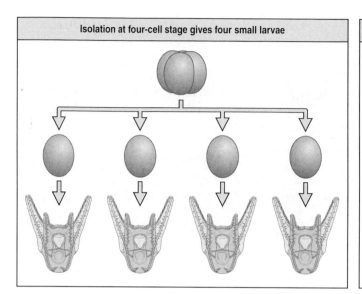

Isolation at four-cell stage gives four small larvae

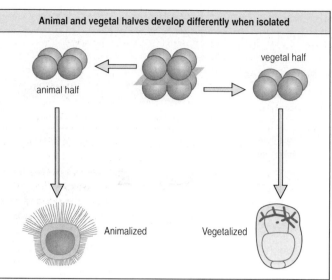

Animal and vegetal halves develop differently when isolated

Fig. 6.20 Development of isolated sea urchin blastomeres.
Left panel: if isolated at the four-cell stage, each blastomere develops into a small but normal larva. Right panel: an isolated animal half from the eight-cell stage forms a hollow sphere of ciliated ectoderm, whereas an isolated vegetal half usually develops into a highly abnormal embryo with a large gut, and some skeletal structures, but with a reduced ectoderm.

The animal–vegetal axis is stable and cannot be altered by centrifugation, which redistributes larger organelles such as mitochondria and yolk platelets. Isolated fragments of egg also retain their original polarity. Blastomeres isolated at the two-cell and four-cell stages, each of which contains a complete animal–vegetal axis, give rise to mostly normal but small pluteus larvae (Fig. 6.20, left panel). Similarly, eggs that are fused together with their animal–vegetal axes parallel to each other form giant, but otherwise normal, larvae.

There is, by contrast, a striking difference in the development of animal and vegetal halves if they are isolated at the eight-cell stage. An isolated animal half merely forms a hollow sphere of ciliated ectoderm, whereas a vegetal half develops into a larva that is variable in form but is usually **vegetalized**, that is it has a large gut and skeletal rods, and a reduced ectoderm lacking the mouth region (see Fig. 6.20, right panel). On occasion, however, vegetal halves from the eight-cell stage can form normal pluteus larvae if the third cleavage is slightly displaced toward the animal pole. These observations show that, as in amphibians, there are maternal cytoplasmic differences present in the egg that specify all three germ layers. Some spatially restricted mRNAs that may encode proteins involved in such specification along the animal–vegetal axis have already been identified.

Even though there are cytoplasmic differences along the animal–vegetal axis it is clear that the sea urchin embryo has considerable capacity for regulation, implying the occurrence of cell–cell interactions. As we see below, important developmental signals are produced by the micromeres in the vegetal region.

6-10   The dorso-ventral axis in sea urchins is related to the plane of the first cleavage.

The dorso-ventral axis of a sea urchin pluteus larva is defined with respect to the mouth, which develops on the ventral side (see Fig. 6.19), but there are also clear differences in skeletal patterning in the ventral and dorsal regions. The ventral side can be recognized well before mouth invagination, because the primary mesenchyme migrates to form two columns of cells on the future ventral face. The migration of these cells is considered in Chapter 8.

Unlike the animal-vegetal axis, the dorso-ventral axis cannot be identified in the egg and appears to be labile up to as late as the 16-cell stage. However, in normal embryos this axis is related to the plane of the first cleavage. In the sea urchin *Strongylocentrotus purpuratus*, a good correlation was found between the dorso-ventral axis and the plane of cleavage: the future dorso-ventral axis lies 45° clockwise from the first cleavage plane as viewed from the animal pole (Fig. 6.21). However, in other species the relationship between the dorso-ventral axis and plane of cleavage is different; the axis may coincide with the plane of cleavage or be at right angles to it. It is worth recalling that in *Xenopus* the first cleavage usually corresponds with the plane of bilateral symmetry.

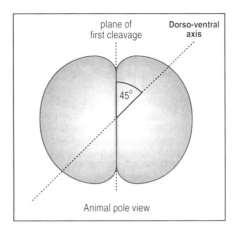

Fig. 6. 21 Position of the dorso-ventral axis of the sea urchin *S. purpuratus* embryo in relation to first cleavage. When viewed from the animal pole, the dorso-ventral axis is usually 45 clockwise from the plane of 1st cleavage. Note that its polarity is not determined at this stage, only the axis.

| 6-11 | **The sea urchin fate map is very finely specified, yet considerable regulation is possible.** |

At the 64-cell stage, four regions of cells can be distinguished in the sea urchin embryo along the animal-vegetal axis and a simplified fate map can be constructed (Fig. 6.22). This is composed of three bands of cells: the micromeres at the vegetal pole, which will give rise to mesoderm (the primary mesenchyme that forms the skeleton); the next tier of cells, which gives rise to the gut, secondary mesoderm and some ectoderm; and the remainder of the embryo, which gives rise to ectoderm and is divided into anterior and posterior regions. Using vital dyes as lineage tracers, the pattern of cleavage has been shown to be invariant but complex, particularly for the ectoderm. However, unlike the nematode, the invariance of cleavage in the sea urchin seems irrelevant to normal development. Compressing an early sea urchin embryo by squashing it with a cover slip, thus altering the pattern of cleavage, still results in a normal embryo after further development. Moreover, since the sea urchin embryo is capable of considerable regulation, specification of fate cannot be closely linked to a cleavage pattern.

When isolated and cultured, some regions of the embryo develop more or less in line with their normal fate: for example, micromeres isolated at the 16-cell stage will form mesenchyme-like cells and may even form skeletal rods. An isolated animal half of presumptive ectoderm forms a ciliated epithelial ball, but there is no indication of mouth formation. By contrast, a vegetal half at the 16-cell stage will develop into a vegetalized embryo with not only a large gut, but also ciliated ectoderm and some skeletogenic mesenchyme. In this tissue, therefore, the fate of some of the cells has been changed from prospective gut endoderm to mesoderm and ectoderm. This capacity for regulation is consistent with the observation that whole vegetal halves isolated at the third cleavage can, on occasion, regulate to give quite normal larvae but only if the third cleavage is displaced toward the animal pole.

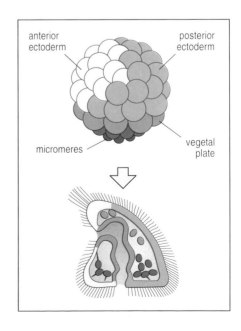

Fig. 6.22 Fate map of the sea urchin embryo. Lineage analysis has shown four main regions at the 64-cell stage. The embryo is divided along the animal-vegetal axis into three bands: the micromeres, which give rise to the primary mesenchyme; the vegetal plate which gives rise to the gut and secondary mesenchyme; and the ectoderm. The ectoderm is divided into anterior and posterior regions. After Ransick, A., *et al.*: 1993.

less is known about how plant cells communicate with each other and transduce hormonal and other signals than is the case in animal cells. The cell wall would seem to impose a barrier to the passage of large molecules, such as proteins, but all known plant growth hormones, such as auxins, gibberellins, cytokinins, and ethylene, are small molecules that easily penetrate cell walls. It may be that plant cells also communicate with each other through the fine cytoplasmic channels—plasmodesmata—which link neighboring plant cells through the cell wall and there is evidence that both proteins and RNA can move between plant cells through the plasmodesmata.

Another important difference between plant and animal cells is that a complete, fertile plant can develop from a single somatic cell and not just from a fertilized egg. This suggests that, unlike the differentiated cells of adult animals, some differentiated cells of the adult plant may retain **totipotency**. Perhaps they do not become fully determined in the sense that adult cells do, or perhaps they are able to escape from the **determined** state, although how either of these could be achieved is as yet unknown. In any case, this difference between plant and animal cells illustrates the dangers of the wholesale application to plant development of concepts derived from animal development. However, the genetic analysis of developmental processes in plants is turning up instances of rather similar genetic strategies for developmental patterning in plants and animals (see Section 7-11).

In this chapter, we first look at the very early development of the zygote in multicellular algae, where it is more easily studied than in higher plants. We then consider the embryogenesis and further development of higher plants, with particular focus on the genetic control of development in flowering plants, and on how cell fate may be determined. The small annual wall cress *Arabidopsis thaliana* has become the *Drosophila* of the plant world (see Section 2-7 for the developmental stages and life-history of *Arabidopsis*), as it is easily grown and is more amenable to genetic analysis than many other plants, and we use it as an example wherever possible.

Embryonic development.

In the same way as an animal zygote, the fertilized plant egg cell undergoes repeated cell division, cell growth, and differentiation to form a multi-cellular embryo. There is enormous diversity in the pattern of early cell divisions in different types of plants but the significance of any variation, in terms of developmental strategy, is not known. Even so, a very common early pattern of cell division is seen with the first division of the zygote. This occurs at right angles to the long axis of the embryo and so divides the zygote into apical and basal regions (see Box 7A, page 209). In many species this first division is unequal, but in most of these cases it is not known to what extent this reflects developmentally **asymmetric cell divisions** that result in daughter cells with different identities.

We start by looking at some aspects of embryogenesis in a multicellular alga that illustrate the importance of asymmetric cell division. In the brown alga *Fucus*, the body axis (apical-basal) is determined by environmental signals while the zygote is still a single cell. Determination of the axis leads to asymmetric cleavage of the zygote transverse to the axis, which delineates the two main regions of the algal body. Studies of *Fucus* illustrate the importance of the cell wall as a potential source of developmental signals.

| 7-1 | **Electrical currents are involved in polarizing the *Fucus* zygote.** |

The eggs of *Fucus* are fertilized externally, and the zygote floats through the water until it encounters a suitable surface on which to settle. Once settled, it begins to develop. At first cleavage, the zygote is divided into two cells of different size, the smaller of which gives rise to the root-like rhizoid, and the other to the thallus, or 'leafy' part of the alga (Fig. 7.1). This asymmetric cell division reflects an apical-basal polarity that is set up in the zygote before cleavage begins. A similar transverse asymmetric division indicates the direction of apical-basal polarity in the cleaving zygote of many higher plants. The first sign of asymmetry in the fertilized *Fucus* egg appears before the first cleavage as a protrusion in the region of the future rhizoid. The fertilized egg is initially symmetric and experimental manipulations have shown that this later polarity is determined by environmental signals. In this, *Fucus* resembles those animal embryos such as *Xenopus*, which have their axes specified by external signals.

A wide range of environmental stimuli, including illumination, pH gradients, and water flow can set the polarity. (In default of such signals, there is an intrinsic mechanism for determining polarity by the point of sperm entry, which corresponds to the future rhizoid region.) The effects of light are dramatically demonstrated by an experiment in which fertilized *Fucus* eggs are placed in a narrow capillary tube that prevents their rotation and a light is shone at one end of the tube. The rhizoids all grow out of the shaded sides of the eggs, irrespective of the site of sperm entry.

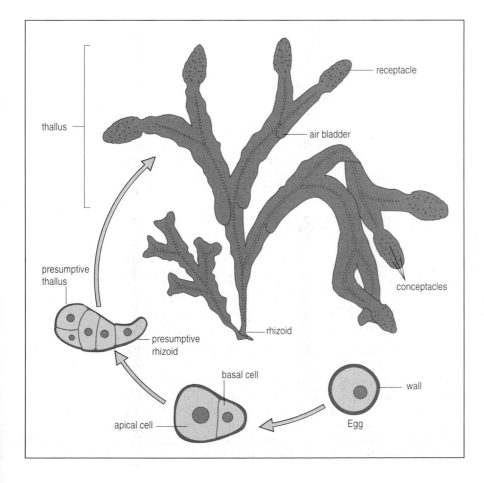

Fig. 7.1 The life history of the multicellular brown alga *Fucus spiralis*. The algal body consists of a flattened thallus divided into fronds, which is anchored to a rock or other hard substrate by the rhizoid or holdfast. Swollen fruiting bodies, the receptacles, develop at the tips of the fronds. They are pitted with small jelly-filled cavities, the conceptacles, in which the gametes are produced. Male and female gametes develop in different receptacles. Mature gametes are released from the conceptacles and the eggs are fertilized externally by the motile sperms. The fertilized egg drifts through the water until it settles on a suitable substrate and begins to develop. The first cleavage of the zygote is unequal, producing a smaller basal cell that gives rise to the rhizoid, and a larger apical cell that gives rise to the thallus.

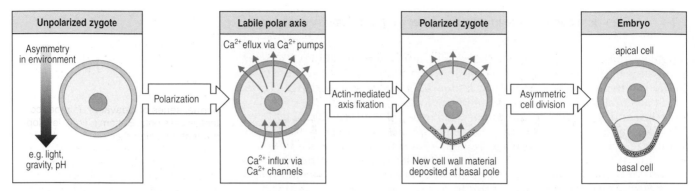

Fig. 7.2 Polarization of the *Fucus* zygote. The fertilized egg is initially unpolarized. Environmental factors, such as illumination from one side, gravity, or pH gradients, impose an asymmetry on the zygote that results in the development of an electrical current through it. This confers a polarity on the zygote. The current is partly carried by calcium ions. The environmental influences result in an increased flow of calcium ions into the cell in one region and an increased outflow on the opposite side. In the case of directional illumination, the shaded side becomes the site of increased calcium influx, which becomes the basal pole of the zygote and the site of future rhizoid formation. The apical-basal axis is fixed by the accumulation of cell vesicles, mitochondria, and Golgi bodies at the site of calcium influx. Cell wall material is deposited at this site and actin filaments become localized there. The first cleavage occurs transverse to the apical-basal axis, and is asymmetric, resulting in a smaller basal cell and a larger apical cell. After Alberts, B., *et al.*: 1989.

Associated with polarization of the egg is the development of an electrical current that flows in the direction of the axis (Fig. 7.2). The current must be partly carried by calcium ions, as local application of a calcium ionophore, which causes a local flow of calcium ions into the cell, causes rhizoid outgrowth at the site of application. All the environmental signals may act by causing calcium channels to open locally, with the resulting current causing further calcium channels to cluster at the site, and calcium pumps (which pump calcium ions out of the cell) to cluster at the other end of the axis. This positive feedback mechanism would act to maintain the direction of flow. The ion flow affects internal components of the egg, which now become localized in the region where calcium ions are entering the cell.

Although ion currents establish the axis, other processes are required to stabilize it. By 12 hours after fertilization the axis is fixed, and it only remains labile—that is, its position can be altered—for up to 7 hours after fertilization. Intracellular components are transported to, and accumulate in, the cytoplasm at the site of rhizoid formation. A particular polysaccharide is transported to the rhizoid site and inserted into the rhizoid wall. It seems that both the cytoskeleton and cell wall are involved in fixation of the axis. Actin filaments soon become localized at the rhizoid pole and the introduction of agents that disrupt them, such as cytochalasin B, prevent axis formation. Removal of the cell wall by enzyme treatment also prevents axis formation. Thus, an attractive hypothesis for polarization of the egg suggests that the calcium currents localize an actin network in the region of the future rhizoid, and set in motion a series of events leading to localized polysaccharide insertion into the cell wall.

7-2   Cell fate in early *Fucus* development is determined by the cell wall.

The pattern of early cleavage in *Fucus* is similar to that in some higher plant embryos. Because the *Fucus* embryo is accessible for microsurgical manipulation, it is possible to gain insight into mechanisms specifying early cell fate, in particular the development of the thallus from apical cells and the rhizoid from the basal cell. The pattern of cell division in these regions is different. The fates of both presumptive rhizoid and

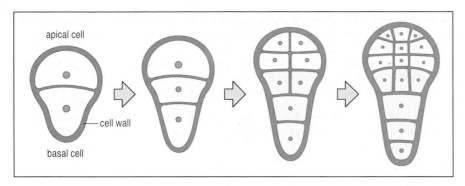

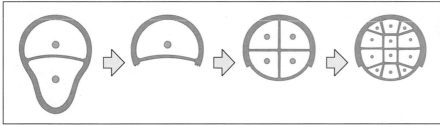

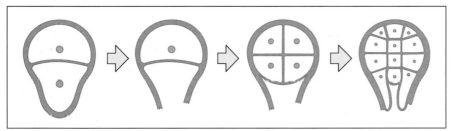

Fig. 7.3 Cell walls can determine cell fate in *Fucus*. Top panel: normal embryonic development. Presumptive rhizoid cells are shown in yellow. Middle panel: development of a complete apical cell when isolated from a two-cell embryo. The isolated apical cell develops by a normal series of cleavages to produce thallus cells only. Bottom panel: development of an isolated apical cell with part of the basal cell wall left attached. As the presumptive thallus cells divide, those that come into contact with the former basal cell wall develop differently, becoming rhizoid rather than thallus cells.

thallus cells can be changed by experimental manipulation of the two-cell embryo. If the wall of the basal cell is destroyed with a laser microbeam, the presumptive rhizoid cell is extruded as a spherical protoplast. Within one hour, the isolated protoplast regenerates a cell wall and then goes on to develop as a normal complete embryo, giving rise to both rhizoid and thallus cells. This shows quite clearly that the fate of the basal cell is not fixed at this stage and can be altered. It also implies that components of the cell wall may be involved in directing the basal cell's future development.

More evidence for the role of the cell wall in directing cell fate comes from experiments with apical cells (Fig. 7.3). An intact apical cell, complete with its cell wall, isolated from the two-cell embryo, continues to divide in a normal 'thallus' pattern until at least the eight-cell stage. But if the wall of the basal cell is still attached, those thallus cells that come into contact with it develop into rhizoid cells, while the rest of the cells develop as thallus cells. So development as rhizoid cells is dependent on contact with the basal cell wall. This provides strong evidence that the position-dependent differentiation of the two cells of the early *Fucus* embryo may be linked to factors on the inner face of the cell wall.

7-3   Differences in cell size resulting from unequal divisions could specify cell type in the *Volvox* embryo.

Many cell divisions in early plant development are unequal. Studies in the green alga *Volvox* show that these differences in size could be important in specifying cell fate. *Volvox* is an extremely simple multicellular organism. An asexual adult colony is made up of just two cell types

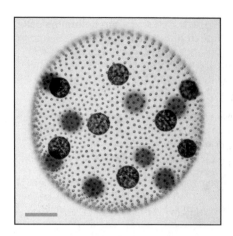

Fig. 7.4 Asexual colony of *Volvox carteri*. This colonial green alga consists of 2000 biflagellate somatic cells surrounding 16 larger gonidia. Each gonidium can give rise to a new colony by asymmetric cell division. Scale bar = 10 μm. Photograph courtesy of D. Kirk.

arranged in an orderly pattern. Around 2000 biflagellate somatic cells are uniformly distributed over the surface of a transparent gelatinous sphere, with 16 larger asexual reproductive cells, known as gonidia, lying beneath the surface (Fig. 7.4). When mature, each gonidium undergoes repeated cleavage to form a juvenile *Volvox* colony *in situ*. These are eventually released from the parent colony, which then dies.

The pattern of 2000 small somatic cells and just 16 large reproductive ones has its origin in the asymmetric cleavage of the gonidium. The first five cleavages are symmetric and result in a hollow spherical embryo of 32 cells with a clear asymmetry in cell packing. At the sixth cleavage, the 16 anterior cells divide asymmetrically to give 16 small and 16 larger cells. The larger cells lie posteriorly and give rise to the gonidia; they divide asymmetrically twice more and then stop. The smaller cells continue to divide for another three or four cycles and differentiate as somatic cells. The 16 small cells that were originally at the posterior pole also divide symmetrically and produce somatic cells.

There could be several possible causes for the difference in developmental fate between the large gonidia and the smaller somatic cells. It might, for example, be the result of asymmetric distribution of a localized cytoplasmic determinant. Experimental evidence suggests, however, that the difference in size is the determining factor. In mutants where the number of divisions is reduced, and so all the cells are larger, the number of gonidia is increased. And when heat shock is used to interrupt cleavage in otherwise normal embryos, the presumptive somatic cells remain large and develop as gonidia. Even more striking is the demonstration that when somatic cells are connected by microsurgery to make one larger cell, this cell will develop into a gonidium. How the differences in size are transduced into differences in gene expression so that somatic cells or gonidia develop is not yet known. Whether differences in cell size affects cell fate in other plants remains to be investigated.

7-4   Both asymmetric cell divisions and cell position pattern the early embryos of flowering plants.

There is great diversity in the pattern of cell division in angiosperm embryos of different species. One common principle is that, as in *Fucus*, the first division in angiosperm embryos is transverse with respect to the long axis of the zygote, and it divides the zygote into an apical cell and a basal cell. In some plant embryos the basal cell contributes little to further development of the embryo but divides to give rise to the suspensor, which may be several cells long. The apical cell undergoes a complex pattern of divisions which, in dicotyledons such as *Capsella bursa-pastoris* (shepherd's purse) and *Arabidopsis*, leads in about five days to a globular-stage and then to a heart-stage embryo (Box 7A).

A rough fate map for *Arabidopsis* can be constructed for the very early embryo on the basis of the pattern of cell divisions, by following the fate of the embryonic cells and finding what region(s) each gives rise to (Fig. 7.5). When there are just eight cells at the apical end—the so-called octant stage—one can begin to map the future main regions of the heart stage, at which stage a clear fate map can be drawn. The heart-stage embryo can be divided into three main regions: the apical region, which gives rise to the shoot meristem and cotyledons; the central region, which forms the future hypocotyl (embryonic stem); and the basal region, which gives rise to the root meristem. There is a radial pattern made up of concentric rings of tissue layers: epidermis, ground

Box 7A Angiosperm embryogenesis.

In flowering plants (angiosperms), the egg cell (not shown) is contained within an ovule inside the ovary in the flower (right inset in the bottom panel). At fertilization, a pollen grain deposited on the surface of the stigma puts out a pollen tube, down which two male gametes migrate into the ovule (see Section 2-7). One male gamete fertilizes the egg cell while the other combines with another cell inside the ovule to form a specialized nutritive tissue, the endosperm, which surrounds, and provides the food source for, the development of the embryo.

Capsella bursa-pastoris (shepherd's purse) is a typical dicotyledon. The first, asymmetric cleavage divides the zygote transversely into an apical and a basal cell (bottom panel). The basal cell then divides several times to form a single row of cells—the suspensor—which takes no further part in embryonic development, but may have an absorptive function. Most of the embryo is derived from the terminal cell. This undergoes a series of stereotyped divisions, in which a precise pattern of cleavages in different planes gives rise to the heart-shaped embryonic stage typical of dicotyledons. This develops into a mature embryo which consists of a main axis, with a meristem at either end, and two cotyledons, which are storage organs.

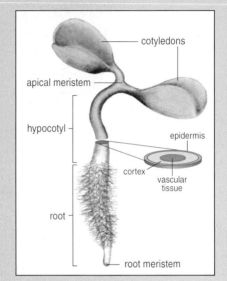

The early embryo becomes differentiated into three main tissues—the outer epidermis, the prospective vascular tissue, which runs through the center of the main axis and cotyledons, and the ground tissue (prospective cortex) that surrounds it. Monocotyledons, such as maize, have a single cotyledon; dicotyledons, such as *Arabidopsis*, have two.

The ovule containing the embryo matures into a seed (left inset in the bottom panel), which remains dormant until suitable external conditions trigger germination and growth of the seedling. A typical dicotyledon seedling (left panel) comprises the shoot apical meristem, two cotyledons, the trunk of the seedling (the hypocotyl) and the root apical meristem. The seedling may be thought of as the phylotypic stage of flowering plants. The seedling body plan is simple. There is one main axis—the apical-basal axis—which defines the polarity of the plant. The shoot forms at the apical pole and the root at the basal pole. The plant axis has radial symmetry in cross-section, which is evident in the hypocotyl and is continued in the root and shoot. In the center is the vascular tissue, which is surrounded by cortex, and an outer covering of epidermis.

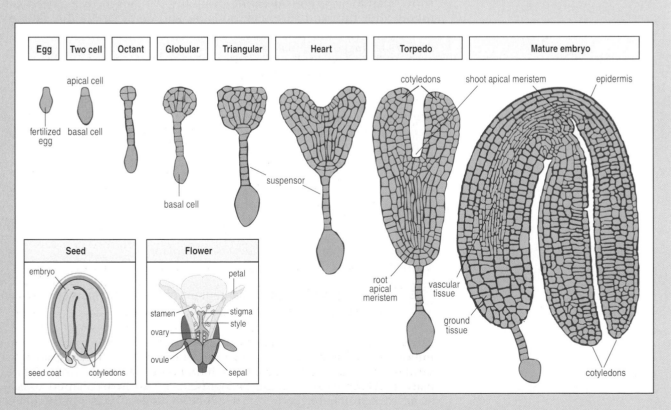

Fig. 7.5 Fate map of the *Arabidopsis* embryo. The stereotyped pattern of cell division in dicotyledon embryos means that at the globular stage it is already possible to map the three regions that will give rise to the cotyledons (dark green) and shoot meristem (red), the hypocotyl (yellow), and the root meristem (purple), in the seedling. After Scheres, B., *et al*.: 1994.

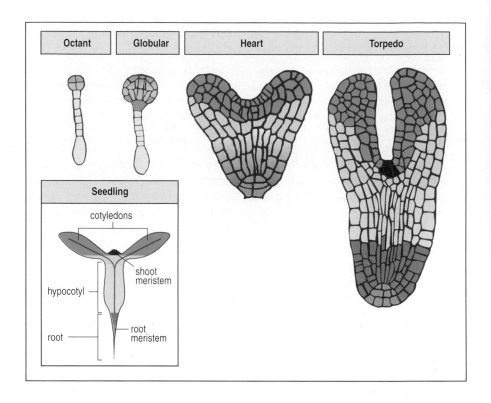

tissue (cortex and endodermis), and vascular tissue. However, it is not known to what extent the fate of the cells is determined at this stage, or whether their fate is in fact dependent on the early pattern of asymmetric cleavages. It seems that cell lineage is not crucial since clonal analysis shows that there is considerable variability. Mutations in the *FASS* gene totally alter the pattern of cell division and disorganize the cells, yet although the resultant seedling is misshapen, it has roots, shoots, and flowers in the correct places.

A very different pattern of cleavage occurs in the embryo of the mono-cotyledon *Zea mays* (maize). Since later cell divisions are irregular, both in orientation and sequence, it is much more difficult to make a fate map at an early stage. The position of the future shoot apical meri-stem can only be identified after about 9 days, when the embryo already consists of many cells. The shoot meristem can then be detected lying at the base of the single cotyledon. The root meristem can be detected several days later in the region near the suspensor. The inability to make a fate map of the early embryo reflects the fact that cell lineages in maize embryos are variable. This suggests that it is relative cell position that determines cell fate in maize embryos. However, no mechanisms for establishing and interpreting positional information have yet been identified in maize, or any other vascular plant.

7-5   The patterning of particular regions of the *Arabidopsis* embryo can be altered by mutation.

The embryos of flowering plants are not easily accessible to experimental microsurgical manipulation, but the course of their development can be altered by mutation. The effects of mutations that alter the basic body plan in *Arabidopsis* can be recognized at the heart-shaped embryo stage. Mutations in zygotic genes, affecting different parts of the pattern along the apical-basal axis, have been found in *Arabidopsis*. In apical mutants the cotyledons and shoot meristem are missing; in central mutants there is no hypocotyl and the cotyledons appear to be attached to the root; in basal mutants there is no hypocotyl or root (Fig. 7.6).

The phenotypes of these mutants suggest that the patterning of particular regions of the embryo along the apical-basal axis is controlled by specific genes. The control mechanisms are not yet known but several of the genes involved have been identified (see Fig. 7.6). The basal mutation *monopterous* eliminates the hypocotyl and the root meristem, and abnormal cell divisions can be observed at the octant stage in the central and basal regions. *fäckel* mutations affect the central region and *gurke* mutants are apical mutants, lacking shoot meristems and cotyledons. Mutations in the gene *SHOOT MERISTEMLESS* completely block the formation of the shoot apical meristem, but have no effect on the root meristem or other parts of the embryo.

Radial pattern formation commences at the eight-cell stage and involves oriented cell division. **Periclinal** (radial) divisions give rise to a new tissue layer and **anticlinal** divisions increase the number of cells in a layer. It seems that tissue fate is determined early in development and then inherited clonally, since in *keule* mutant embryos the epidermal cells are enlarged from an early stage whereas the endodermis is absent when the *short root* mutation is present.

Other mutations have been discovered that affect patterning of the *Arabidopsis* embryo. A mutation in the *LEAFY COTYLEDON* gene, for example, results in what appears to be a homeotic transformation of cotyledons into leaves. A feature of this transformation is the appearance of leaf hairs (trichomes), which are normally only found on leaves and stems on the cotyledons. In addition, the vascular tissue in the transformed cotyledons is intermediate in complexity between that of leaves and cotyledons.

Distinct from the pattern mutations are those that affect morphogenesis —the shape of the embryo and seedling. One such mutation, *fass*, mentioned above, randomizes patterns of cell division and results in fat, short seedlings in which the cells are more rounded and irregularly spaced than the orderly stacks of elongated cells in a wild-type embryo. But although *fass* seedlings are squat, because of an enlarged vascular system and multiple cortical layers, the basic radial pattern is normal. This again suggests that the initial embryonic patterning is not dependent on cell shape or size, but on the relative positions of the cells.

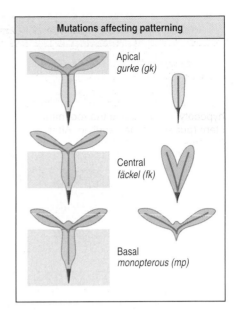

Mutations affecting patterning

Apical
gurke (gk)

Central
fäckel (fk)

Basal
monopterous (mp)

Fig. 7.6 Deletion of pattern by mutation along the apical-basal axis in *Arabidopsis* embryos. Three main classes of mutants have been discovered —apical, central, and basal. The regions deleted in each class of mutants are shown highlighted on the wild-type seedlings on the left, with the resulting mutant seedlings on the right. Examples of genes in which mutations give rise to these phenotypes are given. After Mayer, U., *et al.*: 1991.

7-6 Plant somatic cells can give rise to embryos and seedlings.

As gardeners well know, plants have amazing powers of regeneration. A complete new plant can develop from a small piece of stem or root, or even from the cut edge of a leaf. This reflects an important difference between the developmental potential of plant and animal cells. In animals, with few exceptions, cell determination and differentiation are irreversible. By contrast, many somatic plant cells remain totipotent. Somatic cells from roots, leaves and stems, and even, for some species, a single isolated protoplast, can be grown in culture and induced by treatment with the appropriate growth hormones to give rise to a new plant (Fig. 7.7). Careful observation of plant cells proliferating in culture has revealed that some of the dividing cells give rise to cell clusters that pass through a stage strongly resembling normal embryonic development, although the pattern of cell divisions are not the same as those in the embryo. These 'embryos' can then develop into seedlings. The initial work on plant regeneration in culture was done with carrot cells, but a wide variety

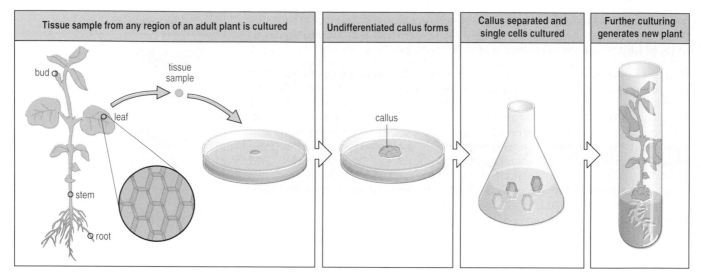

| Tissue sample from any region of an adult plant is cultured | Undifferentiated callus forms | Callus separated and single cells cultured | Further culturing generates new plant |

Fig. 7.7 Cultured somatic cells from a mature plant can form an embryo and regenerate a new plant. The illustration shows generation of a plant from single cells. If a small piece of tissue from a plant stem or leaf is placed on a solid agar medium containing the appropriate nutrients and growth hormones, the cells will start to divide to form a disorganized mass of undifferentiated cells—a callus. The callus cells are then separated and grown in liquid culture, again containing the appropriate growth hormones. In suspension culture, some of the callus cells divide to form small cell clusters. These cell clusters resemble the globular stage of a dicotyledon embryo, and with further culture on solid medium, develop through the heart-shaped and later stages to regenerate a complete new plant.

of plants can form embryos from single cells in this way. The phenomenon has been studied most in dicotyledons, such as carrot, potato, petunia, and tobacco, which are easy to propagate by these means.

The ability of single somatic cells to give rise to whole plants has two important implications for plant development. The first is that maternal determinants may be of little or no importance in plant embryogenesis, as it is unlikely that every somatic cell would still be carrying such determinants. Second, it suggests that many cells in the 'adult' plant body are not fully determined with respect to their fate, but remain totipotent. Of course, this totipotency seems only to be expressed under special conditions, but it is, nevertheless, quite unlike the behavior of animal cells. It is as if such plant cells have no long-term developmental memory.

Summary.

Early embryonic development in many plants is characterized by asymmetric cell division of the zygote, which specifies apical and basal regions. The free-living 'embryos' of the algae *Fucus* and *Volvox* provide useful and accessible models for studying early pattern formation. The apical-basal axis of the *Fucus* zygote, which defines the first plane of cleavage and the future rhizoid and thallus, is specified by environmental signals. It is fixed by electrical currents and stabilized by interactions between the cytoskeleton and cell wall. In *Fucus*, cell wall components seem to be a major determinant of cell fate. In the colonial green alga *Volvox*, cell fate is determined by differences in cell size arising from asymmetric divisions. Embryonic development in flowering plants establishes the shoot and root meristems from which the adult plant develops. Patterning of the embryo of flowering plants seems to rely less on stereotyped patterns of cell divisions than on cell–cell interactions. Particular genes control radial differentiation and patterning of specific regions along the apical-basal axis of the *Arabidopsis* embryo. One major difference between plants and animals is that, in culture, a single somatic plant cell can develop through an embryo-like stage and regenerate a complete new plant, indicating that some differentiated plant cells retain totipotency.

Summary: early development in flowering plants

first asymmetric cell division in embryo establishes apical-basal axis

⇓

embryonic cell fate is determined by position

⇓

shoot and root meristems of seedling give rise to all adult plant structures

Meristems.

The embryonic root and shoot meristems give rise to all the structures of the adult plant. Unlike animals, where the late embryo can usually be regarded as a miniature version of the adult, in plants the adult structures are derived from just two regions of the embryo, the shoot and root meristems. The shoot meristem, for example, gives rise to leaves, internodes, and flowers. As the shoot meristem develops, lateral outgrowths from the meristem give rise to leaves and additional meristems. These outgrowths are initially small but increase in size due to cell proliferation and enlargement. There is usually a time delay between the initiation of two successive leaves, and this results in a plant shoot being formed from repeated modules (Fig. 7.8). Each module consists of an **internode** (the cells produced by the meristem between successive leaf initiations), a **node** and its associated leaf, and an axillary bud. The axillary bud itself contains a meristem, and can form a side shoot when the inhibitory influence of the shoot tip is removed. Root growth is not so obviously modular, but similar considerations apply, as new lateral meristems initiated behind the root apical meristem give rise to lateral roots.

Meristems are small areas, containing relatively small, undifferentiated cells. They are rarely more than 250 μm in diameter in flowering plants. Most, but not all, cell divisions in normal plant development occur within the meristems, or soon after a cell leaves a meristem, and much of the subsequent growth is due to cell enlargement. Since a meristem remains constant in size during growth, cells are continually leaving it, and as they do so they begin to differentiate. The central region of the shoot meristem or the root meristem is commonly referred to as the **promeristem**. This central region contains cells known as **initials**, which behave in the same way as animal stem cells. They are self-renewing and give rise to the cells of the meristem. These initials generally divide rather slowly; their progeny divide more rapidly as they move toward the peripheral parts of the meristem. In the shoot, individual initials eventually get displaced from the meristem and their place is taken by another cell, which then behaves as an initial.

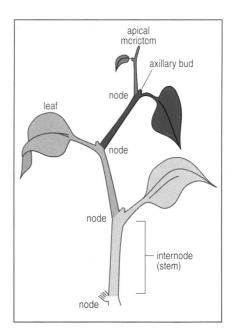

Fig. 7.8 Plant shoots grow in a modular fashion. The shoot apical meristem produces a repeated basic unit of structure called a module. The vegetative shoot module typically consists of internode, node, leaf, and axillary bud (from which a side branch may develop). Successive modules are shown here in different shades of green. As the plant grows, the internodes behind the meristem lengthen and the leaves expand. After Alberts, B., *et al.*: 1989.

The fate of a cell in the shoot meristem is dependent on its position.

The shoot meristem of dicotyledons gives rise to the stalk and leaves. It is made up of three layers (Fig. 7.9). L1 is the outermost layer and is just one cell thick. L2 is also one cell thick and lies beneath L1. In both L1 and L2, cell divisions are **anticlinal**—that is, in a plane perpendicular to the layer—thus maintaining the organization of these layers. The innermost layer is L3, in which the cells can divide in any plane. L1 and L2 are often known as the tunica, and L3 as the corpus.

The fate of the cells in each layer has been determined using plant chimeras. Chimeric tissues are composed of cells of two different genotypes which can be distinguished from each other by distinctive features such as polyploid nuclei (nuclei containing extra sets of chromosomes) or pigmentation. Chimeras can be made by treatment of the meristem with radiation or with chemicals such as colchicine, which induce polyploidy by blocking nuclear division but not chromosomal division. So-called **periclinal chimeras** can be produced, in which one of the three layers of the meristem has a genetic marker that distinguishes its cells and their progeny from those of the other two layers (Fig. 7.10).

Fig. 7.9 Apical meristem of *Arabidopsis*. Top panels: scanning electron micrographs showing the organization of the meristem at the young vegetative apex of *Arabidopsis*. The plant is a *clavata1-1* mutant, which has a broadened apex, allowing for a clearer visualization of the leaf primordia (L) and the meristem (M). Bottom panel: diagram of a vertical section through the apex of a shoot. The three-layered structure of the meristem is apparent in the most apical region. In layer 1 (L1) and layer 2 (L2), the plane of cell division is anticlinal, that is at right angles to the surface of the shoot. Cells in layer 3 (L3, the corpus) can divide in any plane. A leaf primordium is shown forming at one side of the meristem. Scale bar = 10 μm. Photographs courtesy of M. Griffiths.

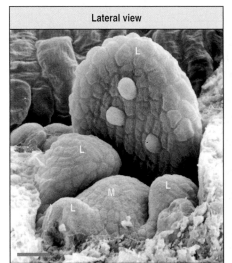

Lateral view

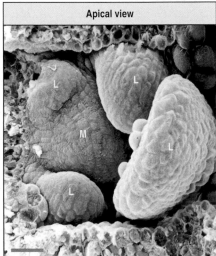

Apical view

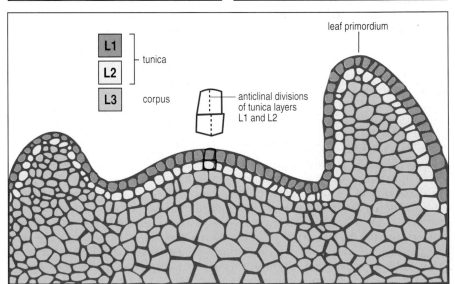

Fig. 7.10 Chimeric meristem (periclinal chimera) composed of cells of two different genotypes. In L1 the cells are diploid whereas the cells of L2 are tetraploid—they have double the normal chromosome number, and are larger and more easily recognized. After Steeves, T.A., *et al.*: 1989.

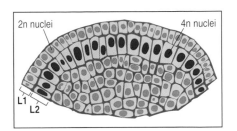

By following the fate of the marked cells one can see what structures each layer gives rise to. In angiosperms, layer L1 gives rise to the epidermis of all the plant structures, while L2 and L3 can both contribute to cortex and vascular structures. Although the three layers maintain their identity in the central region of the meristem over long periods of growth, cells in either L1 or L2 occasionally divide **periclinally** so that new cell walls are formed parallel to the surface of the meristem, and thus one of the new cells invades an adjacent layer. This vagrant cell now develops according to its new position, showing that cell fate is not necessarily determined by the meristem layer in which the cell originated. The L2 layer also becomes disrupted by periclinal divisions as soon as leaves begin to form. At the early stage at which a leaf is determined it is known as a primordium (see Fig. 7.9).

It is important to know what structures in the plant, such as particular internodes and leaves, arise from which regions of the meristem. This can be determined using clonal analysis in a manner similar to that used in *Drosophila* research (see Box 5B, page 153). Single mutant, or otherwise distinguishable meristematic cells, can be generated by X-irradiation or by the activation of a transposable element, which can result, for example, in marked cells of a different color from the rest of the plant. Because there is no cell migration during plant development, all cells deriving from the marked meristem cell form a coherent sector of marked cells as the plant grows. The resulting chimeric plants are known as **mericlinal chimeras**, as only one sector of the plant or of an organ is marked by the clone (Fig. 7.11).

Analysis of maize mericlinal chimeras grown from irradiated seeds shows that the most common pattern of development is of a marked sector that appears at the base of an internode and extends apically, terminating in a leaf. Some **clones** populate just a single internode whereas others populate sectors extending through as many as 13 internodes. This tells us that initials eventually become displaced from the meristem, although some remain as initials for a long time, contributing to several node-internode modules in succession. In similar experiments in sunflowers, marked clones extending through several internodes and up into the **inflorescence** are found, showing that the same initial can contribute to both vegetative and floral structures.

When the varied patterns of the sectors that arise from individual cells are examined, clonal sectors can often be found populating similar but not identical regions of the plant body, showing that the fate of a particular cell is not fixed. Nevertheless, the similarity suggests a predictable pattern of cell division. The clonal analysis experiments indicate that initial cells are simply those that happen to be in the central region of the meristem at any given time. Most initials are displaced from the meristem after a time, and are replaced by other cells. Their self-renewing behavior as initials is a consequence of their position, rather than of some intrinsic difference from other meristem cells.

Using the results of clonal analysis, a rough fate map has been constructed for the shoot apical meristem of the mature maize embryo in the seed. The meristem fate map can only be approximate because it is not possible to mark a cell at a particular location in the embryonic meristem and then follow its development, as it is inaccessible inside the seed. Plant fate maps are thus more probabilistic and are based on indirect methods.

Fig. 7.11 Tobacco plant mericlinal chimera. There is an albino mutation in the L2 layer of the shoot apical meristem. The affected area occupies about one third of the circumference of the shoot, suggesting that there are three apical initial cells in the shoot meristem. Photograph courtesy of S. Poethig.

Box 7B Transgenic plants.

One of the most common ways of generating transgenic plants containing new and modified genes is through infection of plant tissue in culture with the bacterium *Agrobacterium tumefaciens*, the causal agent of crown gall tumors. *Agrobacterium* is a natural genetic engineer. It contains a **plasmid**—the Ti plasmid—that contains the genes required for the transformation and proliferation of infected cells to form a callus. During infection, a portion of this plasmid—the T-DNA (shown in red below)—is transferred into the genome of the plant cell, where it becomes stably integrated.

Genes experimentally inserted into the T-DNA will therefore also be transferred into the plant cell chromosomes. Ti plasmids, modified so that they do not cause tumors but still retain the ability to transfer T-DNA, are widely used as vectors for gene transfer in dicotyledonous plants. The genetically modified plant cells of the callus can then be grown into a complete new transgenic plant that carries the introduced gene in all its cells and can transmit it to the next generation.

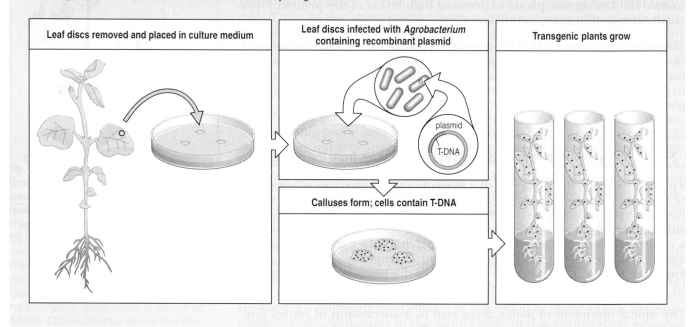

Leaf discs removed and placed in culture medium

Leaf discs infected with *Agrobacterium* containing recombinant plasmid

Transgenic plants grow

plasmid

T-DNA

Calluses form; cells contain T-DNA

7-9 **Leaf positioning and phyllotaxy involves lateral inhibition.**

As the shoot grows, leaves are generated within the meristem at regular intervals and with a particular spacing. The first indication of leaf initiation in the meristem is usually a swelling of a region to the side of the apex to form a leaf primordium. This small protrusion is the result of increased localized cell multiplication and altered patterns of cell division. It also reflects changes in polarized cell expansion. Leaves can be arranged along a shoot in a variety of ways in different plants, and the particular arrangement—or **phyllotaxy**—is reflected in the arrangement of leaf primordia in the meristem. Leaves can occur singly at each node, in pairs, or in whorls of three or more. A common arrangement is the positioning of single leaves spirally up the stalk, which can sometimes form a striking helical pattern in the shoot apex.

An examination of the pattern of leaf initiation in plants, where leaves are borne singly in a spiral arrangement up the stem, shows that a new leaf primordium forms at the center of the first available space outside the promeristem and above the previous primordium (Fig. 7.15). This pattern suggests a mechanism for leaf arrangement based on lateral inhibition, in which each leaf primordium inhibits the formation of a new leaf within a given distance from it. It appears that inhibitory signals emanating from recently initiated primordia prevent leaves from forming

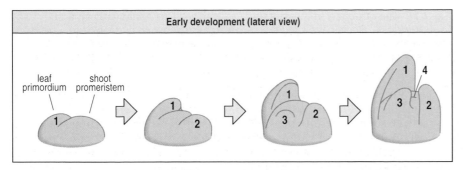

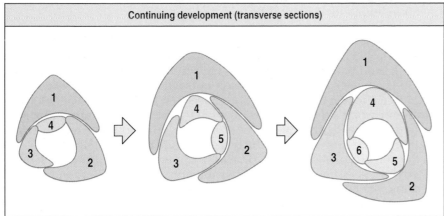

Fig. 7.15 Leaf phyllotaxis. In shoots where single leaves are arranged spirally up the stem, the leaf primordia arise sequentially in a mathematically regular pattern in the meristem. Leaf primordia arise around the sides of the apical dome, just outside the promeristem region. A new leaf primordium is formed slightly above and at a fixed radial angle from the previous leaf, often generating a helical arrangement of primordia visible at the apex. Top panel: lateral views of the shoot apex. Bottom panel: view looking down on cross-sections through the apex near the tip, at successive stages from the top panel. After Poethig, R.S., *et al.*: 1985 (top panel); and Sachs, T.: 1994 (bottom panel).

close to each other. There is some experimental evidence to support such a model in plants. In ferns, leaf primordia are widely spaced, allowing experimental microsurgical interference. Destruction of the site of the next primordium to be formed results in a shift toward that site by the future primordium whose position is closest to it (Fig. 7.16).

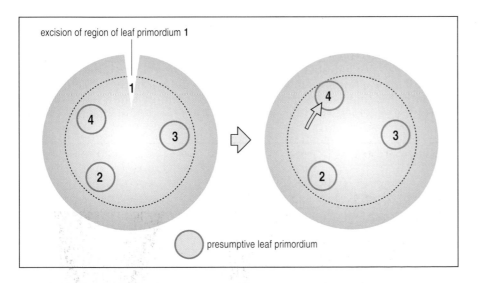

Fig. 7.16 Lateral inhibition may position leaf primordia. In ferns, a cut in the meristem in the region of a presumptive leaf primordium results in the position of the future adjacent primordium shifting closer to that site.

7-10 | **Root tissues are produced from root apical meristems by a highly stereotyped pattern of cell divisions.**

The organization of tissues in the *Arabidopsis* root is shown in Fig. 7.17 The radial pattern comprises single layers of epidermal, cortical, endodermal, and pericycle cells. Root apical meristems resemble shoot apical

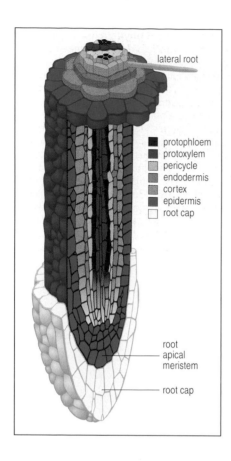

protophloem
protoxylem
pericycle
endodermis
cortex
epidermis
root cap

lateral root

root apical meristem

root cap

Fig. 7.17 The structure of the root tip in *Arabidopsis*. Roots have a radial organization. In the center of the growing root tip is the future vascular tissue (protoxylem and protophloem). This is surrounded by further tissue layers.

meristems in many ways and give rise to the root in a similar manner to that in which the shoot is generated. But there are some important differences between the root and shoot meristems. The shoot meristem is at the extreme tip of the shoot whereas the root meristem is covered by a root cap (which is itself derived from one of the layers of the meristem), and there is no obvious segmental arrangement at the root tip resembling the node–internode–leaf module.

The root is set up early and its origin can be identified in the late heart-stage embryo (Fig. 7.18). Clonal analysis has shown that the root meristem can be traced back to a set of initial cells that come from a single tier of cells in the heart-stage embryo. Each column of cells in the root has its origin in a specific initial cell in the meristem, and each initial cell has a stereotyped pattern of cell divisions that leads to each column. The normal pattern of cell divisions is not obligatory. Ablation with a laser beam of some meristem cells alters the pattern of cell divisions, but the tissues are normal because novel cell divisions replace the cells that have been destroyed. As discussed earlier, *fass* mutants, which have disrupted cell divisions, still have a relatively normal patterning in the root. Such observations show the importance of cell interactions in patterning the root.

Summary.

Meristems are the growing points of a plant. They consist of small regions of undifferentiated cells that are capable of repeated division. The apical meristems, found at the tip of shoots and roots, give rise to all the plant organs—roots, stem, leaves, and flowers. The fate of a cell in the shoot meristem clearly depends upon its position rather than its lineage, because when a cell is displaced from one layer and becomes part of another it adopts the fate of its new layer. Fate maps of the embryonic shoot meristem in maize show that it can be divided into domains, each of which can contribute to the tissues of a particular region of the plant. Cell–cell interactions seem to play a role in determining cell fate and, in line with this, meristems are capable of regulation when parts of them are removed. The shoot meristem gives rise to leaves in species-specific

Fig. 7.18 Fate map of root regions in the heart-stage *Arabidopsis* embryo. The root grows by the division of a set of initial cells. The root meristem comes from a small number of cells in the heart-shaped embryo. Each tissue in the root is derived from the division of a particular initial cell. At the center of the root meristem is a quiescent center, which does not divide. After Scheres, B., *et al.*: 1994.

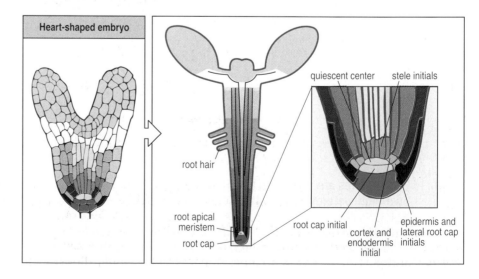

Heart-shaped embryo

root hair

root apical meristem

root cap

quiescent center   stele initials

root cap initial

cortex and endodermis initial

epidermis and lateral root cap initials

patterns—phyllotaxy—which seems best accounted for in terms of lateral inhibition. In the root meristem, the cells are organized rather differently than those in the shoot meristem, and there is a much more stereotyped pattern of cell division. A set of initial cells maintain root structure by dividing along different planes.

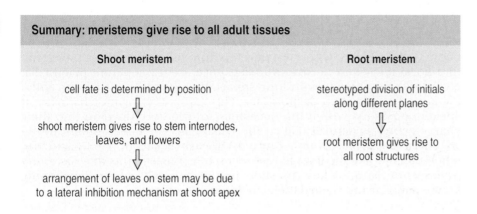

Flower development.

Flowers contain the reproductive cells of higher plants and develop from the shoot meristem. In most plants, the transition from a vegetative shoot meristem to a **floral meristem** that produces a flower is largely or absolutely under environmental control, with day length and temperature being important determining factors. Flowers, with their arrangement of floral organs (sepals, petals, stamens, and carpels), are rather complex, and it is a major challenge to understand how they arise from the floral meristem. Here, we consider the mechanisms that pattern the flower, particularly those genes that specify the identity of the floral organs.

Three classes of genes seem to control the basic patterning of a flower. Organ identity genes specify the identity of the different floral organs, and have a function equivalent to homeotic selector genes in animals (see Section 5-15). Cadastral genes set the boundaries for expression of the organ identity genes and prevent their ectopic expression. Expression of the meristem identity genes converts a vegetative shoot meristem into one that makes flowers.

| 7-11 | Homeotic genes control organ identity in the flower. |

In a flowering shoot, the shoot meristem becomes converted into an inflorescence meristem, which can form one or more floral meristems, each of which develops into a single flower. The **floral organ primordia**, from which the individual parts of the flower develop, arise in the floral meristem by patterned cell division followed by cell differentiation and enlargement. A flower is composed of four concentric whorls of structures (Fig. 7.19), which reflect the arrangement of the floral organ primordia in the meristem. The sepals (whorl 1) form from the outermost ring of meristem tissue, and the petals (whorl 2) from a ring of tissue lying immediately inside it. An inner ring of tissue gives rise to the male reproductive organs—the stamens (whorl 3). The female reproductive

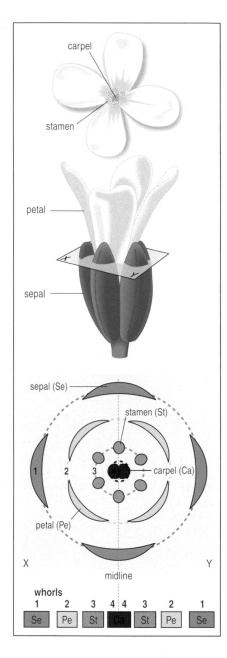

Fig. 7.19 Structure of an *Arabidopsis* flower. Top: *Arabidopsis* flowers are radially symmetrical and have an outer ring of four identical green sepals, enclosing four identical white petals, within which is a ring of six stamens, with two carpels in the center. Bottom: floral diagram of the *Arabidopsis* flower representing a cross-section taken in the plane indicated in the top diagram. This is a conventional representation of the arrangement of the parts of the flower, showing the number of flower parts in each whorl and their arrangement relative to each other. After Coen, E.S., *et al.*: 1991.

organs—the carpels (whorl 4)—develop from the center of the meristem. In a floral meristem of *Arabidopsis*, there are 15 separate primordia, giving rise to a flower with four sepals, four petals, six stamens and a pistil made up of two carpels (see Fig. 7.19). The primordia arise at specific positions within the meristem, where they develop into their characteristic structures (Fig. 7.20). During flower development in the snapdragon (*Antirrhinum*), there is lineage restriction to particular whorls, rather like the lineage restriction to compartments in *Drosophila* (see Section 5-15). A key question is how the cells of the developing flower acquire and record their positional identity.

In *Arabidopsis* and other plants, **homeotic mutations** have been discovered that result in the development of abnormal flowers in which one type of flower part is replaced by another. In the mutant *apetala2*, for example, the sepals are replaced by carpels and the petals by stamens; in the *pistillata* mutant, petals are replaced by sepals and stamens by carpels. These mutations identify the floral organ identity genes, and have enabled their mode of action to be determined.

The homeotic floral mutations in *Arabidopsis* fall into three classes, each of which affects the organs of two adjacent whorls (Fig. 7.21). The first class of mutations, of which *apetala2* is an example, affect whorls 1 and 2, giving carpels instead of sepals in whorl 1, and stamens instead of petals in whorl 2. The phenotype of the flower, going from the outside to the center, is therefore carpel, stamen, stamen, carpel. The second class of homeotic floral mutations affect whorls 2 and 3. In this class, *apetala3* and *pistillata* give sepals instead of petals in whorl 2 and carpels instead of stamens in whorl 3, with a phenotype sepal, sepal, carpel, carpel. The third class of mutations affects whorls 3 and 4 and gives petals instead of stamens in whorl 3 and sepals or variable structures in whorl 4. The mutant *agamous*, which belongs to this class, has an extra set of sepals and petals in the center instead of the reproductive organs.

These mutant phenotypes can be accounted for by quite a simple model of gene activity. Let us assume that the floral meristem is divided into three overlapping regions, A, B, and C, each region corresponding to the site of action of one of the three classes of homeotic mutations (Fig. 7.22). Region A thus covers whorl 1 and 2; B, whorls 2 and

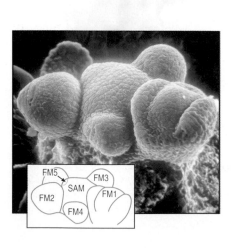

Fig. 7.20 Scanning electron micrograph of an *Arabidopsis* inflorescence meristem. The central inflorescence meristem (shoot apical meristem, SAM) is surrounded by a series of floral meristems (FM) of varying developmental ages. The inflorescence meristem grows indeterminately, with cell divisions providing new cells for the stem below, and new floral meristems on its flanks. The floral meristems (or primordia) arise one at a time in a spiral pattern. The most mature of the developing flowers is on the right (FM1), showing the initiation of four sepal primordia surrounding a still-undifferentiated floral meristem. Eventually such a floral meristem would form petal, stamen, and carpel primordia. Photograph from Meyerowitz, E.M., *et al.*: 1991.

Fig. 7.21 Homeotic floral mutations in *Arabidopsis*. Left panel: an *apetala2* mutant has whorls of carpels and stamens in place of sepals and petals. Center panel: an *apetala3* mutant has two whorls of sepals and two of carpels. Right panel: *agamous* mutants have a whorl of petals and sepals in place of stamens and carpels. Transformations of whorls are shown inset, and can be compared to the wild-type arrangement, as shown in Fig. 7.19. Photographs from Meyerowitz, E.M., *et al.*: 1991 (left panel), and Bowman, J.L., *et al.*: 1989 (center panel).

3; and C, whorls 3 and 4. We next assume that there are three regulatory functions—*a, b,* and *c*—which function in regions A, B, and C, respectively, and which, combinatorially, can give each whorl a unique identity and so specify organ identity (Fig. 7.23, left panel). *a* is expressed in whorls 1 and 2, *b* in 2 and 3, and *c* in whorls 3 and 4. In addition, *a* function inhibits *c* function in whorls 1 and 2 and *c* function inhibits *a* function in whorls 3 and 4—that is, *a* and *c* functions are mutually exclusive. Any floral meristem in which *a* activity alone is present (that is, whorl 1) will develop sepals; *a* and *b* acting together direct petal development in whorl 2; *b* and *c* acting together specify stamens in whorl 3; and *c* alone specifies carpel formation in whorl 4. What the homeotic mutations do is eliminate the functions of *a, b,* or *c*. So mutations in the first class (such as *apetala2* mutations) disrupt function *a*; expression of *c* is then not prevented in whorls 1 and 2, and so *c* is expressed in all whorls, giving the phenotype carpel, stamen, stamen, carpel (see Fig. 7.23, center panel). Mutations in *b*, the second class of mutations (such as *apetala3*), result in only *a* functioning in whorls 1 and 2, and *c* in whorls 3 and 4, giving sepal, sepal, carpel, carpel. Mutations in *c* function genes (such as *agamous*), result in *a* activity in all whorls, giving the phenotype sepal, petal, petal, sepal (see Fig. 7.23, right panel).

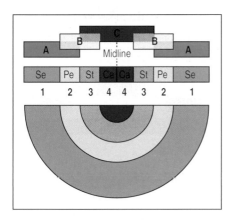

Fig. 7.22 The three overlapping regions of the *Arabidopsis* floral meristem that have been identified by the homeotic floral identity mutations. Region A corresponds to whorls 1 and 2, B to whorls 2 and 3, and C to whorls 3 and 4.

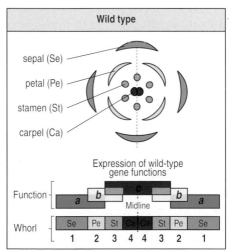
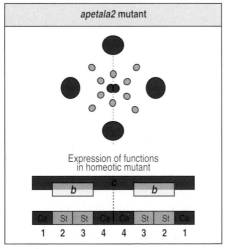
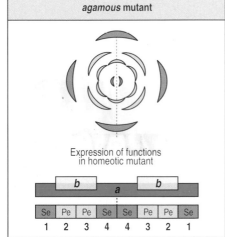

Fig. 7.23 Model for the patterning of the *Arabidopsis* flower. In the wild-type flower (left panel), it is assumed that three regulatory functions, *a, b,* and *c*, are expressed in whorls 1 and 2, 2 and 3, and 3 and 4, respectively. *a* alone specifies sepals, *a* and *b* together specify petals, *b* and *c* stamens, and *c* alone carpels. Mutations alter the regions within the meristem where these functions are expressed.

In the *apetala2* mutant (center), function *a* is absent and *c* spreads throughout the meristem, resulting in the half-flower pattern of carpel, stamen, stamen, carpel. In the *agamous* mutant (right) there is no *c* function and expression of *a* spreads throughout the meristem, resulting in the half-flower pattern of sepal, petal, petal, sepal. After Dennis, E., *et al.*: 1993.

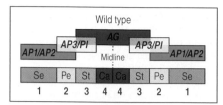

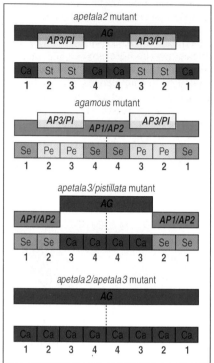

Fig. 7.24 Model of gene action controlling flower pattern in *Arabidopsis*. The gene *APETALA1* (*AP1*) is expressed in whorls 1 and 2; *APETALA3* (*AP3*) and *PISTILLATA* (*PI*) in whorls 2 and 3; and *AGAMOUS* (*AG*) in whorls 3 and 4. Sepals require *APETALA1* and APETALA2, (*APETALA2* is expressed in all whorls, but acts in organ identity specification only along with the whorl 1- and 2-limited *APETALA1* gene); petals a combination of *APETALA1* and *APETALA2* together with *APETALA3* and *PISTILLATA*; stamens *APETALA3* and *PISTILLATA* with *AGAMOUS*; and carpels require *AGAMOUS* alone. Mutations that alter the pattern of expression of one or more of these genes result in differing patterns of expression of the others and homeotic transformations of the floral parts. After Meyerowitz, E.M., *et al.*: 1991.

All the floral homeotic mutants discovered so far can be quite satisfactorily accounted for by this model, and particular genes can be assigned to each of the controlling functions. Function *a* corresponds to the activity of *a* function genes such as *APETALA2*, *b* to *APETALA3* and *PISTILLATA*, and *c* to *AGAMOUS*. The model also accounts for the phenotype of double mutants, such as *apetala2* with *apetala3*, and *apetala3* with *pistillata* (Fig. 7.24).

Another way of examining the action of these homeotic genes is to look at the phenotype of the flower when the action of all of them is lacking, and then study the effects when they are added back one by one (Fig. 7.25). A similar approach was taken with the homeotic genes that specify segment identity in *Drosophila* (see Fig. 5.37). In the absence of all three classes of gene, the flower consists of whorls of identical leaf-like organs; this can be regarded as a developmental ground state. Addition of *a* function genes to the ground state results in all sepals, while addition of *c* function to the ground state gives all carpels. Addition of both *a* and *c* results in flowers with half-flower phenotype sepal, sepal, carpel, carpel, showing that, in accord with the above model, expression of *a* and *c* are now restricted to regions A and C, respectively, as a result of their mutually inhibitory actions.

The model clearly predicts particular spatial patterns of gene activity in the meristem. For example, *b* class genes should be expressed only in the B region, that is, in whorls 2 and 3. Expression of the *b* class gene *APETALA3* is indeed first detected by *in situ* hybridization (see Box 3B, page 65) in the floral meristem at the time the sepal primordia begin to form, and is restricted to whorls of cells that will give rise to petals and stamens. By contrast, the *c* class gene *AGAMOUS* is expressed in the meristem region that gives rise to whorls 3 and 4 (Fig. 7.26).

From the DNA sequence, one can deduce that the homeotic proteins encoded by floral identity genes such as *APETALA1* and *AGAMOUS* contain a conserved sequence of 58 amino acids, known as the MADS box, which is thought to be able to bind to DNA. The MADS box is also present in some transcription factors from yeast and animals. This makes it highly likely that, as with the homeotic selector genes of *Drosophila*, the floral identity genes are transcription factors. The MADS box transcription factors are expressed in regions of the flower that exhibit homeotic transformations when one gene is absent. In line with the model of expression described above, organ identity genes have been found to be sufficient to specify new organ identity. Thus, expression of *AGAMOUS*—a *c* function gene—with a constitutive promoter represses *a* function in whorls 1 and 2; the result is phenotypically the same as loss-of-function *a* mutants. The flower has carpels in place of sepals and stamens in place of petals.

Further evidence for this model comes from expressing *APETALA3* and *PISTILLATA* throughout the flower. The result is that the two outer whorls are petals and the two inner ones are stamens. This shows that

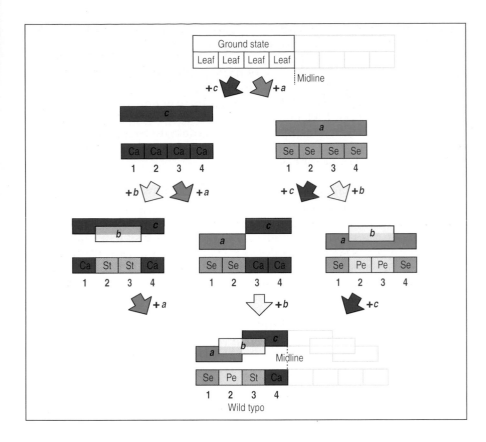

Fig. 7.25 The combinatorial action of the *a*, *b*, and *c* functions in patterning the flower. The floral organs can be considered to be a modification of a ground state in which only leaves are present. Leaves are modified into sepals, petals, stamens, and carpels by the expression of the floral identity functions *a*, *b*, and *c*. Addition of function *a* alone transforms the ground state of leaf into all sepals (Se), while addition of function *c* alone results in all organs becoming carpels (Ca). Addition of different combinations of functions to specific whorls results in the formation of different structures. This model assumes that *a* and *c* restrict each other's expression. After Coen, E.S., *et al.*: 1991.

expression of these two genes is sufficient to provide *b* organ identity —petals and stamens—in combination with class *a* and *c* genes. All that these results emphasize the similarity in function between the homeotic genes in animals and those controlling organ identity in flowers. The similarity with the HOM complex of *Drosophila* is further illustrated by the role of the *CURLY LEAF* gene of *Arabidopsis*, which is necessary for the stable maintenance of homeotic gene activity. *CURLY LEAF* is related to the Polycomb family of *Drosophila* and is similarly required for stable repression of homeotic genes.

Despite the enormous variation in the flowers of different species, the mechanisms underlying flower development seem to be very similar. For example, there are striking similarities between the floral control genes of *Arabidopsis* and the snapdragon, *Antirrhinum*. In developing *Antirrhinum* flowers, the patterns of activity of the corresponding genes fit well with the cell lineage restriction to whorls, seen above for *Arabidopsis*. Before the emergence of organ primordia, cells have not acquired organ identity and there is no lineage restriction between whorls. This only occurs at the time when the pentagonal symmetry of the flower becomes visible and organ identity genes are expressed.

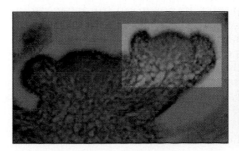

Fig. 7.26 Expression of *APETALA3* and *AGAMOUS* during flower development. *In situ* hybridization shows *AGAMOUS* is expressed in the central whorls (left panel), whereas *APETALA3* is expressed in the outer whorls that give rise to petals and stamens (right panel).

How are the spatial patterns of expression of these homeotic genes controlled? We do not yet know, but genes that restrict the expression of the homeotic genes have been discovered in *Arabidopsis*. These are the cadastral genes, of which *SUPERMAN* is an example. Plants with a mutation in this gene have stamens instead of carpels in the fourth whorl. The action of *SUPERMAN* seems to prevent the expression of *APETALA3* and *PISTILLATA* in the fourth whorl. *SUPERMAN* is expressed in the third whorl, and maintains the boundary between the third and fourth whorls.

7-12   The transition to a floral meristem is under environmental and genetic control.

Flowering in *Arabidopsis* is controlled by internal and external factors, with day length playing an important role (Fig. 7.27, top panels). Most flowering plants have a vegetative phase during which the apical meristem generates leaves. Triggered by environmental signals such as day length, they switch to a reproductive phase and give rise to flowers. There are two types of transition from vegetative growth to flowering. In the determinate type, the inflorescence meristem becomes a terminal flower, whereas in the indeterminate type the inflorescence meristem gives rise to a number of floral meristems. The transition in *Arabidopsis* is of the indeterminate type. During vegetative growth, the shoot meristem generates leaves, together with a new potential meristem at the base of each leaf. A primary response to floral inductive signals in *Arabidopsis* is the transcription of floral meristem identity genes such as *LEAFY* and *APETALA1*. Mutations in these genes partly transform flowers into shoots. In a *leafy* mutant, the flowers are transformed into spirally arranged sepal-like organs along the stem, whereas expression of *LEAFY* throughout the plant is sufficient to determine floral fate in lateral shoot meristems (see Fig. 7.27, bottom panels). The gene in *Antirrhinum* that functions in a similar way to *LEAFY* is *FLORICAULA*. The abnormal expression of *FLORICAULA* in just one layer of a shoot meristem can allow flower development and induce the genes required for flower development in layers in which *FLORICAULA* is normally not expressed. This clearly shows that one layer of the meristem can induce development in adjacent layers.

7-13   The *Antirrhinum* flower is patterned dorso-ventrally as well as radially.

Like *Arabidopsis* flowers, those of *Antirrhinum* consist of four whorls, but unlike *Arabidopsis* flowers, they have five sepals, five petals, four stamens, and two united carpels (Fig. 7.28, left). Floral homeotic mutations similar to those in *Arabidopsis* occur in *Antirrhinum*, and floral organ identity is almost certainly specified in the same way. Several of the *Antirrhinum* homeotic genes have extensive homology with those of *Arabidopsis*, the MADS box in particular being well conserved.

However, an extra element of patterning is required in the *Antirrhinum* flower, which has a bilateral symmetry imposed on the basic radial pattern common to all flowers. In whorl 2, the upper two petal lobes have a shape quite distinct from the lower three, giving the flower its characteristic snapdragon appearance. In whorl 3, the uppermost stamen is absent, as its development is aborted early on. The *Antirrhinum* flower, therefore, has a distinct dorso-ventral axis. Another group of homeotic genes, different from those that govern floral organ identity, appear to act in this dorso-ventral patterning. Mutations in the gene *CYCLOIDEA*, for example, abolish dorso-ventral polarity and produce flowers that are more radially symmetrical (see Fig. 7.28, right).

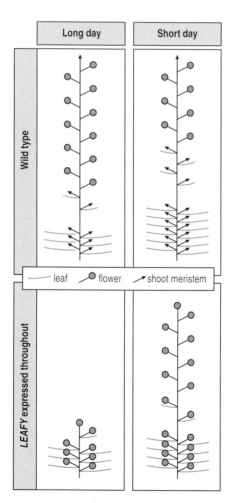

Long day   Short day

Wild type

— leaf   flower   shoot meristem

LEAFY expressed throughout

Fig. 7.27 Day length and *LEAFY* expression can control flowering. If *Arabidopsis* is grown under long day length conditions, few lateral shoots are formed before the apical shoot meristem begins to form floral meristems. Lateral shoots can also give rise to flowers. When grown under short day length conditions, flowering is delayed and there are more lateral shoots and flowers. Expression of the *LEAFY* gene throughout the plant results in the lateral shoot meristems that would normally form being converted to floral meristems.

Dorso-ventral axis

Wild type

CYCLOIDEA mutant

Fig. 7.28 Mutations in *CYCLOIDEA* make the *Antirrhinum* flower symmetrical. In the wild-type flower (left) the petal pattern is different along the dorso-ventral axis. In the mutant (right) the flower is symmetrical. All the petals are like the most ventral one in the wild type and are folded back. Photograph courtesy of E. Coen, from Coen, E.S., *et al.*: 1991.

7-14   The internal meristem layer can specify floral meristem patterning.

Although all three layers of a floral meristem (Fig. 7.29) are involved in organogenesis, the contribution of cells from each layer to a particular structure may be variable. Cells from one layer can become part of another layer without disrupting normal morphology, suggesting that a cell's position in the meristem is the main determinant of its future behavior. Some insight into positional signaling and patterning in the floral meristem can be obtained by making periclinal chimeras (see Section 7-7) from cells that have different genotypes and that give rise to different types of flower. From such chimeras, one can find out whether the cells develop autonomously according to their own genotype, or whether their behavior is controlled by signals from other cells.

Chimeras can be generated by grafting between two plants of different genotypes. A new shoot meristem forms at the junction of the graft, and sometimes contains cells from both genotypes. Such chimeras can be made between wild-type tomato plants and tomato plants carrying the mutation *fasciated*, in which the flower has an increased number of floral organs per whorl. Similar flowers are found in chimeras in which only layer L3 contains *fasciated* cells (Fig. 7.30). The increased number of floral organs is associated with an overall increase in the size of the floral meristem, and this cannot be achieved unless the *fasciated* cells of layer L3 induce the wild-type L1 cells to divide more frequently than normal. The mechanism of intercellular signaling between L3 and L1 is not yet known. These results, together with those described for *FLORI-CAULA* above, illustrate the importance of signaling between layers in flower development.

Summary.

Before flowering, the vegetative shoot apical meristem becomes converted into an inflorescence meristem which either then becomes a flower or produces a series of floral meristems, each of which develops into a single flower. Genes involved in the initiation of flowering and patterning of the flower have been identified in both *Arabidopsis* and *Antirrhinum*. Expression of meristem identity genes is required for the formation of floral meristems from the inflorescence meristem. Homeotic floral organ identity genes, which specify the organ types found in the flowers have been identified from mutations that transform one flower part into another. On the basis of these mutations, a model has been

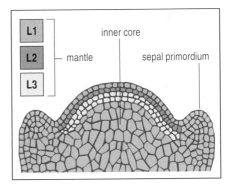

L1

L2

L3

inner core

mantle

sepal primordium

Fig. 7.29 Floral meristem. The meristem is composed of layers L1, L2, and L3. The inner core cells are derived from L3. The sepal primordia are just beginning to develop. After Drews, G.N., *et al.*: 1989.

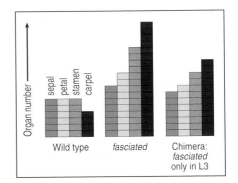

Organ number

sepal petal stamen carpel

Wild type

fasciated

Chimera: *fasciated* only in L3

Fig. 7.30 Floral organ number in chimeras of wild-type and *fasciated* tomato plants. In the *fasciated* mutant there are more organs in the flower than in wild-type plants. In chimeras in which only layer L3 of the floral meristem contains *fasciated* mutant cells, the number of organs per flower is still increased, showing that L3 can control cell behavior in the outer layers of the meristem.

proposed in which the floral meristem is divided into three concentric overlapping regions, in each of which certain floral identity genes act in a combinatorial manner to specify the organ type appropriate to each whorl. Studies with chimeric plants have shown that different meristem layers communicate with each other during flower development.

Summary: flower development in *Arabidopsis*

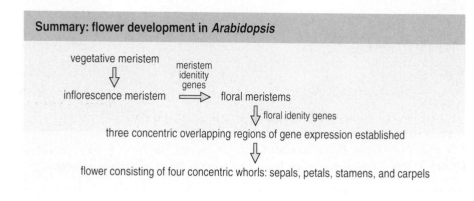

Summary to Chapter 7.

A distinctive feature of plant development is the presence of relatively rigid walls and the absence of any cell migration. Another is that a single isolated somatic cell from a plant can regenerate into a complete new plant. Early embryonic development is characterized by asymmetric cell division of the fertilized egg, which specifies the future apical and basal regions. In the alga *Fucus*, this plane of cleavage is specified by environmental signals. During early development of flowering plants, both asymmetric cell division and cell–cell interactions are involved in patterning the body plan. During this process, the shoot and root meristems are specified and these meristems give rise to all the organs of the plant—stems, leaves, flowers, and roots. The shoot meristem gives rise to leaves in well-defined positions, a process involving lateral inhibition. The shoot meristem eventually becomes converted to an inflorescence meristem, which either becomes a floral meristem (in determinate inflorescences) or gives rise to a series of floral meristems, retaining its shoot meristem identity indefinitely (in indeterminate inflorescences). In floral meristems, each of which develops into a flower, homeotic floral organ identity genes act in combination to specify the floral organ types.

General references.

Weigel, D., Meyerowitz, E.M.: **The ABCs of floral homeotic genes**. *Cell* 1994, **78**:203–209.

Zambryski, P.: **Plasmodesmata: plant channels for molecules on the move**. *Science* 1995, **270**:1943–1944.

Section references.

7-1 Electrical currents are involved in polarizing the *Fucus* zygote.

Goodner, B., Quatrano, R.S.: ***Fucus* embryogenesis: A model to study the establishment of polarity**. *Plant Cell* 1993, **5**:1471–1481.

Robinson, K.R., Cone, R.: **Polarization of fucoid eggs by a calcium ionophore gradient**. *Science* 1980, **207**:77–78.

7-2 Cell fate in early *Fucus* development is determined by the cell wall.

Berger, F., Taylor, A., Brownlee, C.: **Cell fate determination by the cell wall in early *Fucus* development**. *Science* 1994, **263**:1421–1423.

Shaw, S.L., Quatrano, R.S.: **The role of targeted secretion in the establishment of cell polarity and the orientation of the division plane in *Fucus* zygotes**. *Development* 1996, **122**:2623–2630.

7-3 Differences in cell size resulting from unequal divisions could specify cell type in the *Volvox* embryo.

Kirk, M.M., Ransick, A., McRae, S.E., Kirk, D.L.: **The relationship between cell size and cell fate in *Volvox carteri***. *J. Cell Biol.* 1993, **123**:191–208.

7-4 Both asymmetric cell divisions and cell position pattern the early embryos of flowering plants.

Meyerowitz, E.M.: **Plant development: local control, global patterning**. *Curr. Opin. Genet. Dev.* 1996, **6**:475–479.

Meyerowitz, E.M.: **Genetic control of cell division patterns in developing plants**. *Cell* 1997, **88**:299–308.

Torres-Ruiz, R.A., Jürgens, G.: **Mutations in the *FASS* gene uncouple pattern formation and morphogenesis in *Arabidopsis* development**. *Development* 1994, **120**:2967–2978.

7-5 The patterning of particular regions of the *Arabidopsis* embryo can be altered by mutation.

Jürgens, G., Torres-Ruiz, R.A., Berleth, T.: **Embryonic pattern formation in flowering plants**. *Ann. Rev. Genet.* 1994, **28**:351–371.

Jürgens, G.: **Axis formation in plant embryogenesis: cues and clues**. *Cell* 1995, **81**:467–470.

Lloyd, C.: **Plant morphogenesis: life on a different plane**. *Curr. Biol.* 1995, **5**:1085–1087.

Long, J.A., Moan, E.I., Medford, J.I., Barton, M.K.: **A member of the knotted class of homeodomain proteins encoded by the STM gene of *Arabidopsis***. *Nature* 1995, **379**:66–69.

7-6 Plant somatic cells can give rise to embryos and seedlings.

Zimmerman, J.L.: **Somatic embryogenesis: a model for early development in higher plants**. *Pl. Cell* 1993, **5**:1411–1423.

7-7 The fate of a cell in the shoot meristem is dependent on its position.

Irish, V.F., Sussex, I.M.: **A fate map of the *Arabidopsis* embryo shoot apical meristem**. *Development* 1992, **115**:745–753.

Irish, V.F.: **Cell lineage in plant development**. *Curr. Opin. Gen. Dev.* 1991, **1**:169–173.

Langdale, J.A.: **Plant morphogenesis: more knots untied**. *Curr. Biol.* 1994, **4**:529–531.

Turner, I.J., Pumfrey, J.E.: **Cell fate in the shoot apical meristem of *Arabidopsis thaliana***. *Development* 1992, **115**:755–764.

Sinha, N.R., Williams, R.E., Hake, S.: **Overexpression of the maize homeobox gene *knotted-1*, causes a switch from determinate to indeterminate cell fates**. *Genes Dev.* 1993, **7**:787–795.

7-8 Meristem development is dependent on signals from the plant.

Irish, E.E., Nelson, T.M.: **Development of maize plants from cultured shoot apices**. *Planta* 1988, **175**:9–12.

Sachs, T.: *Pattern Formation in Plant Tissues*. Cambridge: Cambridge University Press, 1994.

Sussex, I.M.: **Developmental programming of the shoot meristem**. *Cell* 1989, **56**:225–229.

7-9 Leaf positioning and phyllotaxy involves lateral inhibition.

Mitchison, G.J.: **Phyllotaxis and the Fibonacci series**. *Science* 1977, **196**:270–275.

Smith, L.G., Hake, S.: **The initiation and determination of leaves**. *Plant Cell* 1992, **4**:1017–1027.

7-10 Root tissues are produced from root apical meristems by a highly stereotyped pattern of cell divisions.

Benfey, P.N., Linstead, P.J., Roberts, K., Schiefelbein, J.W., Hauser, M-T., Aeschbacher, R.A.: **Root development in *Arabidopsis*: four mutants with dramatically altered root morphogenesis**. *Development* 1993, **119**:57–70.

Benfey, P.N., Schiefelbein, J.W.: **Getting to the root of plant development: genetics of *Arabidopsis* root formation**. *Trends Genet.* 1994, **10**:84–88.

Doerner, P.: **Radicle development(s)**. *Curr. Biol.* 1995, **5**:110–112.

Dolan, L., Jammaat, K., Willemsen, V., Linstead, P., Poethig, S., Roberts, K., Scheres, B.: **Cellular organization of the *Arabidopsis thaliana* root**. *Development* 1993, **119**:71–84.

Dolan, L., Roberts, K.: **Plant development: Pulled up by the roots**. *Curr. Opin. Genet. Dev.* 1995, **5**:432–438.

Hauser, M-T., Morikami, A., Benfey, P.N.: **Conditional root expansion mutants of *Arabidopsis***. *Development* 1995, **121**:1237–1252.

Scheres, B., McKhann, H.I., van den Berg, C.: **Roots redefined: anatomical and genetic analysis of root development**. *Plant Phys.* 1996, **111**:959–964.

7-11 Homeotic genes control organ identity in the flower.

Bowman, J.L., Sakai, H., Jack, T., Weigel, D., Mayer, U., Meyerowitz, E.M.: ***SUPERMAN*, a regulator of floral homeotic genes in *Arabidopsis***. *Development* 1992, **114**:599–615.

Coen, E.S., Meyerowitz, E.M.: **The war of the whorls: genetic interactions controlling flower development**. *Nature* 1991, **353**:31–37.

which cell division and mitosis succeed each other repeatedly without any intervening periods of cell growth. During cleavage therefore, the mass of the embryo does not increase. Early cleavage patterns can vary widely between different groups of animals (Fig. 8.4). The simplest pattern of division is radial, in which successive symmetric cleavages divide the embryo into equal-sized cells. This pattern is seen in the first three cleavages in the sea urchin. In contrast, the first cleavage of the nematode egg produces two cells of unequal size. The spirally cleaving eggs of molluscs and annelids illustrate yet another pattern of cleavage, in which successive divisions of each blastomere, in different planes, produce a spiral arrangement of cells; many of these divisions are unequal.

Fig. 8.4 Different patterns of early cleavage are found in different animal groups. Radial cleavages are equal and symmetric (sea urchin). Unequal cleavage (nematodes) results in one daughter cell being large. In spiral cleavage, the mitotic apparatus is oriented at an oblique angle to the long axis of the cell.

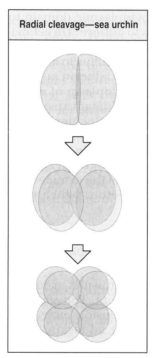

Radial cleavage—sea urchin

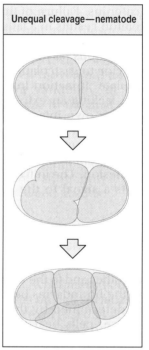

Unequal cleavage—nematode

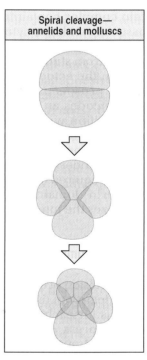

Spiral cleavage—annelids and molluscs

The amount of yolk in the egg can influence the pattern of cleavage. In yolky eggs undergoing symmetric cleavage, a cleavage furrow develops in the least yolky region and gradually spreads across the egg, but its progress is slowed or even halted by the presence of the yolk. Cleavage may thus be incomplete for some time. This effect is most pronounced in the heavily yolked eggs of birds and zebrafish, where complete divisions are restricted to a region at one end of the egg, and the embryo is formed as a cap of cells sitting on top of the yolk (see Fig. 2.10). Even in moderately yolky eggs, such as those of amphibians, the presence of the yolk can influence cleavage patterns. In frogs, for example, the later cleavages are unequal and asynchronous, resulting in an animal half composed of a mass of small cells and a yolky vegetal region composed of fewer and larger cells.

Two key questions arise relating to early cleavage: how are the positions of the cleavage planes determined; and how can cleavage lead to a hollow blastula (or its equivalent), which has a clear inside–outside polarity?

8-3   The asters of the mitotic apparatus determine the plane of cleavage at cell division.

The orientation of the plane of cleavage at cell division is important in a variety of situations. We have already seen its role in relation to unequal cleavages in early invertebrate development, where unequal divisions produce cells with differently specified fates. This is because of the asymmetric distribution of cytoplasmic determinants (see Chapter 6). The plane of cleavage can also be of great importance in later morphogenesis and growth since, for example, it determines whether an epithelial sheet containing dividing cells remains as a single cell layer or becomes multilayered.

Experiments show that the plane of cleavage in animal cells is not specified by the mitotic spindle itself, but by the asters at each pole. To disrupt its normal cleavage, a sea urchin egg is deformed by a glass bead. This displaces the mitotic spindle and interrupts the first cleavage furrow, creating a horseshoe-shaped cell in which the two new nuclei are segregated into the separate arms of the horseshoe (Fig. 8.5). At the next division, two cleavage furrows bisect the spindles formed in the arms of the horseshoe, but a third furrow also forms between the two adjacent asters not connected by a spindle. How the asters specify a furrow between them is not known.

The asters in a dividing cell are, in fact, microtubules that are radiating out from a centrosome, which acts as an organizing center for microtubule growth. In most animal cells the centrosome contains a pair of centrioles, which consist of an array of microtubules. Before mitosis, the centrosome becomes duplicated and, in the cells of the early embryo, the daughter centrosomes, which move to opposite sides of the nucleus and form asters, usually take up positions that cause the plane of cleavage of the cell to be at right angles to the plane of the previous cleavage. An example of this is cleavage of the AB cell, one of the pair of cells formed by cleavage of the nematode zygote (Fig. 8.6). By contrast, the other member of the pair, the P_1 cell, exemplifies the fact that there are many cases where successive cleavages are not at right angles to each other, because local cytoplasmic factors or cytoskeletal components of the cell can cause some centrosome pairs to take up different orientations than others with respect to the original plane of cleavage.

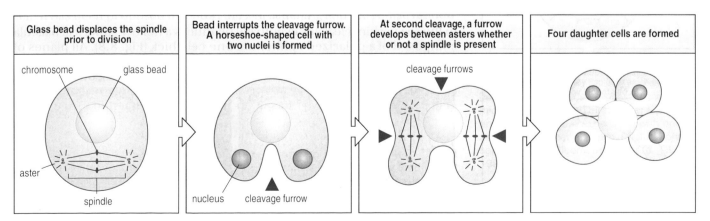

Fig. 8.5 The plane of cleavage in animal cells is determined by the asters and not the mitotic spindle. If the mitotic apparatus of a fertilized sea urchin egg is displaced at the first cleavage by the introduction of a glass bead, the cleavage furrow forms only on the side of the egg to which the mitotic apparatus has been moved. At the next cleavage, furrows bisect each mitotic apparatus, and a furrow also forms between the two adjacent asters, even though there is no spindle between them.

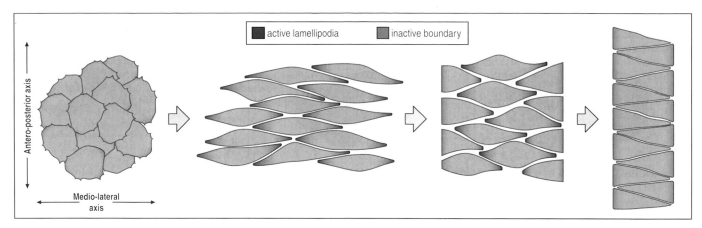

Fig. 8.26 Cell movements during convergent extension. During convergent extension of the mesoderm in amphibians, the cells become elongated in a medio-lateral direction, that is at right angles to the antero-posterior axis. Active movement is confined to the ends of these bipolar cells, which have active lamellipodia, and they shuffle past each other and intercalate. At the tissue boundary, movement ceases. After Keller, R., *et al.*: 1992.

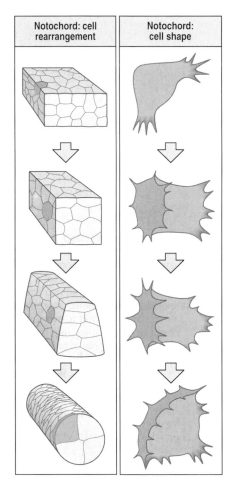

Fig. 8.27 Changes in cell arrangement and cell shape during notochord development. During convergent extension of the notochord, its height increases. At the same time, the cells become elongated. The bottom of the figure shows the eventual arrangement of cells, and their pizza slice shape. After Keller, R., *et al.*: 1989.

monopolar (see Fig. 8.26), with only the tip facing into the tissue still active. It is as if the cell tip that has reached the boundary has become fixed there. Precisely what defines a boundary is not clear, but boundaries between mesoderm, ectoderm, and endoderm seem to be specified before gastrulation commences. As gastrulation proceeds, a further boundary appears within the mesoderm, demarcating the notochord from the future somites (see Section 4-1).

Convergent extension is caused by contractile forces generated at the tips of the elongated cells. These forces both move the cells to the midline and extend the tissue. One can visualize the process of convergent extension in terms of lines of cells pulling on one another in a direction at right angles to the direction of extension. Because the cells at the boundary are trapped at one end, this tensile force causes the tissue to narrow and hence extend anteriorly. This explanation does not, of course, account for the way the cells initially become oriented in a medio-lateral direction; the mechanism by which this occurs remains completely unknown.

All in all, amphibian gastrulation may seem complex, but it is, in essence, nothing more than a special pattern of cell movement and cell adhesiveness. What is not yet understood is how the genes that are expressed in the Spemann organizer, such as *goosecoid*, control these processes (see Section 3-20).

8-11   Notochord elongation is caused by cell intercalation.

The presumptive notochord mesoderm is among the first of the tissues to involute during gastrulation, and in the amphibian it is the first of the dorsal structures to differentiate. Initially it can be distinguished from the adjacent somitic mesoderm by the packing of its cells and by the slight gap that forms the boundary between the two tissues, possibly reflecting differences in cell adhesiveness. The notochord elongates considerably and develops into a stiff rod composed of a stack of thin flat cells shaped like pizza slices. The two main mechanisms involved in its morphogenesis are medio-lateral intercalation leading to convergent extension, and directed dilation.

After the initial elongation of the notochordal mesoderm by convergent extension, it undergoes a further dramatic narrowing in width accompanied by an increase in height (Fig. 8.27). At the same time, the cells become elongated at right angles to the main antero-posterior axis, foreshadowing the later pizza-slice arrangement of cells within the notochord (see Fig. 8.27, bottom). The cells intercalate between their neighbors again, thus causing convergent extension. The later stage in notochord elongation, involving directed dilation, is discussed later in this chapter.

Summary.

At the end of cleavage the animal embryo is essentially a closed sheet of cells, which is often in the form of a sphere enclosing a fluid-filled interior. Gastrulation, strictly the formation of the gut, converts this sheet into a solid three-dimensional embryonic animal body. During gastrulation, cells move into the interior of the embryo, and the regions of endoderm and mesoderm, which were originally adjacent in the cell sheet, take up their appropriate positions in the embryo. Gastrulation results from a well defined spatio-temporal pattern of change in cell shape, cell movement, and change in cell adhesiveness, the main forces of which are generated by localized contractions. In the sea urchin, gastrulation occurs in two phases. In the first, changes in cell shape and adhesion result in migration of mesoderm cells into the interior, and invagination of the endodermal part of the cell sheet to form the gut. In the second phase, extension of the gut to reach the mouth region on the opposite side of the embryo occurs by cell rearrangement within the endoderm, which causes the tissue to narrow and lengthen (convergent extension), and by traction of filopodia extending from the gut tip against the blastocoel wall. Invagination of the mesoderm in *Drosophila* occurs by a similar mechanism to that of the invagination of endoderm in the sea urchin. Gastrulation in *Xenopus* involves more complex movements of cell sheets and also results in elongation of the embryo along the antero-posterior axis. It involves three processes: involution, in which a double-layered sheet of endoderm and mesoderm rolls into the interior over the lip of the blastopore; convergent extension of endoderm and mesoderm in the antero-posterior direction to form the roof of the gut, and the notochord and somitic mesoderm, respectively; and epiboly—spreading of the ectoderm from the animal cap region to cover the whole outer surface of the embryo. Both convergent extension and epiboly are due to cell intercalation, the cells becoming rearranged with respect to their neighbors.

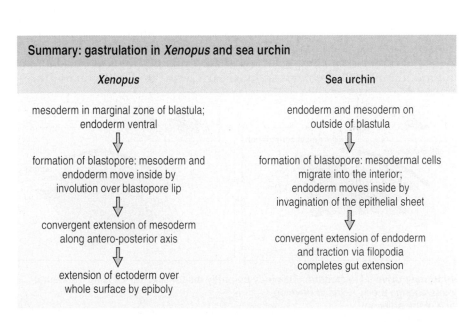

Summary: gastrulation in *Xenopus* and sea urchin

Xenopus	Sea urchin
mesoderm in marginal zone of blastula; endoderm ventral	endoderm and mesoderm on outside of blastula
⇩	⇩
formation of blastopore: mesoderm and endoderm move inside by involution over blastopore lip	formation of blastopore: mesodermal cells migrate into the interior; endoderm moves inside by invagination of the epithelial sheet
⇩	⇩
convergent extension of mesoderm along antero-posterior axis	convergent extension of endoderm and traction via filopodia completes gut extension
⇩	
extension of ectoderm over whole surface by epiboly	

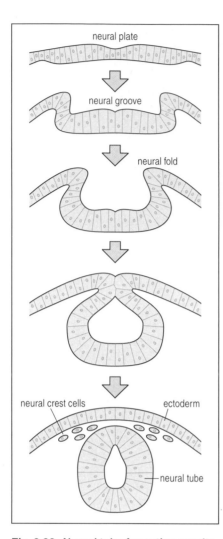

Fig. 8.28 Neural tube formation results from formation and fusion of neural folds. The tube then detaches from the ectoderm.

Neural tube formation.

Neurulation in vertebrates results in the formation of the **neural tube**, an epithelial structure composed of ectoderm, which develops into the brain and spinal cord (see Section 2-2). The tissue that gives rise to the neural tube initially appears as a thickened plate of tissue—the **neural plate**—following induction by the mesoderm (see Section 4-7).

In amphibians, there is convergent extension of the future neural plate during gastrulation. There is then a change in the shape of the cells within the plate, so that the edges of the neural plate become raised above the surface, forming two parallel neural folds with a depression —the neural groove—between them (Fig. 8.28). The neural folds eventually come together along the dorsal midline of the embryo and fuse at their edges to form the neural tube, which then separates from the adjacent ectoderm. This surface layer of ectoderm becomes epidermis. Neurulation in birds and mammals proceeds in a similar manner, although the posterior region of the neural tube in both birds and mammals initially forms as a solid rod, and develops a lumen later. In fishes, the neural tube initially forms as a solid rod throughout its length.

8-12   Neural tube formation is driven by both internal and external forces.

A very early sign of neurulation in *Xenopus* is the formation of the neural groove along the midline. As with gastrulation, changes in cell shape are associated with the curvature of the neural plate and the formation of the folds, but the mechanism is not well understood. During gastrulation, the cells of the neural plate become longer and narrower than the cells of the adjacent ectoderm (Fig. 8.29). As neurulation commences and the plate begins to roll up, the cells at the edge of the plate, where it is bending, become constricted at the apical surface, making them wedge-shaped. This change in cell shape could, in principle, draw the edges of the neural plate up into folds. A mechanism that may be involved in initiating neural fold formation is the crawling of the cells at the edge of the plate down the inner surface of the adjacent ectoderm, thus helping to bring about a local change in curvature (see Fig. 8.29).

In birds and mammals, neurulation does not occur along the whole of the neural plate at the same time, as it does in amphibians, but starts near the anterior end of the embryo, in the region of the midbrain, and proceeds anteriorly and posteriorly (Fig. 8.30). Changes in the shape of neural furrow cells in chick embryos are associated with the folding of

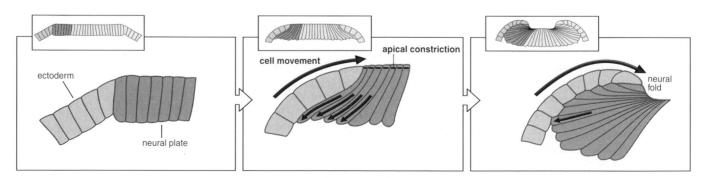

Fig. 8.29 Change in cell shape and cell crawling may drive neural folding. Cells at the edge of the neural plate change their shape and appear to crawl along the undersurface of the adjacent ectoderm. This may be partly responsible for causing the neural folds to develop.

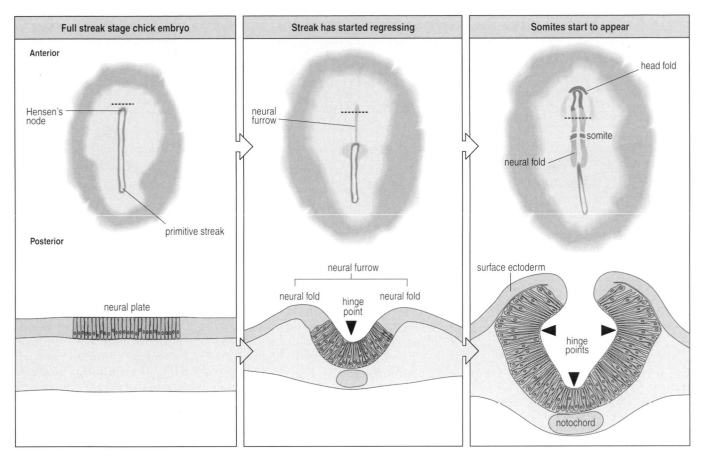

| Full streak stage chick embryo | Streak has started regressing | Somites start to appear |

Fig. 8.30 Change in cell shape in the neural plate during chick neurulation. Top: surface views of the chick epiblast. Bottom: cross-sections taken through the epiblast at the sites indicated by the dotted lines in the top diagrams. Cells anterior to Hensen's node become elongated to form the neural plate.

Cells in the center of the neural plate become wedge-shaped, defining the 'hinge point' at which the neural plate bends. Additional hinge points with wedge-shaped cells also form on the sides of the furrow. After Schoenwolf, G.C., *et al*.: 1990.

the neural tube, but whether they are a cause or a result of folding is not yet clear. Cells in the midline of the chick neural furrow, the so-called hinge point, are wedge-shaped. Later, wedge-shaped cells are seen at additional hinge points on the sides of the furrow, where the furrow curves round further to form a tube (see Fig. 8.30). The detailed cellular mechanism underlying the shape changes is not yet established, but both actin filaments and microtubules are thought to be involved in generating and stabilizing the changes.

The forces supplied by changes in cell shape and convergent extension are intrinsic to the neural plate tissue. But formation of the neural tube may also depend on external forces applied by the surrounding tissue. Thus, when the neural plate is isolated from newt embryos it still rolls up, but in the wrong direction, so that the apical face is now outermost. This suggests that forces intrinsic to the neural plate are not alone sufficient to cause the correct folding of the neural tube.

8-13   Changes in the pattern of expression of cell adhesion molecules accompany neural tube formation.

The neural tube, which is initially part of the ectoderm, separates from the presumptive epidermis after its formation. This involves changes in cell adhesion. The cells of the neural plate, like the rest of the ectoderm

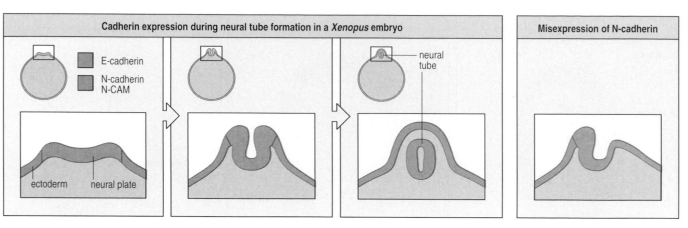

Fig. 8.31 Expression of cell adhesion molecules during neural tube formation in *Xenopus*. The cells of the neural plate express N-cadherin and the adjacent ectoderm expresses E-cadherin. If N-cadherin is misexpressed in the ectoderm on one side, there is a failure of the neural tube to separate at that site (right panel).

in the chick, initially express the adhesion molecule L-CAM on their surface. However, as the neural folds develop, the neural plate ectoderm begins to express both N-cadherin and N-CAM, whereas the adjacent ectoderm expresses E-cadherin only (Fig. 8.31). These changes in adhesiveness may enable the neural tube to separate from the surrounding ectoderm and allow it to sink beneath the surface, the rest of the ectoderm reforming over it in a continuous layer. The changes in adhesiveness could themselves provide the mechanism for neural tube formation, in a manner similar to that described in Section 8-1 for the sorting out of cells. Support for this view comes from altering the pattern of expression of adhesion molecules in *Xenopus* embryos. Injection of N-cadherin mRNA into the gastrula ectoderm in the region adjacent to one side of the future neural tube, results in the persistence of a continuous layer of cells between the ectoderm and neural tube, and a failure of the neural tissue to separate at that site (see Fig. 8.31).

Summary.

In vertebrates, neurulation results in neural tube formation following induction by the mesoderm. Neurulation in mammals, birds, and amphibians results from the folding of the neural plate into a tube, the neural folds fusing in the midline of the embryo. Separation of the neural tube from the adjacent ectoderm requires changes in cell adhesiveness. The formation of the neural folds and their coming together in the midline is apparently driven by changes in cell shape within the tube itself, as well as by forces generated by adjacent tissues.

Cell migration.

Cell migration is a major feature of animal morphogenesis, with cells moving over relatively long distances, from one site to another. In this section, we consider two examples, the migrations of the primary mesenchyme cells from the interior of the sea urchin blastula and of the neural crest cells in the chick embryo. Each of these involve interactions between the migrating cells and the substratum, which control the pattern of migration. As a contrast, we also look at the aggregation of individual amebae of the slime mold *Dictyostelium* into a multicellular fruiting body, which provides an example of how chemotaxis and signal propagation can control migration pattern. We leave until Chapter 11 the migration of immature neurons that is so fundamental to the morphogenesis of the nervous system. Other important examples of cell migration that are dealt with in later chapters include muscle cell migration in vertebrate limbs (see Chapter 10), and germ cell migration (see Chapter 12).

8-14 The directed migration of sea urchin primary mesenchyme cells is determined by the contacts of their filopodia to the blastocoel wall.

After the primary mesenchyme cells have entered the blastocoel, as described in Section 8-7, they move within the blastocoel to become distributed in a characteristic pattern on the inner surface of the blastocoel wall. The mesenchyme cells become arranged in a ring around the gut in the vegetal region at the ectoderm-endoderm border. Some then migrate to form two extensions toward the animal pole on the ventral (oral) side (Fig. 8.32). The migration path of individual cells varies considerably from embryo to embryo, but their final pattern of distribution is fairly constant. The primary mesenchyme cells later lay down the skeletal rods of the sea urchin endoskeleton by secretion and, as these develop, the distribution of the cells changes, so their pattern is a dynamic one.

The primary mesenchyme cells move over the inner surface of the blastocoel wall by means of fine filopodia, which can be up to 40 μm long and may extend in several directions. Each cell at any one time has, on average, six filopodia, most of which are branched. When filopodia make contact with, and adhere to, the blastocoel wall, they contract, drawing the cell body toward the point of contact. Because each cell extends several filopodia (Fig. 8.33), some or all of which may contract on contact with the wall, there is competition between the filopodia, the cell being drawn toward that region of the wall where the filopodia make the most stable contact. The movement of a primary mesenchyme cell therefore resembles a random search for the most stable attachment.

Analysis of video films of migrating cells suggests that the most stable contacts are made in the regions where the cells finally accumulate, namely the vegetal ring and the two ventro-lateral clusters. Thus, the pattern of adhesiveness of the inner surface of the blastocoel wall determines the pattern of migration of the cells, but the molecular basis for

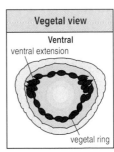

After the primary mesenchyme cells have entered the blastocoel, as

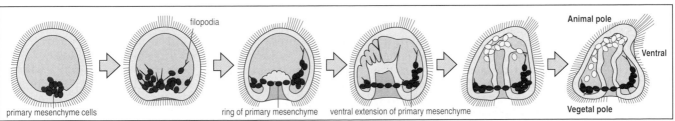

Fig. 8.32 Migration of primary mesenchyme in early sea urchin development. The primary mesenchyme cells enter the blastocoel at the vegetal pole and migrate over the blastocoel wall by filopodial extension and contraction. Within a few hours they take up a well defined ring-like pattern in the vegetal region with extensions along the ventral side.

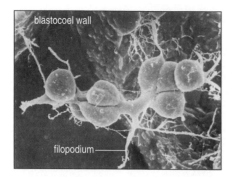

Fig. 8.33 Filopodia of sea urchin mesenchyme. The scanning electron micrograph shows a group of primary mesenchyme cells, some of which have fused together, moving over the blastocoel wall by means of their numerous filopodia, which can extend and contract. Photograph courtesy of J. Morrill, from Morrill, J.B., et al.: 1985.

this adhesion is not known. The surfaces of the cells over which the mesenchyme cells move is covered with a basal lamina, which has been implicated in influencing the contacts between the filopodia and the blastocoel wall.

Primary mesenchyme cells introduced by injection at the animal pole move in a directed manner to their normal positions in the vegetal region. This suggests that guidance cues, possibly graded, are globally distributed over the blastocoel wall. Even cells that have already migrated migrate again to form a similar pattern when introduced into a younger embryo.

8-15   Neural crest migration is controlled by environmental cues and adhesive differences.

Neural crest cells of vertebrates have their origin at the edges of the neural folds and first become recognizable during neurulation. When the neural tube closes, they undergo an epithelial to mesenchymal transition (see Section 8-7) and leave the epithelial sheet in the midline, migrating away from it on either side.

The epithelial to mesenchymal transition in vertebrates involves the *slug* gene, which controls the process by which non-motile epithelial cells become migrating cells. The *slug* gene is related to the *Drosophila snail* gene and is expressed in all migratory neural crest cells. Inhibition of *slug* expression inhibits migration.

Neural crest cells migrate away from the neural tube, giving rise to a wide variety of different cell types that include cartilage in the head, pigment cells in the dermis, the medullary cells of the adrenal gland, glial Schwann cells, and the neurons of both the peripheral and the autonomic nervous systems. Here, we focus on the migration of the crest cells in the trunk region of the chick embryo. We have already discussed the migration of neural crest cells from the hindbrain region into the branchial arches (see Section 4-12).

Various strategies have been employed to follow the migration of neural crest cells. For example, because quail cells have a nuclear marker that distinguishes them from chick cells, grafting a neural tube from a quail embryo into a chick embryo allows the subsequent migration pathways of the quail neural crest cells in the chick embryo to be followed (Fig. 8.34). It is also possible to identify migrating chick neural crest cells by tagging them with labeled monoclonal antibodies, or by labeling them with the dye DiI. The initiation of neural crest migration seems to involve disruption of the basement membrane surrounding the neural tube, allowing the crest cells to escape. Migration also requires loss of adhesion of the cells to the neural tube, and both N-cadherin and E-cadherin (see Section 8-13) are lost from neural crest cells at about the time of migration.

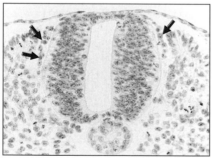

Fig. 8.34 Following cell migration pathways by grafting a piece of quail neural tube to a chick host. A piece of neural tube from a quail embryo is grafted to a similar position in a chick host. The photograph shows migration of the quail neural crest cells (red arrows). Migration can be followed since quail cells have a nuclear marker that distinguishes them from chick cells. Photograph courtesy of N. Le Douarin.

There are two main migratory pathways for neural crest cells in the trunk of the chick embryo (Fig. 8.35). One goes dorso-laterally under the ectoderm and over the somites; cells that migrate this way mainly give rise to pigment cells, which populate the skin and feathers. The

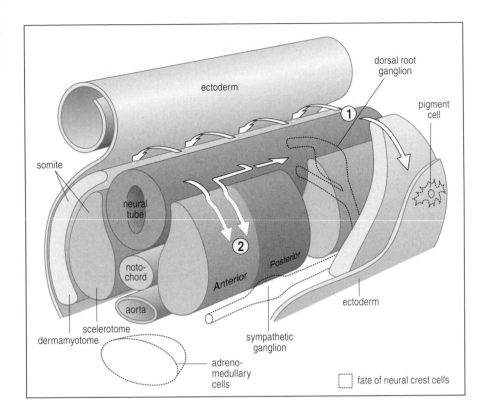

Fig. 8.35 Neural crest migration in the trunk of the chick embryo. One group of cells (1) migrates under the ectoderm to give rise to pigment cells (shown in outline). The other group of cells (2) migrates over the neural tube and then through the anterior half of the somite; the cells do not migrate through the posterior half of the somite. Those that migrate along this pathway give rise to dorsal root ganglia, sympathetic ganglia, and cells of the adrenal cortex, and their future sites are also shown in outline. Those cells opposite the posterior regions of a somite migrate in both directions along the neural tube until they come to the anterior region of a somite. This results in a segmental pattern of migration and is responsible for the segmental arrangement of ganglia.

other pathway is more ventral, primarily giving rise to sympathetic and sensory ganglion cells. Some crest cells move into the somites to form dorsal root ganglia; others migrate through the somites to form sympathetic ganglia and adrenal medulla, but appear to avoid the region around the notochord. Trunk neural crest selectively migrates through the anterior (rostral) half of the somite and not through the posterior (caudal) half. Within each somite, neural crest cells are found only in the anterior half, even when they originate in neural crest adjacent to the posterior half of the somite. This behavior is unlike that of the neural crest cells taking the dorsal pathway, which migrate over the whole dorso-lateral surface of the somite. The anterior migration pathway results in the distinct segmental arrangement of spinal ganglia in vertebrates, with one pair of ganglia corresponding to one pair of somites—one segment—in the embryo (Fig. 8.36). The segmental pattern of migration is due to the adhesive properties of the somites. If the somites are rotated through 180° so that their antero-posterior axis is reversed, the crest cells still migrate through the original anterior halves only. Two members of the Eph family of transmembrane ligands are expressed in the posterior halves of the somites, while crest cells have receptors for these ligands. The interactions between the ligands and receptors could result in the expulsion of the crest cells from the posterior halves of the somites. This would provide a molecular basis for the segmental arrangement of the spinal ganglia.

The neural tube and notochord also both influence neural crest migration. If the early neural tube is inverted through 180°, before neural crest migration starts, so that the dorsal surface is now the ventral surface, one might think that the cells that normally migrate ventrally, now being nearer to their destination, would move ventrally. But this is not the case, and many of the crest cells move upward through the sclerotome in a ventral to dorsal direction, staying confined to the anterior half of each somite. This suggests that the neural tube somehow influences the direction of migration of neural crest cells. The notochord also exerts

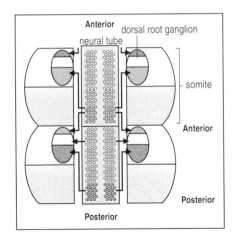

Fig. 8.36 Segmental arrangement of dorsal root ganglia is due to the migration of crest cells through the anterior half of the somite only. Neural crest cells cannot migrate through the posterior (gray) half of a somite. They thus accumulate in the anterior half. The dorsal root ganglion that develops from these cells is made up of crest cells from anterior (orange), adjacent (white), and posterior (red) regions.

an influence, inhibiting neural crest cell migration over a distance of about 50 µm, and thus preventing the cells from approaching it.

Many different extracellular matrix molecules have been detected along neural crest migratory pathways, and the neural crest cells may interact with their molecules by means of their cell-surface integrins. Neural crest cells cultured *in vitro* adhere to, and migrate efficiently on, fibronectin, laminin, and various collagens. Blocking the adhesion of the neural crest cells to fibronectin or laminin by blocking the integrin β_1 subunit *in vivo* causes severe deficiencies in the head region but not in the trunk, suggesting that the crest cells in these two regions adhere by different mechanisms, probably involving other integrins. It is striking how neural crest cells in culture will preferentially migrate along a track of fibronectin, although the role of this molecule in guiding the cells in the embryo is still unclear.

8-16   Slime mold aggregation involves chemotaxis and signal propagation.

The myxamebae of the cellular slime mold *Dictyostelium discoideum* feed on bacteria. When a local food source is exhausted, they enter the multicellular stage of the slime mold life cycle (see Fig. 6.30). The first phase of this stage is aggregation, which when observed in time-lapse videos presents a dramatic spectacle, the cells streaming toward a center of aggregation like small rivers of cells converging into a lake. As they approach the focus of aggregation, the cells in the streams adhere to each other at their anterior and posterior ends through a membrane glycoprotein that is expressed at this stage. Eventually the cells collect into a compact multicellular mound (Fig. 8.37), which develops into a stalked fruiting body (see Section 6-16).

The cells move intermittently rather than continuously, in pulses of inward movement. The mechanism of aggregation involves both chemotaxis of individual cells, and the propagation of the chemotactic signal from one ameba to another. In *Dictyostelium*, cyclic AMP (cAMP) is the chemoattractant. Amebae respond to an increasing concentration of cAMP, and will thus move up a gradient of cAMP by extending a pseudopod in the direction of the source (Fig. 8.38). Chemotaxis up such a gradient only operates over short distances, much less than 1 mm, because it is difficult to establish a reliable diffusion gradient of a

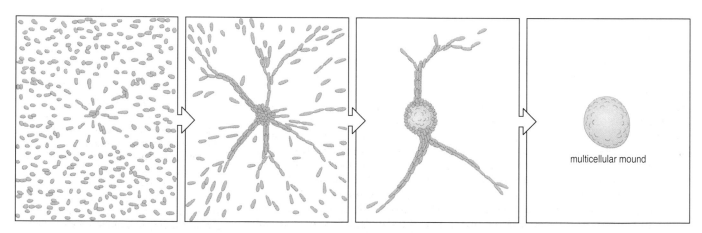

multicellular mound

Fig. 8.37 Aggregation in the cellular slime mold *Dictyostelium*. Aggregating amebae of *D. discoideum* stream toward a focal point, eventually forming a multicellular mound that develops into a fruiting body (not shown). In the cell streams, the normally ameboid cells become bipolar and adhere to each other at either end.

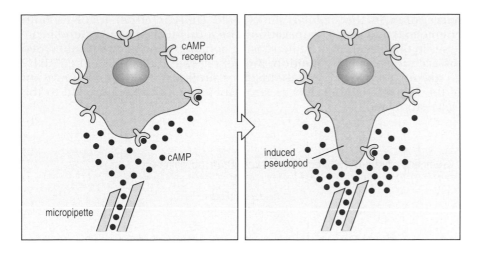

Fig. 8.38 Chemotactic response of slime mold cells. Amebae will move up a gradient of cyclic AMP (cAMP). A localized source of cAMP binds to membrane receptors on the side of the cell facing the source, resulting in the ameba extending a pseudopod toward the source. After Alberts, B., *et al.*: 1989.

chemoattractant over larger distances. Yet amebae can aggregate from distances up to 5 mm from the center. This is achieved by the propagation of the chemotactic signal, in a manner rather similar to the conduction of a nerve impulse.

When an ameba receives a pulse of cAMP, the cAMP binds to membrane receptors that stimulate the cell not only to respond chemotactically by moving towards the source, but also to propagate the signal by producing a pulse of cAMP itself. By this relay mechanism, a wave of cAMP is propagated across the field of aggregating amebae (Fig. 8.39). The pulsatory nature of the signal is shown by introducing a micropipette containing cAMP into a field of amobae ready to aggregate: it can act as an aggregation center only if it provides cAMP pulses of the correct frequency. An important feature of the propagation of the signal is that the cells are refractory to cAMP for a short time immediately after they have given out a pulse of cAMP. This ensures that the pulse only propagates outward. The presence of the enzyme phosphodiesterase, which breaks down cAMP, prevents its concentration rising to a level where it saturates the system, which would prevent propagation.

Aggregation of slime mold cells is the best understood example of chemotaxis in a developing system. We come across further examples of chemotaxis in the development of the nervous system in Chapter 11. However, the signal propagation relay used by the slime mold has not yet been found in any other system.

Summary.

Directed cell migration in animal embryos is controlled mainly by interactions with the substratum over which the cells move, although signals from other cells may also play a part. The directed migration of the primary mesenchyme cells in the sea urchin blastula is due to filopodia on these cells exploring their environment, and then drawing the cell to that region of the blastocoel wall where the filopodia make the most stable attachment. Thus, the final position of a mesenchyme cell is determined by the adhesive properties of the basal lamina and ectoderm over which it moves. Similarly, migration pathways of vertebrate neural crest cells are determined by the adhesive properties over which they move. Differences in adhesiveness between the anterior and posterior halves of the somites result in neural crest being prevented from migrating over or through posterior halves. Thus, presumptive dorsal ganglia cells collect adjacent to anterior halves, giving them a segmental arrangement.

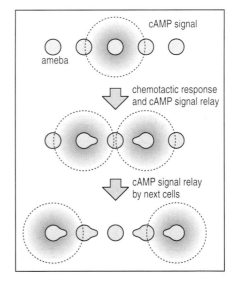

Fig. 8.39 Propagation of a cAMP signal and chemotaxis by slime mold amebae. A cell at the center of the aggregate gives off a pulse of cAMP. This induces pseudopod formation and chemotactic movement in the surrounding cells in the direction of the cAMP source. It also causes the responding cells to produce a pulse of cAMP themselves, and the signal is thus relayed outward. After releasing a pulse of cAMP, a cell becomes refractory to cAMP for a short time. This ensures the signal is propagated outward only.

Aggregation in the cellular slime mold *Dictyostelium* involves both chemotaxis and signal propagation. The individual amebae extend processes in the direction of an increasing concentration of cAMP, and move toward the source. In addition, the cells respond to a pulse of cAMP by giving out a pulse of cAMP themselves, and this results in propagation of the signal, enabling cells as far away as 5 mm to be attracted to the aggregation center.

Summary: directed cell migration of vertebrate neural crest

neural crest cells

migration through anterior of somites   migration over somites

dorsal ganglion cells   pigment cells

Directed dilation.

Hydrostatic pressure can provide the force for morphogenesis in a variety of situations. An increase in hydrostatic pressure inside a spherical sheet of cells causes the sphere to increase in volume. We have already seen how hydrostatic pressure is involved in blastula formation in both the mouse and amphibian. Here, we consider examples of **directed dilation** where the increase in pressure causes an asymmetric change in shape. If the circumferential resistance to pressure in a tube is much greater than the resistance to longitudinal extension, then an increase in the internal pressure causes an increase in length (Fig. 8.40).

After the *Xenopus* notochord has formed, as discussed in Section 8-12, its volume increases threefold, and there is considerable further lengthening as it straightens and becomes stiffer. At this stage, the notochord has become surrounded by a sheath of extracellular material, which restricts circumferential expansion but does allow expansion in the antero-posterior direction. The cells within the notochord develop fluid-filled vacuoles and expand in volume as a result. They thus exert hydrostatic pressure on the notochord sheath, resulting in directed dilation: circumferential expansion of the notochord is prevented by the resistance of the sheath, and this ensures that the increase in volume (dilation) is directed along the notochord's long axis.

The vacuoles in the notochord cells are filled with glycosaminoglycans which, because of their high carbohydrate content, tend to attract water into the vacuoles by osmosis. It is this that produces the hydrostatic pressure that causes the increase in cell volume, and the consequent stiffening and straightening of the notochord.

Changes in the structure of the sheath during the period of notochord elongation fit well with the proposed hydrostatic mechanism. The sheath contains both glycosaminoglycans, which have little tensile strength, and the fibrous protein collagen, which has a high tensile strength; during notochord dilation the density of collagen fibers increases, providing resistance to circumferential expansion. The crucial role of

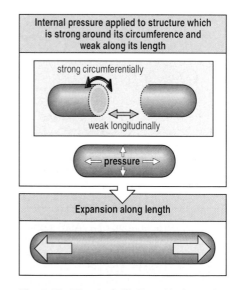

Internal pressure applied to structure which is strong around its circumference and weak along its length

strong circumferentially

weak longitudinally

pressure

Expansion along length

Fig. 8.40 Directed dilation. Hydrostatic pressure inside a constraining sheath or membrane can lead to elongation of the structure. If the circumferential resistance is much greater than the longitudinal resistance, as it is in the notochord sheath, the rod of cells inside the sheath lengthens.

the sheath in dilation and elongation is shown by the fact that if it is digested away, the notochord buckles and folds and the notochord cells, instead of being flat, become rounded.

8-17 **Circumferential contraction of hypodermal cells elongates the nematode embryo.**

During the early development of the nematode there is little change in body shape from the spherical form of the fertilized egg, even during gastrulation. After gastrulation, about 5 hours after fertilization, the embryo begins to elongate rapidly along its antero-posterior axis. Elongation takes about 2 hours, during which time the nematode embryo decreases in circumference about threefold and undergoes a fourfold increase in length.

This elongation is brought about by a change in shape of the hypodermal (epidermal) cells that make up the outermost layer of the embryo; their destruction by laser ablation prevents elongation. During embryo elongation, these cells change shape so that, instead of being elongated in the circumferential direction, they become elongated along the antero-posterior axis (Fig. 8.41). Throughout this elongation the hypodermal cells remain attached to each other by desmosome cell junctions. The desmosomes are also linked within the cells by actin-containing fibers that run circumferentially, and these fibers appear to shorten as the cells elongate. The disruption of actin filaments by cytochalasin D treatment blocks elongation, and so it is very likely that their contraction brings about the change in cell shape. Circumferential contraction of the hypodermal cells causes an increase in hydrostatic pressure within the embryo, forcing an extension in an antero-posterior direction. Circumferentially oriented microtubules may also play a mechanical role in constraining the expansion, in the same way as the sheath of the *Xenopus* notochord described above. Increase in nematode body length is thus another example of directed dilation.

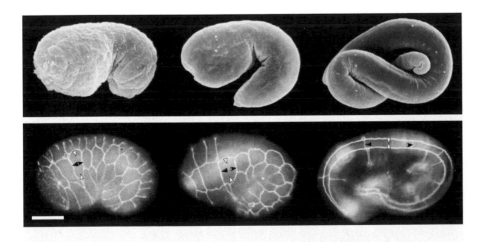

Fig. 8.41 Increase in nematode body length by directed dilation. The change in body shape over 2 hours is illustrated in the top panel. The increase in length is due to circumferential contraction of the hypodermal cells, as shown in the bottom panel. The change in shape of a single cell can be seen in the cell marked with arrows. Scale bar = 10 µm. Photographs courtesy of J. Priess, from Priess, J.R., *et al*.: 1986.

8-18 **The direction of cell enlargement can determine the form of a plant leaf.**

Cell enlargement is a major process in plant morphogenesis, providing up to a 50-fold increase in the volume of a tissue. The driving force for expansion is the hydrostatic pressure—turgor pressure—exerted on the cell wall as the protoplast swells, as a result of water entry into the cell vacuoles by osmosis (Fig. 8.42). Plant cell expansion involves synthesis

Fig. 8.42 Enlargement of a plant cell.
Plant cells expand as water enters the
cell vacuoles and thus causes an
increase in intracellular hydrostatic
pressure. The cell elongates in a
direction perpendicular to the orientation
of the cellulose fibrils in its cell wall.

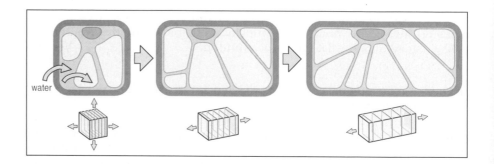

and deposition of new cell wall material, and is an example of directed
dilation. The direction of cell growth is determined by the orientation
of the cellulose fibrils in the cell wall. Enlargement occurs primarily in a
direction at right angles to the fibrils, where the wall is weakest. The orientation of cellulose fibrils in the cell wall is thought to be determined by the
microtubules of the cell's cytoskeleton, which are responsible for positioning the enzyme assemblies that synthesize cellulose at the cell wall. Plant
growth hormones, such as ethylene and giberellins, alter the orientation in
which the fibrils are laid down and so can alter the direction of expansion.
Auxin aids expansion by loosening the structure of the cell wall.

The development of a leaf involves a complex pattern of cell division
and cell elongation, with cell elongation playing a central role in the
expansion of the leaf blade. Two mutations that affect the shape of the
blade by affecting the direction of cell elongation have been identified.
Leaves of the *Arabidopsis angustifolia* mutant are similar in length to
the wild type but are much thinner (Fig. 8.43). In contrast, the *rotundifolia* mutations reduce the length of the leaf relative to its width. Neither
of these mutations affects the number of cells in the leaf. Examination of
the cells in the developing leaf shows that these mutations are affecting
the direction of elongation of the enlarging cells.

Fig. 8.43 The shape of the leaves of
***Arabidopsis* are affected by mutations
affecting cell elongation.** The *angustifolia*
mutation results in less elongation in
width, while mutations in *rotundifolia*
cause short fat leaves to develop.
Photograph courtesy of H. Tsukaya,
from Tsuge, T. *et al.*: 1996.

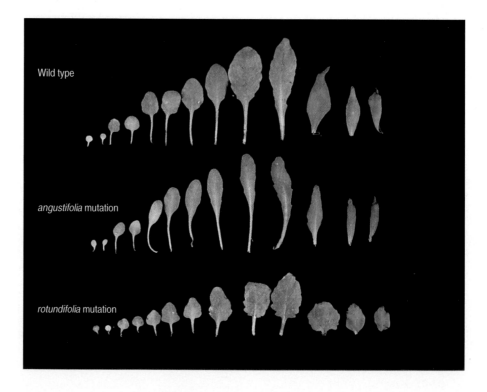

Summary.

Directed dilation results from an increase in hydrostatic pressure and unequal peripheral resistance to this pressure. Extension of the notochord is brought about by directed dilation, in which the notochord increases in volume while its circumferential expansion is constrained by the notochord sheath, forcing it to elongate. Similarly, the nematode embryo elongates after gastrulation due to a circumferential contraction in the outer hypodermal cells that generates pressure on the internal cells, forcing the embryo to extend in an antero-posterior direction. In plants, the direction of cell enlargement determines the shape of leaves. In plant cell enlargement, the direction of elongation is yet another example of directed dilation.

Summary: directed dilation

circumferential restraint
by sheath
⇩
notochord elongation

circumferential restraint
by plant cell wall
⇩
cell elongation

circumferential restraint
⇩
hydrostatic pressure
⇩
nematode elongation

Summary to Chapter 8.

Cells are held together by specific adhesion molecules. Changes in the shape of the embryo and cell migration are due to changes in cell adhesion and to forces generated by the cell. Formation of the blastula results from cell division and cell polarization, and in some cases from the movement of water into the blastocoel. Gastrulation involves major movements of cell sheets so that the future endoderm and mesoderm move inside the embryo to their appropriate positions in relation to the main body plan. Invagination of the mesoderm in both sea urchins and *Drosophila* involves similar changes in cell shape. Convergent extension plays a key role in both sea urchin and amphibian gastrulation, and is due to particular patterns of cell intercalation. Epiboly, the spreading of a multilayered sheet, is due to intercalation. Convergent extension, together with directed dilation, also plays a key role in notochord formation. Directed dilation causes the elongation of the nematode embryo and the direction of enlargement of plant cells. Changes in cell shape and cell adhesion are responsible for neural tube formation. Directed cell migration of sea urchin mesenchyme and vertebrate neural crest cells is controlled by the adhesiveness of the substratum over which the cells move. Migration and aggregation of slime mold cells is by propagation of a chemotactic signal.

References.

Introduction.

Wang, Y., Jones, F.S., Krushel, L.A., Edelman, G.: **Embryonic expression patterns of the neural cell adhesion molecule gene are regulated by homeodomain binding sites.** *Proc. Natl. Acad. Sci.* 1996, **93**:1892–1896.

8-1   Sorting out of dissociated cells demonstrates differences in cell adhesiveness in different tissues.

Steinberg, M.S.: **Does differential adhesion govern self-assembly processes in histogenesis? Equilibrium configurations and the emergence of a hierarchy among populations of embryonic cells.** *J. Exp. Zool.* 1970, **173**:395–433.

Townes, P., Holtfreter, J.: **Directed movements and selected adhesions of embryonic amphibian cells.** *J. Exp. Zool.* 1955, **128**:53–120.

Box 8A   Cell adhesion molecules.

Clark, E.A., Brugge, J.S.: **Integrins and signal transduction pathways: the road taken.** *Science* 1995, **268**:233–239.

Cunningham, B.A.,: **Cell adhesion molecules as morphoregulators.** *Curr. Opin. Cell Biol.* 1995, **7**:628–633.

Gumbiner, B.M.: **Cell adhesion: the molecular basis of tissue architecture and morphogenesis.** *Cell* 1996, **84**:345–357.

Gumbiner, B.M.: **Signal transduction by catenin.** *Curr. Opin. Cell Biol.* 1995, **7**:634–640.

Hynes, R.O: **Integrins: versatility, modulation, and signalling in cell adhesion.** *Cell* 1992, **69**:11–25.

Klymkowsky, M.W., Parr B.: **The body language of cells: the intimate connection between cell adhesion and behavior.** *Cell* 1995: **83**:5–8.

Takeichi, M.: **Cadherins: a molecular family important in selective cell–cell adhesion.** *Ann. Rev. Biochem.* 1990, **59**:237–252.

8-2   Cadherins can provide adhesive specificity.

Levine, E., Lee, C.H., Kintner, C., Gumbiner, B.M.: **Selective disruption of E-cadherin function in early Xenopus embryos by a dominant negative mutant.** *Development* 1994, **120**:901–909.

Steinberg, M.S., Takeichi, M.: **Experimental specification of cell sorting, tissue spreading, and specific spatial patterning by quantitative differences in cadherin expression.** *Proc. Natl. Acad. Sci.* 1994, **91**:206–209.

Takeichi, M.: **Morphogenetic roles of classic cadherins.** *Curr. Opin. Cell Biol.* 1995, **7**:619–627.

8-3   The asters of the mitotic apparatus determine the plane of cleavage at cell division.

Strome, S.: **Determination of cleavage planes.** *Cell* 1993, **72**:3–6.

Staiger, C., Doonan, J.: **Cell division in plants.** *Curr. Opin. Cell Biol.* 1993, **5**:226–231.

8-4   Cells become polarized in early mouse and sea urchin blastulas.

Fleming, T.P., Johnson, M.H.: **From egg to epithelium.** *Ann. Rev. Cell Biol.* 1988, **4**:459–485.

Sobel, J.S.: **Membrane–cytoskeletal interactions in the early mouse embryo.** *Semin. Cell Biol.* 1990, **1**:341–348.

Sutherland, A.E., Speed, T.P., Calarco, P.G.: **Inner cell allocation in the mouse morula: the role of oriented division during fourth cleavage.** *Dev. Biol.* 1990, **137**:13–25.

Winkel, G.K., Ferguson, J.E., Takeichi, M., Nuccitelli, R.: **Activation of protein kinase C triggers premature compaction in the four-cell stage mouse embryo.** *Dev. Biol.* 1990, **130**:1–15.

8-5   Ion transport is involved in fluid accumulation in the blastocoel.

Warner, A.E.: **Physiological approaches to early development.** *Recent Adv. Physiol.* 1984, **10**:87–123.

8-6   Internal cavities can be created by cell death.

Coucouvanis, E., Martin, G.R.: **Signals for death and survival: a two-step mechanism for cavitation in the vertebrate embryo.** *Cell* 1995, **83**:279–287.

8-7   Gastrulation in the sea urchin involves cell migration and invagination.

Hardin, J.: **Local cell interactions and the control of gastrulation in the sea urchin embryo.** *Dev. Biol.* 1994, **5**:77–84.

Hardin, J., McClay, D.R.: **Target recognition by the archenteron during sea urchin gastrulation.** *Dev. Biol.* 1990, **142**:86–102.

Odell, G.M., Oster, G., Alberch, P., Burnside, B.: **The mechanical basis of morphogenesis. I. Epithelial folding and invagination.** *Dev. Biol.* 1981, **85**:446–462.

Ingersoll, E.P., Ettensohn, C.A.: **An N-linked carbohydrate-containing extracellular matrix determinant plays a key role in sea urchin gastrulation.** *Dev. Biol.* 1994, **163**:359–366.

Nakajima, Y., Burke, R. D.,: **The initial phase of gastrulation in sea urchins is accompanied by the formation of bottle cells.** *Dev. Biol.* 1996, **179**:436–446

8-8   Mesoderm invagination in *Drosophila* is due to changes in cell shape, controlled by genes that pattern the dorso-ventral axis.

Leptin, M.: ***Drosophila* gastrulation: from pattern formation to morphogenesis.** *Ann. Rev. Cell Biol.* 1995, **11**:189–212.

Leptin, M.: **Morphogenesis: control of epithelial cell shape changes.** *Curr. Biol.* 1994, **4**:709–712.

Leptin, M., Casal, J., Grunewald, B., Reuter, R.: **Mechanisms of early *Drosophila* mesoderm formation.** *Develop. Suppl.* 1992, 23–31.

Costa, M., Wilson, E.T., Wieschaus, E.: **A putative cell signal encoded by the *folded gastrulation* gene coordinates cell shape changes during *Drosophila* gastrulation.** *Cell* 1994, **76**:1075–1089.

8-9   *Xenopus* gastrulation involves several different types of tissue movement.

Shih, J., Keller, R.: **Gastrulation in *Xenopus laevis*: involution—a current view.** *Dev. Biol.* 1994, **5**:85–90.

8-10   Convergent extension and epiboly are due to cell intercalation.

Keller, R., Shih, J., Sater, A.: **The cellular basis of the convergence and extension of the *Xenopus* neural plate.** *Dev. Dynam.* 1992, **193**:199–217.

Keller, R.: **Early embryonic development of *Xenopus laevis***. *Meth. Cell Biol.* 1991, **36**:61–113.

Keller, R., Shi, J., Domingo, C.: **The patterning and functioning of protrusive actively during convergence and extension of the *Xenopus* organizer**. *Develop. Suppl.* 1992, 81–91.

8-11 Notochord elongation is caused by cell intercalation.

Keller, R., Cooper, M.S., D'Anilchik, M., Tibbetts, P., Wilson, P.A.: **Cell intercalation during notochord development in *Xenopus laevis***. *J. Exp. Zool.* 1989, **251**:134–154.

Adams, D.S., Keller, R., Koehl, M.A.: **The mechanics of notochord elongation, straightening, and stiffening in the embryo of *Xenopus laevis***. *Dev.* 1990, **100**:115–130.

8-12 Neural tube formation is driven by both internal and external forces.

Alvarez, I.S., Schoenwolf, G.C.: **Expansion of surface epithelium provides the major extrinsic force for bending of the neural plate**. *J. Exp. Zool.* 1992, **261**:340–348.

Schoenwolf, G.C., Smith, J.L.: **Mechanisms of neurulation: traditional viewpoint and recent advances**. *Development* 1990, **109**:243–270.

8-13 Changes in the pattern of expression of cell adhesion molecules accompany neural tube formation.

Detrick, R.J., Dickey, D., Kintner, C.R.: **The effect of N-cadherin misexpression on morphogenesis in *Xenopus* embryos**. *Neuron* 1990, **4**:493–506.

Bok, G., Marsh, J.: (eds) *Neural Tube Defects*. Ciba Foundation Symposium 181. John Wiley: Chichester, 1994.

8-14 The directed migration of sea urchin primary mesenchyme cells is determined by the contacts of their filopodia to the blastocoel wall.

Ettensohn, C.A., McClay D.R.: **The regulation of primary mesenchyme cell migration in the sea urchin embryo: transplantations of cells and latex beads**. *Dev. Biol.* 1986, **117**:380–391.

Fink, R.D., McClay, D.R.: **Three cell recognition changes accompany the ingression of sea urchin primary mesenchyme cells**. *Dev. Biol.* 1985, **107**:66–74.

Malinda, K.A., Ettensohn, C.A.: **Primary mesenchyme cell migration in the sea urchin embryo: distribution of directional cues**. *Dev. Biol.* 1994, **164**:562–578.

Malinda, K.M., Fisher, G.W., Ettensohn, C.A.: **Four-dimensional microscopic analysis of the filopodial behavior of primary mesenchyme cells during gastrulation in the sea urchin embryo**. *Dev. Biol.* 1995, **172**:552–566.

8-15 Neural crest migration is controlled by environmental cues and adhesive differences.

Delannet, M., Martin, F., Bussy, B., Chersh, D.A., Reichardt, L.F., Duband, J.L.: **Specific roles of the $\alpha V\beta 1$, $\alpha V\beta 3$ and $\alpha V\beta 5$ integrins in avian neural crest cell adhesion and migration on vitronectin**. *Development* 1994, **120**:2687–2702.

Bronner-Fraser, M.: **Mechanisms of neural crest migration**. *BioEssays* 1993, **15**:221–230.

Erickson, C.A., Perris, R.: **The role of cell–cell and cell–matrix interactions in the morphogenesis of the neural crest**. *Dev. Biol.* 1993, **159**:60–74.

Nieto, M.A., Sargent, M.G., Wilkinson, D.G., Cooke, J.: **Control of cell behaviour during vertebrate development by *Slug*, a zinc finger gene**. *Science* 1994, **264**:835–839.

Wang, H.U., Anderson, D.: **Eph family transmembrane ligands can mediate repulsive guidance of trunk neural crest migration and motor axon outgrowth**. *Neuron* 1997, **18**:383-396.

8-16 Slime mold aggregation involves chemotaxis and signal propagation.

Gerisch, G.: **Cyclic AMP and other signals controlling cell development and differentiation in *Dictyostelium***. *Ann. Rev. Biochem.* 1987, **56**:853–879.

Siu, C.H.: **Cell–cell adhesion molecules in *Dictyostelium***. *BioEssays* 1990, **12**:357–362.

Sager, B.M.: **Propagation of traveling waves in excitable media**. *Genes Dev.* 1996, **10**:2237–2250.

Adams, D.S., Keller, R., Koehl, M.A.: **The mechanics of notochord elongation, straightening and stiffening in the embryo of *Xenopus laevis***. *Development* 1990, **110**:115–130.

8-17 Circumferential contraction of hypodermal cells elongates the nematode embryo.

Priess, J.R., Hirsh, D.I.: ***Caenorhabditis elegans* morphogenesis: the role of the cytoskeleton in elongation of the embryo**. *Dev. Biol.* 1986, **117**:156–173.

8-18 The direction of cell enlargement can determine the form of a plant leaf.

Tsuge, T., Tsukaya, H., Uchimiya, H.: **Two independent and polarized processes of cell elongation regulate leaf blade expansion in *Arabidopsis thaliana* (L.) Heynh**. *Development* 1996, **122**:1589–1600.

Jackson, D.: **Designing leaves. Plant morphogenesis**. *Curr. Biol.* 1996, **6**:917–919.

Cell Differentiation

9

The reversibility and inheritance of patterns of gene activity.

Control of specific gene expression.

Models of cell differentiation.

"Considering that we all started off being similar, the diversity of our characteristics is remarkable."

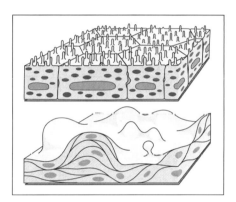

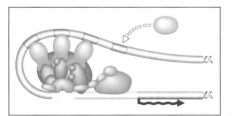

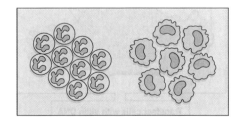

Cellular differentiation describes the process by which embryonic cells become different from one another, acquiring distinct identities and specialized functions. As we have seen in previous chapters, many developmental processes, such as the early specification of the germ layers, involve transient changes in cell form, in patterns of gene activation, and in the molecules synthesized by the cells. However, **cell differentiation** involves the emergence of cell types that have a clear-cut identity in the adult, such as muscle cells, nerve cells, skin cells, and fat cells. In mammals, there are more than 200 clearly recognizable differentiated cell types, such as blood, epidermal, muscle, and nerve cells.

Differentiated cells serve specialized functions and have achieved a terminal and stable state, in contrast with many of the transitory differences in cell state that are characteristic of earlier stages in development. The specified precursors of cartilage and muscle cells have no obvious structural differences from each other and so look the same, but

Fig. 9.16 Interaction between the transcription factors Fos and Jun in gene activation. The proteins Fos and Jun form a heterodimer, which binds to sites known as AP-1 sites in DNA, where it acts as a transcriptional regulator. On its own, Jun can form homodimers, which bind to the AP-1 site less strongly than the Fos:Jun dimer. The homodimers can activate gene expression, but less efficiently than the Fos:Jun heterodimer. Fos itself cannot form homodimers or bind to the AP-1 site alone, and so cannot activate transcription at all.

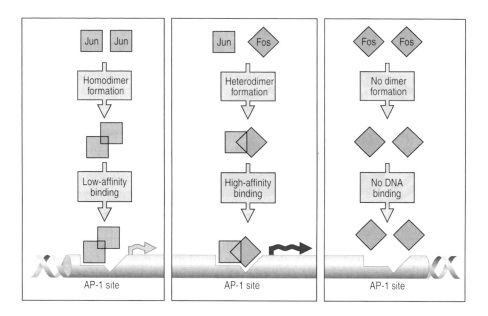

An important feature in the control of transcription by protein factors is the interaction between the factors themselves and with other proteins and small molecules. An example is provided by the two ubiquitous transcription factors Fos and Jun, which increase in activity following the treatment of cells with growth factors. They bind to the AP-1 site, which is present in the control regions of many genes, often in several copies. Fos and Jun bind to DNA as a dimer in the same way as many other gene regulatory proteins. Unlike Fos, Jun can form a homodimer which can also bind to the AP-1 site, but much less strongly than the Fos and Jun heterodimer (Fig. 9.16). This simple case illustrates a general feature of transcription factors: they can interact with each other in a variety of ways, so as to increase or decrease their binding affinity for specific control sites in DNA. The relative concentrations of Fos and Jun in the cell determine which dimeric form is preferentially formed and thus whether the AP-1 sites become activated.

Interactions between transcription factors can become extremely complicated, but we next look at a simple example of the regulation of transcription factor activity, and thus gene expression, by a class of small molecules with an established role in development and cell differentiation. These are the steroid hormones. The action of steroid hormones illustrates in a simple and direct way how an external signal can induce tissue-specific gene expression.

9-7   External signals can activate genes.

The pattern of gene expression in cells can be influenced by external signals. Some signals, such as those provided by steroid hormones, can enter the cell, whereas others, such as protein growth factors, do not enter the cell and act on receptors located at the cell membrane. As we shall see in Chapter 12, steroid hormones produced in the testis are responsible for the secondary sexual characteristics that make male mammals different from females. In insects, the steroid hormone ecdysone is responsible for metamorphosis (see Chapter 14) and induces differentiation in a wide variety of cells. In these cases, the hormone can turn a whole variety of genes in the cell on and off. An example of tissue-specific expression of a single gene is provided by the steroid hormone estrogen, which causes the chick oviduct to produce the protein ovalbumin, a major component of egg white. The continued presence of estrogen is required for transcription of the ovalbumin gene; when estrogen is withdrawn, ovalbumin mRNA and protein disappear.

Unlike protein hormones and growth factors, which bind to receptors on the plasma membrane and exert their effects through intracellular signaling pathways, steroid hormones are lipid soluble; they cross the plasma membrane unaided and enter the cell. Once inside the cell, steroids activate gene expression by binding to receptor proteins; the complex of steroid and receptor is then able to act as a transcriptional regulator, binding directly to control sites in the DNA to activate (or in some cases repress) transcription. In many cases, the control sites, known generally as steroid response elements, act as enhancers.

Some steroid hormones like glucocorticoid bind to a cytoplasmic receptor protein that is then translocated into the nucleus. For example, binding of a steroid to the glucocorticoid receptor in the cytoplasm results in the dissociation of the receptor from a cytoplasmic protein that keeps it inactive in the absence of steroids (Fig. 9.17). Two steroid-receptor complexes then join to form dimers, which move into the nucleus, bind to DNA, and activate specific genes. Other steroids have receptors that are already bound to their target DNA sequences even when the hormone is absent, but hormone binding is necessary to enable them to activate transcription.

The action of steroid hormones in the chick provides an excellent example of the tissue-specificity of the response to these hormones. For example, estrogen activates the ovalbumin gene in the cells of the oviduct, but the ovalbumin gene in liver cells is unaffected. This tissue-specific response in the oviduct cannot easily be accounted for by additional proteins binding to the regulatory regions. Rather, the explanation for the differential response in these two tissues is thought to be some preceding heritable change in the structure of the chromatin. This may, for example, allow transcription factors such as the steroid-hormone receptor complexes to have access to the ovalbumin gene in one cell type but not in the other. Other agents that act through this type of transcription factor complex are thyroid hormone and retinoic acid, which may be a developmental morphogen (see Section 10-4 and Chapter 13).

Most of the signaling molecules that act during development are peptides or proteins. These bind to receptors in the cell membrane and do not themselves enter the cell. The signal is relayed to the cell nucleus by a process known as **signal transduction**. This can be a very complex process but the bare essentials of such a signaling pathway are shown in Fig. 9.18. Like many other intracellular signal transduction pathways, this involves the sequential activation of protein kinases. Binding of the

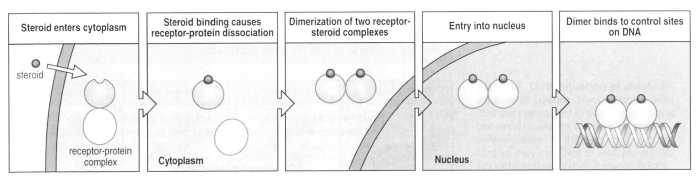

Fig. 9.17 Steroid hormones control transcription by binding to an intracellular receptor to form a transcriptional regulator. In the case of the glucocorticoid receptor shown here, the receptor is present in the cytoplasm as a complex with another protein. Steroid binding causes the receptor to dissociate from this protein and form a dimer with another steroid-bound receptor. This dimer enters the nucleus, where it binds to control sites on DNA and activates transcription.

| 9-13 | Globin gene expression is controlled by distant upstream regulatory sequences. |

Dillon, N., Grosveld, F.: **Transcriptional regulation of multigene loci: multilevel control**. *Trends Genet.* 1993, **9**:134–137.

Dillon, N., Grosveld, F.: **Chromatin domains as potential units of eukaryotic gene function**. *Curr. Opin. Genet. Dev.* 1994, **4**:260–264.

Orkin, S.: **Development of the hematopoietic system**. *Curr. Opin. Genet. Dev.* 1996, **6**:597–602.

Wood, W.G.: **The complexities of β-globin gene regulation**. *Trends Genet.* 1996, **12**:204–206.

| 9-14 | Neural crest cells differentiate into several cell types. |

Le Douarin, N.M., Dupin, E., Ziller, C.: **Genetic and epigenetic control in neural crest development**. *Curr. Opin. Genet. Dev.* 1994, **4**:685–695.

Selleck, M.A., Bronner-Fraser, M.: **The genesis of avian neural crest cells: classic embryonic induction**. *Proc. Natl. Acad. Sci.* 1996, **93**:9352–9357.

Shah, N.M., Groves, A.K., Anderson, D.J.: **Alternative neural crest cell fates are instructively promoted by TGF-β superfamily members**. *Cell* 1996, **85**:331–343.

| 9-15 | Steroid hormones and polypeptide growth factors specify chromaffin cells and sympathetic neurons. |

Stemple, D.L., Anderson, D.J.: **Lineage diversification of the neural crest:** *in vitro* **investigations**. *Dev. Biol.* 1993, **159**:12–23.

Patterson, P.H.: **Control of cell fate in a vertebrate neurogenic lineage**. *Cell* 1990, **62**:1035–1038.

| 9-16 | Neural crest diversification involves signals for both specification of cell fate and selection for cell survival. |

Shah, N.M., Groves, A.K., Anderson, D.J.: **Alternative neural crest cell fates are instructively promoted by TGF-β superfamily members**. *Cell* 1996, **85**:331–343.

Le Douarin, N.M., Ziller, C., Couly, G.F.: **Patterning of neural crest derivatives in the avian embryo:** *in vivo* **and** *in vitro* **studies**. *Dev. Biol.* 1993, **159**:24–49.

Fleischman, R.A.: **From white spots to stem cells: the role of the Kit receptor in mammalian development**. *Trends Genet.* 1993, **9**:285–290.

| 9-17 | Programmed cell death is under genetic control. |

Ellis, R.E., Yuan, J.Y., Horvitz, H.R.: **Mechanisms and functions of cell death**. *Ann. Rev. Cell. Biol.* 1991, **7**:663–698.

Chinnaiyan, A.M., Dixit, V.M.: **The cell-death machine**. *Curr. Biol.* 1996, **6**:555–562.

Jacobson, M.D., Weil, M., Raff, M.C.: **Programmed cell death in animal development**. *Cell* 1997, **88**:347–354.

Steller, H.: **Mechanisms and genes of cellular suicide**. *Science* 1995, **267**:1445–1449.

Organogenesis

10

The development of the chick limb.

Insect imaginal discs.

The insect compound eye.

The nematode vulva.

Development of the kidney.

"When we were older we made use of what we had learnt when young to make quite new patterns and structures."

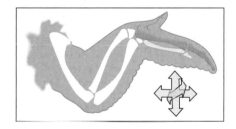

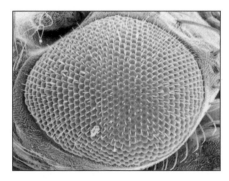

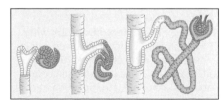

So far, we have concentrated almost entirely on the broad aspects of development involved in laying down the basic body plan in a variety of organisms. We now turn to the development of specific organs and structures—**Organogenesis**, which is a crucial phase of development in which the embryo at last becomes a fully functioning organism, capable of independent survival.

The development of certain organs has been studied in great detail and they provide excellent models for looking at developmental processes such as pattern formation, the specification of positional information, induction, change in form, and cellular differentiation. In this chapter, we consider the development of four organ systems—the chick limb, insect appendages such as legs and wings in *Drosophila*, the insect compound eye, and the nematode vulva. We finish by looking at an example of epithelial patterning and morphogenesis in kidney development. We do not deal with other important organs such as the heart, lung, liver, or pancreas, because their development is, in general, less well understood, and there is no evidence that different principles are involved.

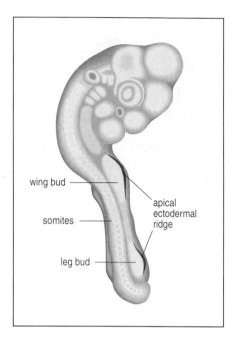

wing bud

somites

apical
ectodermal
ridge

leg bud

Fig. 10.1 The limb buds of the chick embryo. Limb buds appear on the flanks of the embryo on the third day of incubation (only limbs on the right side are shown here). They are composed of mesoderm, with an outer covering of ectoderm. Along the tip of each runs a thickened ridge of ectoderm, the apical ectodermal ridge.

The cellular mechanisms involved in organogenesis are essentially similar to those encountered in earlier stages of development; they are merely employed in different spatial and temporal patterns. There are, for example, striking similarities between the development of the vertebrate limb and the wings and legs of *Drosophila*, and many of the genes and signaling molecules involved will be familiar from earlier chapters.

The development of the chick limb.

The vertebrate embryonic limb is a particularly good system in which to study pattern formation. It has the advantage of having a basic pattern that is initially quite simple, and in the chick embryo, the limbs are easily accessible for surgical manipulation. The limb provides a good model for studying cellular interactions within a structure containing a large number of cells, and for elucidating the role of intercellular signaling in development. In chick embryos, the limbs begin to develop on the third day after laying, when the structures of the main body axis are already well established. The limbs develop from small protrusions—the limb buds—which arise from the body wall of the embryo (Fig. 10.1). By 10 days the main features of the limbs are well developed (Fig. 10.2). These are the cartilaginous elements (which are later replaced by bone), the muscles and tendons, and the epidermally derived surface structures, such as feathers. The limb has three developmental axes: the proximo-distal axis runs from the base of the limb to the tip; the antero-posterior axis runs parallel with the body axis (in the human hand it goes from the thumb to the little finger, and in the chick limb from digit 2 to digit 4); the dorso-ventral axis runs from the back of the human hand to the palm.

Fig. 10.2 The embryonic chick wing. The photograph shows a stained whole mount of the wing of a chick 10 days after laying. By this time, the main cartilaginous elements (e.g. humerus, radius, and ulna) have been laid down. They later become ossified to form bone. The muscles and tendons are also well developed at this stage but cannot be seen in this type of preparation (see Fig. 10.19). The three developmental axes of the limb are proximo-distal, antero-posterior, and dorso-ventral, as shown in the top panel. Note that the chick wing only has three digits which have been called 2, 3, and 4. Scale bar = 1 mm.

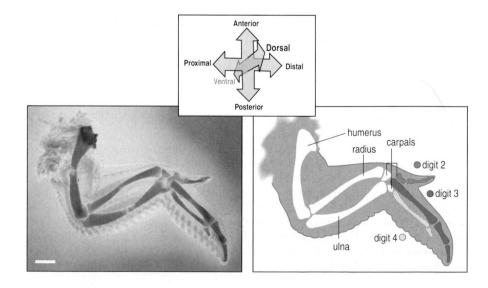

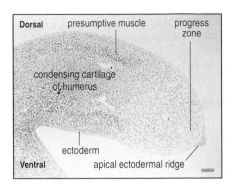

Fig. 10.3 Cross-section through a developing limb bud. The thickened apical ectodermal ridge is at the tip. Beneath the ridge is the progress zone where cells proliferate. Proximal to the progress zone, mesenchyme cells condense and differentiate into cartilage. Presumptive muscle cells migrate into the limb from the adjacent somites. Scale bar = 0.1 mm.

10-1 | The vertebrate limb develops from a limb bud.

The early limb bud has two major components—a core of loose mesenchymal mesoderm cells and an epithelial ectodermal layer (Fig. 10.3). Most of the limb develops from the mesenchymal core, although the muscle cells of the limb have a separate lineage and migrate into the bud from the somites (see Section 4-2). At the tip of the limb bud is a **progress zone** of rapidly dividing and proliferating undifferentiated cells. The progress zone lies directly beneath a thickening in the ectoderm—the **apical ectodermal ridge** (Fig. 10.4). It is only when cells leave the progress zone that they begin to differentiate. As the bud grows, the cells start to differentiate and cartilaginous structures begin to appear in the mesenchyme. The part of the limb nearest to the body wall is the first to differentiate, and differentiation proceeds distally as the limb extends. Among the structures found in the developing limb, the patterning of the cartilage has been the best studied, as this can be stained and seen easily in whole mounts of the embryonic limb. The disposition of muscles and tendons is more intricate, and requires histological examination of serial sections through the limb.

The first clear sign of cartilage differentiation is the increased packing of groups of cells, a process known as condensation. The cartilage elements are laid down in a proximo-distal sequence in the chick wing—the humerus, radius, and ulna, the wrist elements (carpals), and then three easily distinguishable digits, 2, 3, and 4 (see Fig. 10.2). Fig. 10.5 compares this sequence in the development of a chick wing with the similar sequence in the development of a mouse forelimb.

The chick limb bud at 3 days is about 1 mm wide by 1 mm long, but by 10 days it has grown some tenfold, mostly in length. Even then, it is still small compared with both the newborn and adult limb; as the basic pattern is laid down when it is still small, growth occupies the major period of the limb's development. During the later growth phase, the cartilaginous elements are largely replaced by bone. Nerves only enter the limb after the cartilage has been laid down (see Section 11-8). The problem of limb patterning is to understand how cartilage, muscle, and tendons are formed in the right place and how they make the right connections with each other.

10-2 | Patterning of the limb involves positional information.

Patterning in the limb is dependent on cell–cell interactions. The early limb bud has considerable powers of regulation and pieces of it can generally be removed, rotated, or put in a different position without perturbing the final pattern. But there are two regions in which this general rule does not apply. These are crucial organizing regions, and their removal or transplantation has profound effects. One of these regions is the apical ectodermal ridge at the tip of the limb; the other is a region at the posterior margin of the mesenchyme known as the **polarizing region**

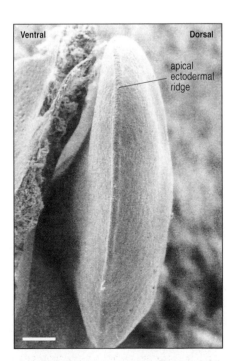

Fig. 10.4 Scanning electron micrograph of a chick limb bud at 4½ days after laying, showing the apical ectodermal ridge. Scale bar = 0.1 mm.

Fig. 10.5 The development of the chick wing and mouse forelimb are similar. The limb cartilage elements are laid down in a proximo-distal sequence as the limb bud grows outward. The cartilage of the humerus is laid down first, followed by the radius and ulna, wrist elements, and digits. Scale bar = 1 mm.

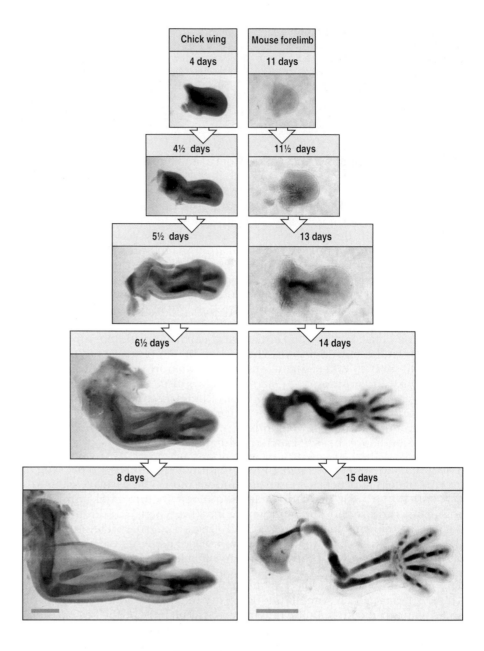

or **zone of polarizing activity** (**ZPA**) (Fig. 10.6, left panel). In the chick embryo, removal of the anterior half of a limb bud has no effect on development and a normal limb is produced. But removal of the posterior half, which contains the polarizing region, leads to abnormal limb development.

Some aspects of vertebrate limb development fit very well with a model of pattern formation based on positional information. The developing chick limb behaves as if its cells' future development is determined by their position with respect to the main axes while they are in the progress zone (see Fig. 10.6). The position of cells along the proximo-distal axis may involve a timing mechanism, their fate being determined by how long they remain in the progress zone. Patterning along the antero-posterior axis is specified by a signal, or signals, emanating from the polarizing region at the posterior margin of the limb bud, with the dorso-ventral axis being specified by a signal, or signals, from the over-lying ectoderm. Patterning along these three axes will be considered in detail later.

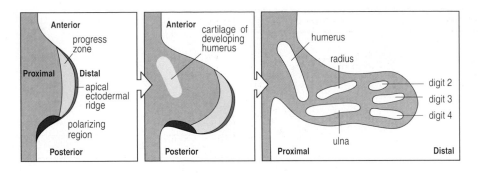

Fig. 10.6 Cells acquire their positional value in the progress zone. The progress zone is specified at the distal end of the bud by the apical ectodermal ridge. As the limb bud grows outward, cells in the progress zone proliferate and acquire a positional value. When they leave the zone, cartilage can begin to differentiate and the cartilaginous elements are laid down in a proximo-distal sequence, starting with the humerus. The position of the cells along the antero-posterior axis is specified by a signal from the polarizing region, which is located at the posterior side of the bud.

Central to the idea of positional information is the distinction between positional specification and interpretation. Cells acquire positional inform-ation first and then interpret these positional values according to their genetic constitution and developmental history. It is the difference in their developmental history that makes wings and legs different, for posi-tional information is signaled in wing and leg buds in the same way. An attractive hypothesis for limb patterning is that a single three-dimensional positional field controls the development of the cells that give rise to all the limb elements—cartilage, muscle, and tendons. The specification of all three axes is, as we shall see, linked by molecular signals.

10-3 | The apical ectodermal ridge induces the progress zone.

The apical ectodermal ridge consists of closely packed columnar cells, which are linked by extensive gap junctions. Their tight packing gives the ridge a mechanical strength that probably keeps the limb flattened along the dorso-ventral axis; the length of the ridge controls the width of the bud. The progress zone, lying beneath the apical ectodermal ridge, is a region of rapidly proliferating mesenchymal cells, which produces the initial outgrowth of the limb bud. It is also the region where limb cells acquire their positional information. Surprisingly, the establishment of the progress zone does not reflect a local increase in cell division in the limb bud region but rather a decrease from a previously high rate in cell proliferation along the rest of the flank. The localization of the apical ectodermal ridge appears to involve the gene *radical fringe*, which is a homolog of the *Drosophila* gene *fringe* that is similarly involved in specifying a dorso-ventral boundary (see Section 10-14). *radical fringe* is expressed in the dorsal limb ectoderm before the formation of the apical ectodermal ridge. The apical ectodermal ridge develops at the boundary between cells that express *radical fringe* and those that do not.

Because of its influence on the underlying progress zone, the apical ectodermal ridge is essential for both the outgrowth and the proximo-distal patterning of the limb. Removal of the ridge from a chick limb bud by micro-surgery results in a significant reduction in growth and in truncation of the limb, with distal parts missing. The proximo-distal level at which the limb is truncated depends on the time at which the ridge is removed (Fig. 10.7). The earlier the ridge is removed, the greater the effect; removal at a later stage only results in loss of the distal parts of the digits. Proximo-distal patterning occurs progressively over time in the progress zone. Follow-ing apical ridge removal, cell proliferation in the progress zone is greatly reduced.

It can be shown that the apical ectodermal ridge signals to the underlying mesenchyme by grafting an isolated ridge to the dorsal surface of an early limb bud. The result is an ectopic outgrowth, which may even develop cartilaginous elements and digits. In the chick, a major signal

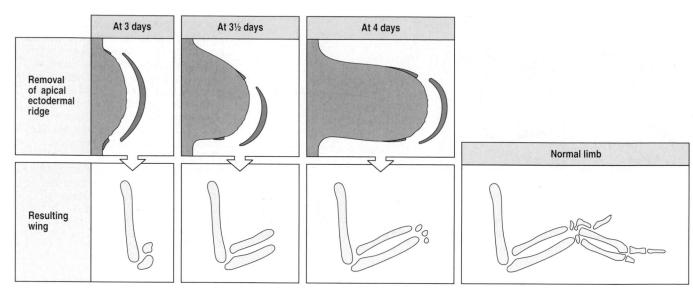

Fig. 10.7 The apical ectodermal ridge is required for proximo-distal development. Limbs develop in a proximo-distal sequence. Removal of the ridge from a developing limb bud leads to truncation of the limb; the later the ridge is removed, the more complete the resulting limb.

from the ridge is provided by proteins of the fibroblast growth factor family. FGF-8 is expressed throughout the ridge and FGF-4 in the posterior region. FGF-4 protein can act as a functional substitute for an apical ridge. If the ridge is removed and beads that release the growth factor grafted into the tip of the limb in its place, more or less normal outgrowth of the limb continues. If sufficient FGF-4 is provided to the outgrowing cells, a fairly normal limb develops (Fig. 10.8).

The ridge itself is maintained by signals from the progress zone and the polarizing region. When all the elements of the limb have been laid down, the ridge disappears, probably because the progress zone no longer provides a maintenance signal. We now consider patterning along each of the three axes in the limb.

10-4   The polarizing region specifies position along the antero-posterior axis.

The polarizing region of a vertebrate limb bud has signaling properties that are almost as striking as those of the Spemann organizer in amphibians. When the polarizing region from one early wing bud is grafted to the anterior margin of another early wing bud, a wing with a mirror-image pattern develops: instead of the normal pattern of digits—2 3 4—the pattern 4 3 2 2 3 4 develops (Fig. 10.9). The pattern of muscles and tendons in the limb shows similar mirror-image changes.

The additional digits come from the host limb bud and not from the graft, showing that the grafted polarizing region has altered the developmental fate of the host cells in the anterior region of the limb bud. The limb bud widens in response to the polarizing graft, which enables the additional digits to be accommodated; widening of a limb bud is always associated with an increase in the extent of the apical ectodermal ridge and the rate of cell division is higher than in the anterior region of a normal bud.

One way that the polarizing region could specify position along the antero-posterior axis is by producing a diffusible morphogen. The concentration

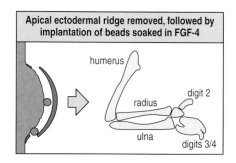

Fig. 10.8 The growth factor FGF-4 can substitute for the apical ectodermal ridge. After the ridge is removed, grafting beads that release FGF-4 into the limb bud results in almost normal limb development.

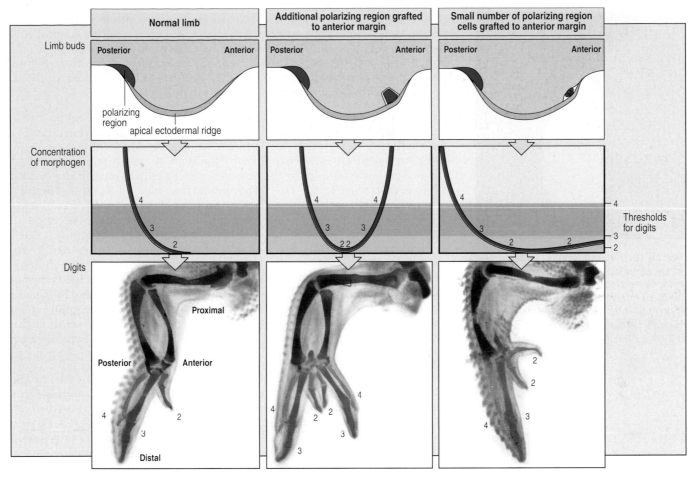

Fig. 10.9 The polarizing region can specify pattern along the antero-posterior axis. If the polarizing region is the source of a graded morphogen, the different digits could be specified at different threshold concentrations of signal (left panels). Grafting an additional polarizing region to the anterior margin of a limb bud (center panels) would result in a mirror-image gradient of signal, and thus the observed mirror-image duplication of digits. The signal from a grafted polarizing region can be attenuated by grafting only a small number of polarizing region cells to the anterior margin of the limb bud (right panels), so that only an extra digit 2 develops.

of morphogen could specify the position of cells along the antero-osterior axis with respect to the polarizing region located at the posterior margin of the limb. Cells could then interpret their positional values by developing specific structures at particular threshold concentrations of morphogen. Digit 4, for example, would develop at a high concentration, digit 3 at a lower one, and digit 2 at an even lower one. According to this model, a graft of an additional polarizing region to the anterior margin would set up a mirror-image gradient of morphogen, which would result in the 4 3 2 2 3 4 pattern of digits that is observed (see Fig. 10.9).

If the action of the polarizing region in specifying the character of a digit is due to a diffusible signal, then when the signal is weakened, the pattern of digits should be altered in a predictable manner. Grafting small numbers of polarizing region cells to a limb bud results only in the development of an additional digit 2 (see Fig. 10.9). An analogous result can be obtained by leaving the polarizing region graft in place for some time and then removing it. When left for 15 hours an additional digit 2 develops, while development of a digit 3 requires the graft to be in place for up to 24 hours.

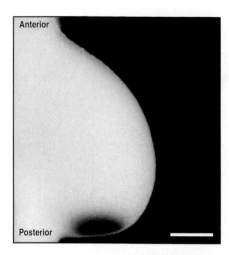

Fig. 10.10 *Sonic hedgehog* **is expressed in the polarizing region of a chick limb bud.** *Sonic hedgehog* is expressed at the posterior margin of the limb bud in the polarizing region and provides a positional signal along the antero-posterior axis. Scale bar = 0.1 mm. Photograph courtesy of C. Tabin.

The consistency of the diffusible morphogen model with the experimental results is, however, not sufficient to prove it. Other models suggest that the same results could be obtained by a relay of short-range signals from the polarizing region. There is also evidence that other mechanisms operate to lay down the cartilage pattern, which is then modified by a signal from the polarizing region, as discussed later.

A persuasive case can be made that the Sonic hedgehog protein is the key component of the natural polarizing signal. The *Sonic hedgehog* gene is expressed in the limb polarizing region (Fig. 10.10) and Sonic hedgehog protein is known to be involved in numerous patterning processes, for example, in the somites (see Section 4-2) and the neural tube (see Section 11-6). The *Drosophila* homolog of Sonic hedgehog, the hedgehog protein, is a key signaling molecule in the patterning of the segments in the *Drosophila* embryo (see Section 5-16), as well as *Drosophila* legs and wings. Chick fibroblast cells transfected with a retrovirus containing the *Sonic hedgehog* gene acquire the properties of a polarizing region; they cause the development of a mirror-image limb when grafted to the anterior margin of a limb bud, as will beads soaked in Sonic hedgehog protein. The bead must be left in place for 24 hours for extra digits to be specified and the pattern of digits depends on the concentration of Sonic hedgehog protein in the bead. A higher concentration is required for a digit 4 to develop than for a digit 2.

Further evidence for the key role of *Sonic hedgehog* comes from the mouse mutation *extra-toes*, in which there is an additional anterior digit and additional anterior expression of *Sonic hedgehog*. Polydactyly, the development of additional digits, occurs in humans (Fig. 10.11) and this could result from anterior expression of the human version of *Sonic hedgehog* or, as we see below, from mutations in the Hox genes. Mice in which the *Sonic hedgehog* gene has been knocked out develop proximal limb structures but lack distal ones, showing that the gene is not required for initiation of limb development. The loss of distal structures probably reflects a requirement for Sonic hedgehog in order to maintain the apical ectodermal ridge and thus the progress zone.

Several other molecules may also be involved in positional signaling in the vertebrate limb. The growth factors BMP-2 and BMP-4 are present in a gradient above the progress zone, with the high point at the posterior margin of the limb bud. Local application of Sonic hedgehog protein induces expression of the BMP growth factors. Retinoic acid is also present at a higher concentration in posterior regions and, when applied locally to the anterior margin of a chick wing bud, causes formation of extra digits in a mirror-image pattern similar to that obtained by grafting a polarizing region. Retinoic acid induces a new polarizing region by inducing *Sonic hedgehog* expression, so is unlikely to act as a positional signal. However, it is possible that retinoic acid is required for limb bud initiation, as inhibition of retinoic acid synthesis blocks limb bud outgrowth.

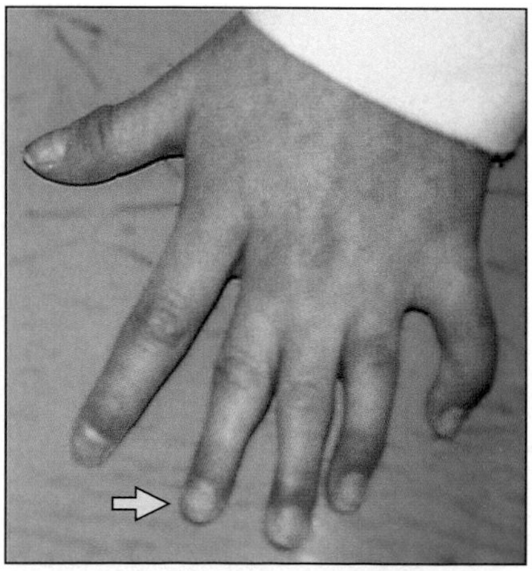

The polarizing region is involved in the maintenance of the apical ridge, probably via BMP-2 and BMP-4, and there is a positive feedback loop between Sonic hedgehog protein in the mesoderm and FGF-4 expression in the ridge. This interaction between FGF-4 and Sonic hedgehog can be shown by the induction of new limbs at ectopic sites. Localized application of FGF-4 to the flank of a chick embryo, between the wing and leg buds, induces the production of FGF-8 in the ectoderm, and then the ectopic expression of *Sonic hedgehog*. The Sonic hedgehog protein then feeds back to induce expression of the embryo's own gene for FGF-4, resulting in the maintenance of the ridge and the outgrowth of an additional limb bud at this site. The type of limb is determined by

Fig. 10.11 Polydactyly in a human hand. The additional digit (arrowed) resembles the adjacent digit. Photograph courtesy of R. Winter.

Fig. 10.12 Result of local application of FGF-4 to the flank of a chick embryo. FGF-4 is applied to a site close to the hindlimb bud, where it induces outgrowth of an additional hindward limb. The dark regions indicate expression of *Sonic hedgehog*. Scale bar = 1 mm. Photograph courtesy of J-C. Izpisúa-Belmonte, from Cohn, M.J., *et al.*: 1995.

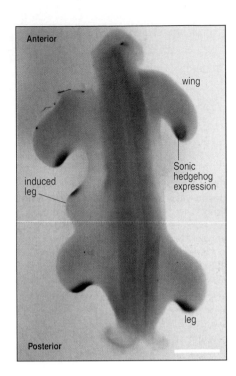

the site of *Sonic hedgehog* expression. In general, additional wing buds develop from application of FGF-4 to the anterior flank, whereas leg buds develop from application to the posterior flank (Fig. 10.12).

| 10-5 | **Position along the proximo-distal axis may be specified by a timing mechanism.** |

In contrast to the signal specifying pattern along the limb's antero-posterior axis, patterning along the proximo-distal axis is less well understood but may be specified by the time cells spend in the progress zone. As the limb bud grows, cells are continually leaving the progress zone. In the forelimb, for example, cells leaving first develop into the humerus and those leaving last form the tips of the digits. If cells could measure the time they spend in the progress zone, for example by counting cell divisions, this would give them a positional value along the proximo-distal axis (Fig. 10.13). A timing mechanism of this sort is consistent with the experimental observation that removal of the apical ridge results in a distally truncated limb. Removal means that cells in the progress zone no longer proliferate, so more distal structures cannot form.

Evidence for a mechanism based on time comes from killing cells in the progress zone or blocking their proliferation at an early stage by, for example, X irradiation. The result is that proximal structures are absent but distal ones are present, and can be almost normal. Because many cells in the irradiated progress zone do not divide, the number of cells that leave the progress zone during each unit of time is much smaller than normal. Proximal structures are thus very small, or even absent. As the bud grows out the progress zone becomes repopulated by the surviving cells, so that only distal structures are formed. If a wing bud is treated in the same way, therefore, only digits will develop.

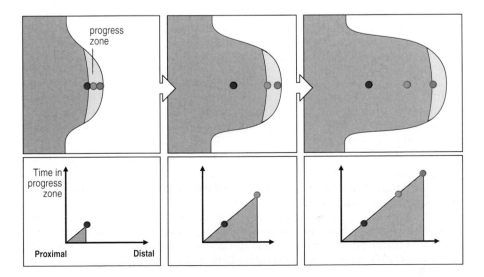

Fig. 10.13 A cell's proximo-distal positional value may depend on the time it spends in the progress zone. Cells are continually leaving the progress zone. If the cells could measure how long they spend in the progress zone, this could specify their position along the proximo-distal axis could be specified. Cells that leave the zone early (red) form proximal structures whereas cells that leave it last (blue) form the tips of the digits.

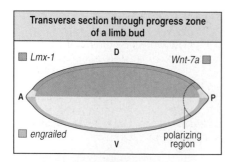

Fig. 10.14 The ectoderm controls dorso-ventral pattern in the developing limb. The gene *Wnt-7a* is expressed in the dorsal ectoderm and *engrailed* (the vertebrate version of the *Drosophila* gene *engrailed*) is expressed in the ventral ectoderm. The gene *Lmx-1* is induced by *Wnt-7a* in the dorsal mesoderm and is involved in specifying dorsal structures.

10-6   The dorso-ventral axis is controlled by the ectoderm.

In the chick wing, there is a well-defined pattern along the dorso-ventral axis; large feathers are only present on the dorsal surface and the muscles and tendons have a complex dorso-ventral organization. The development of pattern along this axis has been studied by recombining ectoderm taken from from left limb buds with mesoderm from right limb buds, so that the dorso-ventral axis of the ectoderm is reversed with respect to the underlying mesoderm.

The right and left wing buds are removed from a chick embryo and the ectoderm separated from the mesenchyme. The wing buds are then recombined with the dorso-ventral axis of the ectoderm opposite to that of the mesoderm. The reconstituted limb buds are then grafted to the flank of a host embryo, with just the dorso-ventral axis of the ectoderm inverted with respect to the enclosed mesenchyme, and allowed to develop. In general, the proximal regions of the limbs that develop have the dorso-ventral polarity of a normal limb: that is their dorso-ventral pattern corresponds to the source of the mesoderm. Distal regions, however, particularly the 'hand' region, have a reversed dorso-ventral axis, with the pattern of muscles and tendons reversed and corresponding to that of the dorso-ventral axis of the ectoderm. The ectoderm can therefore specify dorso-ventral patterning in the limb.

A dorso-ventral pattern is easily seen in the digits, where ventral surfaces —the palm or the paw pads—normally have no fur, whereas dorsal surfaces do. Genes controlling the dorso-ventral axis in vertebrate limbs have been identified from mutations in mice. For example, mutations that inactivate the gene *Wnt-7a*, which encodes a secreted signaling protein of the Wnt family (see Section 3-18), result in limbs in which many of the dorsal tissues adopt ventral fates to give a double ventral limb, the two halves being mirror images. The *Wnt-7a* gene is expressed in the dorsal ectoderm (Fig. 10.14) and this suggests that the ventral pattern is the ground state and is modified dorsally by the dorsal ectoderm, the *Wnt-7a* gene playing a key role in patterning the dorsal mesoderm. Expression of the gene *engrailed* characterizes ventral ectoderm. Mutations in *engrailed* result in *Wnt-7a* being expressed ventrally, giving a double dorsal limb.

One function of *Wnt-7a* is to induce expression of the LIM homeobox gene *Lmx-1* in the underlying mesenchyme. This gene encodes a transcription factor, the expression of which specifies a dorsal pattern in the mesoderm. Ectopic expression of *Lmx-1* in the ventral mesoderm results in the cells adopting a dorsal fate, resulting in a mirror-image dorsal limb. A structural relative of *Lmx-1* in *Drosophila, apterous* is involved in specifying the dorsal surface of the insect wing (see Section 10-14).

In many *Wnt-7a* mutant mice, posterior digits are lacking, suggesting that *Wnt-7a* is also required for normal antero-posterior patterning. Similar results are observed in chick embryos when dorsal ectoderm is removed. This has led to the suggestion that development along all three axes is integrated by interactions between the signals Wnt-7a, FGF-4, and Sonic hedgehog.

10-7   Different interpretations of the same positional signals give different limbs.

The positional signals controlling limb patterning are the same in chick forelimbs and hindlimbs as well as in different vertebrates, but they are interpreted differently. A polarizing region from a wing bud, for example,

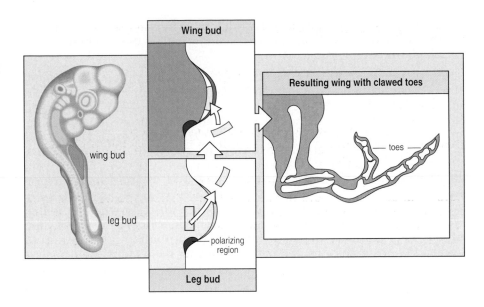

Fig. 10.15 Chick proximal leg bud cells grafted to a distal position in a wing bud acquire distal positional values. Proximal tissue from a leg bud that would normally develop into a thigh, is grafted to the tip of a wing bud. In the progress zone, it acquires more distal positional values and interprets these as leg structures, forming clawed toes at the tip of the wing.

will specify additional mirror-image digits if grafted into the anterior margin of a leg bud, where these digits are toes, not wing digits, as the signal is interpreted by leg cells. Similarly, a polarizing region from a mouse or human limb bud specifies additional wing structures when grafted into the anterior margin of a chick wing bud. A mouse apical ectodermal ridge grafted in place of a chick apical ectodermal ridge can provide the appropriate signal for the chick during early limb development. The signals from the different polarizing regions are the same, as are the signals from the different apical ridges; the difference in the structures formed by the limbs is a consequence of how the signals are interpreted and thus depends on the genetic constitution and developmental history of the responding cells in the limb bud.

Forelimbs and hindlimbs differ because of the difference in their developmental histories, which are related to the position of the limb bud along the antero-posterior axis of the animal's body. This is analogous to the specification of the different thoracic segments in insect development (see Chapter 5). The differences between the homologous limbs of different vertebrates, by contrast, reflect the differences in the activation of the genes that control the interpretation of positional information.

A further demonstration of the interchangeability of positional signals comes from grafting limb bud tissue to different positions along the proximo-distal axis of the bud. If tissue that would normally give rise to the thigh is grafted from the proximal part of an early chick leg bud to the tip of an early wing bud, it develops into toes with claws (Fig. 10.15). The tissue has acquired a more distal positional value after transplantation, but interprets it according to its own developmental program, which is to make leg structures.

10-8 | Homeobox genes are involved in patterning the limbs and specifying their position.

Vertebrate Hox genes specify position along the antero-posterior axis of the body (see Section 4-3) and may also provide positional values in the limbs. At least 23 different Hox genes are expressed during chick limb development. Attention has been largely focused on the genes of the Hoxa and Hoxd gene clusters, which are related to the *Drosophila Abdominal-B* gene (see Box 4A, page 104). These sets of genes are

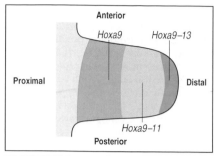

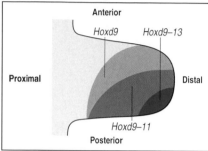

Fig. 10.16 Pattern of Hox gene expression in the chick wing bud. The Hoxa genes (top panel) are expressed in a nested pattern along the proximo-distal axis, *Hoxa13* being expressed most distally, while the Hoxd genes (bottom panel) are expressed in a similarly nested pattern along the antero-posterior axis, *Hoxd13* being expressed most posteriorly.

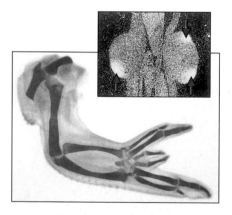

Fig. 10.17 Change in expression of Hoxd genes following a polarizing region graft. Grafting a polarizing region to the anterior margin of a limb bud results in mirror-image expression of the *Hoxd13* genes and a mirror-image duplication of the digits (main panel). The inset shows the expression of *Hoxd13* (arrowed) at the posterior margin in a normal limb bud (inset left). In the operated limb (inset right) *Hoxd13* is expressed both at the anterior and posterior margins.

expressed in both forelimbs and hindlimbs, whereas the expression of Hoxb and Hoxc gene clusters is restricted to either the forelimb or the hindlimb bud.

The expression of Hox genes during limb development is dynamic; the pattern of expression of a particular gene can undergo significant changes as the limb bud grows. Here, we confine our attention to the expression of the Hoxa and Hoxd genes at just one stage of wing development. At this stage, expression of *Hoxd9* through *Hoxd13* is sequentially initiated at the posterior margin of the limb in the progress zone, resulting in a nested pattern of expression of different Hoxd genes along the antero-posterior limb axis. That is, *Hoxd9* expression covers the complete region where the Hoxd genes are expressed, and successive genes occupy more and more posterior regions so that *Hoxd9–Hoxd13* are all expressed in a small posterior region. Expression of the Hoxa genes is similarly sequentially initiated in the progress zone and ends up as a nested pattern of expression along the proximo-distal axis (Fig. 10.16).

At the stage of limb development depicted in Fig. 10.16, the proximo-distal sequence of expression of Hoxa genes corresponds to the three main proximo-distal regions of the limb, which are: upper limb, where the humerus (or femur in the hindlimb) forms, *Hoxa9* expressed; the lower limb, where the radius and ulna (or tibia and fibula) develop, *Hoxa9–11* expressed; and the wrist and digits, *Hoxa9–13* expressed.

If the Hox genes are involved in recording positional information, then experimental manipulations that lead to changes in the pattern of the limb skeletal elements should be preceded by a corresponding change in Hox expression domains. If a polarizing region is grafted to the anterior margin of a wing bud, there is indeed a change to a mirror-image pattern of Hoxd expression (Fig. 10.17). This occurs within 24 hours of grafting, which is about the time required for the polarizing region to exert its effect.

Does alteration of Hox gene expression in the limb lead to homeotic changes, similar to those seen in the vertebrae in Section 4-4. The results of gene knock-out experiments designed to answer this question are complicated, and there is evidently no simple Hox code for the cartilaginous elements of the limb. Knock-outs of individual Hox genes in the mouse do not transform one digit into another. Instead, many bones in the hand are affected in both size and shape, and new elements may even develop. When more than one Hox gene is knocked out at the same time, the effects can be much more severe, and it seems that the Hox genes have an important influence on growth of the cartilaginous elements. Thus, knock-out of both *Hoxa11* and *Hoxd11* results in the absence of both the radius and ulna. Overexpression of *Hoxa13*, which is normally expressed in the distal region of the limb bud, results in limbs in which the radius and the ulna are reduced in size. This suggests that they are transformed into small wrist-like elements, probably by a change in the control of cell multiplication. Overexpression of *Hoxd13* results in the shortening of the long bones of the leg owing to *Hoxd13* affecting the rate of cell proliferation in the growing cartilaginous elements. These results show that expression of the Hox genes can control the size of the cartilaginous elements in the limbs at both early and later stages.

We have seen that limb bud initiation involves FGF and the formation of a polarizing region. It is not difficult to envisage how Hox gene coding along the antero-posterior axis of the body would be able to specify position of the limb bud by specifying where FGF is expressed along the flank, and thus where an apical ridge and polarizing region develop. The pattern of Hox gene expression in the mesoderm, where wing and leg buds form in the chick, is different. That these patterns specify whether a wing or leg develops is supported by the changes in Hox gene expression when an FGF bead is placed in the flank, inducing a wing or a leg (see Section 10-3). The new pattern of Hox gene expression corresponds to that normally found in either the wing or the leg.

Evidence that Hox genes are involved in specifying the position of the polarizing region comes from mouse embryos that express a transgene of *Hoxb8* in more anterior regions of the embryo than normal. In these mice, an extra polarizing region forms at the anterior margin of the forelimb buds, causing extra digits to develop. Evidence for the combined action of Hox genes in determining antero-posterior position of the forelimb bud comes from knock-out mice lacking *Hoxb5* expression, in which the forelimbs develop at a more anterior level.

The involvement of Hox genes in human limb development is shown by the phenotype of human Hox gene mutations. A mutation in the human *Hoxd13* gene results in polydactyly and fusion of digits (see Fig. 10.11), whereas a mutation in human *Hoxa13* results in anterior and posterior digits that are reduced in size.

<table>
<tr><td>10-9</td><td>**Self-organization may be involved in pattern formation in the limb bud.**</td></tr>
</table>

The development of a pattern of cartilaginous elements along the antero-posterior axis may involve mechanisms other than signaling by the polarizing region. Evidence for this comes, for example, from observing the development of reaggregated limb buds after the mesenchymal cells of early chick limb buds have first been disaggregated and thoroughly mixed to disperse the polarizing region. They are then reaggregated, placed in an ectodermal jacket, and grafted to a site where they can acquire a blood supply, such as the dorsal surface of an older limb. Limb-like structures develop from these grafted buds even though they have developed in the absence of a polarizing region. Several long cartilaginous elements may form in the more proximal regions of these abnormal limbs, although none of the proximal elements can be easily identified with normal structures. More distally, however, reaggregated hindlimb buds develop identifiable toes (Fig. 10.18). The fact that well-formed cartilaginous elements can develop at all in the absence of a discrete polarizing region shows that the bud has a considerable capacity for self-organization. In the digits that develop in hindlimb reaggregates there is no sign of the correlation between Hoxd gene expression and antero-posterior position seen in normal development.

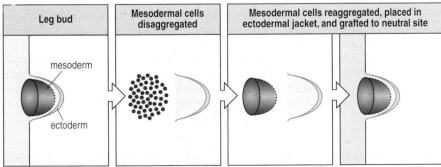

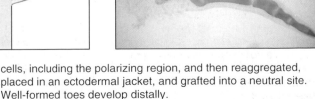

Fig. 10.18 Reaggregated limb bud cells form digits in the absence of a localized polarizing region. Mesodermal cells from a chick leg bud are separated, mixed to disperse all the cells, including the polarizing region, and then reaggregated, placed in an ectodermal jacket, and grafted into a neutral site. Well-formed toes develop distally.

There may therefore be a mechanism in the limb bud that generates a basic pattern—a prepattern—of equivalent cartilaginous elements. These elements could then be given their identities and further refined by response to positional information involving signals like Sonic hedgehog and the Hox genes. The mechanism for generating the prepattern could be based on a **reaction-diffusion mechanism** (Box 10A). In the wing, for example, a reaction-diffusion or related mechanism could result in a single peak in some morphogen forming in the proximal region of the limb, which would specify a prepattern for the humerus. More distally, alterations in the reaction-diffusion conditions, due to changes in proximo-distal positional information, could give rise to three peaks of the morphogen, giving the cartilaginous elements of the three wing digits. These prepatterns could then be modified by signals specifying antero-posterior and dorso-ventral positional information.

In this reaction-diffusion model, polydactyly in humans could simply result from a chance widening of the limb bud. If a reaction-diffusion mechanism generates a periodic pattern of cartilage elements across the limb as the digits are forming, merely widening the limb bud by some small developmental accident would enable a further digit to develop.

| 10-10 | **Limb muscle is patterned by the connective tissue.** |

If quail somites are grafted into a chick embryo at a site opposite to where the wing bud will develop, the wing that subsequently forms will have muscle cells of quail origin, but all other cells will be of chick origin. This demonstrates that limb muscle cells have a different lineage to that of limb connective tissues (cartilage and tendons). Cells that give rise to muscle migrate into the limb bud from the somites at a very early stage (see Section 4-2). After migration, the future muscle cells multiply and initially form a dorsal and a ventral block of presumptive muscle (Fig. 10.19). These blocks undergo a series of divisions to give the final muscle masses. Unlike the cartilage and connective tissue cells, which acquire positional information in the progress zone, the presumptive muscle cells, at least initially, do not acquire positional values, and are all equivalent.

Evidence for this equivalence comes from experiments where somites are grafted from the future neck region of the early embryo in order to replace the normal wing somites. The muscle cells that develop thus come from neck somites, but a normal pattern of limb muscles still develops. This shows that the muscle pattern is determined by the connective tissue into which the muscle cells migrate, rather than by the muscle cells themselves. A mechanism that could pattern the muscle is based on the prospective muscle-associated connective tissue having surface or adhesive properties that the muscle cells recognize, resulting in the migration of muscle cells to these regions. The pattern of muscle could thus be determined by the pattern of muscle-associated connective tissue, which is presumably specified by mechanisms similar to those that produce the pattern of cartilage. If the pattern of connective tissue changed with time, the presumptive muscle cells would migrate to the new sites, and this could account for the splitting of the muscle masses.

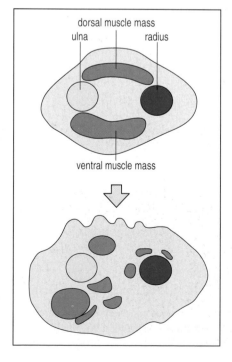

Fig. 10.19 Development of muscle in the chick limb. A cross-section through the chick limb in the region of the radius and ulna shortly after cartilage formation shows presumptive muscle cells present as two blocks—the dorsal muscle mass and the ventral muscle mass. These blocks then undergo a series of divisions to give rise to individual muscles.

Box 10A Reaction-diffusion mechanisms.

There are some self-organizing chemical systems that spontaneously generate spatial patterns of concentration of some of their molecular components. The initial distribution of the molecules is uniform, but over time the system forms wave-like patterns. The essential feature of such a self-organizing system is the presence of two or more types of diffusible molecules that interact with one another. For that reason such a system is known as a reaction-diffusion system. For example, if the system contains an activator molecule that stimulates both its own synthesis and that of an inhibitor molecule, which in turn inhibits synthesis of the activator, a type of lateral inhibition will occur so that synthesis of activator is confined to one region.

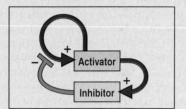

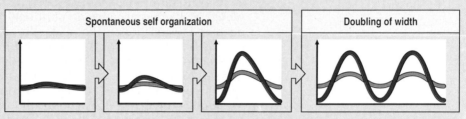

Spontaneous self organization | Doubling of width

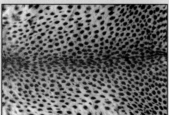

Increasing concentration

Under appropriate conditions, which are determined by the reaction rates and diffusion constants of the components, a closed system of a certain size can spontaneously develop a spatial pattern of activator with a single concentration peak. If the size of the system is increased, two peaks will eventually develop, and so on. Such a mechanism could thus generate periodic patterns such as the arrangement of digits, or the sepals and petals of flowers. When the chemicals can diffuse in two dimensions, the system can give rise to a number of peaks that are somewhat irregularly spaced.

Such a system may underlie some of the patterns of pigmentation that are common throughout the animal kingdom, such as the patterns of spots and stripes seen in the zebra and cheetah (see left). How these patterns are generated is not yet known, but one possibility is a reaction-diffusion mechanism. Assuming that pigment is synthesized in response to some activator, and that synthesis only occurs at high activator concentrations, some animal color patterns can be mimicked by a reaction-diffusion system in computer simulations.

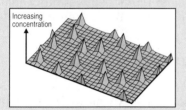

A characteristic feature of reaction-diffusion patterns is that new intermediate peaks appear as the system grows in size. The angelfish *Pomocanthus semicirculatus* provides a remarkable example of striping that could be generated by a reaction-diffusion mechanism (see below). Juvenile *P. semicirculatus*, less than 2 cm long, have only three dorso-ventral stripes. As the fish grow, the intervals between the stripes increase until the fish is around 4 cm long. New stripes then appear between the original stripes and the stripe intervals revert to those present at the 2 cm stage. As the fish grows larger, the process is repeated. This type of dynamic patterning is what would be expected of a reaction-diffusion mechanism. Computer simulations of reaction-diffusion mechanisms can also generate the patterns seen on a wide variety of mollusc shells. Nevertheless, there is as yet no direct evidence for a reaction-diffusion system patterning any developing organism. Top figure after Meinhardt, H., *et al.*: 1974.

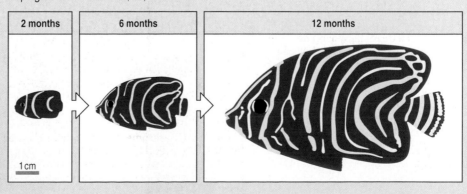

2 months | 6 months | 12 months

1 cm

10-11 | The initial development of cartilage, muscles, and tendons is autonomous.

The pattern of cartilage, tendon, and muscle in the limb may be specified by the same signals, as a polarizing region graft causes the development of a mirror-image pattern of all these elements. Each of the elements develops in its final position and there is little interaction between them. Thus, for example, if just the tip of an early chick wing bud is removed and grafted to the flank of a host embryo, it initially develops into normal distal structures with a wrist and three digits. The long tendon that normally runs along the ventral surface of digit 3 starts to develop in this situation, even though both its proximal end and the muscle to which it attaches are absent, although the tendon does not continue to develop because it does not make the necessary connection to a muscle, and so is not put under tension. The mechanism whereby the correct connections between tendons, muscles, and cartilage are established has still to be determined. It is clear, however, that there is little or no specificity involved in making such connections; if the tip of a developing limb is inverted dorso-ventrally, dorsal and ventral tendons may join up with inappropriate muscles and tendons. They simply make connections with those muscles and tendons nearest to their free ends.

10-12 | Separation of the digits is the result of programmed cell death.

Programmed cell death by apoptosis plays a key role in molding the form of the chick and mammalian limb, especially the digits. The region where the digits form is initially plate-like, as the limb is flattened along the dorso-ventral axis. The cartilaginous elements of the digits develop from the mesenchyme at the correct positions within this plate. Separation of the digits then depends on the death of the cells between these cartilaginous elements (Fig. 10.20). There is evidence for the involvement of BMP-4 in cell death: if the function of the BMP-4 receptor is blocked in the developing chick leg, cell death does not occur and the digits are webbed.

This cell death is a normal programmed part of patterning and cell differentiation (see Chapter 9). The fact that ducks and other waterfowl have webbed feet whereas chickens do not, is simply the result of there being less cell death between the digits of waterfowl. If chick limb mesoderm is replaced with duck limb mesoderm, cell death between the digits is reduced and the chick limb develops 'webbed' feet (Fig. 10.21). It is the mesoderm that determines the patterns of cell death both within the mesoderm and the overlying ectoderm. In amphibians, digit separation is not due to cell death, but results from digits growing more than the interdigital region. That the mesoderm specifies the fate of the overlying epithelium is a general developmental principle.

Fig. 10.20 Cell death during leg development in the chick. Programmed cell death in the interdigital region results in separation of the toes. Scale bar = 1 μm. Photographs courtesy of V. Garcia-Martinez, from Garcia-Martinez, V., et al.: 1993.

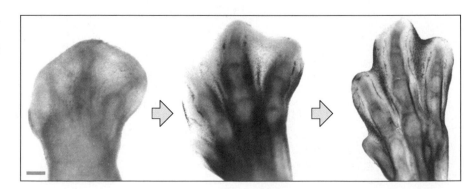

Fig. 10.21 The mesoderm determines the pattern of cell death. The webbed feet of ducks and other water birds form because there is less cell death between the digits than in birds without webbed feet. When the mesoderm and ectoderm of embryonic chick and duck limb buds are exchanged, webbing develops only when duck mesoderm is present.

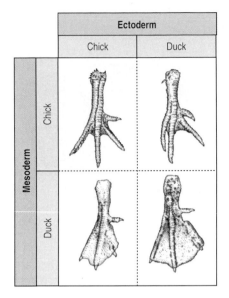

		Ectoderm	
		Chick	Duck
Mesoderm	Chick		
	Duck		

Programmed cell death also occurs in other regions of the developing limb, such as the anterior margin of the limb bud, between the radius and ulna, and in the wing polarizing region at a later stage. It was the investigation of cell death in this region by transplanting it to the anterior margin of the limb bud, that led to the discovery that it acts as a polarizing region.

Summary.

Positioning and patterning of the vertebrate limb is largely carried out by intercellular interactions. The position of the limb buds along the antero-posterior axis of the body is probably related to Hox gene expression. There are two key signaling regions within the limb bud. One is the apical ectodermal ridge, which specifies the progress zone in the underlying mesenchyme where cells acquire their positional identity; the second is the polarizing region at the posterior margin of the limb, which specifies pattern along the antero-posterior axis. The signals from the apical ridge are essential for limb outgrowth; one of these signals is probably a fibroblast growth factor. Sonic hedgehog protein is expressed in the polarizing region and may provide a graded positional signal. The dorso-ventral axis is specified by the ectoderm. Hox genes are expressed in a well-defined spatio-temporal pattern within the limb bud and may provide the molecular basis of positional identity. The limb muscle cells are not generated within the limb but migrate in from the somites and are patterned by the limb connective tissue. Patterning of the cartilaginous elements may involve a self-organizing mechanism. Programmed cell death can result in separation of digits.

Summary: vertebrate limb development

expression of Hox genes and FGF in flank establishes position of limb bud

apical ectodermal ridge in limb bud

signals from polarizing region help maintain apical ridge

polarizing region in limb bud

ectoderm specifies dorso-ventral axis

signals from progress zone help maintain apical ridge

signals from apical ridge (FGF) establish progress zone

signals from polarizing region (Sonic hedgehog) signal position along antero-posterior axis

progress zone

↓ proximo-distal specification

produces cartilage elements

↓ muscle and nerve cells migrate into developing limb

complex pattern of Hox gene expression controls pattern of cartilage and muscle

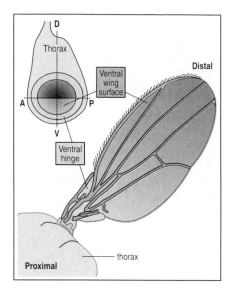

Fig. 10.22 Fate map of the wing imaginal disc of *Drosophila*. The wing disc prior to metamorphosis is an ovoid epithelial sheet bisected by the boundary between the anterior (A) and the posterior (P) compartments. There is also a compartment boundary between the future ventral (V) and dorsal (D) surfaces of the wing. At metamorphosis, the ventral surface will fold under the dorsal surface (see Fig. 10.23). The disc also contains part of the dorsal region of thoracic segment 2 in the region of wing attachment. After French, V., *et al.*: 1994.

Insect imaginal discs.

The appendages of the adult *Drosophila*, such as legs and wings, develop from **imaginal discs**, which have provided excellent systems for analyzing pattern formation because of the power of *Drosophila* genetics. The discs invaginate from the embryonic ectoderm as simple pouches of epithelium during embryonic development and remain as such until **metamorphosis** (see Fig. 2.34). In the leg and wing discs, the specification as leg or wing and the basic patterning occur within the embryonic epithelium, at the time when the segments are being patterned and given their identity. During the larval stages (instars), the discs grow by cell division and patterning continues.

Both leg and wing discs are divided by a compartment boundary that separates them into anterior and posterior developmental regions (see Section 5-15). In the wing disc, a second compartment boundary between the dorsal and ventral regions develops during the second larval instar (Fig. 10.22). When the wing forms at metamorphosis, the future ventral surface folds under the dorsal surface in the distal region to form the double-layered insect wing.

Insect legs have a quite different structure from those of vertebrates, the former being essentially jointed tubes of epidermis. The epidermal cells secrete the hard outer cuticle that forms the exoskeleton. Internally, there are muscles, nerves, and connective tissues. The adult wing is similarly a largely epidermal structure in which two epidermal layers —the dorsal and ventral surfaces—are close together. At metamorphosis of the insect, by which time patterning of the imaginal discs is largely complete, each disc undergoes a series of profound anatomical changes to produce a leg or a wing. Essentially, the invaginated epithelial pouch is turned inside out, as its cells differentiate and change shape. In the case of the wing, one surface folds under the other to form the double-layered wing structure (Fig. 10.23).

Although they differ so greatly in appearance, insect wings and legs are developmentally homologous structures, and the strategy of their patterning is very similar. Moreover, the mechanisms whereby they are patterned, and even the genes involved, show some amazing similarities to the patterning of the vertebrate limb.

Although all imaginal discs superficially look rather similar, they develop according to the segment in which they are located. We look later at how the segment-specific character of a particular disc is specified; first we discuss the patterning of the wing disc.

Fig. 10.23 Schematic representation of the development of the wing blade from the imaginal disc. Initially, the dorsal and ventral surfaces are in the same plane. At metamorphosis, the sheet folds and extends, so the dorsal and ventral surfaces come into contact with each other.

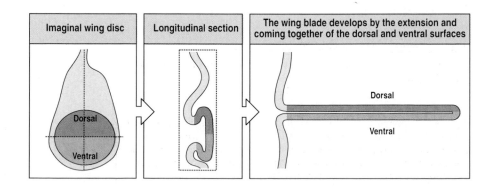

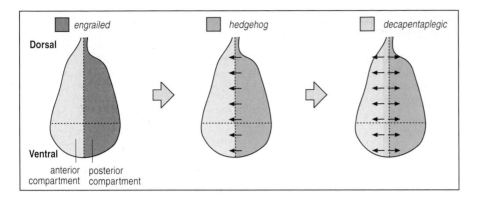

Fig. 10.24 Establishment of a signaling region in the wing disc at the antero-posterior compartment boundary. The disc is divided into anterior and posterior compartments. The gene *engrailed* is expressed in the posterior compartment where the cells also express the gene *hedgehog* and secrete the hedgehog protein. Where hedgehog protein interacts with anterior compartment cells, the gene *decapentaplegic* is activated, and decapentaplegic protein is secreted into both compartments, as indicated by the arrows.

10-13   Signals from the antero-posterior compartment boundary pattern the wing imaginal disc.

Imaginal discs are essentially sheets of epidermis. The epidermis of segments, wings, and legs is divided into anterior and posterior compartments—regions of lineage restriction (see Section 5-15). The wing is also divided into dorsal and ventral compartments, which are discussed in the next section.

In the wing disc, cells at the antero-posterior compartment boundary form a signaling region that specifies pattern along the antero-posterior axis of the wing. A cascade of events sets up this signaling center. It begins with the expression of the *engrailed* gene in the posterior compartment of the disc, which reflects the pattern of gene expression in the embryonic parasegment from which the disc derives.

The cells that express *engrailed* also express the segment polarity gene *hedgehog* (see Section 5-16). At the compartment boundary, the secreted hedgehog protein induces adjacent cells in the anterior compartment to express the *decapentaplegic* gene. The hedgehog protein activates *decapentaplegic* by inhibiting the activity of proteins that normally keep *decapentaplegic* expression repressed. The decapentaplegic protein, which is a member of the TGF-β family of signaling proteins, is secreted at the compartment boundary (Fig. 10.24), and is probably the positional signal for patterning both the anterior and posterior compartments along the antero-posterior axis.

There is evidence that the decapentaplegic protein provides a long-range signal controlling the localized expression of the gene *spalt* in a band overlapping the expression of decapentaplegic (Fig. 10.25), providing a further level of patterning in the wing. The expression of *spalt* occurs where a threshold level of decapentaplegic protein is present (Fig. 10.26).

The role of the *hedgehog* and *decapentaplegic* genes in wing patterning is shown by ectopically expressing *hedgehog* in genetically marked cell clones generated at random in wing discs. When such clones form in the posterior compartment they have little effect and development is more or less normal. However, with *hedgehog* clones in the anterior compartment a mirror-symmetric repeated pattern is produced along

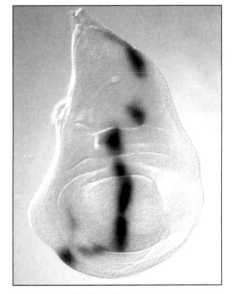

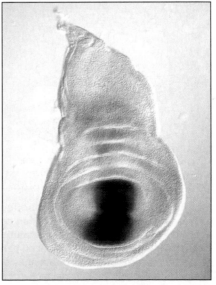

Fig. 10.25 *decapentaplegic* and *spalt* expression in *Drosophila* wing disc. The top panel shows the expression of *decapentaplegic*, which is expressed at the antero-posterior compartment border of the wing blade (dotted line). The bottom panel shows the expression of *spalt* in the wing blade. Photographs courtesy of K. Basler, from Nellen, D., *et al.*: 1996.

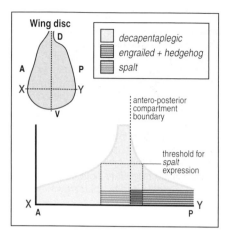

Fig. 10.26 A model for *decapentaplegic* patterning the antero-posterior axis. A schematic cross-section of the wing disc along the antero-posterior axis is shown at the bottom of the figure. *engrailed* and *hedgehog* are expressed in the posterior compartment and *decapentaplegic* is activated at the compartmental boundary. The presumed gradient in decapentaplegic in both anterior and posterior compartments activates the gene *spalt* at threshold concentration.

the antero-posterior axis in that compartment (Fig. 10.27). If we describe the normal wing pattern from anterior to posterior as 123/45, where the numbers describe the wing veins and / indicates the compartment boundary, then the pattern in the experimental wings can be 123h321123/45, where h indicates the *hedgehog* clone. One can interpret these results in terms of decapentaplegic protein being secreted wherever the hedgehog protein interacts with anterior compartment cells. In a normal wing this interaction is localized at the anterior side of the compartment boundary, but in an experimental wing, the ectopic expression of *hedgehog* sets up new sites of *decapentaplegic* expression in the anterior compartment. This results in new gradients of decapentaplegic protein being formed.

10-14   The dorso-ventral boundary of the wing acts as a pattern-organizing center.

The wing disc is subdivided into dorsal and ventral compartments that correspond to the dorsal and ventral surfaces of the adult wing (see Fig. 10.22). These compartments develop after the wing disc is formed —during the second larval instar. The dorsal and ventral compartments were originally identified by cell lineage studies, but they are also defined by expression of the homeotic selector gene *apterous*, which is confined to the dorsal compartment, and which defines the dorsal state. Expression of *apterous* activates the secretion of proteins encoded by the genes *fringe* and *Serrate*. Dorsal cells secreting Serrate protein interact with ventral cells via a receptor (the Notch protein) to form a wing

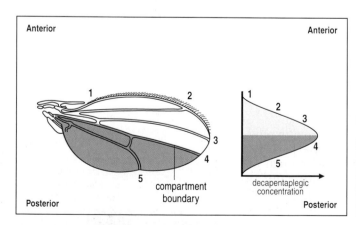

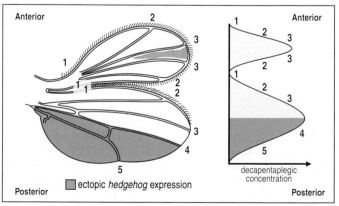

Fig.10.27   Alteration of wing patterning due to ectopic expression of *decapentaplegic*. When *hedgehog* is expressed in a clone of cells in the anterior compartment a new source of decapentaplegic protein is set up. Left panel: in the normal wing, decapentaplegic protein is made at the compartment boundary. The decapentaplegic protein may act as a morphogen, patterning both anterior and posterior compartments, the different veins being specified at threshold concentrations. Right panel: the ectopic expression of *hedgehog* in the anterior compartment results in a new source of decapentaplegic protein and a new wing pattern develops in relation to this new source with veins appearing at threshold concentrations.

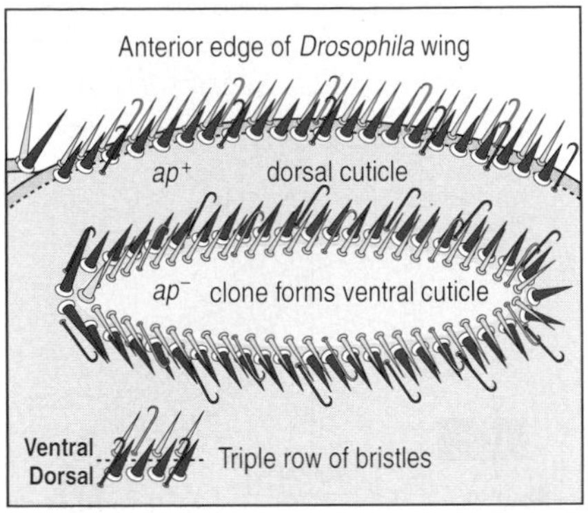

Fig. 10.28 Formation of a new wing margin by ectopic expression of *apterous⁻* cells on the wing dorsal surface. When a clone of cells which do not express the *apterous* gene (*ap⁻*) develops on the dorsal surface of the wing, the cells adopt a ventral character and do not express the gene *fringe*. At the junction of the clone and wild-type dorsal cells (*ap⁺*), a new wing margin with a triple row of characteristic bristles develops. The margin is made up of cells from the clone, which form one type of bristle, and adjacent wild-type cells, which form another type of bristle. After Diaz-Benjumea, F.J., et al.: 1993.

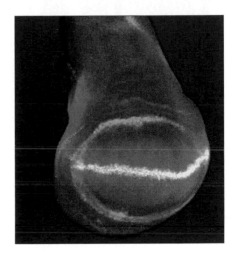

margin at the boundary between them. Ectopic expression of *fringe* in the ventral compartment results in the formation of a new margin. If clones of cells not expressing the *apterous* gene are made in the dorsal compartment of a normal wild-type wing, they adopt a ventral fate, not expressing *fringe* or *Serrate*. In these genetic mosaics, a new wing margin, which can be identified by its characteristic bristle pattern, forms around the clone (Fig. 10.28).

Like the antero-posterior compartment boundary, the dorso-ventral boundary also acts as an organizing center, and the signaling molecule is the protein wingless (see Section 5-16), which belongs to the same protein family as the vertebrate Wnt proteins, and is expressed at the dorso-ventral boundary in the disc. The wingless protein serves a role analogous to decapentaplegic in the antero-posterior patterning system. Another gene expressed at the dorso-ventral boundary is *vestigial*, which is required for cell proliferation (Fig. 10.29).

10-15 The leg disc is patterned in a similar manner to the wing disc, except for the proximo-distal axis.

The easiest way of visualizing a leg imaginal disc is to think of it as a collapsed cone. Looking down on the disc, one can imagine it as a series of concentric rings, each of which will form a proximo-distal segment of the leg. A change in the shape of the epithelial cells is responsible for the outward extension of the leg at metamorphosis. For the leg, the process is rather like turning a sock inside out, with the result that the center of the disc ends up as the distal end, or tip, of the leg (Fig. 10.30). The outermost ring gives rise to the base of the leg, which is attached to the body, and those nearer the center give rise to the more distal structures (Fig. 10.31).

Fig. 10.29 Expression of *wingless* and *vestigial* in the wing disc. *wingless* expression is shown in green, while *vestigial* is red. Both *wingless* and *vestigial* are expressed at the dorso-ventral boundary as indicated by the yellow stripe. Photograph courtesy of K. Basler, from Zecca, M., *et al.*: 1996.

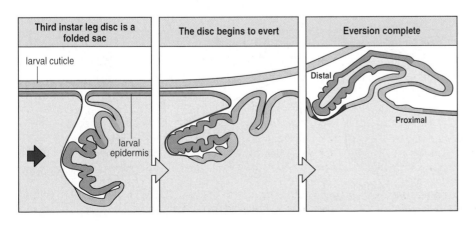

Fig. 10.30 *Drosophila* leg disc extension at metamorphosis. The disc epithelium, which is an extension of the epithelium of the body wall, is initially folded internally. At metamorphosis it extends outward, rather like turning a sock inside out. The red arrow in the first panel is the viewpoint that produces the concentric rings shown in Fig. 10.31.

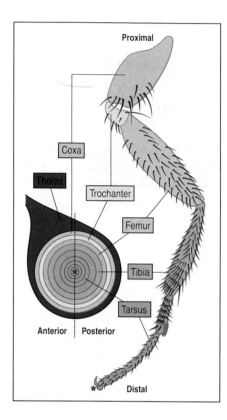

Proximal

Coxa

Thorax

Trochanter

Femur

Tibia

Tarsus

Anterior | Posterior

* Distal

Fig. 10.31 Fate map of the leg imaginal disc of *Drosophila*. The disc is a roughly circular epithelial sheet, which becomes transformed into a tubular leg at metamorphosis. The center of the disc becomes the distal tip of the leg and the circumference gives rise to the base of the leg —this defines the proximo-distal axis. The presumptive regions of the adult leg, such as the tibia, are thus arranged as a series of circles with the future tip at the center. A compartment boundary divides the disc into anterior (A) and posterior (P) regions. After Bryant, P.J.: 1993.

The first steps in the patterning of the leg disc along its antero-posterior axis are the same as in the wing. The *engrailed* gene is expressed in the posterior compartment and induces expression of *hedgehog*. The hedgehog protein induces a signaling region at the compartment border. In the dorsal region of the leg disc, the expression of *decapentaplegic* is induced, as it is in the wing; in the ventral region, however, hedgehog induces expression of the *wingless* gene instead of *decapentaplegic* along the border and *wingless* protein acts as the positional signal (Fig. 10.32). The complementary patterns of *wingless* and *decapentaplegic* expression are maintained by mutual inhibition of each other's expression.

The setting up of the proximo-distal axis of the leg disc also involves interactions between the wingless and decapentaplegic proteins. The future distal tip of the leg is specified at the center of the leg disc and the gene *Distal-less* is expressed there. This site of *Distal-less* expression corresponds to where the wingless and decapentaplegic proteins meet. The activity of *Distal-less* seems to be necessary to specify the proximo-distal axis within the disc. Ectopic expression of *decapentaplegic* or *wingless* can lead to duplication of the proximo-distal axis: this is due to *Distal-less* being expressed at the new site where wingless and decapentaplegic interact, thus specifying another proximo-distal axis. The gene *aristaless* is involved in specifying the very distal-most elements of the leg and is expressed at the same site as *Distal-less*.

| 10-16 | **Butterfly wing markings are organized by additional positional fields.** |

The variety of color markings on butterfly wings is remarkable: more than 17,000 species can be distinguished. Many of these patterns are variations on a basic 'groundplan' consisting of bands and concentric eyespots (Fig. 10.33). The wings are covered with overlapping cuticular scales, which are colored by pigment synthesized and deposited by the epidermal cells. How are these patterns specified? Butterfly wings develop from imaginal discs in the caterpillar in a similar way to *Drosophila*

Fig. 10.32 Establishment of signaling centers in the antero-posterior compartment of the leg disc, and the specification of the distal tip. The gene *engrailed* is expressed in the posterior compartment and induces expression of *hedgehog*. Where hedgehog protein meets and signals to anterior compartment cells, the gene *decapentaplegic* is expressed in dorsal regions and the gene *wingless* is expressed in ventral regions. Both of these genes encode proteins that are secreted. Expression of the gene *Distal-less*, which specifies the proximo-distal axis, is activated where the wingless and decapentaplegic proteins meet.

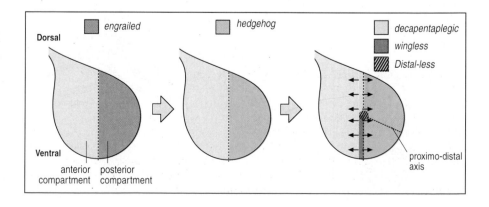

engrailed   hedgehog   decapentaplegic
Dorsal   wingless
                         Distal-less

Ventral
anterior posterior   proximo-distal
compartment compartment   axis

wings. Surgical manipulation has shown that the eyespot is specified at a late stage in the development of the wing disc, and that the pattern is dependent on a signal from the center of the spot. A number of the genes that control wing development in *Drosophila*, such as *apterous*, are expressed in the butterfly in a spatial and temporal pattern similar to that in the fly. Thus, as in *Drosophila*, the shape and structure of the butterfly wing is patterned by a field of positional information. The patterning of the pigmentation, however, involves the establishment of additional fields of positional information.

The expression pattern of *Distal-less* in butterfly wing discs suggests that the mechanism used to delineate the color pattern on the butterfly wing is similar to that which specifies positional information along the proximo-distal axis of the insect leg. In the butterfly wing, *Distal-less* is expressed in the center of the eyespot, whereas in the *Drosophila* leg disc it is expressed in the central region corresponding to the future tip of the limb. Thus, it is possible that eyespot development and distal leg patterning involve similar mechanisms. The eyespot may be thought of as a proximo-distal pattern superimposed on the two-dimensional wing surface. The center of the eyespot may represent the distal-most positional value, with the surrounding rings representing progressively more proximal positions, as in the leg. The site of the eyespot may be specified with reference to the primary wing pattern (i.e. the anterior, posterior, dorsal, and ventral compartments); a secondary coordinate system centered on the eyespot would then be established, with expression of the *Distal-less* gene defining the central focus.

Fig. 10.33 Butterfly wing pattern. Ventral view of a female African butterfly *Bicyclus anynana*, showing wing color pattern and prominent eyespots. Scale bar = 5 mm. Photograph courtesy of V. French and P. Brakefield.

10-17 | **The segmental identity of imaginal discs is determined by the homeotic selector genes.**

The patterning of the leg and wing involves similar signals, such as decapentaplegic protein, yet the actual pattern that develops is very different. This implies that the wing and leg discs interpret positional signals in different ways. This interpretation is under the control of the Hox genes and can be illustrated with respect to the leg and antenna. If the *Antennapedia* gene is expressed in the head region, the antennae develop into legs (Fig. 10.34). Using the technique of mitotic recombination (see Box 5B, page 153) it is possible to generate a clone of mutant cells in the antenna. These cells develop as leg cells but exactly which type of leg cell depends on their position along the proximo-distal axis; if, for example, they are at the tip, they form a claw. It is as if the positional values of the cells in the antenna and leg are similar and that the difference between the two structures lies in the interpretation. This can be illustrated with respect to the French flag and the Union Jack (Fig. 10.35)—cells develop according to their position and genetic constitution. This principle applies also to wing and haltere imaginal discs. Thus, we find a similarity in developmental strategy between insects and vertebrates; both use the same positional information in appendages like legs and wings, but interpret it differently.

The character of a disc and how positional information is interpreted is determined by the Hox genes. Insect legs develop only on the three thoracic segments and not on the abdominal segments and, in *Drosophila*, wings develop only on the second thoracic segment. These adult structures are segment-specific because the particular type of imaginal disc that gives rise to them is only formed by certain parasegments. The absence of appendages on abdominal segments in *Drosophila* is due to the genes required for the formation of the leg and wing discs being suppressed in the abdomen. The type of disc formed in a particular

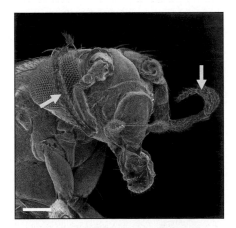

Fig. 10.34 Scanning electron micrograph of *Drosophila* carrying the *Antennapedia* mutation. Flies with this mutation have the antennae converted into legs (arrowed). Scale bar = 0.1 mm. Photograph by D. Scharfe, from Science Photo Library.

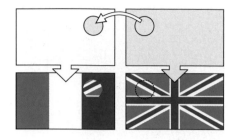

Fig. 10.35 Cells interpret their position according to their developmental history and genetic make-up. For example, if there are two flags which use the same positional information to produce a different pattern, then a graft from one to the other would result in the graft developing according to its new position. This is what happens in imaginal discs.

thoracic segment is typically specified by the action of one of the Hox genes expressed in the segment. Expression of the genes *Antennapedia* and *Ultrabithorax* specify the second and third pair of legs, respectively.

The leg imaginal discs arise from small clusters of ectodermal cells in parasegments 3–6, which contribute to the future thoracic segments of the embryo (Fig. 10.36). In the embryo, each disc initially contains around 25 cells and is formed during growth of the blastoderm. The discs arise at the parasegment boundaries, with anterior and posterior compartments of adjacent parasegments contributing to each disc. In the future second thoracic segment, the leg disc splits off a second disc early in its development, which becomes the wing disc. Similarly, an additional disc, which develops into the haltere, a balancing organ, is formed in the future third thoracic segment. The discs increase in size during larval growth, the wing growing 1000-fold.

Mutations in Hox genes in *Drosophila* adults can cause compartment-specific homeotic transformations of halteres into wings. In normal *Drosophila* adults, the wing is on the second thoracic segment and the haltere on the third thoracic segment; they arise from imaginal discs originating at the boundaries of parasegments 4/5 and 5/6, respectively (see Fig. 10.36). In the normal embryo, the *Ultrabithorax* gene, one of the genes of the bithorax complex (see Box 4A, page 104), is expressed in parasegments 5 and 6 and is involved in specifying their identity. The *bithorax* mutation (*bx*), which causes the *Ultrabithorax* gene to be misexpressed, can transform the anterior compartment of the third thoracic segment, and thus of the haltere, into the corresponding anterior compartment of the second segment: the anterior half of the haltere thus becomes a wing (see Fig. 5.36). The *postbithorax* mutation (*pbx*), which affects a regulatory region of the *Ultrabithorax* gene, transforms the posterior compartment of the haltere into a wing (Fig. 10.37). If both mutations are present in the same fly, the effect is additive and the result is a fly that has four wings but cannot fly (see Fig. 5.36). Another mutation, *Haltere mimic*, causes a homeotic transformation in the opposite direction: the wing is transformed into a haltere.

As with the antenna and leg, it is possible to generate a mosaic haltere in the disc with a small clone of cells containing an *Ultrabithorax* mutation (such as *bithorax*) in the disc; the cells in the clone make wing structures, which correspond exactly to those that would form in a similar position in a wing. It is as if the positional values in haltere and wing discs are identical and all that has been altered in the mutant is how this positional information is interpreted (see Fig. 10.35). In fact, other discs seem to have similar positional fields. This concept of different interpretation of the same positional information in different segments also applies to other structures that develop from imaginal discs, such as antennae and legs.

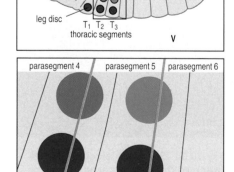

Fig. 10.36 The position in the late *Drosophila* embryo of the imaginal discs that give rise to the adult thoracic appendages. The imaginal discs for the legs, wings, and halteres lie across the parasegment boundaries in the thoracic segments T1, T2, and T3.

Fig. 10.37 **Effects of the *postbithorax* mutation in *Drosophila*.** The mutation *postbithorax* acts on the posterior compartment of a haltere, converting it into the posterior half of a wing.

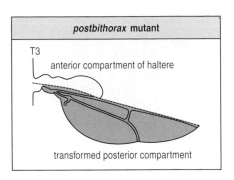

postbithorax mutant

T3

anterior compartment of haltere

transformed posterior compartment

Summary.

The legs and wings of *Drosophila* develop from epithelial sheets —imaginal discs—which are set aside in the embryo. Hox genes acting in the parasegments specify which sort of appendage will form and can direct the interpretation of positional information. The leg and wing imaginal discs are divided at an early stage into anterior and posterior compartments. The boundary between the compartments, where expression of the *decapentaplegic* gene is activated by the hedgehog protein, is a pattern-organizing center and a source of signals that pattern the disc. In the leg disc, wingless protein is the signal in the ventral regions. The dorso-ventral compartment boundary of the wing disc also acts as a pattern-organizing center, with wingless again acting as the signal. The proximo-distal axis of the leg is specified by the interaction between decapentaplegic and wingless proteins, which activates the gene *Distal-less* at the distal end of this axis. The colorful eyespots on butterfly wings may be patterned by a mechanism similar to that used to organize the proximo-distal axis of the insect leg.

Summary: pattern formation in leg and wing imaginal discs of *Drosophila*

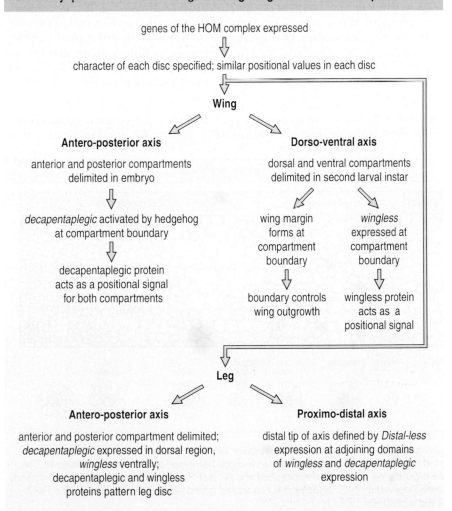

genes of the HOM complex expressed

⇩

character of each disc specified; similar positional values in each disc

Wing

Antero-posterior axis

anterior and posterior compartments delimited in embryo

⇩

decapentaplegic activated by hedgehog at compartment boundary

⇩

decapentaplegic protein acts as a positional signal for both compartments

Dorso-ventral axis

dorsal and ventral compartments delimited in second larval instar

wing margin forms at compartment boundary

⇩

boundary controls wing outgrowth

wingless expressed at compartment boundary

⇩

wingless protein acts as a positional signal

Leg

Antero-posterior axis

anterior and posterior compartment delimited; *decapentaplegic* expressed in dorsal region, *wingless* ventrally; decapentaplegic and wingless proteins pattern leg disc

Proximo-distal axis

distal tip of axis defined by *Distal-less* expression at adjoining domains of *wingless* and *decapentaplegic* expression

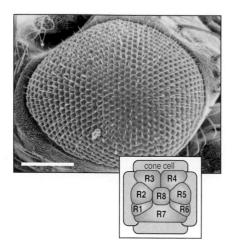

Fig. 10.38 Scanning electron micrograph of the compound eye of *Drosophila*. Each unit is an ommatidium. At the third instar, the ommatidial cluster (inset) is made up of eight photoreceptor neurons (R1–R8) and four cone cells. Scale bar = 50 μm.

The insect compound eye.

The *Drosophila* compound eye is composed of about 800 identical photoreceptor organs (**ommatidia**) arranged in a regular hexagonal array (Fig. 10.38). Each ommatidium is made up of eight photoreceptor neurons (R1–R8), together with four overlying cone cells (which secrete the lens) and additional pigment cells (see Fig. 10.38). It is these red-pigmented cells that give wild-type *Drosophila* its red eyes. The genetic analysis of ommatidia development has provided one of the best model systems for studying the patterning of a small group of cells, and the basis of their intercellular interactions.

The eye develops from the single layered epithelial sheet of the eye imaginal disc, located in the head. Specification and patterning of the cells of the ommatidia begins in the middle of the third larval instar. Patterning starts at the posterior of the eye disc and progresses anteriorly, taking about 2 days, during which time the disc grows eight times larger. One of the earliest events in eye differentiation is the formation of a groove, the **morphogenetic furrow**, in the disc. As the furrow sweeps across the epithelium from posterior to anterior, clusters of cells that will give rise to the ommatidia appear behind it, spaced in a hexagonal array. The furrow moves slowly and takes about 2 days to move across the disc, at a rate of 2 hours per row of ommatidial clusters. As the morphogenetic furrow moves forward, the cells behind it start to differentiate to form regularly spaced ommatidia. The ommatidia are arranged in rows, with each row half an ommatidium out of register with the previous one. This gives the characteristic hexagonal packing arrangement (see Fig. 10.38). The first cells to differentiate are the R8 photoreceptor neurons. These appear as regularly spaced cells in each row, separated from each other by about eight cells. This separation sets the spacing pattern for the ommatidia.

Each R8 cell initiates a cascade of signals that recruits a surrounding cluster of 20 cells to form an ommatidium (Fig. 10.39). R2 and R5 differentiate

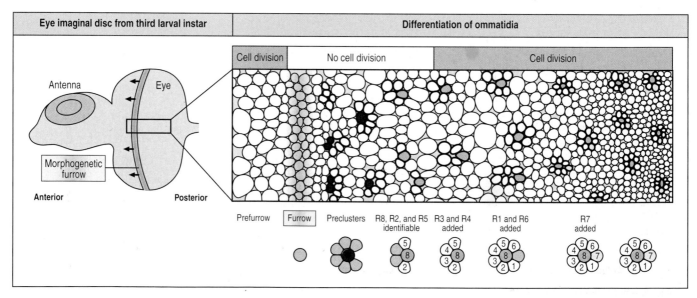

Fig. 10.39 Development of ommatidia in the *Drosophila* compound eye. The compound eye develops from the eye imaginal disc, which is part of a larger disc that also gives rise to an antenna. During the third larval instar, the morphogenetic furrow develops in the eye disc and moves across it in a posterior to anterior direction. Ommatidia develop behind the furrow, the photoreceptor neurons being specified in the order shown, R8 developing first and R7 last. The individual ommatidia are regularly spaced in a hexagonal grid pattern. After Lawrence, P.: 1992.

on either side of it, to form two functionally identical neurons. R3 and R4, which are a slightly different type of photoreceptor, form next. All these cells become arranged in a semi-circle with R8 at the center. R1 and R6 differentiate next and almost complete the circle, which is finally closed by the differentiation of R7 adjacent to R8 (see Fig. 10.39). This arrangement is further modified in the mature ommatidium, but we concentrate here on this earlier pattern.

In the following sections we consider two aspects of eye development: the mechanism that spaces the ommatidia in such a beautifully regular hexagonal array; and the patterning of an individual ommatidium, in particular the patterning of the eight photoreceptor neurons.

10-18 Signals maintain progress of the morphogenetic furrow and the ommatidia are spaced by lateral inhibition.

The morphogenetic furrow is the result of a wave of signals that initiates the development of ommatidia from the disc cells. Its passage across the eye disc is essential for differentiation of the ommatidia, as mutations that block its progress also block the differentiation of new rows of ommatidia, resulting in a fly with abnormally small eyes. Although there are no anterior and posterior compartments in the eye discs, the cells just behind the furrow can be regarded as resembling posterior cells as they secrete hedgehog protein, which triggers the expression of *decapentaplegic* in the furrow and initiates R8 differentiation. However, the system is dynamic, and as the furrow moves forward, *decapentaplegic* expression is turned off and the cells start to express *hedgehog*, thus initiating *decapentaplegic* expression, which draws the furrow forward. The third member of the ubiquitous trio, the gene *wingless*, also plays a part in eye disc patterning. It is expressed at the lateral edges of the eye disc, and prevents the furrow starting in these regions. We thus see that, even though imaginal discs give rise to very diverse structures, the key signals involved in patterning the leg, wing, and eye are similar, although they have different roles.

The question of how the regular spacing of the ommatidia within the eye is achieved can be answered by considering how the R8 cells are spaced (see Fig. 10.39). This involves a lateral inhibition mechanism (see Section 1-14), in which differentiating cells inhibit the differentiation of those cells immediately adjacent to them. All cells in the eye disc initially have the capacity to differentiate as R8 cells, and as the morphogenetic furrow passes they start to do so. But some inevitably gain a lead and are thus able to inhibit the differentiation of another R8 cell over a range of around three cell diameters. A candidate for the inhibitor that spaces the R8 cells is the scabrous protein, a secreted protein related to the vertebrate fibrinogens (proteins involved in blood clotting). Mutations where the *scabrous* gene is inactivated result in reduced spacing between ommatidia, which suggests that the function of the scabrous protein is to keep the ommatidia at the correct distance apart. The Notch protein has also been implicated in the lateral inhibition mechanism.

10-19 The patterning of the cells in the ommatidium depends on intercellular interactions.

There is no constant lineage relationship determining which cells become photoreceptors, for example, and which become pigment cells. By making genetic mosaics in which individual cells in the developing eye are marked, it is found that any pair of cells in the ommatidium could be sister cells. For example, one daughter cell arising from the division of a precursor cell can become a photoreceptor neuron, whereas the other can become a pigment cell. The patterning of the ommatidium is achieved entirely by inductive interactions between the cells forming the photoreceptor cluster. These interactions occur in a fixed sequence of small steps as each set of cells is recruited into the cluster and becomes specified. R8 differentiates first and induces the specification and differentiation of R2 and R5, which in turn induce R3 and R4, and so on. These interactions involve activation of the *Drosophila* epidermal growth factor receptor. After the photoreceptors have differentiated, the four lens-producing cone cells develop, and finally the surrounding ring of accessory cells. Note that in the insect ommatidium (unlike vertebrate limbs and *Drosophila* appendages) cells are being specified and determined individually, not as groups of cells.

| 10-20 | **The development of R7 depends on a signal from R8.** |

Mutations in two key genes involved in eye development—*sevenless* and *bride-of-sevenless*—were instrumental in analyzing the specification of the R7 photoreceptor cell. When either gene is inactivated, the phenotype is the same: R7 does not develop. Instead, an extra cone cell is formed. As we shall see, however, the roles of the two genes in the development of R7 are quite different. This interaction has become a classic example of an intercellular induction that requires direct cell–cell contact.

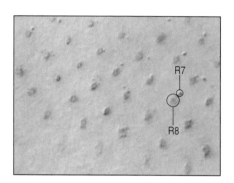

Fig. 10.40 The bride-of-sevenless protein is made in R8. Some of it will migrate into the adjacent R7 cell at the point of contact. Photograph courtesy of L. Zipursky.

Genetic mosaics in which some cells express *sevenless* and some do not show that the *sevenless* gene is the partner involved in responding to the signal rather than producing it. In these mosaics, the only requirement for R7 development is that the prospective R7 cell expresses the *sevenless* gene; it does not matter if all the other cells do not. But if all the surrounding cells express *sevenless* and the prospective R7 cell does not, then it does not develop as R7. It develops as a fifth cone cell instead. Thus, *sevenless* is clearly involved in the cell's response to a signal from other cells. In fact, *sevenless* codes for a transmembrane receptor tyrosine kinase. Other cells in the ommatidium, including lens cells, also express *sevenless*, and so other factors are clearly involved in the specification of R7. Expression of the *sevenless* gene is a necessary, but not sufficient, condition for R7 specification.

We now turn to the *bride-of-sevenless* gene. Again using genetic mosaics, it can be shown that for R7 to develop, only R8 need express the gene. This clearly suggests that *bride-of-sevenless* encodes the signal by which R8 induces R7. This signal is an integral membrane protein, which is present on the apical surface of the R8 cell, where it makes contact with R7. Direct evidence that the sevenless and bride-of-sevenless proteins bind to each other comes from expressing them separately in two cell lines, and then mixing the cells together. Cells of the different lines adhere strongly to each other, showing that the proteins bind to each other. bride-of-sevenless protein then enters the R7 cell (Fig. 10.40).

Binding of bride-of-sevenless protein to the sevenless receptor triggers an intracellular signal transduction pathway; this results in the activation of transcription factors and consequent changes in gene expression that eventually specify the cell as R7. Four other non-neuronal cells also express the sevenless protein and can develop as R7 if the receptor is activated by ectopic expression of bride-of-sevenless. Only one R7 cell normally develops, as the signal from R8 is spatially restricted and thus only reaches the presumptive R7 cell.

10-21 | **Activation of the gene *eyeless* can initiate eye development.**

The gene *eyeless* is required for eye development. Mutations in the *eyeless* gene result in the reduction or complete absence of the fly's compound eye. It is expressed in the region of the eye disc anterior to the morphogenetic furrow. Induction of expression of the *eyeless* gene in other imaginal discs results in the development of ectopic eyes, which have been induced in this way in wings, legs, antennae, and halteres. The fine structure of these ectopic eyes is remarkably normal, and distinct ommatidia are present. Therefore, *eyeless* is clearly a key gene for eye development and its expression switches on the requisite pathway. It is estimated that some 2000 genes are eventually activated as a result of *eyeless* activity, and all of them are required for eye morphogenesis. The action of *eyeless* may be similar to that of the Hox genes in changing the interpretation of positional information in the discs in which it is ectopically expressed (see Section 10-17).

The *eyeless* gene is an example of the remarkable conservation of developmental genes throughout the animal kingdom. Its homolog in vertebrates is called *Pax6*. Like *Drosophila* without *eyeless* function, mice with defective *Pax6* function have smaller eyes than normal, or no eyes at all. And mutations in the human *Pax6* gene are responsible for a variety of eye malformations in the disease known as aniridia, because of the partial or complete absence of the iris. Ectopic expression of *Pax6* in the fly can substitute for *eyeless*, resulting in the development of an ectopic insect eye.

	Summary.

The *Drosophila* compound eye contains around 800 individual ommatidia, arranged in a regular hexagonal pattern. It develops from an imaginal disc, which is patterned at a late larval stage. The early development of the disc involves the signaling molecules wingless, hedgehog, and decapentaplegic. The regular spacing of the ommatidia is achieved by lateral inhibition. The patterning of the eight photoreceptor neurons (R1–R8) of each ommatidium is due to local cell interactions, in which the photoreceptor cells are specified and differentiate in a strict order. The specification of photoreceptor R7 involves direct contact with R8, which carries the bride-of-sevenless protein on its surface. This binds to the membrane receptor sevenless on the prospective R7 cell to stimulate its determination and differentiation as R7. *eyeless* is a key gene in eye development and can induce ectopic eyes in other imaginal discs.

Summary: development of the ommatidia of the *Drosophila* eye

eyeless gene expression required for eye development. Morphogenetic furrow moves ● across eye disc, associated with hedgehog and decapentaplegic signals

⇩

the eight photoreceptors of the future ommatidia begin to develop behind the furrow, R8 developing first and R7 last.
Ommatidia spaced by lateral inhibition

⇩

the development of R7 depends on a signal from R8

The nematode vulva.

The vulva forms the external genitalia of the hermaphrodite of the nematode worm *Caenorhabditis elegans* and connects with the uterus. Its interest as a model developmental structure arises from the small number of cells—just four, one inducing and three responding—involved in its initial specification, and the fact that more than 40 genes involved in its development have been identified. As with the insect ommatidium, we deal here with patterning and induction of the vulva at the level of the individual cell. The vulva is an adult structure that develops in the last larval stage. The mature vulva contains 22 cells, with a number of different cell types. It is derived from three of six P cells of ectodermal origin, which persist in a small row aligned anteroposteriorly on the ventral side of the larva. Three of these cells—$P5_p$, $P6_p$, and $P7_p$—give rise to the vulva, each having a well defined lineage. In discussing the development of the vulva, three distinct cell fates are conventionally distinguished—primary (1°), secondary (2°), and tertiary (3°). The primary and secondary fates refer to different cell types in the vulva, whereas the tertiary fate is non-vulval. $P6_p$ normally gives rise to the primary lineage, whereas $P5_p$ and $P7_p$ give rise to secondary lineages. The other three P cells give rise to the tertiary lineage, in this case the epidermis (Fig. 10.41). Initially however, all six P cells are equivalent in their ability to develop as vulval cells. One of the questions we address here is how the final three cells become selected as the vulval precursor cells.

Cell fate of the three vulval precursor cells is specified by an inductive signal from a fourth cell, the gonadal anchor cell, which confers a primary fate on the cell nearest to it, $P6_p$, and a secondary fate on the cells lying just beyond it, $P5_p$ and $P7_p$ (see Fig. 10.41). In addition, once induced, the primary cell inhibits its immediate neighbors from expressing a primary fate. P cells that do not receive the anchor cell signal adopt a tertiary fate.

The lineage of the P cells is fixed. If one of the daughter cells of a P cell is destroyed, the other does not change its fate. There is no evidence for cell interactions in the further development of these P cell lineages, and cell fate is thus probably specified by asymmetric cell divisions. The

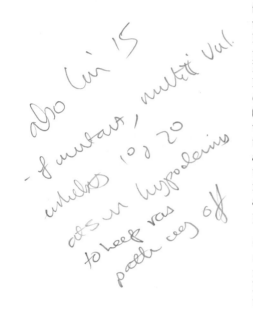

Fig. 10.41 Development of the nematode vulva. The vulva develops from three cells, $P5_p$, $P6_p$, and $P7_p$, in the post-embryonic stages of nematode development. Under the influence of a fourth cell, the anchor cell, $P6_p$, undergoes a primary pathway of differentiation that gives rise to eight vulval cells. $P6_p$ is flanked by $P5_p$ and $P7_p$, which each undergo a secondary pathway of differentiation that gives rise to seven cells of different vulval cell type. Three other P cells nearby adopt a tertiary fate and give rise to epidermis.

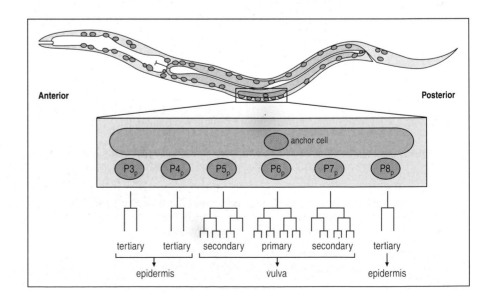

question is how are just three P cells specified, and how is the primary fate of the central cell made different from the secondary fate of its neighbors?

10-22   The anchor cell induces primary and secondary fates.

The six P cells are initially equivalent, in that any of them can give rise to vulval tissue. The key determining signal is that provided by the anchor cell, which normally lies above P6p. The vital role of the anchor cell is shown by cell ablation experiments; when the anchor cell is destroyed with a laser microbeam, the vulva does not develop. If all except one P cell is destroyed, its fate depends on how close it is to the anchor cell. If it is very close it adopts a primary fate, if far away a tertiary fate, and at intermediate distances, a secondary fate. This suggests that a graded signal with thresholds specifies the primary and secondary fates.

The anchor cell's signal is the product of the *lin-3* gene, which is a protein similar to the EGF growth factor of vertebrates. Mutations in *lin-3* result in the same abnormal development that follows removal of the anchor cell, and so no vulva is formed. The receptor for the inductive signal is an EGF receptor-like transmembrane tyrosine kinase, encoded by the *let-23* gene. Mutations that inactivate *let-23* also result in no vulva being formed, as the P cells cannot respond to the inductive signal. Direct evidence for a graded signal comes from placing *lin-3* under the control of a heat-shock promoter, so that its activity can be controlled. All but one P cell are then killed and the effect of increasing *lin-3* expression is followed. At low levels of *lin-3* expression, a secondary fate is induced in the remaining P cell, and at high levels a primary fate is the result.

In addition to a graded signal, there is a separate signal from the primary cell that specifies a secondary fate in the adjacent cells. Evidence for this comes from the observation that in genetic mosaics in which the future secondary cells lack the receptor for the *lin-3* signal, development is normal. This shows that the primary cell must be providing a signal for secondary fate, which is not lin-3 protein. The induction of secondary fate involves a transmembrane receptor, encoded by *lin-12*, which belongs to the Notch family. Thus, two systems pattern early vulval development, perhaps to ensure that it develops correctly.

At least one other signal is involved in establishing the vulva: a signal from the hypodermis—the larval nematode's epidermis—which inhibits adoption of a vulval fate in all the six cells of the equivalence group, and which is overruled by the inductive signal. The interactions that give rise to vulval cells are shown in Fig. 10.42.

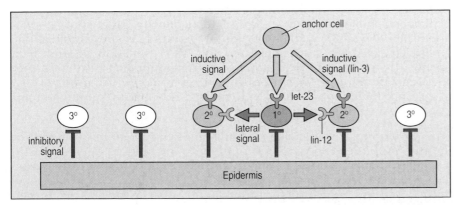

Fig. 10.42 Cell interactions in vulval development. The anchor cell produces a diffusible signal lin-3, which induces a primary fate in the precursor cell closest to it by binding to the receptor let-23. It also induces a secondary fate in the two P cells slightly further away, where the concentration of the signal is lower. The cell adopting a primary fate inhibits adjacent cells from adopting the same fate by a mechanism of lateral inhibition involving lin-12, and also induces a secondary fate in these cells. A constitutive signal from the epidermis inhibits the development of both primary and secondary fates, but is overruled by the initial inductive signal from the anchor cell.

Summary.

The nematode vulva is derived from just three out of six potentially competent cells, and involves both cell interactions and a well-defined cell lineage. The fates of the three precursor P cells are specified by a graded signal from an adjacent anchor cell; this induces a primary fate in the cell nearest to it, and a secondary fate in the two cells adjacent to the primary cell. In addition, the cell adopting a primary fate induces the adjacent cells to develop a secondary fate. The inducing signal (lin-3) is a protein similar to the vertebrate growth factor EGF, and the second signal involves a Notch-like transmembrane receptor—lin-12.

Development of the kidney.

Epithelia are the commonest type of tissue organization found in animals and play a key role in the development of many organs like the kidney, lung, skin, blood vessels, and mammary glands. Cells in epithelia adhere tightly together to form a sheet, which can be single layered, as in the endothelial lining of capillaries and kidney tubules, or multilayered, as in skin. A common feature of epithelia is that they are separated from underlying tissue by a basal lamina of extracellular matrix. It is common for epithelia involved in organogenesis to branch and form tubules, and to be induced by the adjacent mesenchyme. We will illustrate some aspects of epithelial morphogenesis in the mammalian kidney, which provides examples of induction, epithelial branching, mesenchyme to epithelium transition, and tubule formation.

10-23   The development of the ureteric bud and mesenchymal tubules involves induction.

Tubule formation in the mammalian kidney is an example of organogenesis that involves both induction and a **mesenchyme to epithelium transition**. The adult kidney develops from a mass of mesenchyme known as the metanephric blastema. This gives rise to the renal tubules and glomeruli—the basic functional units of the kidney—and an epithelial bud, the ureteric bud, which buds off the embryonic ureter and branches to form the collecting system for urine (Fig. 10.43). The ureteric bud induces the mesenchymal cells to condense around it and form epithelial structures that develop into renal tubules. Each tubule elongates to form a glomerulus at one end, where filtration of the blood will occur, while the other end fuses to a collecting duct connecting to the ureter. The intermediate portion of the tubule is convoluted and its epithelial cells become specialized for the resorption of ions from urine. The mammalian kidney is a good developmental system to study as it can develop in organ culture; over a period of about 6 days explanted metanephric mesenchyme will form many glomeruli and tubules, even though a blood supply is absent.

The development of the ureteric bud and the mesenchyme depends on mutual inductive interactions, neither being able to develop in the absence of the other. The mesenchyme causes the ureteric bud to grow and bifurcate, and the bud induces the mesenchyme to differentiate into nephrons. The transcription factor WT1 is necessary for the

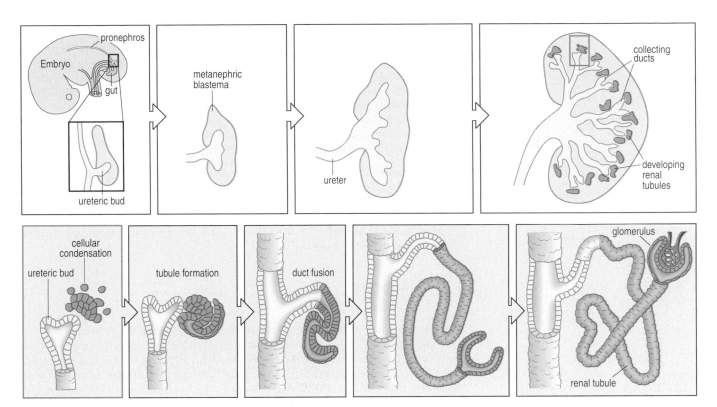

Fig. 10.43 Kidney development from mesenchyme. The kidney develops from a loose mass of mesenchyme, the metanephric blastema, which is induced to form tubules by the ureteric bud. The ureteric bud is itself induced by the mesenchyme to grow and branch to form the collecting ducts of the kidney that connect to the ureter. The mesenchyme cells form cellular condensations that become epithelial tubules, which open into the collecting system formed from the ureteric bud. Each tubule develops a glomerulus, through which waste products are filtered out of the blood.

competence of the mesenchyme to be induced. Mutations in the *WT1* gene are associated with a cancer of the kidney in children known as Wilm's tumor. Glial-derived neurotrophic factor (GDNF) is a likely inducer of ureteric bud growth. It is expressed in the mesenchyme and its receptor Ret is on the bud. Knock-out of the gene for either GDNF or Ret results in the absence of ureter bud outgrowth. Another factor involved in kidney induction is hepatocyte growth factor; antibodies to hepatocyte growth factor prevent bud development in culture. It can also stimulate a line of kidney cells in culture to form tubules. The requirements for induction of the mesenchyme can be studied by placing mesenchyme on one side of a coarse filter and the ureteric bud on the other. Tubule formation in the mesenchyme is taken as a sign that induction has occurred. The nature of the inducing signal is unknown. Several tissues other than the ureteric bud, such as the neural tube, can also induce tubule formation in the kidney mesenchyme. This suggests an innate capacity of the mesenchyme for self-organization into tubules that simply requires a relatively nonspecific permissive signal for development to proceed.

The inducing signal for the mesenchyme is probably fibroblast growth factor-2, and both BMP-7 and Wnt-4 are also involved. A very early response to induction is the expression of *Pax2* in the mesenchyme, which is essential for subsequent tubule development. There is then an increase in the expression of *WT1*, the loose mass of mesenchyme cells condense, and the cells start to express a matrix glycoprotein, syndecan, on their surface. This condensing stage is followed by the formation of distinct cellular aggregates, in which the mesenchymal cells become polarized and acquire an epithelial character. Each aggregate then forms an S-shaped tube, which elongates and differentiates to form the functional unit of a renal tubule and glomerulus. During this transition, the composition of the extracellular matrix secreted by the cells changes: mesenchymal collagen I is replaced by basal lamina proteins, such as collagen IV and laminin, which are typically secreted by epithelial cells. The adhesion molecules (see Box 8A, page 233) expressed by the cells also change; for example, the N-CAM expressed by the mesenchymal cells is replaced by E-cadherin in the epithelial cells. Integrins are also involved in the epithelial–mesenchyme interactions.

Wnt-4 is a mammalian member of the family of signaling proteins that includes the *Drosophila* wingless protein. Knockout mice lacking Wnt-4 activity fail to form the pretubular cell aggregates. Wnt-4 itself is not a cell adhesion molecule but it may be involved in modifying the adhesive properties of the cells so that they can undergo the mesenchyme to epithelium transformation.

Summary.

The development of the mammalian kidney illustrates aspects of epithelial morphogenesis. It involves both mesenchyme to epithelium transition and reciprocal induction. The ureteric bud is induced by the prospective kidney mesenchyme to grow and branch, while the ureteric bud induces the cells in the mesenchyme to condense and form tubules. These tubules then form a glomerulus and fuse with the ureteric system.

Summary to Chapter 10.

The development of organs involves similar processes, and in some cases the same genes, as those used earlier in development. The pattern of the vertebrate limb is specified along three axes, at right angles to each other. Signaling molecules provide positional information, and interpretation of this information involves Hox genes. The development of *Drosophila* wings and legs from imaginal discs depends on signals generated at compartment boundaries, and some of these signal proteins are similar to those used in vertebrate limb patterning. The insect eye also develops from an imaginal disc, and the individual ommatidia in it are spaced by lateral inhibition, with the pattern of photoreceptors depending on local cell–cell interactions. Lateral inhibition and cell–cell signaling also underlies vulva development in the nematode. The development of the kidney involves epithelial morphogenesis and reciprocal induction.

References.

10-1   The vertebrate limb develops from a limb bud.

Tickle, C., Eichele, G.: **Vertebrate limb development**. *Ann. Rev. Cell Biol.* 1994, **10**:121–152.

10-2   Patterning of the limb involves positional information.

Cohn, M.J., Tickle, C.: **Limbs: a model for pattern formation within the vertebrate body plan**. *Trends Genet.* 1996, **12**:253–257.

10-3   The apical ectodermal ridge induces the progress zone.

Niswander, L., Tickle, C., Vogel, A., Booth, I., Martin, G.R.: **FGF-4 replaces the apical ectodermal ridge and directs outgrowth and patterning of the limb**. *Cell* 1993, **75**:579–587.

10-4   The polarizing region specifies position along the antero-posterior axis.

Chiang, C., Litingtung, Y., Lee, E., Young, K.E., Corden, J.L., Westphal, H., Beachy, P.A: **Cyclopia and defective axial patterning in mice lacking *Sonic hedgehog* gene function**. *Nature* 1996, **383**:407–413.

Riddle, R.D., Johnson, R.L., Laufer, E., Tabin, C.: *Sonic hedgehog* mediates polarizing activity of the ZPA. *Cell* 1993, **75**:1401–1416.

Tabin, C.: **The initiation of the limb bud: growth factors, Hox genes, and retinoids.** *Cell* 1995, **80**:671–674.

Tickle, C.: **The number of polarizing region cells required to specify additional digits in the developing chick wing.** *Nature* 1981, **289**:295–298.

Tickle, C.: **Retinoic acid and chick limb development.** *Development Suppl.* 1991, **1**:113–121.

Wanek, N., Gardiner, D.M., Muneoka, K., Bryant, S.V.: **Conversion by retinoic acid of anterior cells into ZPA cells in the chick wing bud.** *Nature* 1991, **350**:81–83.

10-5 Position along the proximo-distal axis may be specified by a timing mechanism.

Wolpert, L., Tickle, C., Sampford, M.: **The effect of cell killing by X-irradiation on pattern formation in the chick limb.** *J. Embryol. Exp. Morph.* 1979, **50**:175–193.

10-6 The dorso-ventral axis is controlled by the ectoderm.

Geduspan, J.S., MacCabe, J.A.: **The ectodermal control of mesodermal patterns of differentiation in the developing chick wing.** *Dev. Biol.* 1987, **124**:398–408.

Parr, B.A., McMahon, A.P.: **Dorsalizing signal Wnt-7a is required for normal polarity of D-V and A-P axes of mouse limb.** *Nature* 1995, **374**:350–353.

Riddle, R.D., Ensini, M., Nelson, C., Tsuchida, T., Jessell, T.M, Tabin, C.: **Induction of the LIM homeobox gene *Lmx1* by Wnt-7a establishes dorsoventral pattern in the vertebrate limb.** *Cell* 1995, **83**:631–640.

Tickle, C.: **Vertebrate limb development.** *Curr. Opin. Genet. Dev.* 1995, **5**:478–484.

Yang, Y., Niswander, L.: **Interaction between the signaling molecule Wnt-7a and Shh during vertebrate limb development: dorsal signals regulate antero-posterior patterning.** *Cell* 1995, **80**:939–947.

10-7 Different interpretations of the same positional signals give different limbs.

Krabbenhoft, K. M., Fallon, J. F.: **The formation of leg or wing specific structures by leg bud cells grafted to the wing bud is influenced by proximity to the apical ridge.** *Dev. Biol.* 1989, **131**:373–382.

10-8 Homeobox genes are involved in patterning the limbs and specifying their position.

Charité, J., De Graaff, W., Shen, S., Deschamps, J.: **Ectopic expression of *Hoxb-8* causes duplication of the ZPA in the forelimb and homeotic transformation of axial structures.** *Cell* 1994, **78**:589–601.

Davis, A.P., Capecchi, M.R.: **A mutational analysis of the 5′ HoxD genes: dissection of genetic interactions during limb development in the mouse.** *Development* 1996, **122**:1175–1185.

Goff, D.J., Tabin, C.J.: **Analysis of *Hoxd-13* and *Hoxd-11* misexpression in chick limb buds reveals that *Hox* genes affect both bone condensation and growth.** *Development* 1997, **124**:627–636.

Morgan, B.A., Izpisúa-Belmonte, J.C., Duboule, D., Tabin, C.I: **Targeted misexpression of Hox-4.6 in the avian limb bud causes apparent homeotic transformations.** *Nature* 1992, **358**: 236–239.

Nelson, C.E., Morgan, B.A., Burke, A.C., Laufer, E., DiMambro, E., Muytaugh, L.C., Gonzales, E., Tessarollo, L., Parada, L.F., Tabin, C.: **Analysis of Hox genes expression in the chick limb bud.** *Development* 1996, **122**:1449–1466.

Scott, M.P.: **Hox genes, arms, and the man.** *Nature Genet.* 1997, **15**:117–118.

Yokouchi, Y., Nakazato, S., Yamamoto, M., Goto, Y., Kameda, T., Iba, H., Kuroiwa, A.: **Misexpression of *Hoxa-13* induces cartilage homeotic transformation and changes cell adhesiveness in chick limb buds.** *Genes Dev.* 1995, **9**:2509–2522.

10-9 Self-organization may be involved in pattern formation in the limb bud.

Hardy, A., Richardson, M.K., Francis-West, P.N., Rodriguez, C., Izpisúa-Belmonte, J.C., Duprez, D., Wolpert, L.: **Gene expression, polarising activity and skeletal patterning in reaggregated hind limb mesenchyme.** *Development* 1995, **121**:4329–4337.

Box 10A Reaction-diffusion mechanisms.

Kondo, S., Asai, R.: **A reaction-diffusion wave on the skin of the marine angelfish *Pomocanthus*.** *Nature* 1995, **376**:765–768.

Murray, J.D.: **How the leopard gets its spots.** *Sci. Amer.* 1988, **258**:80-87.

Richardson, M.K., Hornbruch, A., Wolpert, L.: **Pigment patterns in neural crest chimaeras constituted from quail and guinea fowl embryos.** *Dev. Biol.* 1991, **143**:303–319.

10-10 Limb muscle is patterned by the connective tissue.

Robson, L.G., Kara, T., Crawley, A., Tickle, C.: **Tissue and cellular patterning of the musculature in chick wings.** *Development* 1994, **120**:1265–276.

10-11 The initial development of cartilage, muscles, and tendons is autonomous.

Ros, M.A., Rivero, F.B., Hinchliffe, J.R., Hurle, J.M.: **Immunohistological and ultrastructural study of the developing tendons of the avian foot.** *Anat. Embryol.* 1995, **192**:483–496.

10-12 Separation of the digits is the result of programmed cell death.

Garcia-Martinez, V., Macias, D., Gañan, Y., Garcia-Lobo, J.M., Francia, M.V., Fernandez-Teran, M.A., Hurle, J.M.: **Internucleosomal DNA fragmentation and programmed cell death (apoptosis) in the interdigital tissue of the embryonic chick leg bud.** *J. Cell. Sci.* 1993, **106**:201–208.

Zou, H., Niswander, L.: **Requirement for BMP signaling in interdigital apoptosis and scale formation.** *Science* 1996, **272**:738–741.

10-13 Signals from the antero-posterior compartment boundary pattern the wing imaginal disc.

Blair, S.S.: **Compartments, and appendage development in *Drosophila*.** *Bioessays* 1995, **17**:299–309.

Brook, W.J., Diaz-Benjumea, F.J., Cohen, S.M.: **Organizing spatial pattern in limb development.** *Ann. Rev. Cell Dev. Biol.* 1996, **12**:161–180.

Lawrence, P.A., Struhl G.: **Morphogens, compartments, and pattern: lessons from *Drosophila*?** *Cell* 1996, **85**:951–961.

Lecuit, T., Brook, J.W., Ng, M., Calleja, M., Sun, H., Cohen, S.: **Two distinct mechanisms for long-range patterning by Decapentaplegic in the *Drosophila* wing.** *Nature* 1996, **381**:387–393.

Nellen, D., Burke, R., Struhl, G., Basler, K.: **Direct and long-range action of a DPP morphogen gradient.** *Cell* 1996, **85**:357–368.

Smith, J.: **How to tell a cell where it is.** *Nature* 1996, **381**:367–368.

Vincent, J.P., Lawrence, P.A.: **Developmental genetics. It takes three to distalize.** *Nature* 1994, **372**:132–133.

10-14   The dorso-ventral boundary of the wing acts as a pattern-organizing center.

Diaz-Benjumea, F.J., Cohen, S.M.: **Interaction between dorsal and ventral cells in the imaginal disc directs wing development in *Drosophila*.** *Cell* 1993, **75**:741–752.

Irvine, K.D., Wieschaus, E.: ***fringe*, a boundary-specific signaling molecule, mediates interactions between dorsal and ventral cells during *Drosophila* wing development.** *Cell* 1994, **79**:595–606.

Kim, J., Sebring, A., Esch, J.J., Kraus, M.E., Vorwerk, K., Magee, J., Casroll, B.: **Integration of positional signals and regulation of wing formation and identity by *Drosophila vestigial* gene.** *Nature* 1996, **382**:133–138.

Williams, J.A., Paddock, S.W., Vorwerk, K., Carroll, S.B.: **Organization of wing formation and induction of a wing-patterning gene at the dorso-ventral compartment boundary.** *Nature* 1994, **368**:299–305.

10-15   The leg disc is patterned in a similar manner to the wing disc, except for the proximo-distal axis.

Brook, W.J., Cohen, S.M.: **Antagonistic interactions between wingless and decapentaplegic responsible for dorsal-ventral pattern in the *Drosophila* leg.** *Science* 1996, **273**:1373–1377.

Campbell, G., Tomlinson, A.: **Initiation of the proximo-distal axis in insect legs.** *Development* 1995, **121**:619–625.

10-16   Butterfly wing markings are organized by additional positional fields.

Brakefield, P.M., Gates, J., Keys, D., Kesbeke, F., Wijngaarden, P.J., Monteiro, A., French, V., Carroll, S.B.: **Development, plasticity, and evolution of butterfly eyespot patterns.** *Nature* 1996, **384**:236–242.

Carroll, S.B., Gates, J., Keys, D.N., Paddock, S.W., Panganiban, G.E., Silegue, J.E., Willliams, J.A.: **Pattern formation and eyespot determination in butterfly wings.** *Science* 1994, **265**:109–114.

North, G., French, V.: **Insect wings. Patterns upon patterns.** *Curr. Biol.* 1994, **7**: 611–614.

Nijhout, H.F.: *The Development and Evolution of Butterfly Wing Patterns.* Washington. Smithsonian Institution Press, 1991.

10-17   The segmental identity of imaginal discs is determined by the homeotic selector genes.

Carroll, S.B.: **Homeotic genes and the evolution of arthropods and chordates.** *Nature* 1995, **376**:479–485.

Vachon, G., Cohen, B., Pfeifle, C., McGuffin, M.E., Botas, J., Cohen, S.M.: **Homeotic genes in the Bithorax complex repress limb development in the abdomen of the *Drosophila* embryo through the target gene *Distal-less*.** *Cell* 1992, **71**:437–450.

The insect compound eye.

Bonini, N.M., Choi, K.W.: **Early decisions in *Drosophila* eye morphogenesis.** *Curr. Opin. Genet. Dev.* 1995, **5**:507–515.

Halder, E., Callaerts, P., Gehring, W.J.: **Induction of ectopic eyes by targeted expression of the *eyeless* gene in *Drosophila*.** *Science* 1995, **267**:1788–1792.

10-18   Signals maintain progress of the morphogenetic furrow and the ommatidia are spaced by lateral inhibition.

Baker, N.E., Mlodzik, M., Rubin; G.M.: **Spacing differentiation in the developing *Drosophila* eye: a fibrinogen-related lateral inhibitor encoded by *scabrous*.** *Science* 1990, **250**:1370–1377.

Heberlein, U., Singh, C.M., Huk, A.Y., Donohoe, T.J.: **Growth and differentiation in the Drosphila eye coordinated by hedgehog.** *Nature* 1995, **373**: 709–711.

10-19   The patterning of the cells in the ommatidium depends on intercellular interactions.

Krämer, H., Cagan, R.L.: **Determination of photoreceptor cell fate in the *Drosophila* retina.** *Curr. Opin. Neurobiol.* 1994, **4**:14–20.

10-20   The development of R7 depends on a signal from R8.

Domínguez, M., Hafen, E.: **Genetic dissection of cell fate specification in the developing eye of** Drosophila. *Cell Dev. Biol.* 1996, **7**:219–226.

10-21   Activation of the gene *eyeless* can initiate eye development.

Halder, G., Callaerts, P., Gehring, W.J.: **Induction of the ectopic eyes by targeted expression of the *eyeless* gene in *Drosophila*.** *Science* 1995, **267**:1788–1792.

10-22   The anchor cell induces primary and secondary fates.

Grunwald, I., Rubin, G.M.: **Making a difference: the role of cell–cell interactions in establishing separate identities for equivalent cells.** *Cell* 1992, **68**:271–281.

Kenyon, C.: **A perfect vulva every time: gradients and signaling cascades in *C. elegans*.** *Cell* 1995, **82**:171–174.

Newman, A.P., Sternberg, P.W.: **Coordinated morphogenesis of epithelia during development of the *Caenorhabditis elegans* uterine-vulval connection.** *Proc. Natl. Acad. Sci.* 1996, **93**:9329–9333.

Sundaram, M., Han, M.: **Control and integration of cell signaling pathways during *C. elegans* vulval development.** *BioEssays* 1996, **18**:473–480.

10-23   The development of the ureteric bud and mesenchymal tubules involves induction.

Bard, J.B., Davies, J.A., Karavanova, I., Lehtonen, E., Sariola, H., Vainio, S.: **Kidney development: the inductive interactions.** *Seminars in Cell Dev.* 1996, **7**:195–202.

Müller, U., Wang, D., Denda, S., Meneses, J.J., Pederson, R.A., Reichardt, L.F.: **Integrin α8β1 is critically important for epithelial-mesenchymal interactions during kidney morphogenesis.** *Cell* 1997, **88**:603–613.

Sorokin, L., Klein, G., Mugrauer, G., Eecker, L., Ekblom, M., Ekblom, P.: **Development of kidney epithelial cells.** In *Epithelial Organization and Development* (Fleming, TP, Ed) 163–190. Chapman & Hall: London, 1992.

Stark, K., Vainio, S., Vassileva, G., McMahon, A.P.: **Epithelial transformation of metanephric mesenchyme in the developing kidney regulated by Wnt-4.** *Nature* 1994, **372**:679–683.

Woolf, A.S., Cale, C.M.: **Roles of growth factors in renal development.** *Curr. Opin. Neph. Hyp.* 1997, **6**:10–14.

Development of the Nervous System

11

Specification of cell identity in the nervous system.

Axonal guidance.

Neuronal survival, synapse formation, and refinement.

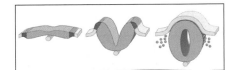

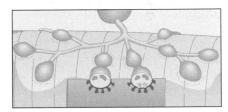

"Special individuals were selected as communicators and they made many connections, which often required them to go on long journeys."

The nervous system is the most complex of all the organ systems in the animal embryo. In mammals, for example, billions of neurons develop a highly organized pattern of connections, creating the neuronal network that makes up the functioning brain and the rest of the nervous system. Yet the nervous system is the product of the same kind of cellular and developmental processes that are involved in the development of other organ systems.

All nervous systems, vertebrate and invertebrate, provide a system of communication through a network of neurons of varying shapes and functions (Fig. 11.1). Each neuron connects with other neurons or target cells through a process of action potential propagation and neurotransmitter release. There are also supporting tissues in the nervous system known collectively as **glia**; for example, the Schwann cells surround the axons of peripheral neurons, and oligodendrocytes and astrocytes ensheath the central neurons. In complex nervous systems, such as those of vertebrates, it has been estimated that there are hundreds of different types of neurons. The **dendrites** and **axon** terminals of individual neurons can be extensively branched; a single neuron can receive as many as 100,000 different inputs. As the nervous system can only function properly if the neurons are correctly connected to one another, a central question surrounding neural system development is how the connections among neurons develop the necessary specificity. The overall process of nervous system development can be divided up into four major stages: the specification of neural (neuron or glial) cell identity; the migration of neurons and the outgrowth of axons to their targets; the formation of **synapses** with the target cells—which can be other neurons, muscle or gland cells; and the refinement of synaptic connections, through the elimination of axon branches and by cell death.

We have already considered some aspects of the development of the vertebrate nervous system, including neural induction, the formation of the neural tube, the segmental arrangement of sensory dorsal root ganglia, and the differentiation of neurons from neural crest cells. We will re-examine some of these key steps when we consider the specification of neural cells. The general focus of this chapter is on the mechanisms that control the identity of cells in the nervous system and that pattern their synaptic connections.

Specification of cell identity in the nervous system.

We start by considering how neural cells acquire their individual identities. As in other developing systems, nerve cell specification is governed both by external signals and by intrinsic differences generated through asymmetric cell divisions. We first consider neural specification in *Drosophila*, which has revealed many key developmental processes in neurogenesis, and then turn to vertebrates, where similar mechanisms are involved.

11-1   Neurons in *Drosophila* arise from proneural clusters.

In *Drosophila*, the presumptive central nervous system is specified at an early stage of development as two longitudinal stripes of cells along the dorso-ventral axis of the embryo, just dorsal to the ventral meso-derm. This region is called the neurogenic zone, or neurectoderm (Fig. 11.2), and comprises ectodermal cells with the potential to form neural cells and epidermis.

Expression of genes known as proneural genes gives the neurectodermal cells the potential to become neural precursors. Of particular impor-tance as proneural genes are the genes of the achaete-scute complex. These encode transcription factors, notably the proteins achaete and scute, which have a basic helix-loop-helix DNA-binding motif. These transcription factors form homodimers and heterodimers with each other, and bind to target genes to initiate neural specification. Within the neurectoderm, cells expressing these genes form groups called proneural clusters, which are evident when one looks at the pattern of *achaete* gene expression in the early *Drosophila* embryo. The achaete protein first appears after gastrulation in a segmentally repeated patt-ern. In each future segment, some clusters are formed in the posterior region, where *engrailed* is also expressed, whereas others form more anteriorly (Fig. 11.3).

The correct pattern of expression of these genes depends on a long stretch of regulatory DNA sequence lying adjacent to the *achaete* and *scute* genes. This region contains regulatory modules that control the pattern of neurogenic gene expression at specific sites. Site-specific expression of *achaete* and *scute* depends on combinations of transcrip-tion factors binding to the regulatory modules. The pair-rule genes, together with genes expressed along the dorso-ventral axis and the segment polarity gene *wingless* (see Section 5-16), are involved in patterning the proneural clusters and distinguishing them from the rest of the neurecto-derm, which will become epidermis. The process is very similar to that involved in the early antero-posterior and dorso-ventral regionaliz-ation of *Drosophila*: combinations of transcription factors, which are

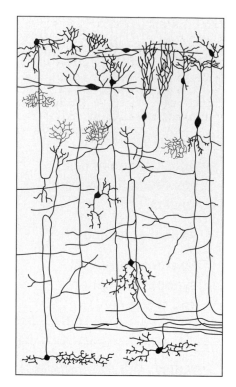

Fig. 11.1   A network of varied neurons in the optic lobe of a bird. Axons and dendrites extend from the cell bodies, which are usually small and rounded, and contain the nucleus.

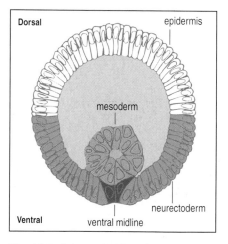

Fig. 11.2   Cross-section of a *Drosophila* early gastrula showing the location of the neurectoderm (pale blue). The meso-derm (red), which originally lies along the ventral midline, has been internalized.

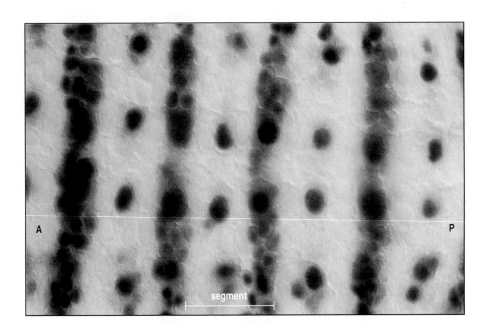

Fig. 11.3 Pattern of expression of the gene _achaete_ in the neurectoderm of the _Drosophila_ embryo. Within each future segment, there are eight proneural clusters expressing _achaete_ (black). Half of these are formed in the region of _engrailed_ expression (purple) at the posterior end of the segment, and half in a more anterior region. Photograph from Skeath, J.B, _et al._: 1996.

themselves distributed in a spatially restricted pattern, locally activate or repress gene activity so that the expression of _achaete_ is eventually restricted. The localization of pair-rule gene expression in the patterning of the antero-posterior body axis, discussed in Sections 5-13 and 5-14, provides a very good model for this general process.

Shortly after the formation of a proneural cluster in the embryonic neurectoderm, one cell in each cluster—the neuronal precursor or **neuroblast**—starts to express _achaete_ at a higher level than the other cells. It then leaves the ectoderm, moving into the interior of the neurula, where it stops expressing the gene. The neuroblasts leave the surface layer during germ band extension and eventually give rise to neurons, which will form the ventral nerve cord. The fate of the neurons is determined by the region from which they derive. The mechanism whereby just one cell is singled out to become a neuroblast is considered in the next section.

The achaete-scute complex also has an important role in the specification of the sensory nervous system of the adult fly. The pattern of sensory bristles (Fig. 11.4) and other sensory organs on the adult cuticle is already present in the imaginal discs in the form of a pattern of sensory mother cells that later give rise to these structures. Proneural clusters are first selected by expression of the achaete-scute complex genes in a precise spatial pattern (Fig. 11.5). As with the proneural clusters of the central nervous system, the pattern of prospective sensory cells within the imaginal disc is constructed by the independent action of site-specific regulatory modules distributed along the achaete-scute complex. Again, as in the specification of neuroblasts, only one of the cells in the cluster expressing achaete-scute adopts a neural fate and becomes a sensory mother cell.

Some of the genes responsible for activating the genes of the achaete-scute complex in specific spatial patterns have been identified. One of these is _iroquois_, which encodes a transcription factor. It is expressed in a broad lateral domain in the thorax, and mutations in _iroquois_ result in the loss of sensory bristles in that domain. In imaginal discs, _iroquois_ is expressed in broad domains, where these patterns of expression are governed by the activity of genes such as _decapentaplegic_ and _wingless_, which provide the discs with positional information.

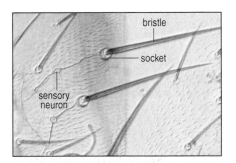

Fig. 11.4 Sensory bristles in _Drosophila_. The bristles respond to external stimuli and stimulate the sensory neurons. Photograph courtesy of Y. Jan.

Fig. 11.5 Specification of sensory bristles in the *Drosophila* wing. The location of proneural clusters in *Drosophila* wing imaginal discs is largely under the control of the achaete-scute gene complex, the expression of which is activated at specific sites through different enhancers. Top panel: the position of the various enhancers that determine the site-specific expression of the *achaete* and *scute* genes. These enhancers are located both upstream and downstream of the *achaete* and *scute* genes. Bottom panel: location of the proneural clusters in the imaginal disc (left) and of the corresponding sensory bristles in the adult wing (right), with colors corresponding to the enhancer responsible for the site-specific expression of the genes. After Campuzano, S., *et al.*: 1992.

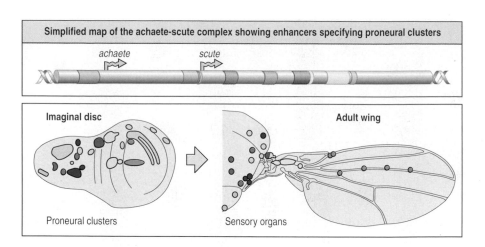

11-2   Lateral inhibition allocates neuronal precursors.

Not all of the cells in the neurectoderm become neurons. The singling out of the prospective neural cells involves lateral inhibition, which we have already seen in action in the spacing of ommatidia in the insect eye (see Section 10-18). Initially, all the cells within a proneural cluster are capable of becoming neuroblasts, but one cell, through an apparently random event, then produces a signal that prevents its immediate neighbors from developing further along the neural pathway (Fig. 11.6). This lateral inhibition eventually leads to just one cell developing as a neuroblast, with the rest of the cluster becoming epithelial epidermal cells. In the grasshopper embryo, if a neuroblast is killed just as it begins to develop, a neighboring cell develops into a neuroblast instead. Once specified, the neuroblasts migrate individually from the superficial cell layer inward and go through a stereotyped series of cell divisions to produce neurons and glia.

The sensory mother cells that give rise to the sensory bristles (sensilla) of the *Drosophila* adult epidermis are thought to be singled out from the proneural clusters in the imaginal discs through a virtually identical mechanism as that outlined above. In the specification of both central nervous system neurons and sensory organs, the *Notch* and *Delta* genes play a key role in lateral inhibition once the proneural clusters have formed. Their activity ensures that just one neuron or sensory organ develops from each cluster at a given time. *Delta* and *Notch* both encode transmembrane proteins; *Delta* encodes a ligand and *Notch* encodes its receptor (Fig. 11.7). Activation of Notch receptor by Delta protein leads to inhibition of the proneural genes and thus to the loss of

Fig. 11.6 The role of lateral inhibition in neural cell development. The proneural cluster gives rise to a single neuronal precursor cell by means of lateral inhibition. The rest of the cells in the cluster become epidermis. In a series of subsequent cell divisions, the neuronal precursor, the neuroblast, gives rise to a single neuron and three glial cells. After Jan, Y.N., *et al.*: 1990.

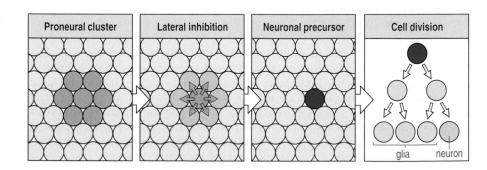

Fig. 11.7 A model for Notch-Delta interaction. Both Notch and Delta are transmembrane proteins. Delta is the ligand for the Notch receptor. Activation of Notch by Delta triggers a signal, which is transmitted to the nucleus of the Notch-expressing cell.

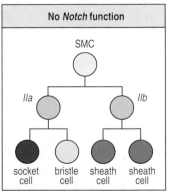

the cell's ability to give rise to a neuroblast. Small differences in *Notch/ Delta* signaling between cells in the cluster allow one cell to advance along the pathway to neural specification sooner than the others; as a result, it sends a signal that prevents the less advanced cells from adopting a neural fate. Such cells go on to develop into epidermal cells and cease expression of *Delta*.

Initially, all of the cells in a proneural cluster can express both *Notch* and *Delta*. One cell then gains an advantage by, for example, starting to express *Delta* sooner or more strongly than the others, and thus interacts with the neighboring cells to inhibit further neural development and also to suppress their production of the Delta protein signal. Hence, the initially equivalent cells of the proneural cluster are resolved by lateral inhibition into one neural precursor cell. If either Notch or Delta function is inactivated, the neurogenic ectoderm makes many more neuroblasts and fewer epidermal cells than normal.

11-3 Asymmetric cell divisions are involved in *Drosophila* sensory organ development.

Most of the sensory organs in the adult *Drosophila* are each formed from the division of a single neuroblast. There are two main types of sensory organs with invariant positions: external sensory organs, which can act as mechanoreceptors or chemoreceptors, and internal chordotonal organs, which monitor stretching. Both types of organ are made up of four cells, only one of which is a sensory neuron (Fig. 11.8, first panel).

The achaete-scute complex is required for the formation of all external sensory organs, but not for the internal chordotonal organs. The proneural clusters that give rise to chordotonal organs express the gene

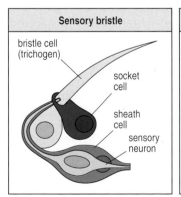
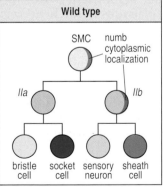
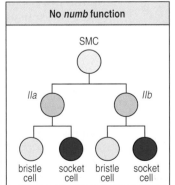

Fig. 11.8 Development of a *Drosophila* sensory bristle, which comprises four cells, depends on asymmetric cell division. The sensory mother cell (SMC) contains numb protein on one side only. It undergoes an asymmetric division in which the numb protein is passed on to just one of the daughter cells (IIb). This cell gives rise to a sensory neuron and a sheath cell.

The cell that does not receive numb protein (IIa) gives rise to a bristle cell and a socket cell. In mutants lacking *numb*, both daughters of the sensory mother cell form bristle and socket cells. In the absence of *Notch* function, a sheath cell develops instead of a neuron. After Guo, M., *et al.*: 1996.

atonal, which codes for a basic helix-loop-helix transcription factor. Ectopic expression of *atonal* results in the formation of additional chordotonal organs, whereas ectopic expression of the achaete-scute genes leads to more external sensory organs. The type of organ that develops also depends on other transcription factors such as *cut*, which is expressed in external sensilla but not in chordotonal organs. Mutations in *cut* can lead to transformation of external sensilla into chordotonal organs, with the reverse transformation occurring when *cut* is expressed ectopically.

The development of a sensory bristle from a sensory mother cell involves two cell divisions to generate the four different cells that form the organ—a sensory neuron and its associated sheath cell, and two support cells, in this case a bristle cell and a socket cell (see Fig. 11.8, second panel). The first cell division gives two daughter cells (IIa and IIb). IIa divides to give a bristle cell and a socket cell; IIb divides to give a neuron and its sheath cell. One determinant of cell fate is the protein product of the *numb* gene, which is asymmetrically localized in the sensory mother cell so that at the first division only one daughter cell receives numb protein. This daughter cell gives rise to the neuron and sheath cell at the second division, whereas the cell lacking numb protein gives rise to the support cells. Numb protein is essential for the formation of the neuron and sheath cell; in mutant embryos lacking *numb* expression these cells do not develop and four outer support cells develop instead (see Fig. 11.8, third panel). Ectopic induction of *numb* expression in both daughter cells of the sensory mother cell results in their developing similarly, each giving a sensory neuron and a sheath cell. *Notch* also plays a role in determining cell fate; in the absence of its function, a sheath cell develops instead of a neuron (see Fig. 11.8, fourth panel).

We have seen many examples of the importance of localization of cytoplasmic determinants in early development. Asymmetric division during sensory cell development in *Drosophila* is a very clear example of cytoplasmic localization involved in cell specification. We now consider related developmental processes in the vertebrate nervous system.

11-4   The vertebrate nervous system is derived from the neural plate.

All the cells of the vertebrate central nervous system derive from the neural plate, a region of columnar epithelium induced from the ectoderm on the dorsal surface of the embryo during gastrulation (see Section 4-7) and from sensory placodes in the head region, which contribute to the cranial nerves. Toward the end of gastrulation, the neural plate begins to fold and form the neural tube (Fig. 11.9). Neural crest cells migrate away from the dorsal half of the tube—undergoing an epithelial to mesenchymal transition—and give rise to both the sensory neurons of the peripheral nervous system and to the autonomic nervous system. The cells that remain within the neural tube give rise to the brain and spinal cord, which together comprise the central nervous system. The mechanisms by which cells are specified as neural precursors in the

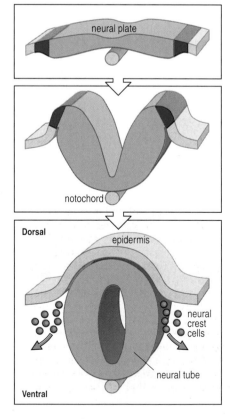

Fig. 11.9 The vertebrate nervous system is derived from the neural plate. The neural tube forms by folding of the neural plate. It sinks beneath the surface and is overlain by the epidermis. The neural tube gives rise to both the brain and the spinal cord. Neural crest cells originating from the line of fusion of the neural folds migrate away from the neural tube to form sensory nerves and the autonomic nervous system (among other structures).

vertebrate neural plate are gradually emerging, and important insights have come from studying the corresponding processes in *Drosophila*.

The neural tube becomes regionalized along the antero-posterior axis, as described in Section 4-8, and this regionalization is particularly clear in the hindbrain region where the Hox genes are expressed in a well-defined pattern. The expression of these genes is thought to give the cells of the neural tube their positional value along the antero-posterior axis of the hindbrain. The neural tube also becomes patterned along the dorso-ventral axis. We next consider the specification of neurons as a class of cells, and then turn to how different types of neurons can be specified in the correct position.

11-5 Specification of vertebrate neuronal precursors involves lateral inhibition.

The specification of cells as neuronal precursors in vertebrates involves processes similar to those already described for *Drosophila* (see Sections 11-1 and 11-2). In early *Xenopus* embryos, neurons are not generated uniformly over the neural plate but are initially confined to three longitudinal stripes on each side of the midline. The medial stripe, which becomes the ventral region of the neural tube, gives rise to motor neurons, whereas the lateral stripes give rise to interneurons and sensory neurons. The *Neurogenin* gene, which codes for a basic helix-loop-helix transcription factor, is related to the achaete and scute proteins of *Drosophila*. Selection of neuronal cells within the stripes involves lateral inhibition, in which the proteins Delta and Notch again play a key role.

Delta is the ligand that binds to the Notch receptor protein, the interaction of which provides a signal that inhibits neuronal differentiation by inhibiting Neurogenin synthesis. Initially, all cells are able to express *Neurogenin*, *Delta*, and *Notch*. When a cell, by chance, starts to express *Delta* more strongly than its neighbors, it both inhibits neural differentiation in the adjacent cells and, by repressing Delta protein expression in these cells, stops them from delivering any reciprocal inhibition. Expression of *Neurogenin* in this one cell leads to the cell expressing *neuroD*, which codes for a transcription factor required for neuronal differentiation (Fig. 11.10). In support of this model, overexpression of Delta leads to a reduction in the number of neurons. If the Notch protein is expressed in an activated form so that it signals continuously, a reduction in neuron number also occurs, while inhibition of Delta function leads to an increase in the number of neurons.

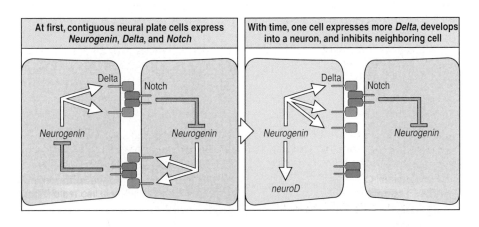

At first, contiguous neural plate cells express *Neurogenin*, *Delta*, and *Notch*

With time, one cell expresses more *Delta*, develops into a neuron, and inhibits neighboring cell

Fig. 11.10 Lateral inhibition specifies single cells as neuronal precursors in the vertebrate nervous system. The *Neurogenin* gene is initially expressed in stripes of contiguous cells in the neural plate, and these cells also express the Delta and Notch proteins. The Delta-Notch interaction between two adjacent cells mutually inhibits expression of the *Neurogenin* gene. When, by chance, one cell expresses more *Delta* than its neighbors, it inhibits the expression of Delta protein in the neighboring cells and so develops as a neuron expressing *Neurogenin* and *neuroD*.

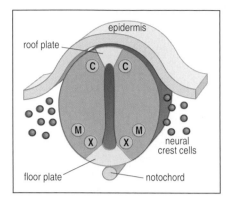

Fig. 11.11 Dorso-ventral organization in embryonic spinal neural tube. In neural tube that develops into spinal cord, a floor plate of non-neuronal cells develops along the ventral midline, and a roof plate of non-neuronal cells along the dorsal midline. Neuronal cells whose phenotype is not known (X) differentiate nearest to the floor plate. Immediately above this region are future motor neurons (M). Commissural neurons (C) differentiate in the dorsal region, near the roof plate.

Fig. 11.12 Formation of sensory and motor neurons in the chick spinal cord. Three days after an egg has been laid, motor neurons are beginning to form within the chick embryo neural tube. One day later, the motor neurons migrate to form the lateral ventral root of the spinal cord. The sensory neurons that migrate from the dorsal part of the spinal cord are derived from neural crest cells and form the segmental dorsal root ganglia (see Section 8-15).

| 11-6 | **The pattern of differentiation of cells along the dorso-ventral axis of the spinal cord depends on ventral and dorsal signals.** |

There is a distinct dorso-ventral pattern in the developing spinal cord. Future motor neurons are located ventrally, whereas commissural neurons, which send axons across the two sides of the cord, differentiate in the dorsal region (Fig. 11.11). In addition to the neuronal cell types, there is, in the ventral midline, a group of non-neuronal cells, which form the floor plate. There are also special roof plate cells in the dorsal midline. Sensory neurons derived from the neural crest cells arise laterally and dorsally and migrate into the dorsal root ganglia (Fig. 11.12). Each cell type is present symmetrically on either side of the midline. Patterning of the cells in the ventral half of the spinal cord neural tube appears to be controlled by the notochord and floor plate cells. The floor plate itself is induced by the underlying notochord. The dorsal epidermal ectoderm, and later the roof plate, patterns the dorsal half of the neural tube.

Before they show signs of visible differentiation, the different regions of the spinal cord can be identified by molecular markers. One of the first signs of dorso-ventral patterning is the expression of homeobox genes of the Pax family in different regions along the dorso-ventral axis of the neural tube; *Pax3* and *Pax7* are, for example, progressively restricted to dorsal regions as the neural tube closes.

The patterning activity of the notochord can be shown by grafting a segment of notochord to a site lateral or dorsal to the neural tube; a second floor plate is induced in neural tube in direct contact with the transplanted notochord and additional motor neurons are generated. Molecular markers of dorsal cell differentiation are suppressed in the region of the graft. Both the short-range and long-range signals are mediated by Sonic hedgehog protein, whose signaling role we have already seen in the development of the vertebrate limb. The *Sonic hedgehog* gene is expressed in both notochord and floor plate, and its misexpression

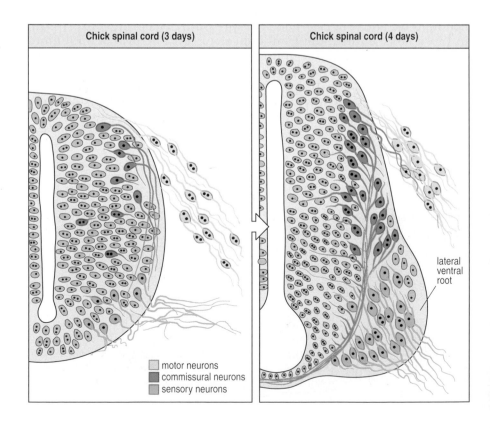

can induce the development of floor plate and motor neurons. An early effect of *Sonic hedgehog* activity is the repression of *Pax3* and *Pax7* expression, which are necessary for generation of ventral cell types. The identity and pattern of cell types generated in the ventral half of the neural tube is controlled by different threshold concentrations of the Sonic hedgehog protein. The threshold concentration of Sonic hedgehog required to induce motor neurons is about three times lower than that required to induce floor plate.

In addition to the signal provided by Sonic hedgehog from the floor plate and notochord in the ventral midline, there are also dorsal signals. Neural plate cells acquire dorsal fates through an inductive signal from the adjacent epidermal ectoderm during neural tube closure. The growth factors BMP-4 and BMP-7 are expressed in the epidermal ectoderm and, in experiments in chick embryos, have been shown to mimic the effect of the dorsalization signal. These BMP signals apparently ensure dorsal differentiation by opposing the action of signals from the ventral region (Fig. 11.13). The induction of dorsal fate results in the expression of the gene for BMP-4 and of another TGF-β family gene, *Dorsalin-1*, in the dorsal region of the neural tube. The expression of BMP-4 and dorsalin appears to be involved in propagating the dorsalizing signal between cells in a contact-dependent fashion.

The pattern of expression of the *Dorsalin-1* gene is strongly influenced by ventral signals. Grafts of notochord to dorsal regions of the neural tube repress *Dorsalin-1* expression, whereas removal of notochord from its normal ventral position results in the area of *Dorsalin-1* expression expanding ventrally. Its normal pattern of expression, therefore, appears to be partly determined by repression by ventral signals. This system bears a very strong similarity to the mechanism of patterning of the dorso-ventral body axis in early *Drosophila* development, where the expression of *decapentaplegic*, another TGF-β family gene, is confined to dorsal regions because its expression is repressed in ventral regions by the nuclear protein, dorsal (see Section 5-8). As in *Drosophila*, one can imagine the expression of dorsalin, or other BMPs, establishing a gradient that could pattern vertebrate dorsal regions.

Thus, in the spinal cord, Sonic hedgehog protein and BMPs may provide positional signals with opposing actions from the two sides of the dorso-ventral axis. The similarity to the patterning of both the vertebrate somite (see Section 4-2) and the early *Drosophila* embryo again emphasizes how patterning mechanisms have been conserved.

While these processes specify where neurons will form, not all neurons are the same. Even among those of the same general type, such as motor neurons, each has a unique identity. Motor neurons in the spinal cord can be classified on the basis of the position of their cell bodies within the cord, and of the muscles that their axons innervate. For example, motor neurons located near the midline of the spinal cord innervate axial muscles of the trunk, whereas motor neurons in the lateral region of the cord innervate limb muscles, with some innervating ventral limb muscles, and others innervating dorsal limb muscles.

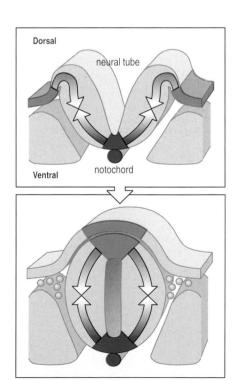

Fig. 11.13 Dorso-ventral patterning of the spinal cord involves both dorsal and ventral signals. As the neural tube forms (top panel), dorsal and ventral signaling occurs. The ventral signal from the floor plate is the Sonic hedgehog protein (red), whereas the dorsal signal (blue) includes BMP-4, which is induced during neural tube formation by the adjacent ectoderm. Ventral and dorsal signals oppose the action of each other.

Fig. 11.14 Different motor neurons in the chick spinal cord express different LIM homeodomain proteins. The photograph (inset) shows a transverse section through a chick embryo at the level of the developing wing. Motor neurons can be seen entering the limb. Initially, all developing motor neurons express the LIM proteins Isl-1 and Isl-2. At the time of axon extension, motor neurons that innervate different muscle regions express different combinations of Isl-1, Isl-2, LIM-1, and LIM-3.

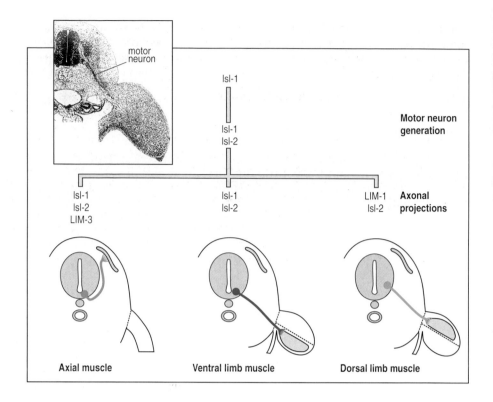

Within the region that innervates a limb, medial motor neurons (those motor neurons nearest the midline) project to ventral muscle, whereas more lateral neurons project to muscles derived from the dorsal muscle mass. Each motor neuron subtype in the chick and zebrafish expresses a combination of proteins encoded by the homeobox genes of the LIM family, which may provide the neurons with positional identity and enable their axons to select a particular pathway (Fig. 11.14). The precise pattern of neuromuscular connections established is impressive and clearly depends on local cues, as discussed later in this chapter.

Motor neurons also acquire a regional identity along the antero-posterior axis. Transplantation of a region of the spinal cord from a site adjacent to a limb to a thoracic region leads to expression of both LIM and Hox genes appropriate to the new site. The signals determining their expression are thought to come from the adjacent mesoderm.

11-7 | **Neurons in the mammalian central nervous system arise from asymmetric cell divisions, then migrate away from the proliferative zone.**

The neural tube generates a large number of different neuronal and glial cell types. In both brain and spinal cord, all neurons and glia arise from a proliferative layer of epithelial cells lining the lumen of the neural tube that is called the ventricular proliferative zone. Once formed, a neuron never divides, and migrates away from the proliferative zone. Here we look at how neurons are generated by cell division in the proliferative zone.

The mammalian cerebral cortex is organized into six layers (I–VI), numbered from the cortical surface inward. They each contain neurons with distinctive shapes and connections. For example, large pyramidal cells are concentrated in layer V, and smaller stellate neurons predominate in layer IV. All of these neurons have their origin in the ventricular proliferative zone lining the inside of the neural tube, and migrate outward to their final positions along radial glial cells—greatly elongated cells that extend across the developing neural tube (Fig. 11.15).

The identity of a cortical neuron is specified before it begins to migrate. The layer to which neurons migrate after their birth in the proliferative layer is related to the time when the neuron is born. The neurons of the mammalian central nervous system do not divide once they have become specified, and so a neuron's time of birth is defined by the last mitotic division that its progenitor cell underwent. The newly formed neuron is still immature; later it extends an axon and dendrite processes, and assumes the morphology of a mature neuron. Birth times of individual neurons can be determined by exposing the neural tube to a short pulse of [³H]thymidine (Fig. 11.16). This is incorporated into the neuron's DNA during the S-phase of the cell cycle, when DNA synthesis takes place. A neuron that is formed at the end of a cell cycle during which [³H]thymidine is incorporated can be identified by its heavily labeled DNA, whereas neurons born later have the label diluted out through further rounds of DNA synthesis in the progenitor cells.

Neurons born at early stages of cortical development migrate to layers closest to their site of birth, whereas those born later end up further away, in more superficial layers (see Fig. 11.16). The younger neurons must migrate past the older ones on their way to their correct position. Thus, there is an inside–outside sequence of neuronal differentiation in the neural tube, which gives rise to the cerebral cortex and other layered brain structures.

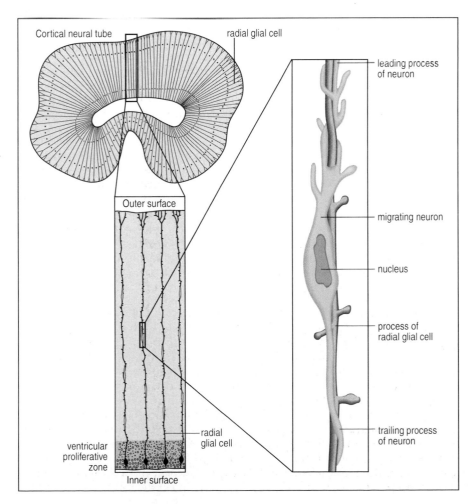

Fig. 11.15 Cortical neurons migrate along radial glial cells. Neurons generated in the ventricular proliferative zone migrate to their final locations in the cortex along radial glial cells that extend right across the wall of the neural tube. After Rakic, P.: 1972.

Fig. 11.16 The time at which a neuron is born determines which cortical layer it becomes part of. In the monkey visual cortex, a neuron's time of birth is determined as the time that the precursor cell stops dividing and leaves the mitotic cycle. This figure shows the results of [³H]thymidine injection into a developing neural tube at various times during development. The red bars represent the distribution of heavily labeled neurons found after each injection. Neurons born first remain closest to their site of birth, the proliferative ventricular zone. Neurons born later become part of successively higher cortical zones. Cortical zones are numbered from the outer surface inward; layer VI is the layer nearest to the ventricular zone.

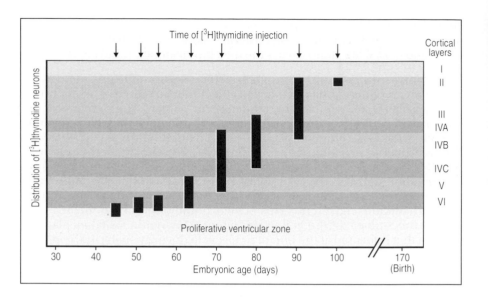

Environmental influences can affect neuronal fate by influencing the character of the progenitor cells. In experiments on the mammalian visual cortex, progenitor cells that produce neurons that would normally migrate to layer VI are transplanted into older embryos. If these cells are transplanted at an early stage in the cell cycle, or undergo further divisions, the neurons they give rise to migrate to layers II and III instead, which indicates an environmental influence on their fate. However, if the cells are transplanted late in their final division cycle, the neurons derived from them still migrate to layer VI, their normal location. Thus, commitment to a particular neuronal fate occurs late in the division cycle and, once it occurs, the neuron becomes unresponsive to environmental influences on its fate. Signals both from the extracellular environment and intrinsic intracellular restrictions therefore have a role in the specification of neuronal identity in vertebrates.

How is the decision made for a cell to leave the cell cycle and become a neuron? Direct observation of slices of cells from developing brain in culture show that the ventricular epithelial cells that give rise to neurons can divide in either of two different planes. After division of a ventricular epithelial cell parallel to the epithelial surface, which leads to formation of distinct apical and basal cells, the basal cell does not undergo any further division and becomes a neuron; the apical cell behaves as a stem cell and can divide again. In contrast, when the plane of division is at right angles to the epithelium, the two daughter cells have a tendency to remain in the proliferative layer and continue to proliferate (Fig. 11.17).

Why do daughter cells behave similarly after vertical divisions but differently after horizontal divisions? By analogy with the asymmetrical divisions in *Drosophila* sensory cell development, the answer may involve two asymmetrically localized cytoplasmic factors, the Notch-1 protein and the numb protein (the vertebrate equivalents of the *Drosophila* Notch and numb proteins). At cell division, Notch-1 protein is localized in the basal domain of the mammalian ventricular epithelial cells, so that in horizontal cell divisions the basal cell, which becomes a neuron and migrates away, contains all the Notch-1. In contrast, numb protein is localized at the dividing cells' apical end and cells inheriting it remain in the proliferative layer (see Fig. 11.17). These two proteins may help to determine whether a cell becomes a neuron or remains a stem cell.

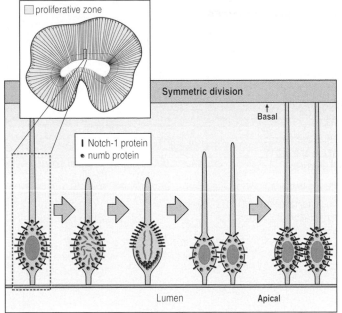

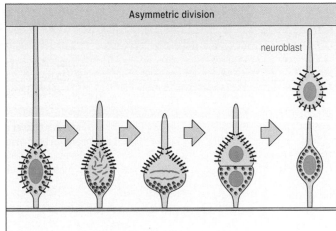

Fig. 11.17 Cortical neurons are formed by asymmetric divisions in the ventricular proliferative zone of the neural tube. Stem cells in the epithelium divide to generate neurons. Divisions at right angles to the epithelial surface give rise to two equal cells that continue to proliferate (left panel). Divisions parallel to the surface give rise to one proliferative cell that remains in the zone and one cell that migrates away to become a neuroblast (right panel). The Notch-1 protein is present on the basal face of dividing ventricular zone cells and is distributed to the basal cell only in horizontal divisions, whereas numb protein is localized in the apical region.

Summary.

In *Drosophila*, prospective neural tissue is specified early in development as a band of neurectoderm along the dorso-ventral axis, above the mesoderm. Within the neurectoderm, expression of proneural genes, such as those of the achaete-scute complex, provides clusters of neurectoderm cells with the potential to form neural cells. Only one cell of the cluster finally gives rise to a neuroblast because of lateral inhibition, involving the Notch receptor protein and its ligand Delta protein, which prevents other cells developing as neuroblasts. Development of the spatially patterned sensory bristles on the adult cuticle involves a similar mechanism of lateral inhibition. Sensory organs, each composed of a single sensory neuron and three associated cells, develop from the neural precursor cell by a mechanism involving asymmetric division, which differentially allocates the asymmetrically distributed numb protein to daughter cells.

The vertebrate nervous system is derived from the neural plate, which is specified during gastrulation. It contains the cells that form the brain and spinal cord, as well as the neural crest cells that contribute to the peripheral nervous system. Neurons are specified within the neural plate by a mechanism of lateral inhibition, similar to that in *Drosophila*. Patterning of neuronal cell types along the dorso-ventral axis of the spinal cord involves signals from both the ventral and dorsal regions. The notochord induces the ventral floor plate and patterns cells in the ventral region, and dorsal signals pattern the dorsal region. The specification of neuronal cell type in the mammalian cortex depends on asymmetric cell divisions and the time when the neurons are formed, as well as on environmental signals. The earliest formed neurons make the innermost cortical layer, while neurons formed later make successive outer layers.

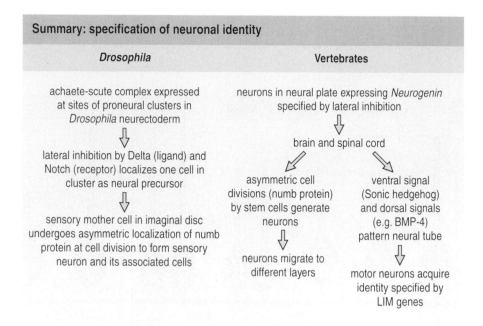

Summary: specification of neuronal identity

Drosophila	Vertebrates
achaete-scute complex expressed at sites of proneural clusters in *Drosophila* neurectoderm	neurons in neural plate expressing *Neurogenin* specified by lateral inhibition

⇓

Drosophila side:

lateral inhibition by Delta (ligand) and Notch (receptor) localizes one cell in cluster as neural precursor

⇓

sensory mother cell in imaginal disc undergoes asymmetric localization of numb protein at cell division to form sensory neuron and its associated cells

Vertebrates side:

brain and spinal cord

⇙   ⇘

asymmetric cell divisions (numb protein) by stem cells generate neurons

⇓

neurons migrate to different layers

ventral signal (Sonic hedgehog) and dorsal signals (e.g. BMP-4) pattern neural tube

⇓

motor neurons acquire identity specified by LIM genes

Axonal guidance.

The working of the nervous system depends on discrete neuronal circuits, in which neurons make numerous connections with each other (Fig. 11.18). Here, we look at how these connections are set up. The functioning network of neurons is established by migration of immature neurons and by the guided outgrowth of axons toward their target cells. What guides the migrating neurons and controls the growth of axons and the contacts they make with other cells?

We have already discussed one example of extensive cell migration in relation to neural development—the migration of neural crest cells to give rise to the segmental arrangement of dorsal root ganglia (see Section 8-15). In that case, the migration is controlled by signals from

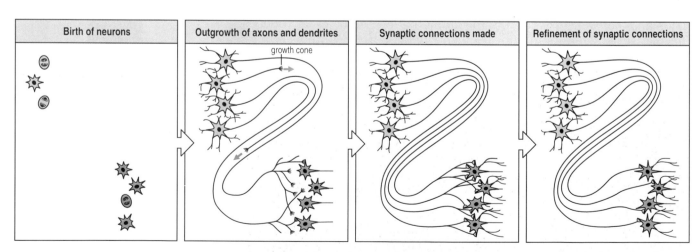

| Birth of neurons | Outgrowth of axons and dendrites | Synaptic connections made | Refinement of synaptic connections |

growth cone

Fig. 11.18 Neurons make precise connections with their targets. Neurons (green) and their target cells usually develop in different locations. Connections between them are established by axonal outgrowth, guided by the movement of the axon tip (growth cone). The initial set of relatively nonspecific synaptic connections are then refined to produce a precise pattern of connectivity. After Alberts, B., *et al.*: 1989.

the adjacent somites. In this chapter, we deal with axon guidance in the central and peripheral nervous system. We first look at the evidence for specificity of guidance pathways, and then consider possible molecular mechanisms.

11-8 | Motor neurons from the spinal cord make muscle-specific connections.

An animal's muscular system enables it to carry out an enormous variety of movements, which all depend on motor neurons having made the correct connections with muscles. The development of the chick limb provides a good model for studying how these connections are made. Motor neurons at different dorso-ventral positions within the spinal cord innervate distinct target muscles within the limb. Those near the midline project to ventral muscle, whereas lateral neurons project to muscles derived from the dorsal muscle mass. Each neuronal subtype expresses a combination of LIM family homeobox genes (see Fig. 11.14), which may provide the neurons with positional identity and enable their axons to select a specific pathway.

The pattern of innervation of the limb muscles of the chick is rather precise (Fig. 11.19). In the chick embryo, the motor neuron axons that enter a developing limb appear to make their connections with particular muscles from an early stage. When the motor axons growing out from the spinal cord first reach the base of the limb, they are all mixed up in a single bundle. At the base of the developing limb, however, the axons separate out and form new nerve tracts containing only those axons that will make connections with particular muscles.

Evidence that local cues in the limb are responsible for this sorting out process comes from inverting along the antero-posterior axis, a section of the spinal cord whose motor neurons will innervate the hindlimb. After inversion, motor axons from lumbo-sacral segments 1–3, which normally enter anterior parts of the limb, initially grow into posterior limb regions, but then take novel paths to innervate the correct muscles. So, even when the axon bundles enter in reverse order, the correct relationship between motor neurons and muscles is achieved. Motor axons can clearly find their appropriate muscles when their displacement relative to the limb, is small. But with a large displacement, this is not the case. A complete dorso-ventral inversion of the limb bud results in the axons failing to find their appropriate targets. In such cases, the axons follow paths normally taken by other neurons.

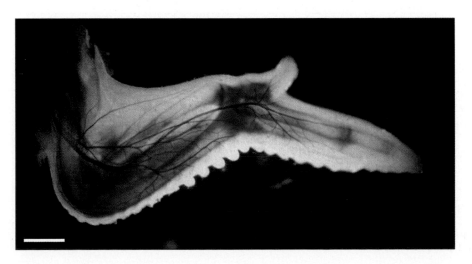

Fig. 11.19 Innervation of the embryonic chick wing. The nerves are stained brown and include both sensory and motor neurons. Scale bar = 1 mm. Photograph courtesy of J. Lewis.

These results show that the mesoderm of the limb provides local cues for the motor axons and that the axons themselves have an identity that allows them to choose the correct pathway. It is necessary to consider how developing neurons extend their axons before understanding how they explore their surroundings and choose a pathway.

11-9   The growth cone controls the path taken by the growing axon.

An early event in the differentiation of a neuron is the extension of its axon by the **growth cone** at the axon tip (Fig. 11.20). The growth cone both moves and senses its environment. As in other cells capable of migration, for example, the primary mesenchyme of the sea urchin embryo (see Section 8-14), the growth cone can continually extend and retract filopodia, which help to pull the axon tip forward over the underlying substratum. Between the filopodia, the edge of the growth cone forms thin ruffles—lamellipodia—similar to those on a moving fibroblast (see Box 8B, page 244). Indeed, in its ultrastructure and mechanism of movement, the growth cone closely resembles the leading edge of a fibroblast crawling over a surface. Unlike a moving fibroblast, however, the extending axon also grows in size, with an accompanying increase in the total surface area of the neuron's plasma membrane. This is achieved by the continual incorporation of additional membrane into the growth cone plasma membrane by fusion of intracellular vesicles.

The activity of the growth cone guides axon outgrowth, and is influenced by the contacts the filopodia make with other cells and with the extracellular matrix. In general, the growth cone moves in the direction where its filopodia make the most stable contacts. In addition, diffusible substances can bind to receptors on the growth cone surface, and so influence its direction of migration. Some of these cues promote axon extension, whereas others inhibit it by causing growth cone collapse.

Axon growth cones are guided by these two main types of cue—attractive and repulsive. In addition, the cues can act either at long range or at short range, thus giving four ways in which the growth cone can be guided (Fig. 11.21). Long-range attraction involves diffusible chemoattractants released from the target cells and is, in principle, similar to the chemotaxis of motile cells, such as those of slime molds (see Section 8-16). Both chemoattractants and chemorepellents have been identified in the developing nervous system and are discussed later. Short-range guidance is mediated by contact-dependent mechanisms involving molecules bound to other cells or to the extracellular matrix; again, such interactions can be either attractive or repulsive.

Fig. 11.20 A developing axon and growth cone. The axon growing out from a neuron's cell body ends in a motile structure called the growth cone. Many filopodia are continually extended and retracted from the growth cone to explore the surrounding environment. Scale bar = 10 μm. Photograph courtesy of P. Gordon-Weeks.

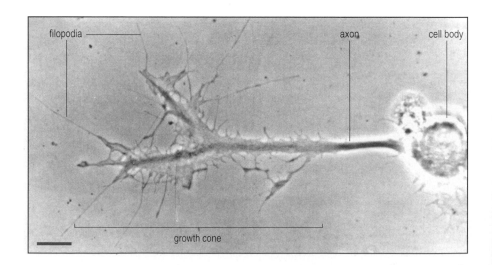

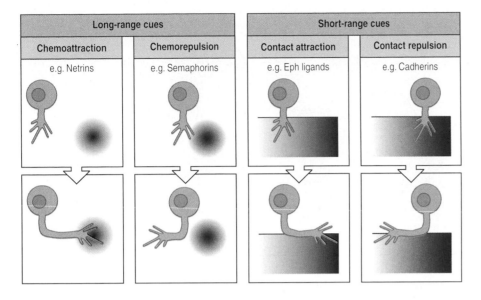

Fig. 11.21 Axon guidance mechanisms. Four types of mechanism can contribute to controlling the direction in which a growth cone moves. Attractant molecules can either be present as a diffusible gradient of chemoattractant, or be bound to the substratum, leading to contact-dependent attraction. Similarly, molecules that repel axon growth can be either bound or diffusible.

Two families of cell adhesion molecules are involved in contact-dependent guidance—the cadherins, and members of the immunoglobulin super-family of proteins (see Box 8A, page 233). In addition, a variety of receptor tyrosine kinases in the axon plasma membrane can regulate axon outgrowth in response to extracellular signals. In the next sections, we consider examples of growth cone guidance during development.

11-10   Choice of axon pathway depends on environmental cues and neuronal identity.

One of the ways in which axons reach a distant target is to make use of 'stepping stones' or 'guide posts' along the pathway. These are localized regions to which the growth cone moves when its filopodia make contact. Arrival at the ultimate target is achieved through a sequence of stages, characterized by arrival at the consecutive guide posts. For example, early in the development of the grasshopper, sensory neurons arise in the epithelium at the distal tip of each developing leg. (The grasshopper does not have a separate larval stage.) These neurons extend axons under the epithelium, along a well-defined route within the leg, to make connections with the central nervous system. Because the behavior of the growth cones can be directly observed, it is known that the pathway they follow, although stereotyped, is not absolutely fixed, and they make frequent small corrections to their course.

The numerous fine filopodia of each growth cone continually explore the surrounding substratum. They migrate in a proximal direction until the growth cone makes contact with the first of three guidepost cells. These cells, which will themselves become neurons, provide a high-affinity substratum for the migrating growth cone. Growth cones in the vicinity of the guidepost cells often turn sharply toward them as the fine filopodia make contact (Fig. 11.22). The axon then continues along its proximal path until it meets the second guidepost cell. At this site, the growth cone extends branches in both dorsal and ventral directions. Eventually, the dorsal branch is withdrawn, and the axon extends ventrally to make contact with the third guidepost cell, after which it moves toward the central nervous system (Fig. 11.23).

If the first two guidepost cells are removed, one observes increased neuronal branching and slower axon movement, but the axon eventually reaches its destination. This indicates the presence of guidance cues in

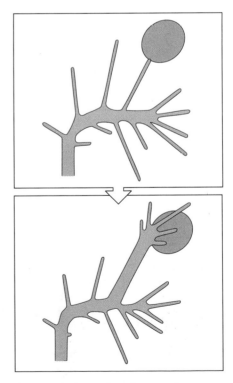

Fig. 11.22 Behavior of a growth cone near a guidepost cell. Contact of a single filopodium with a guidepost cell (red) results in further extension of the growth cone toward it. After O'Connor, T.P., *et al.*: 1990.

Fig. 11.23 Outgrowth and guidance of a peripheral sensory neuron in the developing grasshopper leg. The cell body of the sensory neuron lies beneath the epithelium at the tip of the developing leg. The axon tip migrates over the inner surface of the epithelium covering the limb. The axon extends until it meets a guidepost cell (red), and then continues to extend to make contact with two more guidepost cells in succession. Eventually, it makes contact with the central nervous system (CNS).

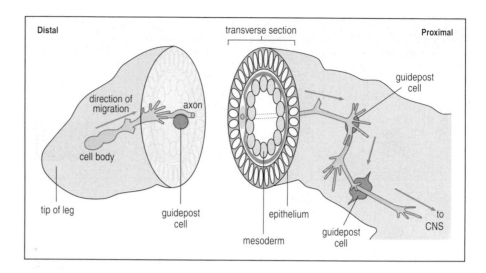

the epithelium over which the growth cones migrate. Local cues of this type are thought to provide the basis of axonal outgrowth in a wide variety of systems, including the vertebrate retino-tectal system, which provided much of the early evidence that suggested the presence of local cues.

11-11   Neurons from the retina make ordered connections on the tectum to form a retino-tectal map.

The vertebrate visual system has been the subject of intense investigation for many years, and the highly organized projection of neurons from the eye to the brain is one of the best models we have to show how specific neural connections are made. A characteristic feature of the vertebrate brain is the presence of topographic maps. That is, the neurons from one region of the nervous system project in an ordered manner to one region of the brain, so that nearest-neighbor relations are maintained. The most studied topographical projection is that of the optic nerve from the retina into the brain.

The retina develops light-sensitive cells that indirectly activate neurons —the retinal ganglion cells—whose axons are bundled together to form the optic nerve, which connects the retina and a region of the brain. This region is called the **optic tectum** in amphibians and birds, and the **lateral geniculate nucleus** in mammals. In amphibians, the optic nerve from the right eye makes connections with the left optic tectum, while the nerve from the left eye makes connections to the right side of the brain. The optic nerve from each eye is made up of thousands of axons, which connect with the optic tectum in a highly ordered manner: there is a point-to-point correspondence between a position on the retina and one on the tectum. Neurons in the dorsal region of the retina project to the ventral region of the tectum, and regions in the anterior (nasal) region of the retina project to the posterior region of the tectum (Fig. 11.24). The development of the projection from the retina to the tectum is initially only reasonably precise, but is fine-tuned later by nerve impulses from the retina.

Remarkably, in some lower vertebrates, such as fish and amphibians, the pattern of connections can be re-established with precision when the optic nerve is cut. The ends of the axons distal to the cut die, new growth cones form, and axon outgrowth reforms connections to the tectum. In frogs, even if the eye is inverted through 180° the axons still

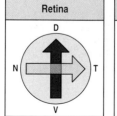
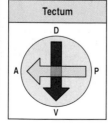

Fig. 11.24 The retina maps onto the tectum. Dorsal retinal neurons connect to ventral tectum and temporal retinal neurons to anterior tectum.

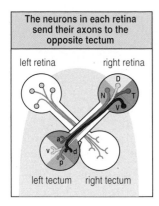

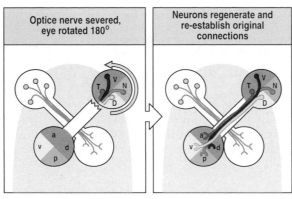

| The neurons in each retina send their axons to the opposite tectum | Optice nerve severed, eye rotated 180° | Neurons regenerate and re-establish original connections |

Fig. 11.25 Retino-tectal connections in amphibians are re-established in the original arrangement after severance of the optic nerve and rotation of the eye. Left panel: neurons in the optic nerve from the left eye mainly connect to right optic tectum and *vice versa*. There is a point-to-point correspondence between neurons from different regions in the retina (nasal (N), temporal (T), dorsal (D), ventral (V)) and their connections in the tectum (posterior (p), anterior (a), ventral (v), dorsal (d), respectively). Center and right panels: if one optic nerve of a frog is cut and the eye rotated dorso-ventrally through 180°, the severed ends of the axons degenerate. When the neurons regenerate, they make connections with their original sites of contact in the tectum.

find their way back to their original sites of contact (Fig. 11.25). However, the animals subsequently behave as if their visual world has been turned upside down: if a visual stimulus, such as a fly is presented above the inverted eye, the frog moves its head downward instead of upward (Fig. 11.26), and can never learn to correct this error.

On the basis of such experiments, it was suggested that each retinal neuron carries a chemical label that enables it to connect reliably with an appropriately chemically labeled cell in the tectum. This is known as the chemoaffinity hypothesis of connectivity. The chemical labels probably do not provide unique lock and key interactions. Rather, it is thought that graded spatial distributions of a relatively small number of factors on the tectum provide positional information, which can be detected by the retinal axons. The spatially graded expression of another set of factors on the retinal axons would provide them with their own positional information. The development of the retino-tectal projection could thus, in principle, result from the interaction between these two gradients.

Such gradients have indeed been found in the developing visual system of the chick embryo. An axon-guiding activity based on repulsion has been detected along the antero-posterior axis of the chick tectum. Normally, the temporal (posterior) half of the retina projects to the anterior part of the tectum and the nasal (anterior) retina projects to the posterior tectum. When offered a choice between growing on posterior

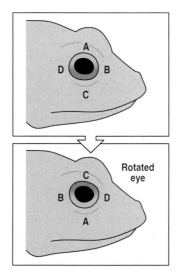

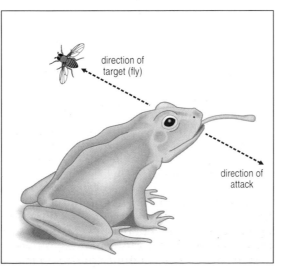

Fig.11.26 The effect of eye rotation on the visual behavior of a frog. After severance of the optic nerve and rotation of the eye, the retina re-establishes its original connections with the optic tectum (see Fig. 11.25). However, because the eye has been rotated, the image falling on the tectum is upside down compared with normal. When the frog sees a fly above its head, it thinks the fly is below it, and moves its head downward to try to catch it.

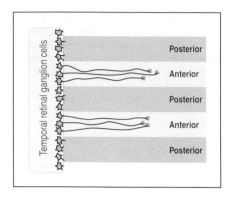

Fig. 11.27 Choice of targets by retinal axons. If pieces of temporal retina are placed next to a 'carpet' of alternating stripes (90 μm wide) of anterior and posterior tectal membranes, the temporal retinal cells only extend axons onto anterior tectal membranes.

or anterior tectal cells, temporal axons from an explanted chick retina show a preference for anterior tectal cells (Fig. 11.27). This choice is mediated by repulsion of the axons—as shown by the collapse of their growth cones—by a factor located on the surface of posterior tectal cells.

Complementary gradients of cell-surface molecules that could be involved in this repulsion of axons have been found in chick tectum and retina. ELF-1 is a cell-surface protein that binds to the receptor protein Mek-4, which is a member of the large Eph family of receptor tyrosine kinases. ELF-1 is expressed in an antero-posterior gradient along the tectum, whereas Mek-4 is expressed in a gradient on the surface of retinal neurons, with the high point at the temporal end of the retina. The high end of the retinal gradient of Mek-4 maps to the low end of the tectal gradient of ELF-1 (Fig. 11.28). Both in culture and *in vivo*, ELF-1 exerts a strong repulsive effect on temporal retinal axons.

11-12   Axons can be guided by gradients of diffusible agents.

An additional mechanism that could enable axons to find their target is directed migration based on chemotaxis. The target cells would set up a gradient of a diffusible substance that could either attract or repel growth cones. We now look at two examples where such a mechanism may be operating.

During the development of the spinal cord, a well-defined pattern of cell differentiation is established along the dorso-ventral axis (see Section 11-6). Commissural neurons develop in the dorsal region (see Fig. 11.11), and send their axons ventrally along the lateral margin of the cord. When the axons reach about halfway down the cord, they make an abrupt turn and project to the floor plate in the ventral midline, bypassing the motor neurons (Fig. 11.29, left panel).

Evidence for chemotaxis toward a diffusible chemoattractant produced by the floor plate comes from experiments *in vitro*. When rat dorsal neural tube containing only commissural cells is cultured, there is hardly any outgrowth of the axons, but when it is cultured with a floor plate explant, there is extensive outgrowth of axons toward the floor plate tissue. If the floor plate explant is placed to the side of the dorsal neural tube explant so that it is parallel to its dorso-ventral axis, most of the axons that emerge from the dorsal explant are oriented toward the floor plate (see Fig. 11.29, right panel). The fact that only the axons close to the floor plate tissue grow toward it is good evidence for the presence of a diffusible chemoattractant.

A diffusible factor responsible for this chemotaxis in the spinal cord is netrin-1, a protein that was originally isolated from embryonic brain and whose mRNA is present in floor plate cells. Cells transfected with mRNA for netrin-1 can mimic the long-range chemoattractant effect of floor plate cells. Similar proteins control neuron migration toward the midline in the nematode, and axon outgrowth toward the midline in *Drosophila*. Knocking-out mouse *netrin-1*, or of one of its receptors, results in abnormal commissural axon pathways (Fig. 11.30).

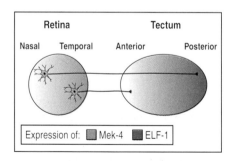

Fig. 11.28 Complementary expression of the proteins ELF-1 and Mek-4 in the chick embryo retino-tectal projection. Mek-4 is a receptor tyrosine kinase and ELF-1 is its ligand. Temporal retinal neurons, which express high levels of Mek-4 protein on their surface, are repulsed by the posterior tectum, where high levels of ELF-1 are also present, but can make connections with the anterior tectum, where ELF-1 is scarce or absent. Nasal neurons from the retina make the best contacts in the posterior tectum, as they express low levels of Mek-4 and adhere to high levels of ELF-1 on the tectum.

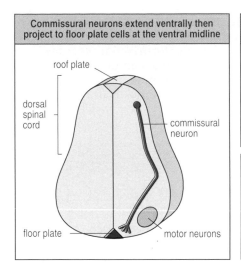

Commissural neurons extend ventrally then project to floor plate cells at the ventral midline

roof plate

dorsal spinal cord

commissural neuron

floor plate

motor neurons

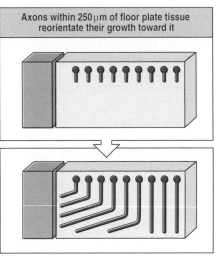

Axons within 250 μm of floor plate tissue reorientate their growth toward it

Fig. 11.29 Chemotaxis can guide commissural axons in the spinal cord. In the chick, spinal cord commissural neurons extend ventrally, and then toward the floor plate cells (left panel). In explants of rat spinal cord cultured together with floor plate tissue, axons that are within 250 μm of the floor plate tissue grow toward it. After Tessier-Lavigne, M., *et al.*: 1991.

The second example of axons probably being guided by a diffusible substance, in this case a chemorepellent rather than a chemoattractant, occurs after the commissural axons have reached the floor plate in the vertebrate spinal cord. The axons then not only cross the midline, but turn sharply upward. Growing evidence suggests that a family of proteins present in the spinal cord and known as the semaphorins are involved in this process. Semaphorins, which can repel growth cones (see Fig. 11.21), are secreted in the ventral region of the developing spinal cord, and may prevent certain sensory axons that come from the dorsal region from entering and making synapses in the ventral region, thus accounting for their sharp turn upward.

Summary.

Growth cones at the tip of the extending axon guide it to its destination. Filopodial activity at the growth cone is influenced by environmental factors, such as contact with the substratum and with other cells, and can also be guided by chemotaxis. In the development of motor neurons that innervate vertebrate limb muscles, the growth cones guide the axons so as to make the correct muscle-specific connections, even when their normal site of entry into the limb is disturbed. The sensory neurons in the limb of the grasshopper are guided both by the substratum and by

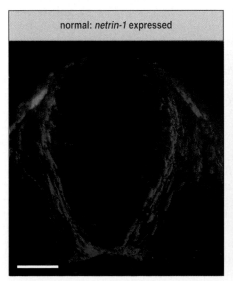

normal: *netrin-1* expressed

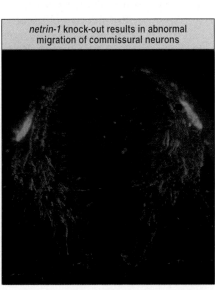

netrin-1 knock-out results in abnormal migration of commissural neurons

Fig. 11.30 Effect of *netrin-1* gene knock-out in mice. In mice lacking *netrin-1*, the commissural axons do not migrate toward the floor plate. Scale bar = 0.1 mm. Photographs courtesy of M. Tessier-Lavigne, from Serafini, T., *et al.*: 1996.

special guidepost cells along the migration pathway. Neuronal guidance of axons from the amphibian retina to make the correct connections with the optic tectum involves gradients in cell-surface molecules both on the tectal neurons and on the retinal axons that can promote or repel growth cone approach. Gradients in diffusible molecules are probably responsible for the directional growth of commissural axons in the spinal cord.

Summary: axonal guidance

axonal outgrowth led by growth cone

⇩

attractive and repulsive cues are both short range and long range

↙   ↘

retinal axons connect with tectal   commissural axons in spinal cord
neurons to make retino-tectal map   guided by chemotaxis—gradient in netrin-1

⇩

connections guided by gradients;
ELF-1 on tectum and Mek-4 on retinal cells

Neuronal survival, synapse formation, and refinement.

When axons reach their targets they make specialized connections —**synapses**—which are essential for signaling between the neurons and their target cells. Formation of synapses in the correct pattern is a basic requirement for any developing nervous system. The connections may be made with other nerve cells, with muscles, and also with certain glandular tissues. Here, we focus mainly on the development and stabilization of synapses at the junctions between nerve cells and muscle cells in vertebrates, the **neuromuscular junctions**.

Setting up the organization of a complex nervous system in vertebrates seems to involve refining an initially rather imprecise organization by extensive programmed cell death (see Section 9-17). The establishment of a connection between a neuron and its target appears to be essential not only for the functioning of the nervous system but for the very survival of many neurons. Neuronal death is common in the developing vertebrate nervous system; it appears that too many neurons are produced initially, and only those that make appropriate connections survive. Survival depends on the neuron receiving neurotrophic factors, such as nerve growth factor, which are produced by the target tissue and for which neurons compete.

A special feature of nervous system development is that fine-tuning of synaptic connections depends on the interaction of the organism with its environment and the consequent neuronal activity. This is particularly true of the vertebrate visual system, where sensory input from the retina in a period immediately after birth modifies synaptic connections so that the animal can perceive fine detail. Again, this refinement seems to involve competition for neurotrophic factors. We return to this topic after next considering the survival of motor neurons that innervate the developing vertebrate limb.

11-13 | Many motor neurons die during limb innervation.

Some 20,000 motor neurons are formed in the segment of spinal cord that provides innervation to a chick leg, but about half of them die soon after they are formed (Fig. 11.31). Cell death occurs after the axons have grown out from the cell bodies and entered the limb, at about the time as the axon terminals are reaching their potential targets—the skeletal muscles of the limb. The role of the target muscles in preventing cell death is suggested by two experiments. If the leg bud is removed, the number of motor neurons surviving sharply decreases. When an additional limb bud is grafted at the same level as the leg, providing additional targets for the axons, the number of surviving motor neurons increases.

Survival of a motor neuron may depend on its establishing a functional synapse with a muscle cell. Once a neuromuscular junction is established, the neuron can activate the muscle, and this is followed by the death of a proportion of the other motor neurons that are approaching the muscle cell. Muscle activation by neurons can be blocked by the drug curare, which prevents neuromuscular transmission, and this blocking results in a large increase in the number of motor neurons that survive. A possible explanation is that in the absence of activation, the muscle produces a trophic factor in amounts sufficient to enable many neurons to survive. Once a neuromuscular junction is established and the muscle is activated, production of the trophic factor may be reduced. Neurons that have already established a connection to the muscle cell will survive, whereas neurons that have not yet reached the cell will die.

Even after neuromuscular connections have been made, some of them are subsequently eliminated. At early stages of development, single muscle fibers are innervated by axon terminals from several different motor neurons. With time, most of these connections are eliminated, until each muscle fiber is innervated by the axon terminals from just one motor neuron. The continued survival of a neuron that has established a connection must therefore be ensured by additional means. It is now clear that this later elimination process is also dependent on neural activity.

The well-established matching of the number of motor neurons to the number of appropriate targets by these mechanisms suggests that a general mechanism for the development of nervous system connectivity in all parts of the vertebrate nervous system is that excess neurons are generated and only those that make the required connections are selected for survival. This mechanism is well suited to regulating cell numbers by matching the size of the neuronal population to its targets. We now look in more detail at the neurotrophic factors that promote neuronal survival, and then consider the role of neural activity in the elimination of neuromuscular synapses.

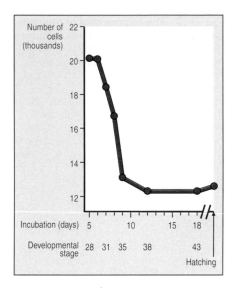

Fig. 11.31 Motor neuron death is a normal part of development in the chick spinal cord. The number of motor neurons innervating a limb decreases by about half before hatching, as a result of programmed cell death during development. Most of the neurons die over a period of 4 days.

11-14 | Neuronal survival depends on competition for neurotrophic factors.

The first factor required for neuronal survival to be identified was nerve growth factor (NGF). The discovery of NGF started with the serendipitous observation that a mouse tumor implanted into a chick embryo evoked extensive growth of nerve fibers toward the tumor. This suggested that the tumor was producing a factor that promotes axonal outgrowth of neurons. The factor was eventually identified as a protein, NGF, using axon outgrowth in culture as an assay. NGF is necessary for the survival of a number of types of neurons, particularly those of the sensory and sympathetic nervous systems.

Fig. 11.32 The requirement of trigeminal ganglion cells for neurotrophins changes during axon outgrowth. The trigeminal nerve innervates different parts of the face. It has motor neurons which control chewing, and sensory neurons which control various muscles in the face. As the axons approach their targets, they require BDNF, NT-3, and NT-4/5. Later, once connections have been established, they require NGF. After Davies, A.: 1994.

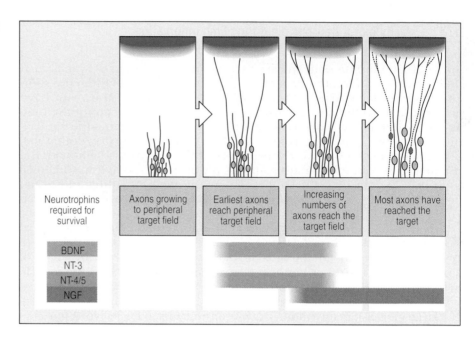

NGF is a member of a family of proteins known as the **neurotrophins**. As well as NGF, these include brain-derived neurotrophic factor (BDNF), neurotrophin-3 (NT-3) and neurotrophin-4/5 (NT-4/5). The receptors for the neurotrophins are receptor tyrosine kinases known as Trk proteins; one of the receptors for NGF, for example, is called TrkA. Different types of neuron require different neurotrophins for their survival, and the requirement for certain neurotrophins also changes during development (Fig. 11.32). Knock-out of genes coding for neurotrophins or their receptors in mice has, in general, confirmed their roles as deduced from *in vitro* culture. In mice lacking TrkA, for example, there is a loss of both sympathetic and sensory neurons.

Mice that lack the Trk receptors for NT-3 and NT-4/5, which may be required for motor neuron survival, show a greater than normal loss of motor neurons. Another neurotrophin, glial cell-line derived neurotrophic factor (GDNF), prevents the death of facial motor neurons, and is also expressed in developing limb buds. There is increasing evidence that individual neurotrophins act on a variety of neural cell types.

11-15 Reciprocal interactions between nerve and muscle are involved in formation of the neuromuscular junction.

The mature neuromuscular junction is a complex structure involving extensive modification of the nerve ending and the muscle cell membrane. Just before the junction, the axon terminal branches into a network-like arrangement. Each branch ends in a swelling, which is in contact with a special endplate region on the muscle fiber. The axon's plasma membrane is separated from the muscle cell's plasma membrane by a narrow cleft (the synaptic cleft) filled with extracellular material (the basal lamina), which is secreted by both the nerve and the muscle cell. The whole structure, comprising the axon terminal plasma membrane, the opposing muscle cell plasma membrane, and the cleft between them, is called a neuromuscular junction or synapse (Fig. 11.33).

Electrical signals cannot pass across the synaptic cleft, and for neuron to muscle signaling, the electrical impulse propagated down the axon is converted at the axon terminal into a chemical signal—the release of a

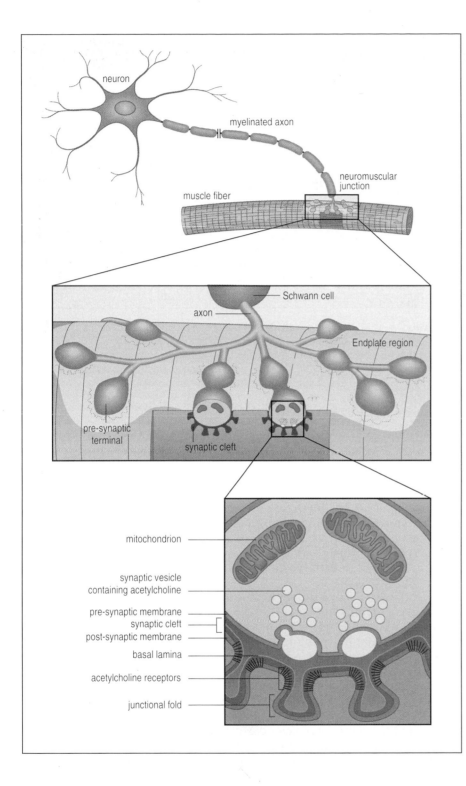

Fig. 11.33 Structure of the vertebrate neuromuscular junction. The motor axon, which is covered in a myelin sheath, innervates a muscle fiber. At the neuro-muscular junction, the axon branches in the endplate region, and makes synaptic connections with the muscle membrane. Communication between nerve and muscle is by release of the neurotrans-mitter acetylcholine from synaptic vesicles into the synaptic cleft. The acetylcholine diffuses across the cleft and binds to acetylcholine receptors on the muscle cell membrane. After Kandell, E.R., *et al.*: 1991.

neurotransmitter into the synaptic cleft from the synaptic vesicles in the terminal. Molecules of the neurotransmitter diffuse across the cleft and interact with receptors on the muscle cell membrane, causing the muscle fiber to contract. The neurotransmitter used by motor neurons making connections with skeletal muscles is the small molecule acetylcholine. Because the signal travels in the direction of nerve to muscle, the axon terminal is called the pre-synaptic part of the junction and the muscle cell is called the post-synaptic partner (see Fig. 11.33).

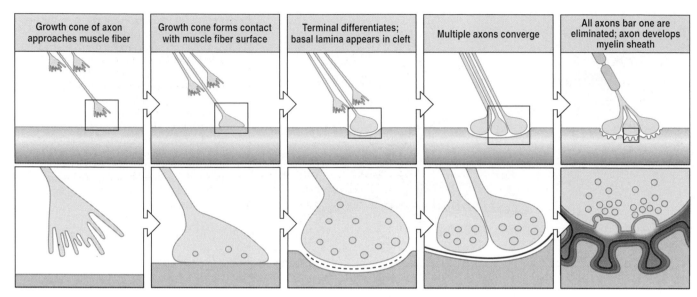

Fig. 11.34 Development of the neuromuscular junction. The boxed areas in the top panels are shown at a higher magnification in the bottom panels. From left to right: a growth cone approaches a muscle fiber and forms an unspecialized, but functional, contact on its surface. The terminal then differentiates and basal lamina (red line) appears in the widened synaptic cleft. As the basal lamina appears extrasynaptically, multiple axons converge on the synaptic site. Finally, all axons but one are eliminated. The surviving axon terminal expands and forms a mature neuromuscular junction, with the axon developing a myelin sheath.

Development of a neuromuscular junction is progressive—in the rat it takes about 3 weeks. How does a single axon terminal become selected and develop, and how do acetylcholine receptors become localized in the right place on the post-synaptic muscle cell membrane opposite the terminal? The axon terminals that make contact with the muscle cell are initially unspecialized, but they soon begin to accumulate synaptic vesicles. Initially, several synapses from different axons are made on the same immature muscle cell (or myotube) but, with time, all but one are eliminated. Synaptic transmission begins shortly after a contact is made. The development of a neuromuscular junction is shown in Fig. 11.34.

The permanent establishment of a neuromuscular junction relies on an exchange of signals between the axon terminal and the muscle cell. A key event in establishment is the aggregation of acetylcholine receptors at the post-synaptic membrane. Initially, acetylcholine receptors are present throughout the plasma membrane of the immature muscle fiber, but soon after axon contact they begin to accumulate at the site of contact. A key signal for clustering of the receptors is provided by the protein agrin (Fig. 11.35). Agrin is secreted by motor neurons at the pre-synaptic terminal and induces membrane specialization. It probably acts by activating a tyrosine kinase receptor named Musk on the muscle cell. Knock-out of the genes for either agrin or Musk in mice results in the absence of functional neuromuscular junctions; there is little or no clustering of acetylcholine receptors and consequently no muscle activity.

The innervation of muscle causes not only local clustering of acetylcholine receptors but also a localized increase in their synthesis. This is due to an increased rate of acetylcholine receptor gene transcription in those muscle cells that lie directly beneath the post-synaptic membrane (Fig. 11.36). Stimulation of receptor gene transcription is thought to be caused by proteins called neuregulins or ARIAs released from the axon terminal. Receptor synthesis elsewhere in the muscle is reduced in response to the enhanced electrical activity in the muscle triggered through the junction.

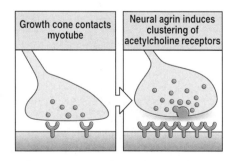

Fig. 11.35 Clustering of acetylcholine receptors during development of the neuromuscular junction. When a growth cone contacts a muscle fiber it releases the protein agrin. This causes acetylcholine receptor clustering in the muscle cell membrane, and localized deposition of basal lamina material.

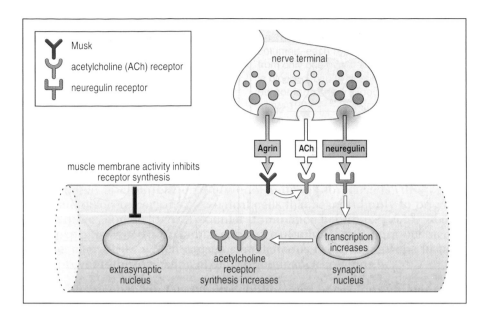

Fig. 11.36 Control of acetylcholine receptor localization and synthesis in a muscle cell membrane by the pre-synaptic axon. Beneath the neuro-muscular junction, acetylcholine receptors cluster because of agrin interacting with the Musk receptor. Release of neuregulin from the axon stimulates transcription of the genes for the acetylcholine receptor proteins in the nuclei nearest to the junction and production of acetylcholine receptors. Away from this region, electrical activity in the muscle cell membrane reduces synthesis of the receptor.

Neural activity also plays a key role in refining the pattern of neuromuscular junctions. Almost all mammalian muscle fibers are innervated by two or more motor axons during the embryonic period. After birth, branches of motor neurons are withdrawn, so that each muscle fiber is finally innervated by a single motor neuron (Fig. 11.37). This change in connectivity is due to competition between synapses. It seems that when a motor neuron stimulates a muscle fiber, the activity between other neurons and the muscle fiber is suppressed, and these synapses are eventually eliminated. As we now see, a similar mechanism seems to be involved in refining synaptic connections between neurons.

11-16 | The map from eye to brain is refined by neural activity.

We have already seen how axons from the amphibian retina make connections with the tectum so that a retino-tectal map is established. This map is initially rather coarse grained, in that axons from neighboring cells in the retina make contacts over a large area of the tectum. This area is much larger early on than at later stages, when the retino-tectal map is more finely tuned. This fine tuning of the map results, as in muscle, from the withdrawal of axon terminals from most of the initial contacts, and requires neural activity. This requirement is seen particularly clearly in the development of visual connections in mammals.

The mammalian visual system is more complex than that of lower vertebrates (Fig. 11.38). Axons from the retina first connect to the lateral geniculate nucleus, onto which they map in an ordered manner. Input from one half of each eye goes to the opposite side of the brain, whereas input from the other half of each eye goes to the same side of the brain. Neurons from the lateral geniculate nucleus then send axons to the visual cortex. When there is a visual stimulus, the inputs from the retinal axons activate neurons in the lateral geniculate nucleus, which then activate neurons in the corresponding region of the visual cortex. There is thus input from both eyes at the same location in the cortex. The adult visual cortex consists of six cell layers, but we need only focus here on layer 4, which is where many axons from the lateral geniculate make connections.

In the lateral geniculate nucleus, as in the visual cortex, the neurons are arranged in layers. Each layer receives input from retinal axons from either the right or the left eye, but not from both. Thus, from the beginning,

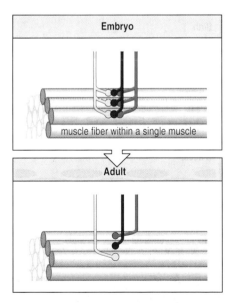

Fig. 11.37 Refinement of muscle innervation by neural activity. Initially, several motor neurons innervate the same muscle fiber. Elimination of synapses means that each fiber is eventually innervated by only one neuron. After Goodman, C.S., *et al.*: 1993.

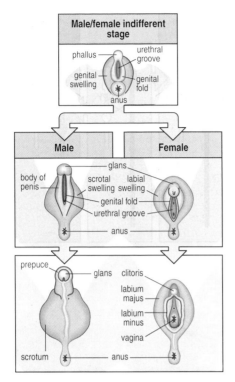

Fig. 12.4 Development of the genitalia in humans. At an early embryonic stage, the genitalia are the same in males and females (top panel). After testis formation in males, the phallus and the genital fold give rise to the penis, whereas in females they give rise to the clitoris and the labia minus. The genital swelling forms the scrotum in males and the labia majus in females.

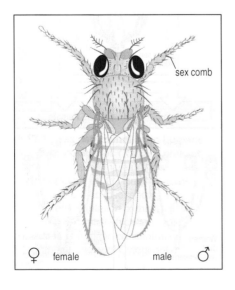

in the absence of the testes, the Müllerian ducts develop into the **oviducts** (Fallopian tubes), which transport eggs from the ovaries to the uterus, while the Wolffian ducts degenerate. In males, the cells of the testes secrete a Müllerian-inhibiting substance, which specifically causes the degeneration and resorption of the Müllerian ducts. The Wolffian duct becomes the vas deferens which carries sperm to the penis.

The main secondary sexual characters that distinguish males and females are the reduced size of mammary glands in males, and the development of a penis and a scrotum in males instead of the clitoris and labia of females (Fig. 12.4). At early stages of embryonic development the genital region of males and females is indistinguishable. Differences only arise after gonad development, as a result of the action of gonadal hormones. For example, in humans, the phallus gives rise to the clitoris in females and the end of the penis in males.

The role of gonadal hormones in sexual development is illustrated by rare cases of abnormal sexual development. Certain XY males develop as phenotypic females in external appearance, even though they have testes and secrete testosterone. They have a mutation that renders them insensitive to testosterone because they lack the testosterone receptor, which is present throughout the body. Conversely, genetic females with a completely normal XX constitution can develop as phenotypic males in external appearance if they are exposed to male hormones during their embryonic development.

Sex-specific behavior is also affected by the hormonal environment because of the effects of hormones on the brain. For example, male rats castrated after birth develop the sexual behavioral characteristics of genetic females.

12-3   In *Drosophila*, the primary sex-determining signal is the number of X chromosomes and is cell autonomous.

The sexual differences between *Drosophila* males and females are mainly in the genital structures, although there are also some differences in bristle patterns and pigmentation, and male flies have a sex comb on the first pair of legs. In flies, sex determination of the somatic cells is cell autonomous, and is therefore specified on a cell-by-cell basis; there is no process resembling the control of somatic cell sexual differentiation by hormones. The pathway of somatic sexual development is the result of a series of gene interactions that are initiated by the primary sex signal, which act on a binary genetic switch. The end result of this cascade is the expression of just a few effector genes, whose activity controls male or female differentiation of the somatic cells.

Like mammals, fruit flies have two unequally sized sex chromosomes, X and Y, and males are XY and females XX. But these similarities are misleading. In flies, sex is not determined by the presence of a Y chromosome, but by the number of X chromosomes. Thus, XXY flies are female and X flies are male. The chromosomal composition of each somatic cell determines its sexual development. This is beautifully illustrated by the creation of genetic mosaics in which the left side of the animal is XX and the right side X: the two halves develop as female and male, respectively (Fig. 12.5).

Fig. 12.5 A *Drosophila* female/male genetic mosaic. The left side of the fly is composed of XX cells and develops as a female, whereas the right side is composed of X cells and develops as a male. The male fly has smaller wings, a special structure, the sex comb, on its first pair of legs, and different genitalia at the end of the abdomen (not shown).

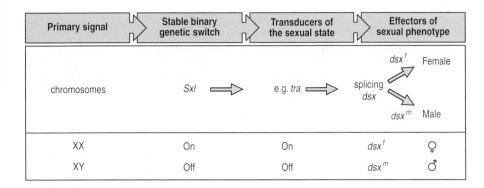

Primary signal	Stable binary genetic switch	Transducers of the sexual state	Effectors of sexual phenotype	
chromosomes	Sxl ⟹	e.g. tra ⟹	splicing dsx	dsx^f Female dsx^m Male
XX	On	On	dsx^f	♀
XY	Off	Off	dsx^m	♂

Fig. 12.6 Outline of the sex determination pathway in *Drosophila*. The number of X chromosomes is the primary sex-determining signal and in females, the presence of two X chromosomes activates the gene *Sex-lethal* (*Sxl*). This produces Sex-lethal protein, whereas no Sex-lethal protein is made in males (who only have one X chromosome). The activity of *Sex-lethal* is transduced via the *transformer* gene (*tra*) and causes sex-specific splicing of *double sex* (*dsx*) RNA, such that the cells follow a female developmental pathway. In the absence of Sex-lethal protein, the splicing of *double sex* RNA leads to male development.

In flies, the presence of two X chromosomes results in the production of the protein Sex-lethal, coded for by the gene *Sex-lethal*, which is located on the X chromosome. This leads to female development via further gene activation involving transducers of the sexual state and finally effectors of the sexual phenotype (Fig. 12.6). The end of the sex determination pathway is the gene *double sex*; this encodes a transcription factor, which ultimately determines the somatic sex. The *double sex* gene is active in both males and females, but different protein products are produced in the two sexes as a result of sex-specific RNA splicing. Males and females thus express similar but distinct double sex proteins, which act in somatic cells to induce expression of sex-specific genes, as well as to repress characteristics of the opposite sex. Sex-specific *double sex* RNA splicing is controlled by the gene *transformer*, which is in turn controlled by the *Sex-lethal* gene. In the presence of Sex-lethal protein, *transformer* RNA is productively spliced and this, together with the transformer-2 protein, leads to the female form of the double-sex protein being made, resulting in female differentiation. Production of the male form of the protein is the default pathway.

Sex-lethal is only turned on in females with two X chromosomes. In the absence of early *Sex-lethal* expression, male development occurs. Once *Sex-lethal* is activated in females, it remains activated through an autoregulatory mechanism. Early expression of *Sex-lethal* in females occurs through activation of a promoter P_e (e stands for establishment), at about the time of syncytial blastoderm formation. At this stage, Sex-lethal protein is synthesized and accumulates in the blastoderm of female embryos. At the cellular blastoderm stage, another promoter for *Sex-lethal*, P_m (m stands for maintenance) becomes active in both males and females, and the P_e promoter is shut off, but the sex is already determined. The *Sex-lethal* RNA transcribed can only be spliced into an mRNA for the Sex-lethal protein if some Sex-lethal protein is already present. Only females contain any Sex-lethal protein, and so only in females is the mRNA productively spliced and more Sex-lethal protein synthesized (Fig. 12.7). This autoregulatory loop at the post-transcriptional level results in Sex-lethal protein being synthesized throughout female development.

How does the number of X chromosomes control these key sex-determining genes? The mechanism in *Drosophila* involves interactions between the products of so-called numerator genes on the X chromosome and those of genes on the autosomes, as well as maternally specified factors. In females, the twofold higher level of numerator protein activates the *Sex-lethal* gene by binding to sites in the P_e promoter. The mechanism is essentially the same as that involved in activating genes such as *hunchback* and *even-skipped* in early development (see Section 5-14).

Fig. 12.7 Production of Sex-lethal protein in *Drosophila* sex determination. When two X chromosomes are present, the early establishment promoter (P_e) of the *Sex-lethal* (*Sxl*) gene is activated at the syncytial blastoderm stage in future females, but not in males. This results in the production of Sxl protein. Later, at the blastoderm stage, the maintenance promoter (P_m) of *Sxl* becomes active in both females and males, and P_e is turned off. The *Sxl* RNA is only correctly spliced if Sxl protein is already present, which is only in females. A positive feedback loop for Sxl protein production is thus established in females. The continued presence of Sxl protein initiates a cascade of gene activity leading to female development. If no Sxl protein is present, male development ensues. After Cline, T.W.: 1993.

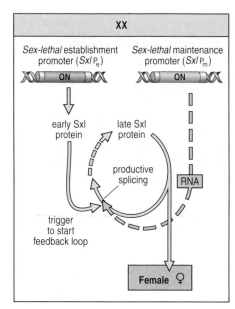
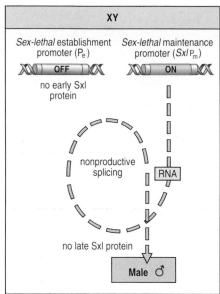

A special feature of *Drosophila* is that there seems to be a mechanism determining sexual behavior, which is not dependent on *double sex*. This involves the *fruitless* gene, whose activity is necessary for male sexual behavior. The *transformer* gene may act directly on *fruitless*, turning it off.

12-4   Somatic sexual development in *Caenorhabditis* is determined by the number of X chromosomes.

In the nematode *C. elegans*, the two sexes are hermaphrodite (essentially a modified female) and male (Fig. 12.8), although in other nematodes they are male and female. Hermaphrodites produce a limited amount of sperm early in development, with the remainder of the germ cells developing into oocytes. Sex in *C. elegans* (and other nematodes) is determined by the number of X chromosomes: the hermaphrodite (XX) has two X chromosomes, whereas the presence of just one X chromosome leads to development as a male (XO). The primary sex signal in *C. elegans* acts on the gene *XO lethal* (*xol-1*). With two X chromosomes, *xol-1* expression is low, resulting in the development of a hermaphrodite.

A cascade of gene activity converts the level of *xol-1* expression into the somatic sexual phenotype (Fig. 12.9). Genes involved include those encoding nuclear proteins, such as *sdc-1*, and secreted proteins, such as

Fig. 12.8 Hermaphrodite and male *C. elegans*. The hermaphrodite has a 'two-armed' gonad and makes both eggs and sperm. The eggs are fertilized internally. The male makes sperm only.

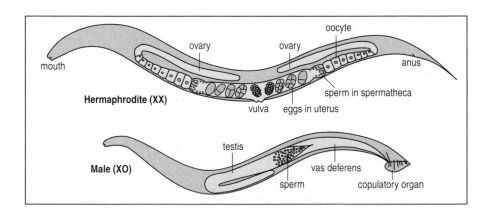

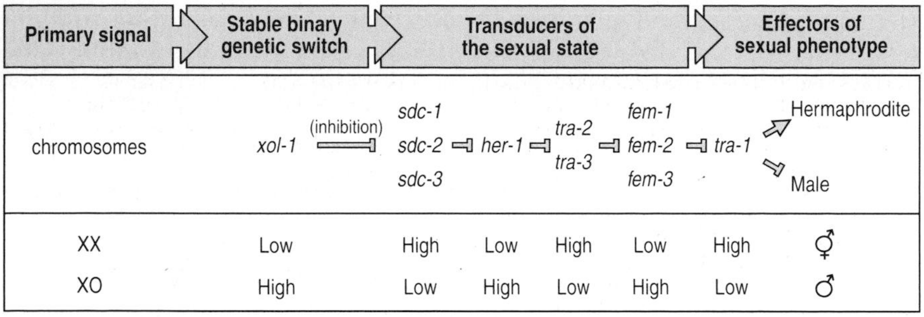

Primary signal	Stable binary genetic switch		Transducers of the sexual state			Effectors of sexual phenotype
chromosomes	xol-1 $\xrightarrow{\text{(inhibition)}}$	sdc-1 sdc-2 ⊣ her-1 ⊣ sdc-3	tra-2 tra-3	⊣ fem-1 fem-2 ⊣ tra-1 fem-3		Hermaphrodite ↗ ↘ Male
XX	Low	High	Low	High	Low	High ♀
XO	High	Low	High	Low	High	Low ♂

Fig. 12.9 Outline of the somatic sex determination pathway in *C. elegans*. The primary signal for sex determination is set by the number of X chromosomes. When two X chromosomes are present, the expression of the gene *XO lethal* (*xol-1*) is low, leading to hermaphrodite development, whereas *xol-1* is expressed at a high level in males. There is a cascade of gene expression starting from *xol-1* that leads to the gene *transformer-1* (*tra-1*), which codes for a transcription factor. If *tra-1* is active, development as a hermaphrodite occurs, but if it is expressed at low levels, males develop. The product of the *hermaphrodite-1* (*her-1*) gene is a secreted protein, which probably binds to a receptor encoded by *transformer-2* (*tra-2*), inhibiting its function.

hermaphrodite-1 (*her-1*). At the end of the cascade is the gene *transformer-1* (*tra-1*), which encodes a transcription factor. Expression of transformer-1 protein is both necessary and sufficient to direct all aspects of hermaphrodite (XX) somatic cell development. A gain-of-function mutation in *tra-1* leads to hermaphrodite development in an XO animal, irrespective of the state of any of the regulatory genes that normally control its activity. Mutations that inactivate *tra-1* lead to complete masculinization of XX hermaphrodites. Unlike in *Drosophila*, sex determination in *C. elegans* requires cell–cell interactions, as the process involves secreted proteins.

Before considering how the sex of the germ cells is determined and how the imbalance of X-linked genes between the sexes is dealt with in animals, we need to digress briefly to touch upon sex determination in flowering plants.

12-5 | **Most flowering plants are hermaphrodites, but some produce unisexual flowers.**

As we have seen, the flowers of angiosperms share a common organization in which the four types of floral organs are arranged in concentric whorls (see Section 7-11). The two inner whorls are the sexual organs —the stamens and carpels. Sepals and petals in the outer whorls are not sexual organs, but may serve to attract pollinators. Stamens produce pollen, which contains the male gametes corresponding to the sperm of animals. The female reproductive structures of flowers are the carpels, which are either free, or are fused to form a compound ovary. Carpels are the site of ovule formation, and each ovule produces an egg cell. Most flowering plants are hermaphrodites, bearing flowers with functional male and female sexual organs. However, not all flowering plants are of this type.

In about 10% of flowering plants, flowers of just one sex are produced. Flowers of different sexes may occur on the same plant, or may be confined to different plants. The development of male or female flowers usually involves the selective resorption of either the stamens or pistil after they have been specified and have started to grow.

In maize, male and female flowers develop at particular sites on the shoot. The tassel at the tip of the main stem (see Fig. 7.12) only bears flowers with stamens; the ears of corn which are at the end of the lateral branches bear female flowers containing pistils. Sex determination becomes visible when the flower is still small, with the stamen primordia being larger in males and the pistil longer in females. The smaller organs eventually degenerate.

The plant hormone gibberellic acid may be involved in sex determination, as differences in gibberellin concentration are associated with the different sexual organs. In the maize tassel, gibberellin concentration is 100-fold lower than in the developing ears. If the concentration of gibberellic acid is increased in the tassel, pistils can develop.

12-6   Germ cell sex determination may depend both on cell signals and genetic constitution.

The determination of the sex of animal germ cells, that is, whether they will develop into eggs or sperm, makes use of different mechanisms from those used for somatic cell sexual development. In the mouse, the primordial germ cells continue to proliferate for a few days after entering the genital ridge, which is where the gonads form. At this stage, male and female germ cells are indistinguishable. Their future development is determined largely by the sex of the gonad in which they reside, and not by their own chromosomal constitution. In the female embryo, diploid germ cells enter prophase of the first meiotic division, and then become arrested until the mouse becomes a sexually mature female. In male embryos, diploid germ cells divide mitotically for some time in the embryo, but then also stop dividing, becoming arrested in the G_1 phase of the cell cycle (Fig. 12.10). They start dividing again after birth and enter meiosis as the mouse attains sexual maturity.

All mouse germ cells that enter meiosis before birth develop as eggs, whereas those not entering meiosis until after birth develop as sperm. Germ cells, whether XX or XY, that fail to enter the genital ridge and end up in adjacent tissues, such as the embryonic adrenal gland or mesonephros, enter meiosis and begin developing as oocytes in both male and female embryos. In XX/XY chimeras, XX germ cells that are surrounded by testis cells develop along the spermatogenesis pathway. However, the later development of germ cells that develop in these inappropriate sites is abnormal. There is no reproductive future for XY germ cells in the ovaries or for XX germ cells in the testes.

In *Drosophila*, the difference in the behavior of XY and XX germ cells depends initially on the number of X chromosomes, as in somatic cells; the *Sex-lethal* gene again plays an important role, although other elements in the sex determination pathway may differ. As in mammals, both chromosomal constitution and cell interactions are involved in the development of germ cell sexual phenotype. Transplantation of genetically marked pole cells (see Section 2-5) into an embryo of the opposite sex shows that male XY germ cells in a female XX embryo become integrated into the ovary and begin to develop as sperm; that is, their behavior is autonomous with respect to their genetic constitution. By contrast, XX germ cells in a testis develop as sperm, showing a role for environmental signals. In neither case, however, are functional sperm produced.

The hermaphrodite of *C. elegans* provides a particularly interesting example of germ cell differentiation, as both sperm and eggs develop within the same gonad. Unlike the somatic cells, which have a fixed lineage and number (see Section 6-1), the number of germ cells in an adult nematode is indeterminate, with about 1000 germ cells in each 'arm' of the gonad. At hatching of the first stage larva, there are just two founder germ cells, which proliferate to produce the germ cells. The germ cells are flanked by distal tip cells on each side, and their proliferation is controlled by a signal from these distal tip cells. The distal tip signal is the lag-2 protein, which is homologous to the Delta protein of *Drosophila* (see Section 11-2). The receptor for lag-2 protein on the germ cells is probably the glp-1 protein, which is similar both to the nematode lin-12 protein (see Section 10-22) and *Drosophila* Notch protein.

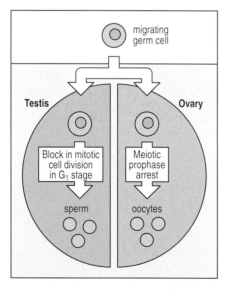

Fig. 12.10 Environmental signals specify germ cell sex in mammals. Migrating germ cells, whether XX or XY, enter meiotic prophase and start developing as oocytes unless they enter a testis. In the testis, the germ cells receive an inhibitory signal which blocks mitotic cell division and prevents them entering meiotic prophase.

In *C. elegans*, entry of germ cells into meiosis from the third larval stage onward is controlled by the distal tip signal. In the presence of this signal, the cells proliferate, but as they move away from it, they enter meiosis and develop as sperm (Fig. 12.11, top). In the hermaphrodite gonad, all the cells that are initially outside the range of the distal tip signal develop as sperm, but cells that later leave the proliferative zone and enter meiosis develop as oocytes (see Fig. 12.11, bottom). The eggs are fertilized by stored sperm as they pass into the uterus. The male gonad has similar proliferative meiotic regions, but all the germ cells develop as sperm.

Sex determination of the nematode germ cells is somewhat similar to that of the somatic cells, in that the chromosomal complement is the primary sex-determining factor and many of the same genes are involved in the subsequent cascades of gene expression. The terminal regulator genes required for spermatogenesis are called *fem* and *fog*. In hermaphrodites, there must be a mechanism for activating the *fem* genes in some of the XX germ cells, so allowing them to develop as sperm.

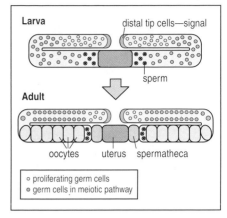

Fig. 12.11 Determination of germ cell sex in the hermaphrodite nematode gonad. Top: during the larval stage, germ cells in a zone close to the distal tips of the gonad multiply; when they leave this zone in the larval stage they enter meiosis and develop into sperm. Bottom: in the adult, cells that leave the proliferative zone develop into oocytes. The eggs are fertilized as they pass into the uterus. After Clifford, R., *et al.*: 1994.

12-7 | Various strategies are used for dosage compensation of X-linked genes.

In each of the animals whose sex determination we have looked at there is an imbalance of X-linked genes between the sexes. One sex has two X chromosomes, whereas the other has one. This imbalance has to be corrected to ensure that the level of expression of genes carried on the X chromosome is the same in both sexes. The mechanism by which the imbalance in X chromosomes is dealt with is known as **dosage compensation**. Failure to correct the imbalance leads to abnormalities and arrested development. Different animals deal with the problem of dosage compensation in different ways (Fig. 12.12).

Mammals, such as mice and humans, achieve dosage compensation by inactivating one of the X chromosomes in females after the blastocyst

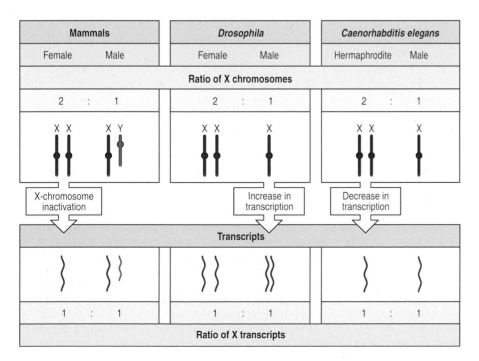

Fig. 12.12 Mechanisms of dosage compensation. In mammals, *Drosophila*, and *C. elegans* there are two X chromosomes in one sex and and only one in the other. Mammals inactivate one of the X chromosomes in females; in *Drosophila* males there is an increase in transcription from the single X chromosome; and in *C. elegans* there is a decrease in transcription from the X chromosomes in hermaphrodites. The result of these different dosage compensation mechanisms is that the level of X chromosome transcripts is approximately the same in males and females.

has implanted in the uterine wall. Once X inactivation is initiated in female embryos, it is maintained in all somatic cells throughout life. The inactive X chromosome can be identified in the nucleus as a Barr body (see Section 9-5) and X inactivation provides an important model for the inheritance of gene expression. Whatever the sex chromosome constitution, for example, XY, XX, XXY, or XXXY, there is only one active X chromosome per somatic cell, with all the other X chromosomes inactivated. During early cleavage stages, both X chromosomes are active and the first X inactivation occurs in the extra-embryonic tissues when the paternal X is exclusively inactivated. Later, at gastrulation, random X inactivation occurs in the cells of the embryo. As we see below, the inactive X becomes reactivated during germ cell development.

X inactivation is dependent on a small region of the X chromosome, the inactivating center. It contains a proposed switch gene, *Xist*, which produces a non-coding RNA. Inactivation seems to be caused by the chromosome being coated by *Xist* RNA, which is expressed by the inactive X but not by an active X chromosome. If the *Xist* gene is introduced into another chromosome, that chromosome becomes silenced. This inactivation is correlated with methylation of the DNA (see Section 9-5).

Dosage compensation in *Drosophila* works in the opposite way to that in the mouse; instead of repression of the 'extra' X activity in females, transcription of the X chromosome in males is increased nearly twofold, and there is also an increase in the translation of mRNA derived from the X chromosome. This increased activity is regulated by the primary sex-determining signal, which results in the dosage compensation mechanism operating when the *Sex-lethal* gene is 'off'. In females, where *Sex-lethal* is 'on', a cascade of gene activity turns off the dosage compensation mechanism.

In *C. elegans*, dosage compensation is achieved by reducing the level of X chromosome expression in XX animals to that of the single X chromosome in XO males. This involves a cascade of gene interactions set off by the primary sex-determining signal. A key gene that acts to reduce expression of both the X chromosomes in hermaphrodites is *dpy-27*. The protein product of this gene, dpy-27, although present in both XX and XO nuclei, becomes associated specifically with the X chromosomes in XX animals and reduces transcription, possibly by causing condensation of the chromatin. The sdc-3 protein, which is also involved in somatic sex determination, directs the dpy-27 (and related proteins) to the X chromosome.

Summary.

The development of early embryos of both sexes is very similar. A primary sex-determining signal sets off development toward one or the other sex, and in mammals, *Drosophila*, and *C. elegans*, this signal is determined by the chromosomal complement of the fertilized egg. In mammals, the *Sry* gene on the Y chromosome is responsible for the embryonic gonad developing into a testis and producing hormones that determine male sexual characteristics. The sexual phenotype of the somatic cells is determined by the gonadal hormones. In *C. elegans* and *Drosophila*, the primary sex-determining signal is the number of X chromosomes. In *Drosophila*, the gene *Sex-lethal* is turned on in females but not in males, in response to this signal. In both cases, this results in further gene activity in which sex-specific RNA splicing is involved. In *C. elegans*, the gene *XO lethal* is turned off in hermaphrodites and on in males, eventually leading to sex-specific expression of the gene *transformer-1*, which determines the sexual phenotype. Somatic sexual differentiation in *Drosophila* is cell-autonomous and is controlled by the number of

X chromosomes; in *C. elegans*, cell–cell interactions are also involved. Most flowering plants are hermaphrodites, having flowers with both male and female organs.

In mammals, signals from the gonads determine whether the germ cells develop into oocytes or sperm. Male germ cells in *Drosophila* develop along the sperm pathway even in an ovary, but female germ cells develop along the sperm pathway when placed in a testis. Most *C. elegans* adults are hermaphrodites, and produce both sperm and eggs from the same gonad.

Various strategies of dosage compensation are used to correct the imbalance of X chromosomes between males and females. In female mammals one of the X chromosomes is inactivated; in *Drosophila* males the activity of the single X chromosome is upregulated; whereas in *C. elegans* the activity of the single X chromosome is upregulated; whereas in *C. elegans* the activity of the X chromosome in XX animals is downregulated.

Summary: determination of sexual phenotype

Mammals

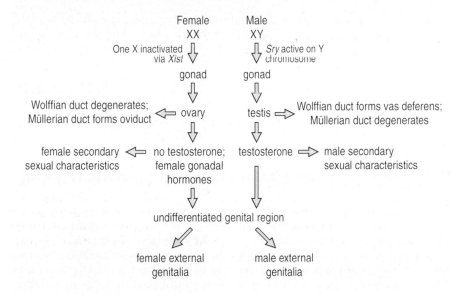

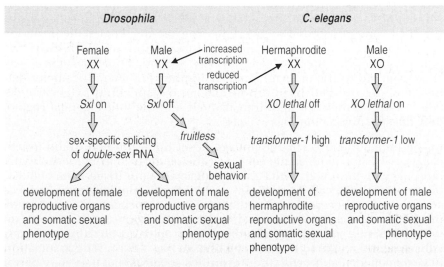

The development of germ cells.

In all but the simplest animals, the cells of the germ line are the only cells that can give rise to a new organism. Unlike the somatic cells, which eventually all die, they can survive for the lifetime of the body that produced them. They are therefore very special cells. The outcome of germ cell development is either a male gamete (the sperm in animals), or a female gamete (the egg). The egg is a particularly remarkable cell, as it is the cell from which all the cells in an organism are derived. In species whose embryos receive no nutrition from the mother after fertilization, the egg must also provide everything necessary for development, as the sperm contributes virtually nothing to the organism, other than its chromosomes.

Germ cells are specified very early in some animals, although not in mammals, by cytoplasmic determinants present in the egg. We therefore start our discussion of germ cell development by looking at the specification of primordial germ cells by special cytoplasm—the **germ plasm**. The localization of this special cytoplasm in the insect egg during oogenesis has been particularly well studied, and many of the genes involved are known.

In plants, there is no early specification of germ cells, as a single somatic plant cell can give rise to a sexually mature plant. The germ cells are specified late in development, during the development of the flower from the floral meristem (see Section 7-11).

In animals, once the primordial germ cells are specified, they migrate into the gonads, somatic structures that usually develop some distance away from the site of germ cell origin. Once within the gonads, the germ cells differentiate as either male or female gametes.

12-8   Germ cell fate can be specified by a distinct germ plasm in the egg.

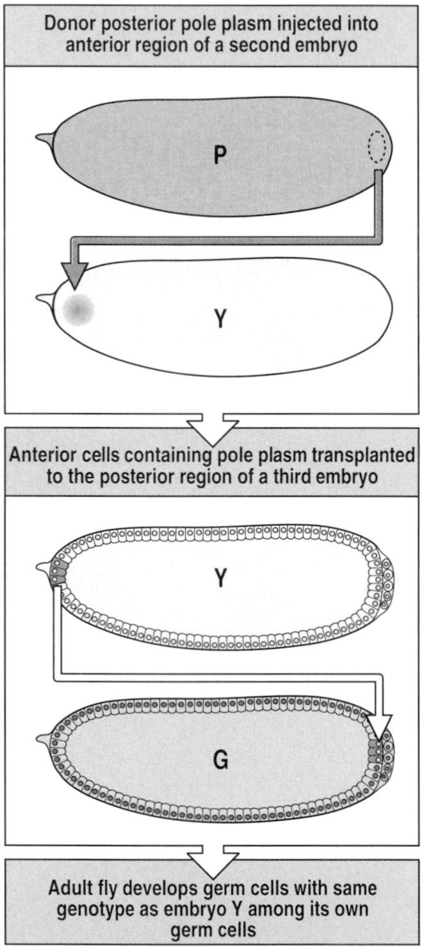

In flies, nematodes, and frogs, molecules localized in special egg cytoplasm are involved in specifying the germ cells. The clearest example of this is provided by *Drosophila*, where primordial germ cells become distinct at the posterior pole of the egg about 90 minutes after fertilization, several hours before cellularization of the rest of the embryo (see Section 2-5). The cytoplasm at the posterior pole is distinguished by large organelles, the polar granules, which contain both proteins and RNAs. That there is something special about this posterior cytoplasm—the so-called **pole plasm**—is demonstrated by two key experiments. First, if the posterior end of the egg is subjected to ultraviolet irradiation, which destroys the pole plasm activity, no germ cells develop. Second, if pole plasm of an egg is transferred to the anterior pole of another embryo, the nuclei that become surrounded by the pole plasm are specified as germ cells (Fig. 12.13). If they are then transplanted into the future genital region, they develop as functional germ cells.

In the nematode, a germ cell lineage is set up at the end of the fourth cleavage division, with all the germ cells being derived from the P_4 blastomere (see Section 6-2). The P_4 cell is derived from three stem-cell like divisions of the P_1 cell. At each of these divisions, one daughter produces somatic cells whereas the other divides again to produce a somatic cell progenitor and a P cell. The egg contains P granules in its cytoplasm which become asymmetrically distributed after fertilization, and are subsequently confined to the P cell lineage (Fig. 12.14). The association of germ cell formation with the P granules suggests that they may play a

Donor posterior pole plasm injected into anterior region of a second embryo

P

Y

Anterior cells containing pole plasm transplanted to the posterior region of a third embryo

Y

G

Adult fly develops germ cells with same genotype as embryo Y among its own germ cells

Fig. 12.13 Transplanted pole plasm can induce germ cell formation. Pole plasm from a fertilized egg of genotype P (pink) is transferred to the anterior end of an early cleavage stage embryo of genotype Y (yellow). After cellularization, cells containing pole plasm induced at the anterior end of embryo Y are transferred to the posterior end (a site from which germ cells can migrate into the gonad) of another embryo, of genotype G (green). The adult fly that develops from embryo G contains germ cells of genotype Y as well as those of G.

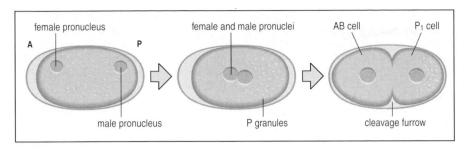

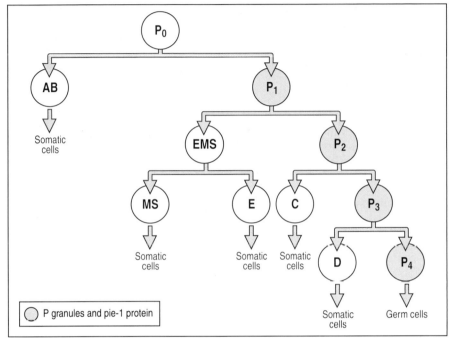

Fig. 12.14 P granules and pie-1 protein become asymmetrically distributed to germ-line cells during cleavage of the nematode egg. Before fertilization, P granules are distributed throughout the egg. After fertilization, P granules become localized at the posterior end of the egg. At the first cleavage, they are only included in the P_1 cell (top panel), and thus become confined to the P cell lineage. The pie-1 protein is only present in P cells. All germ cells are derived from P_4, which is formed at the fourth cleavage.

key role in germ cell specification in *C. elegans*, but this is not yet confirmed. The gene *pie-1* is involved in preserving the stem cell property of the P blastomeres. It encodes a nuclear protein that is expressed maternally; the pie-1 protein is only present in the germ-line blastomeres, and eventually disappears.

There is also evidence for germ plasm in *Xenopus* eggs. After fertilization, distinct yolk-free patches of cytoplasm aggregate at the yolky vegetal pole. When the blastomeres at the vegetal pole cleave, this cytoplasm is distributed asymmetrically so that it is retained only in the most vegetal daughter cells, from which the germ cells are derived. Ultraviolet irradiation of the vegetal cytoplasm abolishes the formation of germ cells, and transplantation of fresh vegetal cytoplasm into an irradiated egg restores germ cell formation. At gastrulation, the germ plasm is located in cells in the floor of the blastocoel cavity, among the cells that give rise to the endoderm. Cells containing the germ plasm are, however, not yet determined as germ cells and can contribute to all three germ layers if transplanted to other sites. At the end of gastrulation, the primordial germ cells are determined, and migrate out of the presumptive endoderm and into the genital ridge, as described below.

There is no evidence for germ plasm being involved in germ cell formation in the mouse or other mammals. Germ cell specification in the mouse involves cell–cell interactions, as cultured embryonic stem cells can, when injected into the inner cell mass, give rise to both germ cells and somatic cells (see Box 3C, page 81). In the mouse, germ cells first become distinguishable from somatic cells in the posterior primitive streak. As in *Xenopus*, they migrate to the genital ridge, which is where the gonads form.

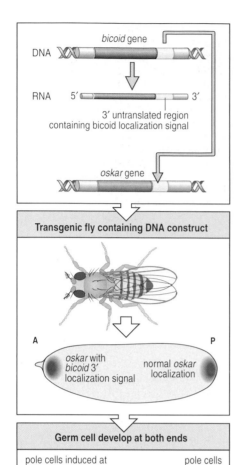

Fig. 12.15 The gene *oskar* is involved in specifying the germ plasm in *Drosophila*. In normal eggs, *oskar* mRNA is localized at the posterior end of the embryo, whereas *bicoid* mRNA is at the anterior end. The localization signals for both *bicoid* and *oskar* mRNA are in their 3' untranslated regions. By manipulating the *Drosophila* DNA, the localization signal of *oskar* can be replaced by that of *bicoid* (top panel). A transgenic fly is made containing the modified DNA. In its egg, *oskar* becomes localized at the anterior end (middle panel). The egg therefore has *oskar* mRNA at both ends, and germ cells develop at both ends of the embryo, as shown in the photograph (bottom panel). Thus, *oskar* alone is sufficient to initiate the specification of germ cells. Photograph courtesy of R. Lehmann.

12-9   Pole plasm becomes localized at the posterior end of the *Drosophila* egg.

In Chapter 5, we saw how the main axes of the *Drosophila* egg are specified by the follicle cells in the ovary, and how the mRNA for proteins such as bicoid and nanos become localized in the egg (see Section 5-7). The pole plasm also becomes localized at the posterior end of the egg under the influence of the follicle cells.

Several maternal genes are involved in pole plasm formation in *Drosophila*. Mutations in any of at least eight genes result in the affected heterozygous individual being 'grandchildless'. Its homozygous offspring lack a proper pole plasm, and so although they may otherwise develop normally, they lack germ cells and are therefore sterile. One of these eight genes is *oskar*, which plays a central role in the organization and assembly of the pole plasm; of the genes involved in pole plasm formation, *oskar* is the only one to have its mRNA localized at the posterior pole. The signal for localization is contained in the 3' untranslated region of the mRNA, which interacts with the microtubule apparatus that directs anterior or posterior localization in the egg (see Section 5-7). The region of the *oskar* gene that codes for the 3' localization signal can be replaced by the *bicoid* 3' localization signal. This DNA construct is then used to make a transgenic fly, and the eggs of this fly have *oskar* mRNA localized both in the anterior and posterior ends of the egg (Fig. 12.15).

12-10   Germ cells migrate from their site of origin to the gonad.

In many animals, germ cells develop at some distance from the gonads, and only later migrate to them, where they differentiate into eggs or sperm. This separation of site of origin from the final destination seems to be a mechanism to exclude germ cells from the general developmental upheaval involved in laying down the body plan.

The vertebrate gonad develops from the mesoderm lining the abdominal cavity, which is known as the **genital ridge**. The primordial germ cells migrate there from distant sites. In *Xenopus*, the primordial germ cells originate in the endoderm (which forms the gut) and migrate to the future gonad along a cell sheet that joins the gut to the genital ridge. Only a small number of cells start this journey, dividing about three times before arrival, so that about 30 germ cells colonize the gonad. The number of primordial germ cells that arrive at the genital ridge of the mouse, by a very similar pathway (Fig. 12.16), is about 2500. In chick embryos, the pattern of migration is different: the germ cells originate at the head end of the embryo, and most arrive at their destination via the blood vessels, leaving the bloodstream at the hindgut and then migrating along the epithelial sheet. In *Drosophila*, prospective germ cells are carried inside

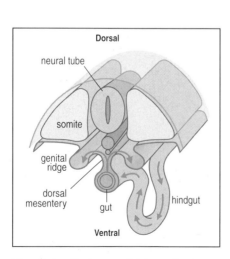

Fig. 12.16 Pathway of primordial germ cell migration in the mouse embryo. In the final stage of migration, the cells move from the gut tube into the genital ridge, via the dorsal mesentery. After Wylie, C.C., *et al*.: 1993.

the embryo during gastrulation, and they then migrate through the gut to the site where the gonads develop.

How migration and proliferation of the germ cells in vertebrates is controlled is not known, but factors produced by the genital ridge are thought to be responsible. In the mouse, two genes that are involved in controlling proliferation of migrating germ cells are *White spotting* (*W*) and *Steel* that we have already met in relation to melanocyte differentiation (see Section 9-16). Mutations that inactivate either of these genes cause a decrease in germ cell numbers. *White spotting* codes for the cell-surface receptor Kit, which is expressed in the migrating germ cells. Its ligand, the Steel protein, is expressed in the cells along which the germ cells migrate. The requirement for these two genes indicates that the migrating cells are continuously receiving signals from the surrounding tissue. Once they arrive at the gonad, the germ cells begin to differentiate into sperm or eggs, as described in the next section.

12-11   Germ cell differentiation involves a reduction in chromosome number.

Germ cells have to reduce their chromosome number by half during gamete formation, so that at fertilization the diploid chromosome number is reinstated exactly. Primordial germ cells are diploid, and reduction from the diploid to the haploid state occurs during meiosis (Fig. 12.17). Meiosis comprises two cell divisions, in which the chromosomes are replicated during the first division, but not during the second division, so that their number is reduced by half.

Development of eggs and sperm follow different courses, even though they both involve meiosis. The development of the egg is known as **oogenesis**. The main stages of mammalian oogenesis are shown in Fig. 12.18 (left panel). In mammals, germ cells undergo a small number of proliferative mitotic cell divisions as they migrate to the gonad. In the case of developing eggs, the diploid oogonia continue to divide mitotically for a short time in the ovary. After entry into meiosis, the primary oocytes become arrested in the prophase of the first meiotic division. They never proliferate again; thus, the number of oocytes at this embryonic stage is the total number of eggs the female mammal ever has. In humans, most of these oocytes degenerate before puberty, leaving about 40,000 out of an original 6 million. In mammals, and in many other vertebrates, oocyte development is held in suspension after birth for months (mice) or years (humans), until the female becomes sexually mature at puberty. After puberty, the oocytes undergo further development, growing in size up to 1000-fold, and meiosis resumes. In mammals, meiosis proceeds as far as the metaphase of the second meiotic division, where it becomes arrested again, and is completed after fertilization.

Fig. 12.17 Meiosis produces haploid cells. Meiosis reduces the number of chromosomes from the diploid to the haploid number. Only one pair of homologous chromosomes is shown here for simplicity. Before the first meiotic division, the DNA replicates, so that each chromosome entering meiosis is composed of two identical chromatids. The paired homologous chromosomes (known as a bivalent) undergo crossing over and recombination, and align on the meiotic spindle at the metaphase of the first meiotic division. The homologous chromosomes separate, and each is segregated into a different daughter cell at the first cell division. There is no DNA replication before the second meiotic division. The daughter chromatids of each chromosome separate and segregate at the second cell division. The chromosome number of the resulting daughter cells is thus halved.

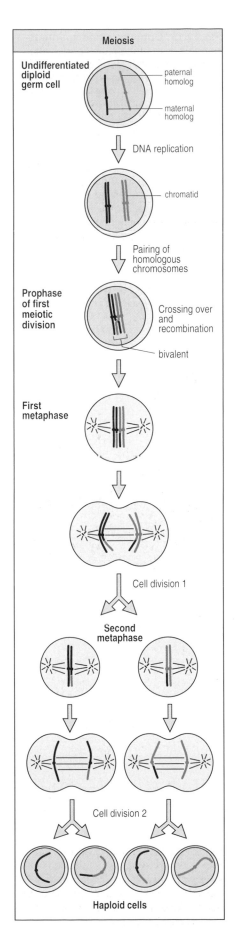

Meiosis

Undifferentiated diploid germ cell
— paternal homolog
— maternal homolog

DNA replication
— chromatid

Pairing of homologous chromosomes

Prophase of first meiotic division
— Crossing over and recombination
— bivalent

First metaphase

Cell division 1

Second metaphase

Cell division 2

Haploid cells

Fig. 12.18 Oogenesis and spermatogenesis in mammals. Left panel: after the germ cells that form oocytes enter the embryonic ovary, they divide mitotically a few times and then enter the prophase of the first meiotic division. No further cell multiplication occurs, but the oocyte increases 100-fold in mass. Further development of the oocyte occurs in the sexually mature adult female. This includes the formation of external cell coats, and the development of a layer of cortical granules located under the oocyte plasma membrane. Eggs continue to mature in the ovary under hormonal influences, but become blocked in the second metaphase of meiosis, which is only completed after fertilization. Polar bodies are formed at meiosis (see Box 2A, page 27). Right panel: germ cells that develop into sperm enter the embryonic testis and become arrested at the G₁ stage of the cell cycle. After birth, they begin to divide mitotically again, forming a population of stem cells (spermatogonia). These give off cells that then undergo meiosis and differentiate into sperm. Sperm can therefore be produced indefinitely.

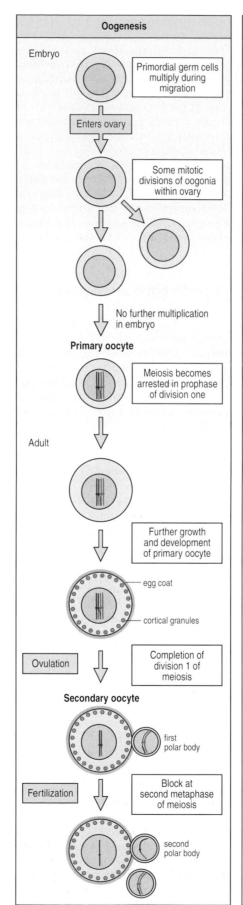

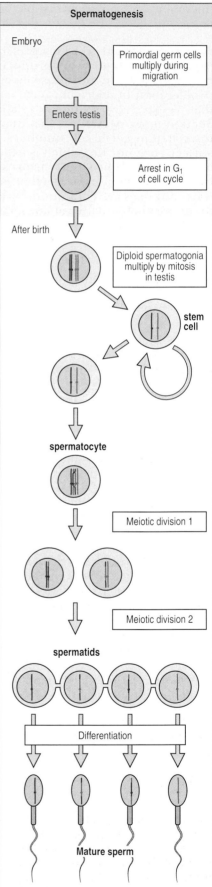

The strategy for **spermatogenesis**—the production of sperm—is different. Diploid germ cells that give rise to sperm do not enter meiosis in the embryo, but become arrested at an early stage of the mitotic cell cycle in the embryonic testis. They resume mitotic proliferation after birth. Later, in the sexually mature animal, the cells, now termed spermatogonia, provide a population of stem cells that give rise to diploid spermatocytes. These undergo meiosis, each forming four haploid spermatids that mature into sperm (see Fig. 12.18, right panel). Thus, unlike the fixed number of oocytes in female mammals, sperm continue to be produced throughout life.

In *Drosophila*, there is a continuous production of both eggs and sperm from a population of stem cells. Oogenesis begins with the division of a stem cell (see Fig. 5.10), and so there is no intrinsic limitation to the number of eggs that a female fly can produce. *Drosophila* oogenesis, and the localization of cytoplasmic determinants, have already been discussed in Section 5-7.

12-12 Oocyte development can involve gene amplification and contributions from other cells.

Eggs vary enormously in size among different animals, but they are always larger than the somatic cells. A typical mammalian egg is about 0.1 mm in diameter, a frog's egg about 1 mm, and a hen's egg about 5 cm (see Fig. 3.1). To achieve these large sizes—and even the mammalian egg has a mass 100-fold greater than that of a typical mammalian somatic cell—a variety of mechanisms have evolved, some organisms using several of them together. One strategy is to increase the overall number of gene copies in the developing oocyte, as this proportionately increases the amount of mRNA that can be transcribed, and thus the amount of protein that can be synthesized. This strategy is generally adopted by vertebrate oocytes, which are arrested in the prophase of the first meiotic division and thus have double the normal diploid number of genes. Transcription and oocyte growth continue while the oocyte is in the arrested stage. Another strategy, adopted by insects and amphibians, is to produce many extra copies of those genes whose products are needed in large quantities in the egg. Thus, the ribosomal RNA genes of amphibians are amplified during oocyte development from hundreds to millions. In insects, the genes that encode the proteins of the egg membrane, the chorion, become amplified in the surrounding follicle cells.

A different strategy is for the oocyte to rely on the synthetic activities of other cells. In insects, the nurse cells adjacent to the oocyte, which are its sister cells, make and deliver many mRNAs and proteins to the oocyte (see Fig. 5.11). Yolk proteins in birds and amphibians are made by liver cells, and are carried by the blood to the ovary, where the proteins enter the oocyte by endocytosis. The proteins become packaged in yolk platelets. From an early stage, the oocyte is polarized, and the yolk platelets accumulate at the vegetal pole.

12-13 Genes controlling embryonic growth are imprinted.

Like all other cells in an animal, mature germ cells have completed a program of differentiation. Yet, at the end of that process they must have a genome that, after fertilization, is capable of controlling the development of the embryo; of all the cells in the body, they are the only ones with genomes that are passed on to future generations. Their genomes therefore have to revert to a state from which all the cells of the organism can be derived, so there must be no permanent alterations in their genetic constitution. However, recent studies on mammals have shown that certain genes in eggs and sperm are programmed to be switched either on or off during development. Evidence for this comes from the different contributions of the maternal and paternal genomes to the development of the embryo.

Mouse eggs can be manipulated by nuclear transplantation to have either two paternal genomes, or two maternal genomes, and can be re-implanted into a mouse for further development. The embryos that result are known, as **androgenetic** and **gynogenetic** embryos, respectively. Although both kinds of embryo have a diploid number of chromosomes, their development is abnormal. The embryos with two paternal genomes have well developed extra-embryonic tissues, but the embryo itself is abnormal, and does not proceed beyond a stage at which several somites are present. By contrast, the embryos with diploid maternal genomes have relatively well developed embryos, but the extra-embryonic tissues—placenta and yolk sac—are poorly developed (Fig. 12.19). These results clearly show that both maternal and paternal genomes are necessary for normal mammalian development: the two parental genomes function differently in development, and are required for the normal development of both the embryo and the placenta. It is for this reason that mammals cannot be produced parthenogenetically, by activation of an unfertilized egg.

From such observations, it is known that the paternal and maternal genomes are modified, or **imprinted**, during germ cell differentiation. Although the paternal and maternal genomes may contain the same genes, the imprinting process can turn certain genes in either the sperm or egg on or off. Some of the genes necessary for yolk sac and placenta development are inactivated in the maternal genome, whereas some of those required for development of the embryo are turned off in the paternal genome. Imprinting implies that the affected genes carry a 'memory' of being in a sperm or an egg.

Imprinting in mammals is a reversible process, whereby modification of the same genes in either the sperm or egg can lead to differences in gene expression in the diploid cells of the embryo. The reversibility requirement is important, because any of the chromosomes may eventually end up in male or female germ cells during development. Inherited imprinting is probably erased during early germ cell development, and imprinting is later established afresh during germ cell differentiation.

Fig. 12.19 Paternal and maternal genomes are both required for normal mouse development. A normal biparental embryo has contributions from both the paternal and maternal nucleus in the zygote after fertilization (left panel). Using nuclear transplantation, an egg can be constructed with two paternal or two maternal nuclei from an inbred strain. Embryos that develop from an egg with two maternal genomes—gynogenetic embryos (center panel)—have under-developed extra-embryonic structures. This results in development being blocked, although the embryo itself is relatively normal and well developed. Embryos that develop from eggs with two paternal genomes—androgenetic embryos (right panel)—have normal extra-embryonic structures, but the embryo itself only develops to a stage where a few somites have formed.

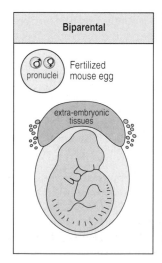
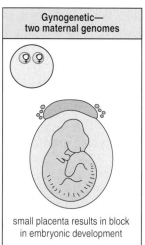
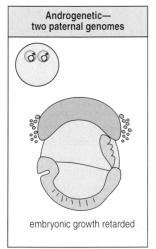

The imprinted genes affect not only early development, but also the later growth of the embryo. Further evidence for imprinted genes being involved in the growth of the embryo comes from studies on chimeras made between normal embryos and androgenetic or gynogenetic embryos. When inner cell mass cells from gynogenetic embryos are injected into normal embryos, growth is retarded by as much as 50%. But when androgenetic inner cell mass cells are introduced into normal embryos, the chimera's growth is increased by up to 50%. The imprinted genes on the male genome thus significantly increase the growth of the embryo.

At least 21 imprinted genes have been identified in mice, some of which are involved in growth control. The insulin-like growth factor (IGF-2) gene, *Igf-2*, is imprinted in the maternal genome—that is, it is turned off, so that only the paternal gene is active; this situation promotes growth. Imprinting of *Igf-2* is probably due to imprinting of the *H19* gene, which regulates expression of *Igf-2*. *H19* is imprinted in the opposite direction: it is expressed in the maternal chromosome but not in the paternal one. The gene for the so-called receptor for IGF-2 protein—*Igf-2r*—is also imprinted in this direction; it is turned off in the paternal genome and turned on in the maternal genome (Fig. 12.20). In fact, *Igf-2r* does not encode the receptor for IGF-2 (the true receptor for IGF-2 is probably the IGF-1 receptor), but a protein that is required to degrade IGF-2 protein; expression of this protein tends to reduce growth by controlling the amount of IGF-2 available. Thus, the activation of the *Igf-2r* gene in the female chromosome reduces growth.

Direct evidence for the *Igf-2* gene being imprinted comes from the observation that when male sperm carrying a mutated, and so defective, *Igf-2* gene fertilize a normal egg, small offspring result. This is because very low levels of IGF-2 are produced from the imprinted maternal gene. If the mutated gene is carried only by the egg's genome, however, development is normal.

A possible evolutionary explanation for the reciprocal imprinting of genes that control growth is that the reproductive strategies of the father and mother are different—that is, paternal imprinting promotes growth whereas maternal imprinting reduces it. The father wants to have maximal growth for his own offspring, and this can be achieved by having a large placenta, as a result of producing growth hormone (whose production is stimulated by IGF-2). The mother, who may mate with different males, benefits more by spreading her resources over all her offspring, and so wishes to prevent too much growth in any one embryo. Paternal genes expressed in the offspring could be selected to extract more resources from mothers, because an offspring's paternal genes are less likely to be present in the mother's other children.

Imprinting occurs during germ cell differentiation, and so a mechanism is required both for maintaining the imprinted condition throughout development and for wiping it out during the next cycle of germ cell development. A possible mechanism for maintaining imprinting is DNA methylation (see Fig. 9.13). Evidence that DNA methylation is required for imprinting comes from transgenic mice in which methylation is impaired. In these mice, the *Igf-2* and *Igf-2r* genes are no longer imprinted.

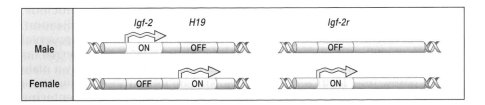

Fig. 12.20 Imprinting of genes controlling embryonic growth. In mouse embryos, the paternal gene for insulin-like growth factor 2 (*Igf-2*) is on, but the gene on the maternal chromosome is off. In contrast, the *Igf-2r* gene is on in the maternal genome and off in the paternal genome. The product of this gene tends to inhibit growth, as it is involved in degrading IGF-2. *H19* may regulate *Igf-2*.

Summary to Chapter 12.

In many animals, the chromosomal constitution of the embryo determines which sex will develop. In mammals, the Y chromosome is male-determining; it specifies the development of a testis, and the hormones produced by the testis cause the development of male sexual characteristics. In the absence of a Y chromosome, the embryo develops as a female. In *Drosophila* and *C. elegans*, sexual development is initially determined by the number of X chromosomes, which sets in train a cascade of gene activity. In mammals, the somatic sexual phenotype is determined by cell–cell interactions; in the fly, somatic cell sexual differentiation is cell autonomous. In many animals, the future germ cells are specified by localized cytoplasmic determinants in the egg. Germ cells in mammals are unusual, as they are specified entirely by cell–cell interactions, and the same is true of flowering plants, in which the germ cells are only specified at a late stage, as the flowers develop. Most flowering plants are hermaphrodites. In animals, the development of germ cells into sperm or egg depends both on chromosomal constitution and cell–cell interactions. At fertilization, fusion of sperm and egg initiates development, and there are mechanisms to ensure that only one sperm enters an egg. Both maternal and paternal genomes are required for normal mammalian development, as some genes are imprinted; for such genes, whether they are expressed or not during development depends on whether they are derived from the sperm or the egg. Animals use a variety of strategies of dosage compensation to correct the imbalance in the number of X chromosomes in males and females.

General references.

Crews, D.: **Animal sexuality**. *Sci. Amer.* 1994, **270**:109–114

Marsh, J., Goodie, J. (eds.): *Germline Development.* (Ciba Symp. 182) Chichester: John Wiley, 1994.

Section references.

12-1 The primary sex-determining gene in mammals is on the Y chromosome.

Goodfellow, P.N., Lovell-Badge, R.: *SRY* **and sex determination in mammals.** *Ann. Rev. Genet.* 1993, **27**:71–92.

Schafer, A.J., Goodfellow, P.N.: **Sex determination in humans**. *BioEssays* 1996, **18**:955–963.

12-2 Mammalian sexual phenotype is regulated by gonadal hormones.

Kelly, D.D.: **Sexual differentiation of the nervous system**. In *Principles of Neural Science*. 3rd edn. Edited by Kandel, E.R., Schwartz, J.H., Jessell, T.M. New York: Elsevier Science Publishing Co. Inc., 1991.

12-3 In *Drosophila*, the primary sex-determining signal is the number of X chromosomes and is cell autonomous.

Hodgkin, J.: **Sex determination compared in** *Drosophila* **and** *Caenorhabditis*. *Nature* 1990, **344**:721–728.

12-4 Somatic sexual development in *Caenorhabditis* is determined by the number of X chromosomes.

Cline, T.W., Meyer, B.J.: **Vive la difference: males vs. females in flies vs. worms**. *Ann. Rev. Genet.* 1996, **30**:637–702.

12-5 Most flowering plants are hermaphrodite, but some produce unisexual flowers.

Irisa, E.N.: **Regulation of sex determination in maize**. *BioEssays* 1996, **18**:363–369.

12-6 Germ cell sex determination may depend both on cell signals and genetic constitution.

McLaren, A.: **Germ cells and germ cell sex**. *Phil. Trans. Roy. Soc. Lond.* 1995, **350**:229–233.

Clifford, R., Francis, R., Schedl, T.: **Somatic control of germ cell development in** *Caenorhabditis elegans*. *Semin. Dev. Biol.* 1994, **5**:21–30.

12-7 Various strategies are used for dosage compensation of X-linked genes.

Lucchesi, J.C. (Ed.): **Dosage compensation**. *Semin. Dev. Biol.* 1993, **4**:91–145.

Willard, H.F., Salz, H.K.: **Remodelling chromatin with RNA**. *Nature* 1997, **386**:228–229.

Panning, B., Jaenisch, R.: **DNA hypomethylation can activate** *Xist* **expression and silence X-linked genes**. *Genes Dev.* 1996, **10**:1991–2002.

Davis, T.L., Meyer, B.J.: **SDC-3 co-ordinates the assembly of a dosage compensation complex on the nematode X chromosome**. *Development* 1997, **124**:1019–1031.

12-8 Germ cell fate can be specified by a distinct germ plasm in the egg.

Mello, C.C., Schubert, C., Draper, B., Zhang, W., Lobel, R., Priess, J.R.: **The PIE-1 protein and germline specification in *C. elegans* embryos.** *Nature* 1996, **382**:710–712.

Williamson, A., Lehmann, R.: **Germ Cell Development in Drosophila.** *Ann. Rev. Cell Dev. Biol.* 1996, **12**:365–391.

12-9 Pole plasm becomes localized at the posterior end of the *Drosophila* egg.

Rongo, C., Gavis, E.R., Lehmann, R.: **Localization of oskar RNA regulates oskar translation and requires oskar protein.** *Development* 1995, **121**:2737–2746.

12-10 Germ cells migrate from their site of origin to the gonad.

Dixon, K.E.: **Evolutionary aspects of primordial germ cell formation.** In *Germline Development* (Ciba Symp. 182). Chichester: John Wiley, 1994; pp92–120.

Wylie, C.C., Heasman, J.: **Migration, proliferation and potency of primordial germ cells.** *Semin. Dev. Biol.* 1993, **4**:161–170.

12-11 Germ cell differentiation involves a reduction in chromosome number.

Metz, C.B., Monroy, A. (eds.): *Biology of the Sperm. Biology of Fertilization.* vol. 2. Orlando, Florida: Academic Press, 1985.

12-12 Oocyte development can involve gene amplification and contributions from other cells.

Spradling, A.: **Developmental genetics of oogenesis.** In *Drosophila Development.* Edited by Bate, M., Martinez-Arias, A. New York: Cold Spring Harbor Laboratory Press, 1993; pp1–69.

Browder, L.W.: *Oogenesis.* New York: Plenum Press, 1985.

12-13 Genes controlling embryonic growth are imprinted.

Haig, D.: **Do imprinted genes have few and small introns?** *BioEssays* 1996, **18**:351–353.

Surani, A.: **Silencing of the genes.** *Nature* 1993, **366**:302–303.

Solter, D.: **Differential imprinting and expression of maternal and paternal genomes.** *Ann. Rev. Genet.* 1988, **22**:127–146.

Razin A, Cedar H: **DNA methylation and genomic imprinting.** *Cell* 1994, **77**:473–476.

12-14 Fertilization involves cell-surface interactions between egg and sperm.

Snell, W.J., White, J.M.: **The molecules of mammalian fertilization.** *Cell* 1996, **85**:629–637.

Longo, F.S.: *Fertilization.* London: Chapman & Hall, 1987.

Miller, D.J., Shur, B.D.: **Molecular basis of fertilization in the mouse.** *Semin. Dev. Biol.* 1994, **5**:255–264.

Romano, C.S., Myles, D.G., Primakoff, P.: **Multiple roles for PH-20 and fertilin in sperm–egg interactions.** *Semin. Dev. Biol.* 1994, **5**:265–271.

12-15 Changes at the egg membrane at fertilization block polyspermy.

Foltz, K.R., Partin, J.S., Lennarz, W.J.: **Sea urchin egg receptor for sperm: sequence similarity of binding domain and hsp70.** *Science* 1993, **259**:1421–1425.

Foltz, K.R.: **The sea urchin egg receptor for sperm.** *Semin. Dev. Biol.* 1994, **5**:243–253.

12-16 A calcium wave initiated at fertilization results in egg activation.

Whitaker, M.: **Cell cycle: Sharper than a needle.** *Nature* 1993, **366**:211–212.

Whitaker, M.: **Lighting the fuse at fertilization.** *Development* 1993, **117**:112.

Swann, K., Lai, F.A.: **A novel signalling mechanism for generating Ca^{2+} oscillations at fertilization in mammals.** *BioEssays* 1997, **19**:371–378.

Regeneration

13

"Even when some of us were removed by force, others took our place."

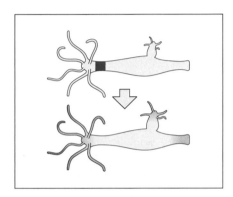

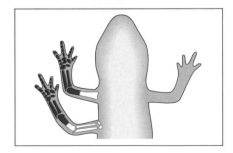

Throughout this book, we have seen many examples of the capacity of the embryo to self-regulate when parts of it are removed or rearranged (see, for example, Sections 3-5 and 6-1). Here we look at the related phenomenon of **regeneration** in adult organisms. Regeneration is the ability of the fully developed organism to replace lost parts by growth or remodeling of somatic tissue. Plants have remarkable powers of regeneration: a single somatic plant cell can give rise to a complete new plant. Some animals also show great ability to regenerate: small fragments of animals such as starfish, planarians (flatworms), and *Hydra* can give rise to a whole animal (Fig. 13.1). The ability of animals like *Hydra* and planarians to regenerate may be related to their ability to reproduce asexually. A remarkable case of regeneration is found in ascidians, whose blood cells alone can give rise to a fully functional organism.

Fig. 13.1 Regeneration in some invertebrate animals. A planarian, *Hydra*, and a starfish all show remarkable powers of regeneration. When parts are removed or a small fragment isolated, a whole animal can be regenerated.

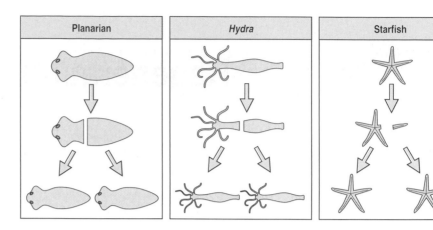

Among vertebrates, newts and other urodele amphibians show a remarkable capacity for regeneration (Fig. 13.2). The lens, for example, regenerates from the pigmented epithelium of the iris (Fig. 13.3). Insects and other arthropods can also regenerate lost appendages, such as legs. The regenerative powers of mammals are much more restricted. The mammalian liver can regenerate if a part of it is removed; and fractured bones can mend by a regenerative process. But mammals cannot regenerate lost limbs, although they do have a limited capacity to replace lost digits, as described later. Nematodes and their relatives cannot regenerate at all.

The issue of regeneration raises several major questions. Why are some animals able to regenerate and others not? What is the origin of the cells that give rise to the regenerated structures? What mechanisms pattern the regenerated tissue and how are these related to the patterning processes that occur in embryonic development? In this chapter we discuss some aspects of regeneration that are most closely linked to developmental processes, but do not consider wound healing, which is essentially a repair process, or the compensatory growth that occurs in mammals when part of the liver or a kidney is lost. We will focus on two systems in which regeneration has been intensively studied: regeneration of the whole animal in *Hydra*, and limb regeneration in insects and amphibians. Plant regeneration will also be discussed briefly.

Fig. 13.2 The capacity for regeneration in urodele amphibians. The emperor newt can regenerate its dorsal crest (1), limbs (2), retina and lens (3 and 4), jaw (5), and tail (not shown).

A distinction can be drawn at the outset between two types of regeneration. In one—**morphallaxis**—there is little new growth, and regeneration occurs mainly by the repatterning of existing tissues and the re-establishment of boundaries. Regeneration in *Hydra* is a good example of morphallaxis. By contrast, regeneration in the newt limb, for example, depends on the growth of new, correctly patterned structures, and this is known as **epimorphosis**. Both types of regeneration can be illustrated with reference to the French flag pattern (Fig. 13.4). In morphallaxis, new boundary regions are first established and new positional values are specified in relation to them; in epimorphosis, new positional values are linked to growth from the cut surface. We first consider morphallaxis in *Hydra*.

Fig. 13.3 Lens regeneration. Removal of the lens from the eye of a newt results in regeneration of a new lens from the pigmented epithelium of the iris.

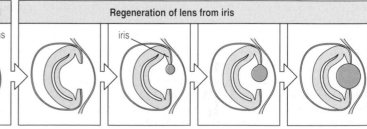

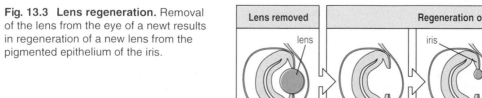

Fig. 13.4 Morphallaxis and epimorphosis. A pattern such as the French flag may be specified by a gradient in positional value (see Fig. 1.22). If the system is cut in half it can regenerate in one of two ways. In regeneration by morphallaxis, a new boundary is established at the cut and the positional values are changed throughout. In regeneration by epimorphosis, new positional values are linked to growth from the cut surface.

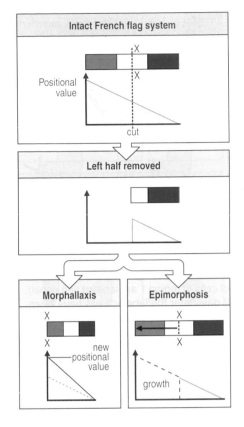

Morphallaxis.

Hydra is a freshwater coelenterate consisting of a hollow tubular body about 0.5 cm long, with a head region at the distal end and a basal region at the other 'proximal' end, with which it can stick to surfaces (Fig. 13.5). The head consists of a small conical hypostome where the mouth opens, surrounded by a set of tentacles, which are used for catching the small animals on which *Hydra* feeds. Unlike most of the animals discussed so far in this book, which have three germ layers, *Hydra* has only two. The body wall is composed of an outer epithelium, which corresponds to the ectoderm, and an inner epithelium, which corresponds to the endoderm. These two layers are separated by a basement membrane. There are about 20 different cell types in *Hydra*, which include nerve cells, muscle cells, and nematocysts that are used to capture prey.

13-1 | *Hydra* grows continuously, with loss of cells from its ends and by budding.

Well-fed *Hydra* are in a dynamic state of continuous growth and pattern formation. The cells of both epithelial layers proliferate steadily and, as the tissues grow, cells are displaced along the body column toward the head or foot (Fig. 13.6). In order for an adult *Hydra* to maintain a constant size, excess cells must be continually lost. Cell loss occurs at the tips of the tentacles and at the basal disc of the foot bud. Most of the excess cell production is taken up by the asexual budding of new *Hydra* from the body column. Budding occurs about two thirds of the way down the body column; the body wall evaginates by a morphogenetic

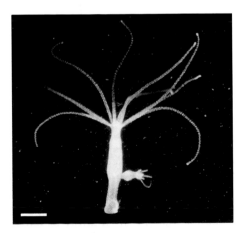

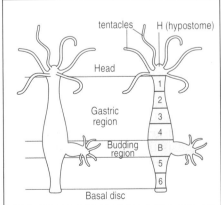

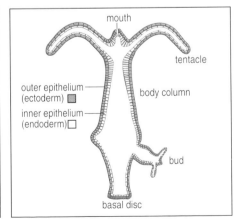

Fig. 13.5 *Hydra*. This freshwater coelenterate (photograph, left panel) has a head with tentacles and mouth at one end and a sticky foot at the other. It can reproduce by budding. For the purposes of grafting experiments, the body column is divided into a series of regions, as indicated in the center panel. The body wall is made up of two epithelial layers, corresponding to the ectoderm and endoderm found in other organisms featured in this book (right panel). Scale bar = 1 mm. Photograph courtesy of W. Müller, from Müller, W.A.: 1989.

Insect limbs intercalate positional values by both proximo-distal and circumferential growth.

Intercalation of missing positional values, as already described for the proximo-distal axis of the amphibian limb (see Section 13-6), seems to be a general property of regenerating epimorphic systems. When cells with disparate positional values are placed next to one another, intercalary growth occurs in order to regenerate the missing positional values. Intercalation is particularly clearly illustrated by limb regeneration in the cockroach.

A cockroach leg is made up of a number of distinct segments, arranged along the proximo-distal axis in the order coxa, femur, tibia, tarsus. Each segment seems to contain a similar set of proximo-distal and circumferential positional values, and will intercalate the missing positional values. When a distally amputated tibia is grafted onto a host tibia that has been cut at a more proximal site, localized growth occurs at the junction between graft and host, and the missing central regions of the tibia are intercalated (Fig. 13.20, left panels). In contrast to amphibian regeneration, there is a considerable contribution from the distal piece. As in the amphibian, however, regeneration is a local phenomenon and the cells are indifferent to the overall pattern of the tibia. Thus, when a proximally cut tibia is grafted onto a more distal site, making an abnormally long tibia, regenerative intercalation again restores the missing positional values, making the tibia even longer (see Fig. 13.20, right panels). The regenerated portion is in the reverse orientation to the rest of the limb, as indicated by the direction in which the bristles point, suggesting

Fig. 13.20 Intercalation of positional values by growth in the regenerating cockroach leg. Left panels: when a distally amputated tibia (5) is grafted to a proximally amputated host (1), intercalation of the positional values 2–4 occurs, irrespective of the proximo-distal orientation of the grafts, and a normal tibia is regenerated. Right panels: when a proximally amputated tibia (1) is grafted to a distally amputated host (4), however, the regenerated tibia is longer than normal and the regenerated portion is in the reverse orientation to normal, as judged by the orientation of surface bristles. The reversed orientation of regeneration is due to the reversal in positional value gradient. The proposed gradient in positional value is shown under each figure. After French, V., et al.: 1976.

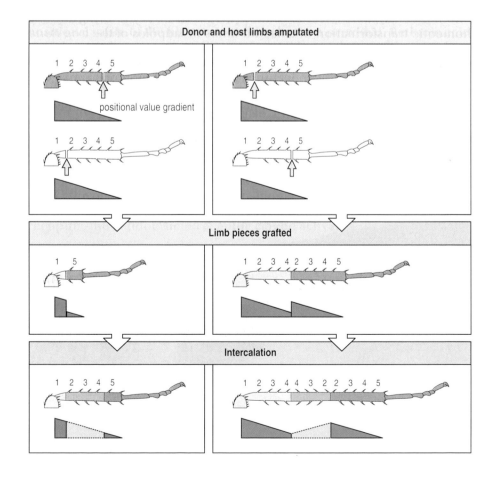

that the gradient in positional values also specifies cell polarity, as in insect body segments (see Chapter 5). These results also show that when cells with non-adjacent positional values are placed next to each other, the missing values are intercalated by growth to provide a set of continuous positional values.

A similar set of positional values is present in each segment of the limb. Thus, a mid-tibia amputation, when grafted to the mid-femur of a host, will heal without intercalation. But grafting a distally amputated femur onto a proximally amputated host tibia results in intercalation, largely femur in type. There must be other factors making each segment different, rather like the segments of the insect larva.

Intercalary regeneration also occurs in a circumferential direction. When a longitudinal strip of epidermis is removed from the leg of a cockroach, normally nonadjacent cells come into contact with one another, and intercalation in a circumferential direction occurs after molting (Fig. 13.21). One can treat positional values in the circumferential direction as a clock face, with values going continuously 12, 1, 2, 3... 6... 9... 11. As in the proximo-distal axis, there is intercalation of the missing positional values.

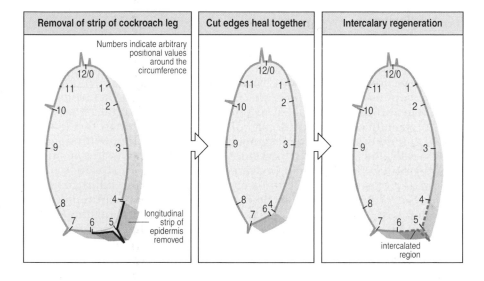

Fig. 13.21 Circumferential intercalation in the cockroach leg. The leg is seen in transverse section. When a piece of cockroach ventral epidermis is removed (left panel), the cut edges heal together (center panel). When the insect molts and the cuticle regrows, circumferential positional values are intercalated (right panel). The positional values are arranged around the circumference of the leg, rather like the hours on a clock face. After French, V., *et al.*: 1976.

| 13-9 | **Polarized regeneration in plants is due to the polarized transport of auxin.** |

We have already described the remarkable ability of single plant cells to generate a whole plant (see Section 7-6), but parts of plants also have considerable powers of regeneration. A piece of stem cut away from a plant can often regenerate a new shoot and roots. In general, roots form from the end of the stem that was originally closest to the root, whereas shoots tend to develop from dormant buds at the end that was nearest to the shoots (Fig. 13.22). This polarized regeneration is related to vascular differentiation and to the polarized transport of the plant hormone auxin. Transport of auxin from its source in the shoot tip toward the root leads to an accumulation of auxin at the root end of a stem cutting, where it induces the formation of roots. One hypothesis suggests that polarity is both induced and expressed by the oriented flow of auxin.

Fig. 13.22 Polarized regeneration in plants. Isolated pieces of stem regenerate roots from the proximal (basal) surface and a shoot from a bud nearest to the distal (apical) surface.

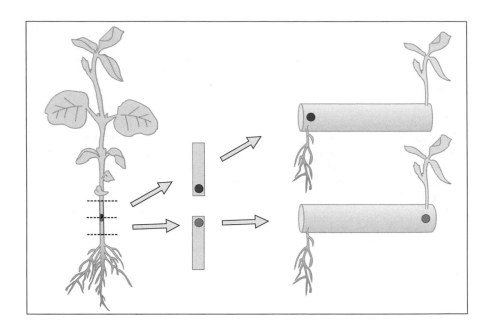

Summary.

Urodele amphibians can regenerate limbs by epimorphosis. Stump tissue at the site of amputation first dedifferentiates to form a blastema, which then grows and gives rise to a regenerated structure. Regeneration is usually dependent on the presence of nerves, but limbs that have never been innervated are capable of regeneration. Regeneration always gives rise to structures with positional values more distal than those at the site of amputation. When a blastema is grafted to a stump with different positional values, proximo-distal intercalation of the missing positional values occurs. Retinoic acid proximalizes the positional values of the cells of the blastema. Insect limbs can also regenerate, intercalation of positional values occurring in both proximo-distal and circumferential directions. Regeneration in plants is polarized, with roots developing from the cut surface originally nearest the root and the shoot developing from the surface originally nearest the shoot in a piece of stem removed from a plant.

Summary: regeneration of an amphibian limb

amputation of a newt limb

⇓

local dedifferentiation of stump tissue to form a blastema

⇓

the blastema grows to form distal structures by epimorphosis,
provided nerves are present

⇓

graft of distal blastema to proximal stump

⇓

intercalation of missing proximo-distal positional values from stump tissue growth

Summary to Chapter 13.

Regeneration is the ability of an adult organism to replace a lost part of its body. The capacity to regenerate varies greatly among different groups of organisms. Mammals have very limited powers of regeneration, whereas a complete plant can be regenerated from a single somatic cell. Regeneration of limbs in animals such as amphibians may reflect the retention or reactivation of embryonic mechanisms. The two main mechanisms involved in regeneration are morphallaxis, in which there is no growth but in which positional values are re-specified, and epimorphosis, in which patterned new growth occurs. *Hydra* regenerates by morphallaxis: when its head is removed, the level of inhibitor normally produced by the head falls, resulting in a new head being specified at the cut end. Regeneration of limbs in amphibians and insects is by epimorphosis: a blastema forms by dedifferentiation of stump cells and the blastemal cells divide and differentiate to regenerate the lost distal structures. Intercalation of positional values can occur when normally non-adjacent values are placed next to each other. Regeneration in plants is polarized; roots and shoots grow from opposite ends of a piece of cut stem.

References.

Introduction.

Goss, R.J.: *Principles of Regeneration*, London & New York: Academic Press, 1969.
Rinkevich, B., Shlemburg, Z., Fishelson, L.: **Whole-body protochordate regeneration from totipotent blood cells.** *Proc. Nat. Acad. Sci.* 1995, **92**:7695–7699.

13-1 *Hydra* grows continuously, with loss of cells from its ends and by budding.

Otto, J.J., Campbell, R.D.: **Tissue economics of *Hydra*: regulation of cell cycle, animal size and development by controlled feeding rates.** *J. Cell Sci.* 1977, **28**:117–132.

13-2 Regeneration in *Hydra* is polarized and does not depend on growth.

Hicklin, J., Wolpert, L.: **Positional information and pattern regulation in Hydra: the effect of gamma-radiation.** *J. Emb. Exp. Morphol.* 1973, **30**:741–752.

13-3 The head region of *Hydra* acts both as an organizing region and as an inhibitor of inappropriate head formation.

Wolpert, L., Hornbruch, A., Clarke, M.R.B.: **Positional information and positional signaling in *Hydra*.** *Am. Zool.* 1974, **14**:647–663.

13-4 Head regeneration in *Hydra* can be accounted for in terms of two gradients.

Hassel, M., Albert, K., Hofheinz, S.: **Pattern formation in *Hydra vulgaris* is controlled by lithium-sensitive process.** *Dev. Biol.* 1993, **156**:362–371.

Müller, W.A.: **Pattern formation in the immortal *Hydra*.** *Trends Genet.* 1996, **12**:91–96.
MacWilliams, W.: **Hydra transplantation phenomena and the mechanism of *Hydra* head regeneration. II. Properties of the head activation.** *Dev. Biol.* 1983, **96**:239–257.
Shenk, M.A., Gee, L., Steele, R.E., Bode, H.R.: **Expression of *Cnox-2*, a Hom/Hox gene, is suppressed during head formation in *Hydra*.** *Dev. Biol.* 1993, **160**:108–118.

13-5 Vertebrate limb regeneration involves cell dedifferentiation and growth.

Muneoka, K., Sassoon, D.: **Molecular aspects of regeneration in developing vertebrate limbs.** *Dev. Biol.* 1992, **152**:37–49.
Lo, D.C., Allen, F., Brockes, J.P.: **Reversal of muscle differentiation during urodele limb regeneration.** *Proc. Nat. Acad. Sci.* 1993, **90**:7230–7234 .
Brockes, J.P.: **Amphibian limb regeneration: rebuilding a complex structure.** *Science* 1997, **276**:81–87.
Tanaka, E.M., Gann, A.A.F., Gates, P.B., Brockes, J.P.: **Newt myotubules re-enter the cell cycle by phosphorylation of the retinoblastoma protein.** *J. Cell Biol.* 1997, **136**:155–165.

13-6 The limb blastema gives rise to structures with positional values distal to the site of amputation.

Reginelli, A.D., Wang, Y.Q., Sassoon, D., Muneoka, K.: **Digit tip regeneration correlates with *Msx-1 (Hox7)* expression in fetal and newborn mice.** *Development* 1995, **121**:1065–1076.
Gardiner, D.M., Bryant, S.V.: **Molecular mechanisms in the control of limb regeneration: the role of homeobox genes.** *Int. J. Devel. Biol.* 1996, **40**:797–805.
Stocum, D.L.: **A conceptual framework for analyzing axial patterning in regenerating urodele limbs.** *Int. J. Dev. Biol.* 1996, **40**:773–783.

13-7   **Retinoic acid can change proximo-distal positional values in regenerating limbs.**

Stocum, D.L.: **Retinoic acid and limb regeneration.** *Semin. Dev. Biol.* 1991, **2**:199–210.

Maden, M.: **The homeotic transformation of tails into limbs in *Rana temporaria* by retinoids.** *Dev. Biol.* 1993, **159**:379–391.

Brockes, J.P.: **New approaches to amphibian limb regeneration.** *Trends Genet.* 1994, **10**:169–173.

Brockes, J.P.: **Introduction of a retinoid reporter gene into the urodele limb blastema.** *Proc. Nat. Acad. Sci.* 1992, **89**:11386–11390.

Bryant, S.V., Gardiner, D.M.: **Retinoic acid, local cell–cell interactions and pattern formation in vertebrate limbs.** *Dev. Biol.* 1992, **152**:125.

Scadding, S.R., Maden, M.: **Retinoic acid gradients during limb regeneration.** *Dev. Biol.* 1994, **162**:608–617.

Pecorino, L.T., Entwistle, A., Brockes, J.P.: **Activation of a single retinoic acid receptor isoform mediates proximo-distal respecification.** *Curr. Biol.* 1996, **6**:563–569.

13-8   **Insect limbs intercalate positional values by both proximo-distal and circumferential growth.**

French, V.: **Pattern regulation and regeneration.** *Phil. Trans. Roy. Soc. Biol. Sci.* 1981, **295**:601–617.

13-9   **Polarized regeneration in plants is due to polarized transport of auxin.**

Sachs, T.: *Pattern Formation in Plant Tissues.* Cambridge: Cambridge University Press, 1991.

Growth and Post-Embryonic Development

14

Growth.

Molting and metamorphosis.

Aging and senescence.

" Growing up was not just about getting bigger, but also about some dramatic changes that made us almost unrecognizable. "

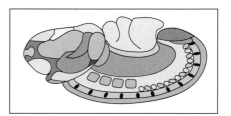

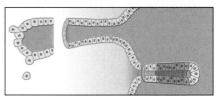

Development does not stop once the embryonic phase is complete. Most, but by no means all, of the growth in animals and plants occurs in the post-embryonic period, when the basic form of the organism has already been established. In many animals, the embryonic phase is immediately succeeded by a free-living larval or immature adult stage. In others, such as mammals, considerable growth occurs during a late embryonic or fetal period, while the embryo is still dependent on maternal resources. Growth then continues after birth. Growth is a central aspect of all developing systems, determining the final size and shape of the organism and its parts. Animals with a larval stage not only grow in size, but may also undergo molting, in which the outer skeleton is shed, and **metamorphosis**, in which the larva is transformed into the adult form. Metamorphosis often involves a radical change in form and the development of new organs.

We first consider the roles of intrinsic growth programs and of extracellular factors such as growth hormones in controlling both embryonic and post-embryonic growth. This is followed by a discussion of metamorphosis in insects and amphibians. Finally, we consider an abnormal aspect of post-embryonic development—aging.

Growth.

Growth is defined as an increase in the mass or overall size of a tissue or organism; this increase may result from cell proliferation, cell enlargement without division, or an increase in extracellular material, such as bone matrix or even water. In early embryonic development, there is little growth during cleavage and blastula formation, and cells get smaller at each cleavage division. However, growth begins in different animals at different stages. In the chick, for example, it occurs in the region anterior to the node, during primitive streak regression, whereas in *Xenopus* it starts after gastrulation.

In animals, the basic body pattern is laid down when the embryo is still small; in humans, all organs are less than 1 cm in their maximum dimension when their pattern is determined. The growth program—that is, how much an organism or an individual organ grows—is also specified at an early stage in development. Overall growth of the organism mainly occurs in the period after the basic pattern of the embryo has been established; there are, however, many examples where earlier organogenesis involves localized growth, as in the vertebrate limb bud (see Chapter 10), and the developing nervous system (see Chapter 11). Different rates of growth in different parts of the body, or at different times, during early development profoundly affect the shape of organs and the organism.

Unlike the situation in animals, where the embryo is essentially a miniature version of the free-living larva or adult, plant embryos bear little resemblance to the mature plant. Most of the adult plant structures are generated after germination by the shoot or root meristems, which have a capacity for continual growth (see Chapter 7). In woody plants, the outer cambial layer, a layer of cells that can give rise to the main tissues of the stem, also retains proliferative capacity, enabling trees to increase in girth year after year.

14-1   Tissues can grow by cell proliferation, cell enlargement, or accretion.

Although growth often occurs through an increase in the number of cells, this is only one of three main growth (Fig. 14.1). A second strategy is growth by cell enlargement—that is, by individual cells increasing their mass and getting bigger. Once differentiated, skeletal and heart muscle cells and neurons never divide again, although they do increase in size. Glial cells, in contrast, do proliferate. Neurons grow by the extension

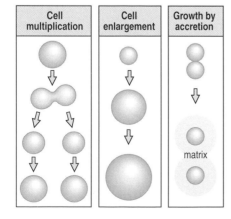

Fig. 14.1 The three main strategies for growth in vertebrates. The most common mechanism is cell proliferation —growth followed by division. A second strategy is cell enlargement, in which cells increase their size without dividing. A third strategy is to increase size by accretionary growth, such as matrix secretion.

and growth of axons and dendrites, whereas muscle growth involves an increase in mass, as well as the fusion of satellite cells to pre-existing muscle fibers to provide new nuclei. Cell enlargement is also a major feature of plant growth, as we shall see. And differences in size between closely related species of *Drosophila* are partly the result of the larger species having larger cells. Some growth is through a combination of cell proliferation and cell enlargement. For example, lens cells are produced from a proliferative zone for an extended period, and their differentiation involves considerable cell enlargement.

The third growth strategy, accretionary growth, involves an increase in the volume of the extracellular space, which is achieved by cell secretion of large quantities of extracellular matrix. This is clearly seen in both cartilage and bone, where most of the tissue mass is extracellular.

Even when they are capable of division, the cells of many adult tissues do not divide, or divide infrequently. Cells can, however, be induced to divide by injury or other stimuli, such as a reduction in mass of the tissue. Liver cells divide relatively infrequently but if, for example, two thirds of a rat's liver is removed, the cells in the remaining third proliferate and restore the liver to its normal size within a few weeks. This restorative capacity indicates the presence of factors in the circulation that control cell proliferation. Removal of a kidney leads to an increase in size of the remaining kidney, but in this case the growth is mainly the result of cell enlargement rather than cell proliferation. There is a significant amount of cell death in many growing tissues, and the overall growth rate is determined by the rates of cell death and cell proliferation.

Certain vertebrate tissues, including the hematopoietic system tissues (see Section 9-11) and epithelia, are continually renewed throughout an animal's lifetime by cell division and differentiation from a stem cell population. The end products of this type of proliferative system, such as mature red blood cells and keratinocytes, are themselves incapable of division and die.

14-2 Cell proliferation can be controlled by an intrinsic program and by external signals.

When a eukaryotic cell duplicates itself it goes through a fixed sequence of events called the **cell cycle**. The cell grows in size, the DNA is replicated, and these replicated chromosomes then undergo mitosis, and become segregated into two daughter nuclei. Only then can the cell divide to form two daughter cells, which can go through the whole sequence again. During cleavage of the fertilized egg, as in *Xenopus* and the mouse, there is no cell growth, and cell size decreases with each division, but in other proliferating cells, the cytoplasmic mass must double in preparation for cell division.

The standard eukaryotic mitotic cell cycle is divided into well-marked phases. At the M-phase, mitosis and cell cleavage give rise to two new cells. The rest of the cell cycle, between one M-phase and the next, is called interphase. Replication of DNA occurs during a defined period in interphase, the S-phase. Preceding S-phase is a period known as G_1, and after it another interval known as G_2, after which the cells enter mitosis again (Fig. 14.2). G_1, S-phase, and G_2 collectively make up interphase, the part of the cell cycle during which cells grow and synthesize proteins, as well as replicate their DNA. Particular phases of the cell cycle are absent in some cells: during cleavage of the fertilized egg, G_1 and G_2 are virtually absent; in meiosis (see Section 12-11) there is no DNA replication at the second division; and in insect salivary glands there is no

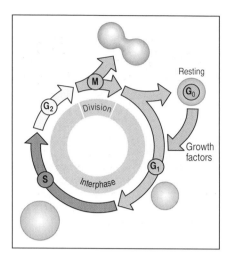

Fig. 14.2 The eukaryotic cell cycle. After mitosis (M), the daughter cells can either enter a resting phase (G_0), in which they effectively withdraw from the cell cycle, or proceed through G_1 to the phase of DNA synthesis (S). This is followed by G_2, and then by mitosis. Cell growth occurs throughout G_1, S, and G_2. The decision to enter G_0 or to proceed through G_1 may be controlled by both intracellular status and extracellular signals, such as growth factors. Cells such as neurons and skeletal muscle cells, which do not divide after differentiation, are permanently in G_0.

M-phase, as the DNA replicates repeatedly without mitosis or cell division, leading to the formation of giant polytene chromosomes, as we see later.

Growth factors and other signaling proteins play a key role in controlling cell growth and proliferation. Studies of cell cultures show that growth factors are essential for cells to multiply, the particular growth factor required depending on the cell type. When somatic cells are not proliferating they are usually in a state known as G_0, into which they withdraw after mitosis (see Fig. 14.2). Growth factors enable the cell to proceed out of G_0 and progress through the cell cycle.

A striking discovery has been the recognition that cells must receive signals, such as growth factors, not only for them to divide, but also for them simply to survive. In the absence of all growth factors, cells commit suicide by **apoptosis**, as a result of activation of their internal cell death program (see Section 8-6). There is a significant amount of cell death in all growing tissues, so that growth rate is dependent on the rates of both cell death and cell proliferation.

The timing of events in the vertebrate cell cycle is controlled by a set of 'central' timing mechanisms. One set of proteins, known as the **cyclins**, control the passage through key transition points in the vertebrate cell cycle. Cyclin concentrations oscillate during the cell cycle, and these oscillations correlate with transitions from one phase of the cycle to the next (see Section 12-16). Cyclins act by forming complexes with, and helping to activate, cyclin-dependent kinases, which in turn phosphorylate proteins that trigger the events of each phase, such as DNA replication in S-phase or mitosis in M-phase. In general, the mechanisms controlling the pattern of cell division in the embryo are poorly understood. But in *Drosophila*, the early cell cycles are under the control of genes that pattern the embryo and they exert their effect by their influence on the cyclins.

The first cell cycles in the *Drosophila* embryo are represented by rapid and synchronous nuclear divisions, without any accompanying cell divisions. This creates a syncytial blastoderm (see Fig. 2.30). There are virtually no G-phases, just alternations of DNA synthesis (S-phase) and mitosis. But at cycle 14, there is a major transition to a different type of cell cycle, a transition similar to the mid-blastula transition in frogs (see Section 3-19). At cycle 14 and subsequently, the cell cycles have a well-defined G_2 phase, and the blastoderm becomes cellularized. After the 17th or 18th cycle, cells in the epidermis and mesoderm stop dividing, and differentiate. This cessation of proliferation is caused by the exhaustion of maternal cyclin E originally laid down in the egg, which is required for progression through the cell cycle.

At the 14th cell cycle, distinct spatial domains with different cell-cycle times can be seen in the *Drosophila* blastoderm (Fig. 14.3). This patterning of cell cycles is produced by a change in the synthesis and distribution of a protein phosphatase called string, which exerts control on the cell cycle by dephosphorylating and activating a cyclin-dependent kinase. In the fertilized egg, the string protein is of maternal origin and is uniformly distributed. It therefore produces a synchronized pattern of nuclear division throughout the embryo. After cycle 13, the maternal string protein disappears and zygotic string protein becomes the controlling factor. Zygotic *string* gene transcription occurs in a complex spatial and temporal pattern. Only cells in which the *string* gene is expressed enter mitosis. This results in a variation in the rate of cell division in different parts of the blastoderm, which ensures that the correct number of cells are generated in different tissues. This pattern of zygotic *string* gene expression is controlled by transcription factors encoded by the early patterning genes, such as the gap and pair-rule genes and those patterning the dorso-ventral axis. Thus, the cell cycles in *Drosophila* development provide a good example of how genes can control the pattern of cell divisions.

In contrast to this intrinsic program of early cell division in the *Drosophila* embryo, the cell proliferation and growth that occurs in the imaginal discs that give rise to adult structures is modulated by cell–cell interactions, and is not closely linked to pattern formation. As discussed in Chapter 10, imaginal disc cells can proliferate to very different extents and still give a normal pattern. In the *Drosophila* wing, the clonal descendants of a single cell can contribute to a tenth of the wing, or to as much as a half. This means that very different patterns of cell proliferation can give rise to a normal wing. In Section 13-8, we saw how intercalary growth is stimulated to replace missing portions of limbs when cells with disparate positional values are placed next to each other. In the *Drosophila* wing, it is possible that proliferation ceases when the wing has reached its final size, and this could depend on the cells recognizing when a complete set of positional values has been established.

In early mammalian embryos, cell proliferation times vary with developmental time and place. The first two cleavage cycles in the mouse last about 24 hours, and subsequent cycles take about 10 hours each. After implantation, the cells of the epiblast proliferate rapidly; cells in front of the primitive streak have a cycle time of just 3 hours, but the signals involved in this proliferation have not yet been identified.

Numerous growth factors that can control cell proliferation have been discovered, but in general their precise roles in normal development are not yet known. Some exceptions include erythropoietin, which promotes proliferation of red blood cell precursors (see Section 9-12), and the growth factors that we will now discuss.

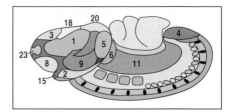

Fig. 14.3 Domains of mitosis in the *Drosophila* blastoderm. Mitotic domains that divide at the same time are indicated by the various colors. The numbers indicate the order in which various regions undergo mitosis at the 14th cycle. After Edgar, B.A., *et al*.: 1994.

14-3   Growth of mammals is dependent on growth hormones.

Human growth during the embryonic, fetal, and post-natal periods provides a good model for mammalian growth. The human embryo increases in length from 150 µm at implantation to about 50 cm over the nine months of gestation. During the first 8 weeks after conception, the embryonic body does not increase greatly in size, but the basic human form is laid down in miniature. The greatest rate of growth occurs at about 4 months, when the embryo grows as much as 10 cm per month. Growth after birth follows a well-defined pattern (Fig. 14.4, left panel). During the first year after birth, growth occurs at a rate of about 2 cm per month. The growth rate then declines steadily until the start of the characteristic adolescent growth spurt at about 11 years in girls and 13 years in boys (see Fig. 14.4, right panel). In pygmies, this adolescent growth spurt does not occur, hence their characteristic short stature.

Growth of the different parts of the body is not uniform, and different organs grow at different rates. At 9 weeks of development, the head of a human embryo is more than a third of the length of the whole embryo, whereas at birth it is only about a quarter. After birth, the rest of the

Fig. 14.4 Normal human growth.
Left panel: an average growth curve
for a human male after birth. Right panel:
comparative growth rates of boys and
girls. There is a growth spurt at puberty
in both sexes, which occurs earlier in girls.

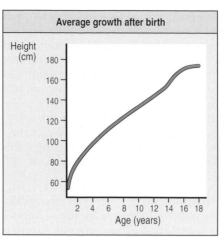

Average growth after birth

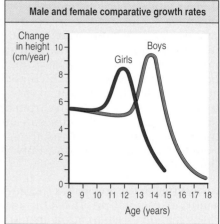

Male and female comparative growth rates

body grows much more than the head, which is only about an eighth of the body length in the adult (Fig. 14.5).

The maternal environment plays an important role in controlling fetal growth. This is well illustrated by crossing a large shire horse with a much smaller Shetland pony. When the mother is a shire mare, the newborn foal is similar in size to a normal shire foal, but when the mother is a Shetland, the newborn is much smaller. However, with growth after birth, the offspring of both these crosses become similar in size, and achieve a final size intermediate between shires and Shetlands.

The amount of nourishment an embryo receives can have profound effects in later life. In early intrauterine life, undernutrition tends to produce small but normally proportioned animals. In contrast, under-nutrition during the post-natal growth period leads to selective organ damage. For example, rats that are undernourished immediately after weaning have normal skeletal growth, but the liver and kidneys do not grow normally and are permanently small. Epidemiological studies in humans have shown that small size at birth is associated with increased

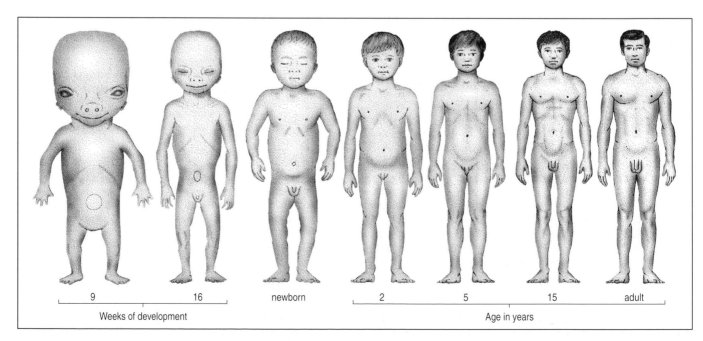

Fig. 14.5 Different parts of the human body grow at different rates. At 9 weeks of development, the head is relatively large but, with time, other parts of the body grow much more than the head. After Gray, H.: 1995.

death rates from cardiovascular diseases and non-insulin dependent diabetes. The mechanisms underlying these long-term effects are not understood, but the studies emphasize the importance of adequate growth for normal development.

Embryonic growth is dependent on growth factors. We have seen that FGFs control cell proliferation in the progress zone of the developing chick limb, and are required for proliferation and outgrowth of the bud (see Section 10-3). Evidence for the role of other growth factors in embryonic growth comes from the technique of gene knock-out (see Box 4B, page 108). Insulin-like growth factors 1 and 2 (IGF-1 and IGF-2), which are single-chain protein growth factors that closely resemble insulin and each other in their amino-acid sequence, have a key role not only in post-natal mammalian growth, but also in growth during embryonic development. Newborn mice lacking a functional *Igf-2* gene develop relatively normally, but weigh only 60% of the normal newborn body weight. Mice in which the *Igf-1* gene has been inactivated are also growth retarded, showing that IGF-1 also has an important role in embryonic growth. Both factors and their receptors are present as early as the eight-cell stage of mouse development. IGF-2 seems to be essential for early embryonic growth, after which the action of IGF-1 becomes dominant. Both proteins act via the IGF-1 receptor; the IGF-2 receptor by contrast reduces the concentration of IGF-2, by enabling it to be degraded. The genes for IGF-2 and its receptor are two of the genes that are imprinted in mammals; they are inactivated in maternal and paternal germ cells, respectively (see Section 12-13).

Growth hormone, which is synthesized in the pituitary gland, is essential for the post-embryonic growth of humans and other mammals and is secreted throughout fetal life. Within the first year of birth, secretion of growth hormone by the pituitary gland commences. A child with insufficient growth hormone grows less than normal, but if growth hormone is given regularly, normal growth is restored. In this case, there is a catch-up phenomenon, with a rapid initial response that tends to restore the growth curve to its original trajectory.

Production of growth hormone is under the control of two hormones produced in the hypothalamus: growth hormone-releasing hormone, which promotes growth hormone synthesis and secretion, and somato-statin, which inhibits its production and release. Growth hormone produces many of its effects by inducing the synthesis of IGF-1 (Fig. 14.6) and, to a lesser extent, IGF-2. Post-natal growth, as well as embryonic growth, is largely due to the actions of these insulin-like growth factors, and complex hormonal regulatory circuits control their production.

As can be seen from measurements of human growth after birth (see Fig. 14.4, left panel), the growth rate decreases until puberty, when there is a sudden growth spurt. This sharp increase is due to secretion of gonadotropins; these cause the increased production of the steroid sex hormones, which in turn cause increased production of growth hormone. This growth spurt occurs a couple of years earlier in girls than in boys. We next look at the growth of some individual organs.

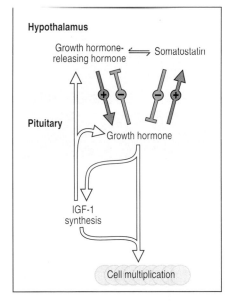

Fig. 14.6 Growth hormone production is under the control of the hypothalamic hormones. Growth hormone is made in the pituitary gland and is secreted. Growth hormone-releasing hormone from the hypothalamus promotes growth hormone synthesis, while somatostatin inhibits it. Growth hormone controls its own release by negative feedback signals to the hypothalamus. Growth hormone causes the synthesis of the insulin-like growth factor IGF-1 and this promotes the production of growth hormone.

| 14-4 | **Developing organs can have their own intrinsic growth programs.** |

Patterning of the embryo occurs while the organs are still very small. For example, human limbs have their basic pattern established when they are less than 1 cm long. Yet, over the years, the limb grows to be at least one hundred times longer. How is this growth controlled? It appears that each of the cartilaginous elements in the limb has its own individual growth program. In the chick wing, the initial size of the cartilaginous

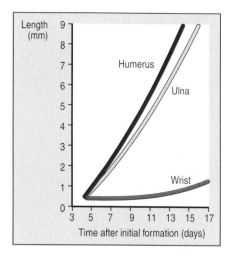

Fig. 14.7 Comparative growth of the cartilaginous elements in the embryonic chick wing. When first laid down, the cartilaginous elements of the humerus, ulna, and wrist are the same size, but the humerus and ulna then grow much more than the wrist element.

elements in the long bones—the humerus and ulna—are similar to the elements in the wrist (Fig. 14.7). Yet, with growth, the humerus and ulna increase many times in length compared with the wrist bones. These growth programs are specified when the elements are initially patterned and involve both cell multiplication and matrix secretion. Each element follows its growth program even when grafted to neutral sites, provided that a good blood supply is established.

A classic illustration of an intrinsic growth program comes from grafting limb buds between large and small species of *Ambystoma*, a genus of salamanders. A limb bud from the larger species grafted to the smaller species initially grows slowly but eventually ends up its normal size, which is much larger than any of the limbs of the host (Fig. 14.8). Whatever the circulatory factors that influence growth, the intrinsic response of different tissues is therefore crucial. Such intrinsically patterned differential growth can affect the overall form of an organism considerably, as can be seen in Fig. 14.5.

14-5   Growth of the long bones occurs in the growth plates.

An important aspect of post-embryonic vertebrate growth is the growth of the long bones of the limbs (humerus, femur, radius, and ulna). The long bones are initially laid down as cartilaginous elements (see Section 10-2) and then become ossified. The early growth of these elements involves both cell proliferation and matrix secretion in a well-defined pattern. In both fetal and post-natal growth, the cartilage is replaced by bone in a process known as **endochondral ossification**, in which ossification starts in the centers (diaphyses) of the long bones and spreads outward (Fig. 14.9). There are also secondary ossification centers at each end of the bones (the epiphyses). The adult long bones thus have a bony shaft with cartilage confined to the articulating surfaces at each end, and to two internal regions near each end, the **growth plates**, in which growth occurs. In the growth plates, the cartilage cells are usually arranged in columns, and various zones can be identified. Just next to the bony epiphysis is a narrow germinal zone, which contains stem cells. Next is a proliferative zone of cell division, followed by a zone of maturation, and a hypertrophic zone, in which the cartilage cells increase in size. Finally, there is a zone in which the cartilage cells die and are replaced by bone laid down by osteoblasts. There is a strong similarity to the development of skin, where basal stem cells give rise to dividing cells, which differentiate into keratinocytes and finally die (see Section 14-7).

The mechanism by which the cartilage cells in the growth plates are controlled so that they stop dividing and enlarge to form the scaffolding for bone has been studied in mice. It involves a signaling molecule called Indian hedgehog, which belongs to the large and widespread family of proteins that includes mammalian Sonic hedgehog and *Drosophila* hedgehog proteins. Indian hedgehog is expressed in the hypertrophic cells of the growth plate. Its target is a protein that is related to parathyroid hormone and is present in the perichondrium surrounding the bone. Indian hedgehog promotes the proliferation of cartilage cells and prevents their hypertrophy. Lack of Indian hedgehog protein in mice results in premature cartilage cell differentiation, and thus short stubby limbs.

Fig. 14.8 The size of limbs is genetically programmed in salamanders. An embryonic limb bud from a large species of salamander, *Ambystoma tigrinum*, grafted to the embryo of a smaller species, *Ambystoma punctatum*, grows much larger than the host limbs—to the size it would have grown in *Ambystoma tigrinum*. Photograph from Harrison, R.G.: 1969.

Growth hormone affects bone growth by acting on the growth plates. The cells in the germinal zone have receptors for growth hormone, and growth hormone is probably directly responsible for stimulating these stem cells to proliferate. Further growth, however, is probably mediated by

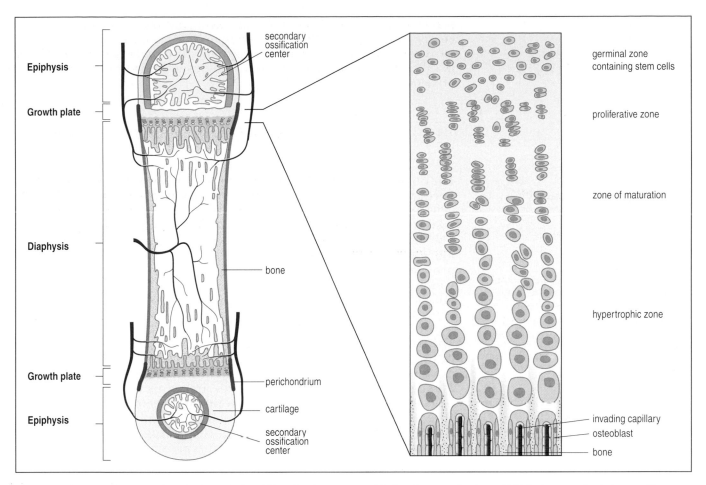

Fig. 14.9 Growth plates and endochondral ossification in the long bone of a vertebrate. The long bones of vertebrate limbs increase in length by growth from cartilaginous growth plates. The growth plates are cartilaginous regions that lie between the epiphysis of the future joint and the central region of the bone, the diaphysis. In the figure, bone has already replaced cartilage in the diaphysis, and more bone is being added at the growth plates. Within the growth plates, cartilage cells multiply in the proliferative zone, then mature and undergo hypertrophy (cell enlargement). They are then replaced by bone, which is laid down by special cells called osteoblasts. Secondary sites of ossification are located within the epiphyses. After Wallis, G.A.: 1993.

IGF-1, whose production in the growth plate is stimulated by growth hormone. Thyroid hormones are also necessary for optimal bone growth; they act both by increasing the secretion of growth hormone and IGF-1, and by stimulating hypertrophy of the cartilage cells. FGF is also important for bone growth; the genetic defect that gives rise to achondroplasia (short-limbed dwarfism) is a dominant mutation in the FGF receptor-3, whose normal function is to limit, rather than promote, bone formation.

The rate of increase in the length of a long bone is equal to the rate of new cell production per column multiplied by the mean height of an enlarged cell. The rate of new cell production depends both on the time cells take to complete a cycle in the proliferative zone, and the size of this zone. Different bones grow at different rates, and this can reflect the size of the proliferative zone, the rate of proliferation, and the degree of cell enlargement in the growth plate.

In view of the complexity of the growth plate, it is remarkable that human bones in the limbs on either side of the body can grow for some 15 years independently of each other, and yet eventually match to an accuracy of about 0.2%. This is achieved by having many columns of cells in each plate, so that growth variations are averaged out. Growth of a bone ceases when the growth plate ossifies, and this occurs at different times for different bones. Ossification of growth plates occurs in a strict order in different bones and can therefore be used to provide a measure of physiological age.

14-6   Growth of vertebrate striated muscle is dependent on tension.

The number of striated (skeletal) muscle fibers in vertebrates is determined during embryonic development. Once differentiated, striated muscle cells lose the ability to divide. Post-embryonic growth of muscle tissue results from an increase in individual fiber size, both in length and girth. The number of myofibrils within the enlarged muscle fiber can increase more than 10-fold. Additional nuclei for the much enlarged cell are provided by the fusion of satellite cells with the fiber. Satellite cells, which are undifferentiated cells lying adjacent to the differentiated muscle, also act as a reserve population of stem cells that can replace damaged muscle.

The increase in the length of a muscle fiber is associated with an increase in the number of sarcomeres, the functional contractile unit. For example, in the soleus muscle of the mouse leg, as the muscle increases in length, the number of sarcomeres increases from 700 to 2300 at 3 weeks after birth. This increase in number seems to depend on the growth of the long bones putting tension on the muscle via its tendons. If the soleus muscle is immobilized by placing the leg in a plaster cast at birth, sarcomere number increases slowly over the next 8 weeks, but then increases rapidly when the cast is removed (Fig. 14.10). One can thus see how bone and muscle growth are mechanically coordinated.

14-7   The epithelia of adult mammalian skin and gut are continually replaced by derivatives of stem cells.

Adult vertebrate skin is composed of three layers—the **dermis**, which mainly contains fibroblasts, the protective outer **epidermis**, which mainly contains **keratinocytes**, and the **basal lamina** or basement membrane, composed of extracellular matrix, which separates the epithelial epidermis from the dermis.

Because of the protective function of the epidermis, cells are continuously lost from its outer surface and must be replaced. Cells are replaced from a basal layer of proliferating epidermal cells in contact with the basal lamina. These proliferating cells are derived from stem cells within the basal layer. The stem cells form a special population in the basal layer that multiplies relatively slowly and retains proliferative potential throughout the life of the organism. They have the property of self-renewal, like hematopoietic stem cells (see Section 9-12). Stem cells divide asymmetrically, with one daughter remaining a stem cell and the other daughter becoming committed to differentiation. After becoming committed, this cell divides a few times more, and its progeny differentiate further after leaving the basal layer. As they move through the skin they mature, until by the time they reach the outermost layer they are completely filled with keratin and their membranes are strengthened with the protein involucrin. The dead cells eventually flake off the surface (Fig. 14.11).

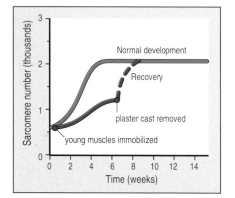

Fig. 14.10 The growth in length of the muscles attached to the long bones of the mouse leg depends on the tension provided by long bone growth. The length of a muscle is related to the number of sarcomeres—the basic contractile unit. If the limb is immobilized by a plaster cast in a position in which there is no tension on the muscle, there is little increase in muscle length until the cast is removed. There is then a rapid increase in length.

Fig. 14.11 Differentiation of keratinocytes in human epidermis. Descendants of stem cells, which will become keratinocytes, become detached from the basal lamina, divide several times, and leave the basal layer before beginning to differentiate. Differentiation of keratinocytes involves the production of large amounts of the intermediate filament protein, keratin. In the intermediate layers, the cells are still large and metabolically active, whereas in the outer epidermal layers, the cells lose their nuclei, become filled with keratin filaments, and their membranes become insoluble, due to deposition of the protein involucrin. The dead cells are eventually shed from the skin surface.

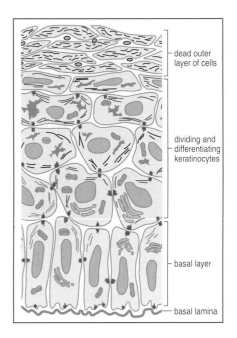

Cell adhesion molecules play a key role in keratinocyte differentiation. The basal cell layer is attached to the basement membrane by specialized cell junctions that include both **hemi-desmosomes** and focal contacts. The epidermal cells in the upper layers are connected to each other through extensive **desmosomal junctions**. The integrin family of cell adhesion molecules (see Box 8A, page 233) plays the most important part in keratinocyte development. The expression of integrins is largely confined to the basal layer, where they mediate attachment of cells to the basal lamina at the hemi-desmosome junctions. Stem cells express higher levels of α2\β1 and α3\β1 integrins (which are receptors for collagen and laminin) than other cells of the basal layer, and these can be used to determine the cells' locations within the layer. It may be that these high levels of integrins keep the cells in the basal layer. It is likely that components of the basal lamina actively inhibit keratinocyte differentiation; differentiation and migration out of the basal layer only begins when a cell becomes detached from the basal lamina. This is accompanied by a decrease in integrins on the cell surface.

The epithelial cells lining the gut are also continuously replaced from stem cells. In the small intestine, the endothelial cells form a single-layered epithelium. This is highly folded into villi that project into the gut, and crypts that penetrate the underlying connective tissue. Cells are continuously shed from the tips of the villi, while the stem cells that produce their replacements lie near the base of the crypts. The new cells generated by the stem cells move upward toward the tips of the villi. En route, they multiply in the lower half of the crypt (Fig. 14.12).

A single crypt in the small intestine of the mouse contains about 250 cells of four main differentiated cell types. About 150 of the cells are proliferative, dividing about twice a day so that the crypt produces 300 new cells each day; about 12 cells leave the crypt each hour. It is estimated that there is just one stem cell per crypt, but some 30–40 potential stem cells that may acquire stem cell properties following injury or other trauma.

14-8 Cancer can result from mutations in genes controlling cell multiplication and differentiation.

Cancer can be regarded as a major perturbation of normal cellular behavior that results from certain mutations in somatic cells. Creating and maintaining tissue organization requires strict controls on cell division, differentiation, and growth. In cancer, cells escape from these normal controls, and proceed along a path of uncontrolled growth and migration that can kill the organism. There is usually a progression from a benign localized growth to malignancy in which the cells **metastasize**—migrate to many parts of the body where they continue to grow.

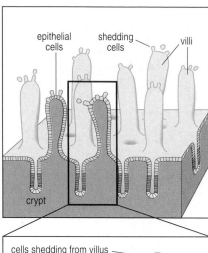

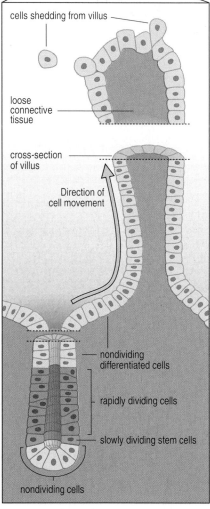

Fig. 14.12 Epithelial cells lining the mammalian gut are continuously replaced. The upper panel shows the villi—epithelial extensions covering the wall of the gut. The lower panel is a detail of a crypt and villus. The base of the crypts contain stem cells. They give rise to cells that proliferate and move upward. These differentiate into epithelium in the projecting villus, and are shed from the tips. The total time of transit is about 4 days. After Alberts, B., *et al.*: 1989.

The cells most likely to give rise to cancer are those that are undergoing continual division. Because they replicate their DNA frequently, they are more likely than other cells to accumulate mutations that arise from errors in DNA replication. In almost all cancers, the cancer cells are found to have a mutation in one or more genes. Not all mutations give rise to a cancer, however, and particular genes in which mutation can contribute to cancer formation have been identified in humans and other mammals. These genes are known as **proto-oncogenes**; when such a gene undergoes mutation it becomes an **oncogene**. In some cases, the presence of a single oncogene is capable of making a cell cancerous. At least 70 proto-oncogenes have been identified in mammals.

Proto-oncogenes are genes that are involved in cell proliferation, differentiation, and migration. We have seen how extracellular signals provided by growth factors and other signal molecules play a key role in the regulation of cell differentiation and cell proliferation. Many proto-oncogenes encode growth factors and other signaling molecules, or their receptors, or proteins involved in the intracellular response to such signals. An altered version of any of these proteins produced by its mutant oncogene can cause permanent activation of the cell cycle, causing the cell and its progeny to divide indefinitely. Such mutations have a dominant effect on the cell as only one copy of the gene needs to be mutated to produce the change that can lead to malignancy.

There is another group of genes in which mutations can also lead to cancer. These are the **tumor suppressor genes**, in which inactivation or deletion of both copies of the gene is required for a cell to become cancerous. The classic example of a tumor caused by the loss of such a gene is the childhood tumor, retinoblastoma, which is a tumor of retinal cells. Although retinoblastoma is normally very rare, there are families in which an inherited predisposition is determined by a single gene, and they have provided the means of identifying the gene responsible for this cancer. The inherited defect in some of these families turns out to be a deletion of a particular region on one of the two copies of chromosome 13. This on its own does not cause the cells to be cancerous. However, if any retinal cell also acquires a deletion of the same region on the other copy of chromosome 13, a retinal tumor develops. The gene in this region that is responsible for susceptibility to retinoblastoma is known as the *retinoblastoma* (*Rb*) gene. Both copies of the *Rb* gene must be lost or inactivated for a cell to become cancerous (Fig. 14.13), and hence *Rb* is regarded as a tumor suppressor gene. The *Rb* gene encodes a protein that is involved in the regulation of the cell cycle. We have seen in Sections 9-9 and 13-5 that the Rb protein plays a role in the normal development of muscle, where it is involved in the withdrawal of muscle precursor cells from the cell cycle.

Another tumor suppressor gene is familiar to us as a gene first identified as a segment polarity gene in *Drosophila*. The gene *patched* codes for a transmembrane protein that is part of the hedgehog signaling pathway (see Section 5-16). In humans, mutations that inactivate or delete both copies of the human homolog of *patched* are associated with a variety of epithelial cell tumors. Loss of just one copy of the homolog leads to some skeletal abnormalities.

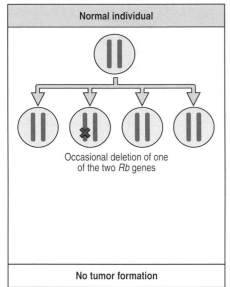
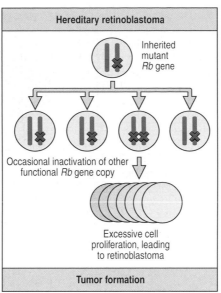

Fig. 14.13 The *retinoblastoma* (*Rb*) gene is a tumor suppressor gene. If only one copy of the *Rb* gene is lost or inactivated, no tumor develops (left panel). In individuals already carrying an inherited mutant *Rb* gene, if the other copy of the gene is lost or inactivated in a cell, that cell will generate a retinal tumor (right panel). Such individuals are thus at a much greater risk risk of developing retinoblastoma, and usually do so at a young age.

A major feature of cancer is the failure of the cells to differentiate properly. The majority of cancers—over 85%—occur in epithelia. This is not surprising when it is recalled that many epithelia (such as the epidermis and the lining of the gut) are constantly being renewed by division and differentiation of stem cells (see Section 14-7). In normal epithelia, cells generated by stem cells continue to divide for a little time until they undergo differentiation, when they stop dividing. By contrast, cancerous epithelial cells continue to divide, although not necessarily more rapidly, and usually fail to differentiate.

This failure of cancer cells to differentiate is also clearly seen in certain leukemias—cancers of white blood cells. All blood cells are continually renewed from a pluripotent stem cell in the bone marrow, by a process in which steps in differentiation are interspersed with phases of cell proliferation. The pathway eventually culminates in terminal cell differentiation and a complete cessation of cell division (see Section 9-11). Several types of leukemia are caused by cells continuing to proliferate instead of differentiating. These cells become stuck at a particular immature stage in their normal development, a stage that can be identified by the molecules expressed on their cell surfaces. Cancers that result from a failure of cell differentiation could possibly be cured by the addition of a factor that could promote differentiation.

A number of other developmental genes that we have considered elsewhere are also involved in cancer. For example, the first mammalian member of the Wnt gene family to be identified was identified as an oncogene, and abnormal expression of this signaling protein can block cell differentiation. E-cadherin is downregulated in cancers that spread; mutations in the Notch pathway can lead to a block in differentiation, and therefore also result in the cancer; and TGF-β members are involved in tumor suppression.

Rarely, cancers can develop without any alteration in the cell's genetic material. The clearest examples are **teratocarcinomas**, solid tumors that spontaneously arise from germ cells. Teratocarcinomas are unusual tumors, in that they can contain a bizarre mixture of differentiated cell types. Spontaneous teratocarcinomas usually occur in the ovary or testis, and are derived from the germ cells. In the mouse ovary, accidental activation of an unfertilized egg results in its development *in situ* to the stage of epiblast formation; this epiblast then gives rise to a tumor. Similarly,

Fig. 14.14 Embryonic stem (ES) cells can develop normally or into a tumor, depending on the environmental signals they receive. Cultured ES cells originally obtained from an inner cell mass of a mouse contribute to a healthy chimeric mouse when injected into an early embryo (bottom left panels), but if injected under the skin of an adult mouse, the same cells will develop into a teratocarcinoma (bottom right panels).

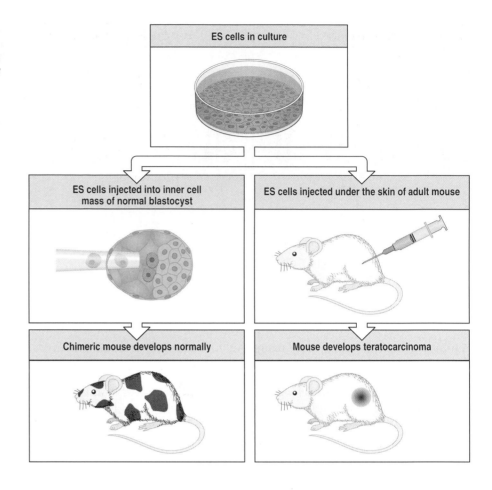

if the epiblast of an early mouse embryo is transplanted to any site in the body of an adult mouse in which it receives a good blood supply, it gives rise to a teratocarcinoma.

There is good reason to believe that teratocarcinomas are not caused by genetic alterations. A mouse inner cell mass placed in culture gives rise to embryonic stem cells (ES cells), which can grow indefinitely in culture (see Box 3C, page 81). These cells maintain their embryonic character, and when put back into the inner cell mass of another embryo, contribute normally to many different tissues, including the germ line to produce a chimeric mouse. However, when the same ES cells are placed under the skin of an adult mouse, they develop into a teratocarcinoma (Fig. 14.14). Transgenic mice containing tissues derived from ES cells do not have an increased probability of forming tumors, yet those same cells consistently form tumors in adult mice. This indicates that the teratocarcinoma must be the result of the inner mass cells receiving the wrong developmental signals, and not of a genetic change.

14-9 Hormones control many features of plant growth.

Unlike the protein growth hormones of animals, plant hormones are typically small molecular weight organic molecules. Auxin (indole-3-acetic acid) is one of the main regulators of plant growth, and is implicated in a large number of developmental processes, including growth toward a light source, tissue polarity, vascular tissue differentiation, and apical dominance, which is the phenomenon of the suppression of the growth

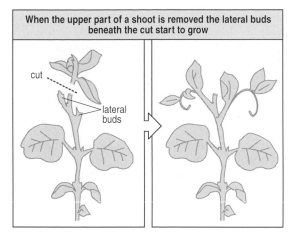

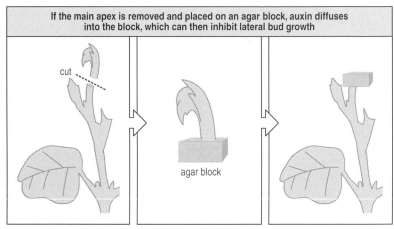

Fig. 14.15 Apical dominance in plants. Left panels: the growth of lateral buds is inhibited by the apical meristem above them. If the upper part of the stem is removed, lateral buds start growing. Right panels: an experiment to show that apical dominance is due to inhibition of lateral bud outgrowth by a substance secreted by the apical region. This substance is the plant hormone auxin.

of lateral buds immediately below the apical bud. Apical dominance is caused by a diffusible inhibitor of bud outgrowth produced by the shoot apex, as shown by putting an excised apex in contact with an agar block, which can then on its own inhibit lateral growth. The inhibitor is auxin. It is produced by the apical bud, transported down the stem, and suppresses the outgrowth of buds that fall within its sphere of influence. If the apical bud is removed, apical dominance is also removed, and lateral buds start to grow (Fig. 14.15). Application of auxin to the cut tip replaces the suppressive effect of the apical bud.

Another family of plant hormones, the gibberellins, regulate stem elongation and have some similar effects to auxin. Cytokinins, which can stimulate cell proliferation in culture, are derivatives of adenine. The chemical nature of plant hormones is very different from those of animals; nevertheless, they are thought to act through specific hormone-binding receptors and to stimulate intracellular signal transduction.

Plant growth is affected by a variety of environmental factors, such as temperature, humidity, and light. A seedling grown in the dark takes on a characteristic etiolated form in which chloroplasts do not develop, there is extensive elongation of internodes, and the leaves do not expand. The effect of light on plant growth (photomorphogenesis) is mediated through a family of intracellular receptor proteins called phytochromes, which respond to red light and regulate many aspects of plant development and growth.

14-10 | Cell enlargement is central to plant growth.

Plant growth is routinely achieved by cell division in meristems and organ primordia, followed by irreversible cell enlargement, which achieves most of the increase in size. Controlled cell enlargement can result in up to a 50-fold increase in the volume of a tissue. The driving force for expansion is the hydrostatic pressure exerted on the cell wall when the protoplast swells as a result of water entry into the cell vacuoles by osmosis (see Fig. 8.42). Cell expansion involves synthesis and deposition of new wall material, and is accompanied by an increase in both protein and RNA synthesis.

The direction of cell expansion is determined by the orientation of the cellulose fibrils in the cell wall, and is an example of directed dilation (discussed in Chapter 8 in relation to notochord expansion, see Fig. 8.40). Enlargement primarily occurs in a direction at right angles to the fibrils, where the wall is weakest. The orientation of cellulose fibrils in the cell wall is thought to be determined by the microtubules of the cell's cytoskeleton, which are responsible for positioning the enzyme assemblies that synthesize cellulose at the cell wall.

Mutations that affect cell expansion can significantly reduce elongation of the root. Plant growth hormones, such as ethylene and gibberellic acid, alter the orientation in which the fibrils are laid down, and so can alter the direction of expansion. Auxin aids expansion, perhaps by activating expansins, cell wall proteins that act by loosening the structure of the cell wall.

Summary.

Growth in both animals and plants mainly occurs after the basic body plan has been laid down. In animals, growth can occur by cell multiplication, cell enlargement, and secretion of large amounts of extracellular matrix. A plant's growth in size is both by cell enlargement and cell division. In mammals, insulin-like growth factors (IGFs) are required for normal embryonic growth, and they also mediate the effects of growth hormone after birth. Human post-natal growth is largely controlled by growth hormone, which is made in the pituitary gland. Transplantation experiments show that individual organs and bones each have their own intrinsic growth program, which determines their final size. Growth in the long bones occurs in response to growth hormone stimulation of the cartilaginous growth plates at either end of the bone. Cancer is the result of the loss of growth control and differentiation. Plant growth is dependent on auxins, gibberellins, and other growth hormones.

Molting and metamorphosis.

Many animals do not develop directly from an embryo into an 'adult' form, but into a larva from which the adult eventually develops by metamorphosis. The changes that occur at metamorphosis can be rapid and dramatic, the classic examples being the metamorphosis of a caterpillar into an adult butterfly, a maggot into a fly, and a tadpole into a frog. In some cases it is hard to see any resemblance between the animal before and after metamorphosis. The adult fly does not resemble the maggot at all, because adult structures develop from the imaginal discs and so are completely absent from the larval stages (see Chapters 5 and 10). Another striking example of metamorphosis is the transformation of the pluteus stage larva of the sea urchin into the adult. In other cases, the changes are somewhat less dramatic, as in amphibian metamorphosis and the various larval stages of nematodes. In frogs, the most obvious changes at metamorphosis are the regression of the tadpole's tail and the development of limbs, although many other structural changes occur. In some insects, the entire body plan is transformed, with most larval tissues undergoing cell death as the adult tissues develop from the imaginal discs and histoblasts. In all arthropods, increase in size in larval and pre-adult stages requires shedding of the external cuticle, which is rigid, a process known as **molting**.

Fig. 14.16 Growth and molting of the tobacco hornworm caterpillar (*Manduca sexta*). The caterpillar goes through a series of molts. The tiny hatchling (1)—indicated by the arrow—molts to become a caterpillar (2), and then undergoes three further molts (3, 4, and 5). The increase in size between molts is about twofold. The caterpillars are sitting on a lump of caterpillar food. Scale bar = 1 cm. Photograph courtesy of S.E. Reynolds.

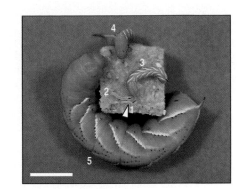

A number of features distinguish early embryogenesis from molting, metamorphosis, and other aspects of post-embryonic development. Whereas the signal molecules in early development act over a short range and are typically protein growth factors, many signals in post-embryonic development are produced by specialized endocrine cells, and include both protein and nonprotein hormones. The synthesis of these hormones is orchestrated by the central nervous system, and there are complex interactions between the endocrine glands.

14-11 Arthropods have to molt in order to grow.

Arthropods have a rigid outer skeleton, the cuticle, which is secreted by the epidermis. This makes it impossible for the animals to increase in size gradually. Instead, increase in body size takes place in steps associated with the loss of the old outer skeleton and the deposition of a new larger one. This process is known as **ecdysis** or **molting**. The stages between molts are known as instars. *Drosophila* larvae have three instars and molts. The increase in overall size between molts can be striking, as illustrated in Fig. 14.16 for the tobacco hornworm.

At the start of a molt, the epidermis separates from the cuticle in a process known as apolysis, and a fluid (molting fluid) is secreted into the space between the two (Fig. 14.17). The epidermis then increases in area by cell multiplication or cell enlargement, and becomes folded. It begins to secrete a new cuticle, and the old cuticle is partly digested away, eventually splits, and is shed.

Molting is under hormonal control. Stretch receptors that monitor body size are activated as the animal grows, and this results in the brain secreting prothoracicotropic hormone. This activates the prothoracic gland to release the steroid hormone ecdysone, which is the hormone that causes molting. The action of ecdysone is counteracted

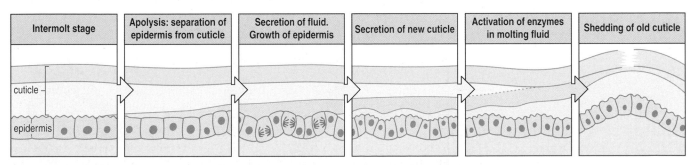

| Intermolt stage | Apolysis: separation of epidermis from cuticle | Secretion of fluid. Growth of epidermis | Secretion of new cuticle | Activation of enzymes in molting fluid | Shedding of old cuticle |

Fig. 14.17 Molting and growth of the epidermis in arthropods. The cuticle is secreted by the epidermis. At the start of molting, the cuticle separates from the epidermis—apolysis—and a fluid is secreted between them. The epidermis grows, becomes folded, and begins to secrete a new cuticle. Enzymes weaken the old cuticle, which is shed.

by juvenile hormone, produced by the corpus allatum. A similar hormonal circuit controls metamorphosis, and is described more fully in the next section.

14-12   Metamorphosis is under environmental and hormonal control.

When an insect larva has reached a particular stage it does not grow and molt any further, but undergoes a more radical **metamorphosis** into the adult form. Metamorphosis occurs in many animal groups other than arthropods, including amphibians. In both insects and amphibians, environmental cues, such as nutrition, temperature, and light, as well as the animal's internal developmental program, control metamorphosis through their effects on neurosecretory cells in the brain. There are two groups of hormone-producing cells, one of which promotes metamorphosis, and another which inhibits it. Metamorphosis occurs when the inhibition, which is predominant in the larval stage, is overcome in response to environmental cues. The signals produced by the two sets of endocrine cells control the development of all the cells involved in metamorphosis.

In insects, temperature and light cues stimulate neurosecretory cells in the larva's central nervous system to release signals that act on a neurosecretory release site behind the brain, which then secretes prothoracicotropic hormone. This acts on the prothoracic gland to stimulate the production of ecdysone. It is ecdysone that promotes metamorphosis, as demonstrated by its ability to induce premature metamorphosis in fly larvae. However, the action of ecdysone can be counteracted by another hormone—juvenile hormone—produced by the corpus allatum, an endocrine gland located just behind the brain. As its name implies, juvenile hormone maintains the larval state. In butterflies, a pulse of ecdysone in the final instar larva triggers the beginning of pupa formation, and another pulse some days later initiates the later stages of metamorphosis (Fig. 14.18).

Fig. 14.18 Insect metamorphosis. The corpus allatum of a butterfly larva secretes juvenile hormone, which inhibits metamorphosis. In response to environmental changes, such as an increase in light and temperature, the corpus allatum of the final instar larva begins to secrete prothoracicotropic hormone (PTTH). This acts on the prothoracic gland to stimulate the secretion of ecdysone, the hormone that overcomes the inhibition by juvenile hormone and causes metamorphosis. After Tata, J.R.: 1993.

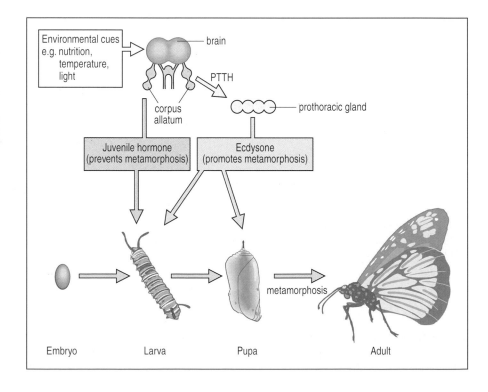

Ecdysone crosses the plasma membrane, where it interacts with intracellular ecdysone receptors that belong to the steroid hormone receptor superfamily. These receptors are gene regulatory proteins, which are activated by binding their hormone ligand (see Section 9-7). The hormone-receptor complex binds to the regulatory regions of a number of different genes, inducing a new pattern of gene activity characteristic of metamorphosis.

In amphibians, in response to nutritional status and environmental cues, such as temperature and light, the neurosecretory cells of the hypothalamus release thyrotropin-releasing hormone, which acts on the pituitary gland, causing it to release thyroid-stimulating hormone. This in turn acts on the thyroid gland to stimulate the secretion of the thyroid hormones that bring about metamorphosis (Fig. 14.19). The thyroid hormones are the iodo-amino acids thyroxine ($T4$), and tri-iodothyronine ($T3$). They are signaling molecules of ancient origin, occurring even in plants. Although very different in chemical structure to ecdysone, they too pass through the plasma membrane and interact with intracellular receptors that belong to the steroid hormone receptor superfamily. The pituitary also produces prolactin, which is an inhibitor of metamorphosis.

A striking feature of the hormones that stimulate metamorphosis is that as well as affecting a wide variety of tissues, they affect different tissues in different ways, their effects varying from the subtle to the gross. In the tadpole limb, for example, thyroid hormones promote development and growth, whereas they cause cell death and degeneration in the tail. Metamorphosis also leads to changes in the responsiveness of cells to other signals; for example, in *Xenopus*, estrogen can only induce the synthesis of vitellogenin, a protein required for the yolk of the egg, after metamorphosis. Each tissue has its own response to the hormones causing metamorphosis, and some of these effects can be reproduced

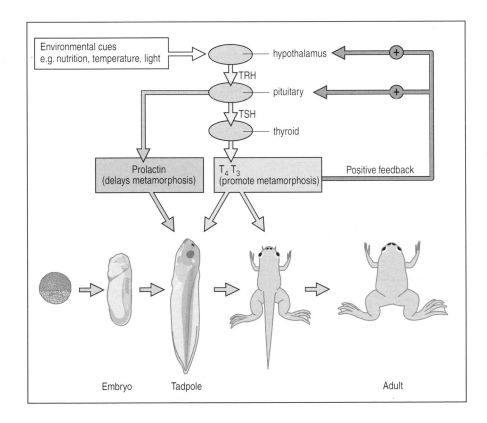

Fig. 14.19 Amphibian metamorphosis. During the early development of a tadpole, the pituitary produces prolactin, which inhibits metamorphosis. Changes in the environment, such as an increase in nutritional levels, cause the secretion of thyrotropin-releasing hormone (TRH) from the hypothalamus, which acts on the pituitary to release thyroid-stimulating hormone (TSH). This in turn acts on the thyroid glands to stimulate secretion of the thyroid hormones thyroxine (T_4) and tri-iodothyronine (T_3), which cause metamorphosis. The thyroid hormones also act on the hypothalamus and pituitary to maintain synthesis of TRH and TSH. After Tata, J.R.: 1993.

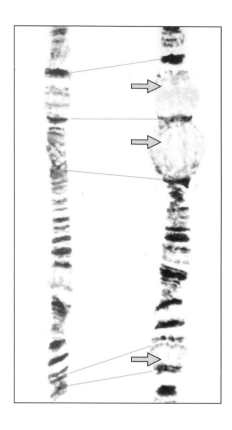

Fig. 14.20 Gene activity seen as puffs on the polytene chromosomes of Drosophila. A region of a chromosome is shown from a young third instar (left), and from an older larva (right), after ecdysone has induced puffs at three loci (arrowed). Photograph courtesy of M. Ashburner.

in culture. When excised *Xenopus* tadpole tails are exposed to thyroid hormones in culture, for example, they cause cell death and complete tissue regression; addition of prolactin prevents this occurring.

In *Drosophila*, it is possible to see the changes in gene activity that occur during metamorphosis because of the special characteristics of polytene chromosomes. Cells in some tissues of the larva grow and repeatedly pass through S-phase without undergoing mitosis and cell division. The cells become very large and can have several thousand times the normal complement of DNA. In salivary gland cells, many copies of each chromosome are packed side by side to form giant polytene chromosomes. When a gene is active, the chromosome at that site expands into a large localized 'puff', which is easily visible (Fig. 14.20). The puff represents the unfolding of chromatin and the associated transcriptional activity. When the gene is no longer active the puff disappears. During the last days of larval life, a large number of puffs are formed in a precise sequence, a pattern that is under the direct influence of ecdysone.

Summary.

Arthropod larvae grow by undergoing a series of molts in which the rigid cuticle is shed. Metamorphosis during the post-embryonic period can result in a dramatic change in the form of an organism. In insects, it is hard to see any resemblance between the animal before and after metamorphosis, whereas in amphibians, the change is somewhat less dramatic. Environmental and hormonal factors control metamorphosis. In both insects and amphibians there are two sets of hormonal signals, one promoting, the other delaying, metamorphosis. Thyroid hormones cause metamorphosis in amphibians, and ecdysone does the same in insects. In *Drosophila*, gene activity during metamorphosis can be monitored by localized puffing on the giant polytene chromosomes of the salivary gland cells.

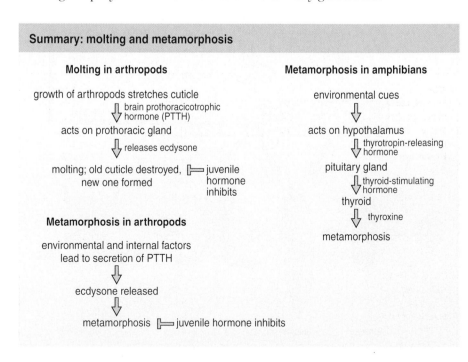

Summary: molting and metamorphosis

Molting in arthropods

growth of arthropods stretches cuticle
⇩ brain prothoracicotrophic hormone (PTTH)
acts on prothoracic gland
⇩ releases ecdysone
molting; old cuticle destroyed, ⊨ juvenile hormone inhibits
new one formed

Metamorphosis in arthropods

environmental and internal factors lead to secretion of PTTH
⇩
ecdysone released
⇩
metamorphosis ⊨ juvenile hormone inhibits

Metamorphosis in amphibians

environmental cues
⇩
acts on hypothalamus
⇩ thyrotropin-releasing hormone
pituitary gland
⇩ thyroid-stimulating hormone
thyroid
⇩ thyroxine
metamorphosis

Aging and senescence.

Organisms are not immortal, even if they escape disease or accidents. With aging—the passage of time—**senescence** occurs. Senescence is the increase of impairment of physiological functions with age, resulting in a decreased ability to deal with a variety of stresses, and an increased susceptibility to disease. This phenomenon raises many unanswered questions as to its underlying mechanisms, but we can at least consider some general questions, such as whether senescence is part of an organism's post-embryonic developmental program or whether it is simply the result of wear and tear.

Although individuals may vary in the time at which particular aspects of aging appear, the overall effect is summed up as an increased probability of dying in most animals, including humans, with increased age. This life pattern, which is illustrated in relation to *Drosophila* (Fig. 14.21), is typical of many animals. There are, however, exceptions, such as the Pacific salmon, in which death does not come after a process of gradual aging, but is linked to a certain stage in the life cycle, in this case to spawning.

One view of senescence is that it is the outcome of an accumulation of damage that eventually outstrips the ability of the body to repair itself, and so leads to the loss of essential functions. For example, some old elephants die of starvation because their teeth have worn out. Nevertheless, there is clear evidence that senescence is under genetic control, as different animals age at vastly different rates, as shown by their different life spans (Fig. 14.22). An elephant, for example, is born after 21 months' embryonic development, and at that point shows few, if any, signs of aging, whereas a 21 month old mouse is already well into middle age and beginning to show signs of senescence. This genetic control of aging can be understood in terms of the 'disposable soma theory', which puts it into the context of evolution. The disposable soma theory proposes that natural selection tunes the life history of the organism so that

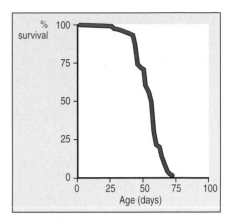

Fig. 14.21 Aging in *Drosophila*. The probability of dying increases rapidly at older ages.

Fig. 14.22 Table showing life span, length of gestation, and age at puberty for various mammals.

Longevity and time to attain reproductive maturity at puberty for various mammals			
	Maximum life span (months)	Length of gestation (months)	Age at puberty (months)
Man	1440	9	144
Finback whale	960	12	—
Indian elephant	840	21	156
Horse	744	11	12
Chimpanzee	534	8	120
Brown bear	442	7	72
Dog	408	2	7
Cattle	360	9	6
Rhesus monkey	348	5.5	36
Cat	336	2	15
Pig	324	4	4
Squirrel monkey	252	5	36
Sheep	240	5	7
Gray squirrel	180	1.5	12
European rabbit	156	1	12
Guinea-pig	90	2	2
House rat	56	0.7	2
Golden hamster	48	0.5	2
Mouse	42	0.7	1.5

sufficient resources are invested in maintaining the repair mechanisms that prevent aging at least until the organism has reproduced and cared for its young. Thus mice, which start reproducing when just a few months old, need to maintain their repair mechanisms for much less time than do elephants, which only start to reproduce when around 13 years old. In most species of animals in the wild, few individuals live long enough to show obvious signs of senescence and senescence need only be delayed until reproduction is complete.

| 14-13 | **Genes can alter the timing of senescence.** |

The maximum recorded life spans of animals show dramatic differences in length (see Fig. 14.22). Humans can live as long as 120 years, some owls 68 years, cats 28 years, *Xenopus* 15 years, and mice 3½ years. Mutations in genes that affect life span have been identified in both *C. elegans* and humans, and may give clues to the mechanisms involved.

A *C. elegans* worm that hatches as a first instar larva in an uncrowded environment with ample food grows to adulthood and dies after 2 weeks. However, in crowded conditions and when food is short, the animal enters a larval state, known as the dauer larval state, where it neither eats nor grows until food becomes available again. This state can last for several months. Mutations in genes that affect dauer formation can also affect the normal life span. Thus, animals carrying a mutation in the gene *daf-2* live twice as long as normal and always enter the dauer state, even when they have ample food. It is possible that the effect of this mutation is related to the fact that the dauer larvae do not eat. Reduction in food intake increases life span in other animals; rats on a minimal diet live about 40% longer than rats allowed to eat as much as they like, which is thought to be partly due to the reduced exposure to free radicals. These are molecules that are formed during metabolism when energy is generated by the breakdown of foods. They are highly reactive, and can damage both DNA and proteins.

Mutations in another *C. elegans* gene, *clk-1*, slow down the cell cycle and both embryonic and post-embryonic development, since the cells do not divide as frequently; these mutant worms have a life span up to 70% longer than normal. There are two other clk genes, and animals with mutations in all these genes as well as *daf-2* can live five times longer than normal. It is not known how the clk genes control senescence, but they may delay it by decreasing metabolic rate and thus reducing exposure to free radicals.

Humans that are homozygous for the recessive gene defect known as Werner's syndrome show striking effects of premature aging. There is growth retardation at puberty, and by their early twenties sufferers from this syndrome have gray hair and suffer from a variety of illnesses, such as heart disease, that are typical of old age. Most die before the age of 50. Fibroblasts taken from Werner's syndrome patients undergo fewer cell divisions in culture before becoming senescent and dying than do fibroblasts from unaffected people of the same age (see next section). The gene affected in Werner's syndrome has been isolated and is thought to encode a protein involved in unwinding DNA. Such unwinding is required for DNA replication, DNA repair, and gene expression. The inability to carry out DNA repair properly in Werner's syndrome patients could subject the genetic material to a much higher level of damage than normal. The link between Werner's syndrome and DNA thus fits with the possibility that aging is closely linked to the accumulation of damage in DNA.

14-14 | Cells senesce in culture.

One might think that when cells are isolated from an animal, placed in culture, and provided with adequate medium and growth factors, they would continue to proliferate almost indefinitely. However, this is not the case. For example, mammalian fibroblasts—connective tissue cells—will only go through a limited number of cell doublings in culture; the cells then stop dividing, however long they are cultured (Fig. 14.23). For normal fibroblasts, the number of cell doublings depends both on the species and the age of the animal from which they are taken. Human fibroblasts taken from a fetus go through about 60 doublings, those from an 80-year old about 30, and those from an adult mouse about 12–15. When the cells stop dividing, they appear to be healthy, but are stuck at some point in the cell cycle, often G_0. Cells taken from patients with Werner's syndrome, who show an acceleration of many features of normal aging, make significantly fewer divisions in culture than normal cells.

A feature shared by senescent cells in culture and *in vivo* is shortening of the telomeres. These are the repetitive DNA sequences at the ends of the chromosomes that preserve chromosome integrity and ensure that chromosomes replicate themselves completely without loss of informational DNA at the ends. The length of the telomeres is reduced in older cells. Telomere length is found to decrease by about 50 base pairs at each DNA replication, suggesting that they are not completely replicated at each cell division and this may be related to senescence.

| Summary.

Aging is largely caused by wear and tear, but is also under genetic control. Normal cells in culture can only undergo a limited number of cell divisions, which correlate with their age at isolation and the normal life span of the animal from which they came. Genes that can affect life span have been identified in *C. elegans*.

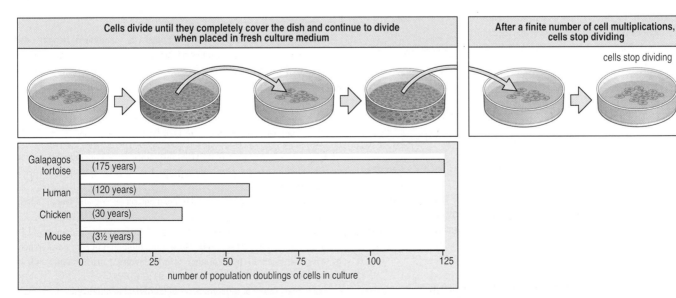

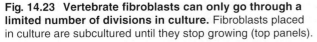

Fig. 14.23 Vertebrate fibroblasts can only go through a limited number of divisions in culture. Fibroblasts placed in culture are subcultured until they stop growing (top panels). The number of cell doublings in culture before they stop dividing is related to their maximum age, as indicated by the figures in brackets on the graph (bottom panel).

Summary to Chapter 14.

The form of many animals is laid down in miniature during embryonic development, and they then grow in size, keeping the basic body form, although different regions grow at different rates. Growth may involve cell multiplication, cell enlargement, and the laying down of extracellular material. In vertebrates, structures have an intrinsic growth program, which is under hormonal control. Arthropod larvae grow by molting and both they and other animals, such as frogs, undergo metamorphosis, which is under hormonal control. Cancer can be viewed as an aberration of growth, as it usually results from mutations that lead to excessive cell proliferation and failure of cells to differentiate. The symptoms of aging appear mainly to be caused by damage to cells accumulating over time, but aging is also under genetic control. In plants, all adult structures are derived by growth from specialized regions called meristems.

References.

14-1   Tissues can grow by cell proliferation, cell enlargement, or accretion.

Goss, R.J.: *The Physiology of Growth*. New York: Academic Press, 1978.

14-2   Cell proliferation can be controlled by an intrinsic program and by external signals.

Follette, P.J., O'Farrell, P.H.: **Connecting cell behavior to patterning: lessons from the cell cycle.** *Cell* 1997, **88**:309–314.

Edgar, B.: **Diversification of cell cycle controls in developing embryos.** *Curr. Opin. Cell Biol.* 1995. **7**:815–824.

Edgar, B.A., Lehner, C.F.: **Developmental control of cell cycle regulators: a fly's perspective.** *Science* 1996, **274**:1646–1652.

Raff, M.C.: **Size control: the regulation of cell numbers in animal development.** *Cell* 1996, **86**:173–175.

14-3   Growth of mammals is dependent on growth hormones.

Barker, D.J.: The Wellcome Foundation Lecture, 1994: The fetal origins of adult disease. *Proc. Roy. Soc. Lond.* 1995, **262**:37–43.

Brook, C.G.D.: *A Guide to the Practice of Paediatric Endocrinology*. Cambridge: Cambridge University Press, 1993.

Baker, J., Liu, J.P., Robertson, E.J., Efstratiadies, A.: **Role of insulin-like growth factors in embryonic and postnatal growth.** *Cell* 1993, **75**:73–82.

Heyner, S., Garside, W.T.: **Biological actions of IGFs in mammalian development.** *BioEssays* 1994, **16**:55–57.

14-4   Developing organs can have their own intrinsic growth programs.

Wolpert, L.: **The cellular basis of skeletal growth during development.** *Brit. Med. Bull.* 1981, **37**:215–219.

14-5   Growth of the long bones occurs in the growth plates.

Kember, N.F.: **Cell kinetics and the control of bone growth.** *Acta Paediatr. Suppl.* 1993, **391**:61–65.

Ohlsson, C., Isgaard, J., Tomell, J., Nilsson, A., Isaksson, O.G., Lindahl, A.: **Endocrine regulation of longitudinal bone growth.** *Acta Paediatr. Suppl.* 1993, **391**:33–40.

Roush, W.: **Putting the brakes on bone growth.** *Science* 1996, **273**: 579.

Deng, C., Wynshaw-Boris, A., Zhou, F., Kuo, A., Leder, P.: **Fibroblast growth factor receptor 3 is a negative regulator of bone growth.** *Cell* 1996, **84**:911–921.

Wallis, G.A.: **Bone growth: coordinating chondrocyte differentiation.** *Curr. Biol.* 1996, **6**:1577–1580.

14-6   Growth of vertebrate striated muscle is dependent on tension.

Williams, P.E., Goldspink, G.: **Changes in sarcomere length and physiological properties in immobilized muscle.** *J. Anat.* 1978, **127**:450–468.

Schultz, E.: **Satellite cell proliferative compartments in growing skeletal muscles.** *Dev. Biol.* 1996, **175**:84–94.

14-7   The epithelia of adult mammalian skin and gut are continually replaced by derivatives of stem cells.

Poten, C.S., Loeffler, M.: **Stem cells: attributes, cycles, spirals, pitfalls and uncertainties. Lessons for and from the crypt.** *Development* 1990, **110**:1001–1020.

Jones, P.H., Harper, S., Watts, F.M.: **Stem cell patterning and fate in human epidermis.** *Cell* 1995, **80**:1–20.

Sellheyer, K., Bickenbach, J.R., Rothnagel, J.A., Bundman, D., Langley, M,A., Krieg, T., Roche, N.S., Roberts, A.B., Roop, D.R.: **Inhibition of skin development by overexpression of transforming growth factor-β1 in the epidermis of transgenic mice.** *Proc. Nat. Acad. Sci.* 1993, **90**:5237–5241.

Watt, F.M.: **Terminal differentiation of epidermal keratinocytes.** *Curr. Opin. Cell Biol.* 1989 1:1107–1115.

Watt, F.M., Jones, P.H.: **Expression and function of the keratinocyte integrins.** *Development Suppl.* 1993, pp185–192.

Gordon, J.I., Hermiston, M.L.: **Differentiation and self-renewal in the mouse gastrointestinal epithelium**. *Curr. Opin. Cell Biol.* 1994, **6**:795–803.

14-8   **Cancer can result from mutations in genes controlling cell multiplication and differentiation.**

Sawyers, C.L., Denny, C.T., Witte, O.N.: **Leukemia and the disruption of normal hematopoiesis**. *Cell* 1991, **64**:337–350.

Rabbitts, T.H.: **Chromosomal translocations in human cancer**. *Nature* 1994, **372**:143–149.

Shilo, B.Z.: **Tumor suppressors. Dispatches from patched**. *Nature* 1996, **382**:115–116.

Hunter, T.: **Oncoprotein networks**. *Cell* 1997, **88**:333–346.

14-9   **Hormones control many features of plant growth.**

Lyndon, R.F.: *Plant Development*. London: Unwin Hymas, 1990.

14-10   **Cell enlargement is central to plant growth.**

Cosgrove, D.J.: **Plant cell enlargement and the action of expansins**. *BioEssays* 1996, **18**:533–540.

Hauser, M.T., Morikami, A., Benfey, P.N.: **Conditional root expansion mutants of Arabidopsis**. *Development* 1995, **121**:1237–1252.

14-11   **Arthropods have to molt in order to grow.**

Reynolds, S.E., Samuels, R.I.: **Physionomy and biochemistry of insect molting fluid**. *Adv. Insect Physical* 1996, **26**:157–232.

14-12   **Metamorphosis is under environmental and hormonal control.**

Thummel, C.S.: **Flies on steroids—*Drosophila* metamorphosis and the mechanisms of steroid hormone action**. *Trends Genet.* 1996, **12**:306–310.

Tata, J.R.: **Gene expression during metamorphosis: an ideal model for post-embryonic development**. *BioEssays* 1993, **15**:239–248.

Aging and senescence.

Kirkwood, T.B.: **Human senescence**. *BioEssays* 1996, **18**:1009–1016.

Orr, W.C., Sohal, R.S.: **Extension of life-span by overexpression of superoxide dismutase and catalase in *Drosophila melanogaster***. *Science* 1994, **263**:1128–1130.

14-13   **Genes can alter the timing of senescence.**

Yu, C.E., Oshima, J., Fu, Y.H., Wijsman, E.M., Hisama, F., Alisch, R., Matthews, S., Nakura. J., Miki, T., Ouais, S., Martin, G.M., Mulligan, J., Schellenberg, G.D.: **Positional cloning of the Werner's syndrome gene**. *Science* 1996, **272**:258–262.

Weindruck, R.: **Caloric restriction and aging**. *Sci. Amer.* 1996, **274**:46–52.

Kenyon, C.: **Ponce d'elegans: genetic quest for the fountain of youth**. *Cell* 1996, **84**:501–504.

Rose, M.R., Archer, M.A.: **Genetic analysis of mechanisms of aging**. *Curr. Opin. Genet. Dev.* 1996, **6**:366–370.

Lithgow, G.J.: **Invertebrate gerontology: the age mutations of *Caenorhabditis elegans***. *BioEssays* 1996, **18**:809–815.

14-14   **Cells senesce in culture.**

Pennisi, E.: **Premature aging gene discovered**. *Science* 1996, **272**:193–194.

Campisi, J.: **Replicative senescence: an old live's tale?** *Cell* 1996, **84**:497–500.

Evolution and Development

Modification of development in evolution.

Changes in the timing of developmental processes during evolution.

"Our communities have a very long history, and our diversity comes from the modifying early schemes by using common mechanisms. Our past is always with us."

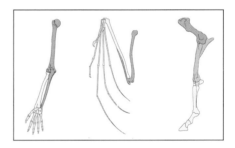

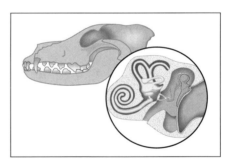

It has been suggested that nothing in biology makes sense unless viewed in the light of evolution. Certainly, it would be very difficult to make sense of many aspects of development without an evolutionary perspective. For example, in our discussion of vertebrate development we saw how, despite different modes of very early development, all vertebrate embryos develop to a similar phylotypic stage, after which their development diverges (see Fig. 3.1). This shared phylotypic stage, which is the embryonic stage after neurulation and the formation of the somites, is probably a stage through which some distant ancestor of the vertebrates passed. It has persisted ever since, to become a fundamental characteristic of the development of all vertebrates, whereas the stages before and after the phylotypic stage have evolved differently in different organisms. Such changes are due to changes in the genes that control development. Heritable developmental changes that generated new adult forms better adapted to their environment, and more successful modes of reproduction, were selected for during evolution.

Throughout this book, we have emphasized the conservation of some developmental mechanisms at the cellular and molecular level among distantly related organisms. The widespread use of the Hox gene complexes and of the same few families of protein signaling molecules provide excellent examples of this. It is this basic similarity in molecular mechanisms that has made developmental biology so exciting in recent years; it has meant that discoveries in one animal, in particular in *Drosophila*, have had important implications for understanding development in other animals. It seems that when a useful developmental mechanism evolved, it was used again and again.

We have already considered the early development of a wide variety of organisms and found some similarities, as well as a number of differences. Here, we mainly confine our attention to two phyla—the chordates, which include the vertebrates, and the arthropods, which include the insects and crustaceans. We focus on those differences that distinguish the members of a large group of related animals, such as the vertebrates or the insects, from each other.

It is generally accepted that all multicellular animals have evolved from a common ancestor, which itself evolved from a single-celled organism. All changes in animal form that have occurred during evolution are ultimately due to changes in DNA. The differences between modern insects and vertebrates, for example, are due to the accumulation of genetic changes over the hundreds of millions of years since they last shared a common ancestor. Changes in form were produced by genetic changes that altered embryonic development. The new adult phenotypes were better adapted to the prevailing conditions, survived better, and so were subject to positive selection. Genetic variability resulting from sexual reproduction and genetic recombination is present in all populations of the organisms we have dealt with in this book, and provides new phenotypes upon which selection can act. The modification of development by changes in gene action is central to the evolution of multicellular organisms.

We look first at the relationship between **ontogeny** (the development of the individual organism) and **phylogeny** (the evolutionary history of the species or group): why, for example, do all vertebrate embryos pass through an apparently fish-like phylotypic stage that has structures resembling gill slits? We then discuss the many variations that occur on the theme of a basic segmented body plan: what determines the different numbers and positions of paired appendages, such as legs and wings, in different groups of segmented organisms? Finally, we consider the timing of developmental events, and how simple alterations in timing, and variations in growth, can have major effects on the shape and form of an organism. In all cases, we ultimately want to understand the changes in the developmental processes and the genes controlling them that have resulted in the extraordinary variety of multicellular animals. This is an exciting area of study in which many problems remain to be solved.

Modification of development in evolution.

Comparisons of embryos of related species has suggested an important generalization about development: the more general characteristics of a group of animals, that is those shared by all members of the group, appear earlier in their embryos than the more specialized ones, and arose earlier in evolution. In the vertebrates, a good example of a general characteristic would be the notochord, which is common to all vertebrates, and is also found in other chordate embryos. Paired appendages, such as limbs, which develop later, are special characters that are not found in other chordates, and that differ in form among different vertebrates. All vertebrate embryos pass through a common phylotypic stage, which then gives rise to the diverse forms of the different vertebrate classes. However, the development of the different vertebrate classes before the phylotypic stage is also highly divergent, because of their very different modes of reproduction; some developmental features that precede the phylotypic stage are evolutionarily highly advanced, such as the formation of a trophoblast and inner cell mass by mammals. This is an example of a special character that developed late in vertebrate evolution, and is related to the nutrition of the embryo via a placenta, rather than a yolky egg.

In this section, we consider the modifications that have occurred to a variety of embryonic structures during evolution, including the basic body plan and the limbs. We start by looking at the branchial arches in vertebrates.

15-1 Embryonic structures have acquired new functions during evolution.

If two groups of animals that differ greatly in their adult structure and habits (such as fishes and mammals) pass through a very similar embryonic stage, this indicates that they are descended from a common ancestor and, in evolutionary terms, are closely related. Thus, an embryo's development reflects the evolutionary history of its ancestors. Structures found at a particular embryonic stage have become modified during evolution into different forms in the different groups. In vertebrates, one good example of this is the evolution of limbs from the embryonic fin-like structures of a fish ancestor, which is discussed in the next section. Another example is the branchial arches and clefts that are present in all vertebrate embryos (see Fig. 2.9), including humans. These are not the relics of the gill arches and gill slits of an adult fish-like ancestor, but of structures that would have been present in the embryo of the fish-like ancestor. During evolution, the branchial arches have given rise both to the gills of the primitive jawless fishes and, in a later modification, to jaws (Fig. 15.1). When the ancestor of land vertebrates left the sea, gills were no longer required, but the embryonic structures that gave rise to them persisted. With time, they became modified, and in mammals, including humans, they now give rise to different structures in the face and neck (Fig. 15.2). The cleft between the first and second branchial arches provides the opening for the Eustachian tube, and endodermal cells in the clefts give rise to a variety of glands, such as the thyroid and thymus.

Evolution rarely generates a completely novel structure out of the blue. New anatomical features usually arise from modification of an existing structure. One can therefore think of much of evolution as a 'tinkering' with existing structures, which gradually fashions something different.

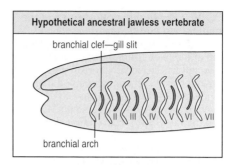

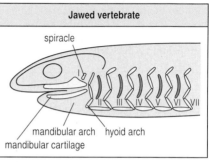

Fig. 15.1 Modification of the branchial arches during the evolution of jaws in vertebrates. The ancestral jawless fish had a series of seven gill slits—branchial clefts—supported by cartilaginous or bony arches. Jaws developed from a modification of the first arch to give the mandibular arch with the mandibular cartilage of the lower jaw and the hyoid arch behind it.

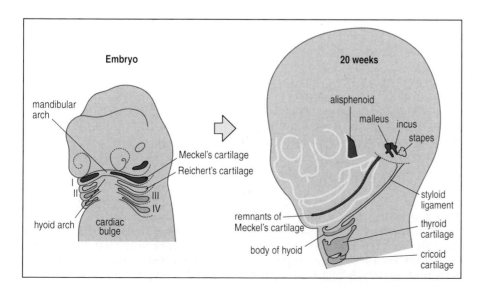

Fig. 15.2 Fate of branchial arch cartilage in humans. In the embryo, cartilage develops in the branchial arches, which gives rise to elements of the three auditory ossicles, the hyoid, and the pharyngeal skeleton. The fate of the various elements is shown by the color coding. After Larsen, W.J.: 1993.

Fig. 15.3 Evolution of the bones of the mammalian middle ear. The articular and quadrate bones of ancestral reptiles (left panel) were part of the lower jaw articulation. Sound was transmitted to the inner ear via these bones and their connection to the stapes. When the lower jaw of mammals became a single bone (the dentary), the articular bone became the malleus, and the quadrate bone the incus of the middle ear, acquiring a new function in transmitting sound to the inner ear from the tympanic membrane (right panel). The eustachian tube forms between branchial arches I and II. After Romer, A.S.: 1949.

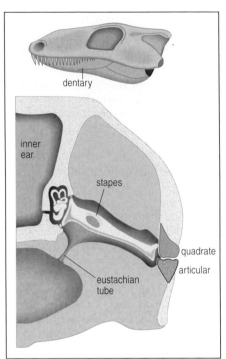

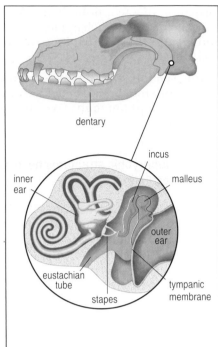

A nice example of a modification of an existing structure is provided by the evolution of the mammalian middle ear. This is made up of three bones that transmit sound from the eardrum (the tympanic membrane) to the inner ear. In the reptilian ancestors of mammals, the joint between the skull and the lower jaw was between the quadrate bone of the skull and the articular bone of the lower jaw, which were also involved in transmitting sound (Fig. 15.3). During mammalian evolution, the lower jaw became just one bone, the dentary, with the articular no longer attached to the lower jaw. By changes in their development, the articular and the quadrate bones in mammals were modified into two bones, the malleus and incus, whose function was now to transmit sound from the tympanic membrane to the inner ear.

Another example of modification of a pre-existing structure is provided by the evolution of the vertebrate kidney. In birds and mammals, three kidney-like structures—metanephros, mesonephros, and pronephros—appear during development. The pronephros and mesonephros are transitory and the functional kidney develops from the metanephros. However, as discussed in Section 12-2, the mesonephros plays a key role in the development of the gonads, giving rise to the somatic cells of the testis and ovary. In lower vertebrates, such as fish and amphibians, the pronephros acts as the functional kidney in the immature juvenile stages, but the mesonephros is the functional kidney in the adult. Thus, in birds and mammals the embryonic kidneys of their ancestors have persisted as embryonic structures, but have been modified to provide structures essential to the development of the gonad.

15-2   Limbs evolved from fins.

The limbs of tetrapod vertebrates are special characters that develop after the phylotypic stage. Amphibians, reptiles, birds, and mammals have limbs, whereas fish have fins. The limbs of the first land vertebrates evolved from the pelvic and pectoral fins of their fish-like ancestors. The basic limb pattern is highly conserved in both the forelimbs and hindlimbs

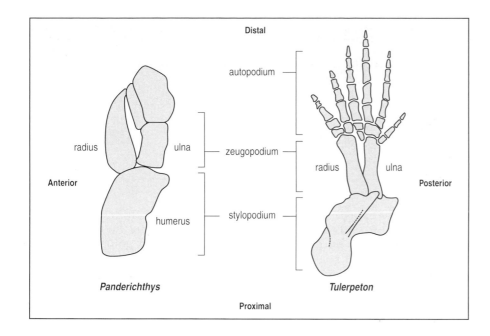

Fig. 15.4 The fin to limb transition. In the lobe-like fin of the Devonian fish *Panderichthys*, there were proximal elements corresponding to the humerus (the stylopodium), radius, and ulna (the zeugopodium), but no distal elements, of the Devonian tetrapod *Tulerpeton* has similar proximal elements, but has also developed digits.

of all tetrapods, although there are some differences both between forelimbs and hindlimbs, and between different vertebrates.

The fossil record suggests that the transition from fins to limbs occurred in the Devonian period, between 400 and 360 million years ago. The transition probably occurred when the fish ancestors of the tetrapod vertebrates living in shallow waters moved onto the land. The fins of Devonian lobe-finned fishes, such as *Panderichthys*, are probably ancestral to tetrapod limbs, an early example of which is the limb of the Devonian tetrapod *Tulerpeton* (Fig. 15.4). The proximal skeletal elements corresponding to the humerus, radius, and ulna of the tetrapod limb are present in the ancestral fish, but there are no structures corresponding to digits. How did digits evolve? Some insights have been obtained by examining the development of fins in a modern fish, the zebrafish *Danio*.

The fin buds of the zebrafish embryo are initially similar to tetrapod limb buds, but important differences soon arise during development. The proximal part of the fin bud gives rise to skeletal elements, which are homologous to the proximal skeletal elements of the tetrapod limb. There are four main proximal skeletal elements in a zebrafish fin, which arise from the subdivision of a cartilaginous sheet (Fig. 15.5). The essential difference between fin and limb development is in the distal skeletal elements. In the zebrafish fin bud, an ectodermal fin fold develops at the distal end of the bud and fine bony fin rays are formed within it. These rays have no relation to anything in the vertebrate limb.

As in the tetrapod limb bud (see Section 10-4), the key signaling gene *Sonic hedgehog* is expressed at the posterior margin of the zebrafish fins and the expression pattern of Hoxd and Hoxa genes is similar to that in tetrapods. However, in the later stages of bud development, the Hox

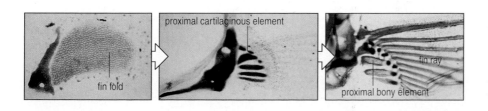

Fig. 15.5 The development of the pectoral fin of the zebrafish *Danio*. Left panel: the pectoral girdle and fin fold. Middle panel: four proximal cartilaginous elements and distal fin rays. Right panel: four proximal bony elements supporting the distal fin rays in the adult fish. Photographs courtesy of D. Duboule, from Sordino, P., *et al.*: 1995.

Fig. 15.6 Regions of Hox gene expression in the chick hindlimb and the zebrafish pectoral fin. Left panel: in the zebrafish fin bud, an apical ectodermal fold extends out from the underlying mesoderm. *Hoxd12* remains expressed in the mesoderm, which gives rise to the proximal cartilaginous elements. Right panel: in the chick leg, the mesoderm grows extensively and *Hoxd11* is expressed in the early bud and, additionally, more distally at later stages. After Coates, M.I.: 1995.

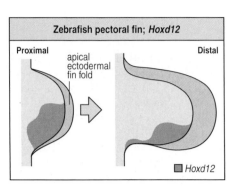
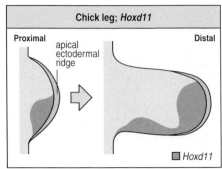

genes are only expressed in the posterior part of the fin bud and there is no distal expression where the fin rays develop. On the other hand, in the tetrapod limb bud, an additional domain of Hox gene expression occurs in the distal region that gives rise to the digits (Fig. 15.6). If zebrafish fin development reflects that of the primitive ancestor, then tetrapod digits are novel structures, whose appearance is correlated with a new domain of Hox gene expression. However, they may have evolved from the distal recruitment of the same developmental mechanisms and processes that generate the radius and ulna. There are, as discussed in Chapter 10, mechanisms in the limb for generating periodic cartilaginous structures such as digits. It is likely that such a mechanism was involved in the evolution of digits by an extension of the region in which the embryonic cartilaginous elements form, together with the establishment of a new pattern of Hox gene expression in the more distal region.

The great range of anatomical specializations in the limbs of mammals (Fig. 15.7) is due to changes both in limb patterning and in the differential growth of parts of the limbs during embryonic development, but the basic underlying pattern of skeletal elements is maintained. If one compares the forelimb of a bat and a horse, one can see that although both retain the basic pattern of limb bones, it has been modified to provide a specialized function in each. In the bat, the limb is adapted for flying: the digits are greatly lengthened to support a membranous wing. In the horse, the limb is adapted for running: in the forelimb lateral digits are reduced, the central metacarpal (a hand bone in humans) is lengthened, and the radius and ulna are fused for greater strength. The role of differential growth rates and the loss of skeletal elements

Fig. 15.7 Diversification of mammalian limbs. The basic pattern of bones in the forelimb is conserved throughout the mammals, but there are changes in the proportions of the different bones, as well as fusion and loss of bones. This is seen particularly in the horse limb, in which the radius and ulna have become fused into a single bone, and the central metacarpal (a hand bone in humans) is greatly elongated. In addition, there has been loss and reduction of the digits in the horse. In the bat wing, by contrast, the digits have become greatly elongated to support the membranous wing.

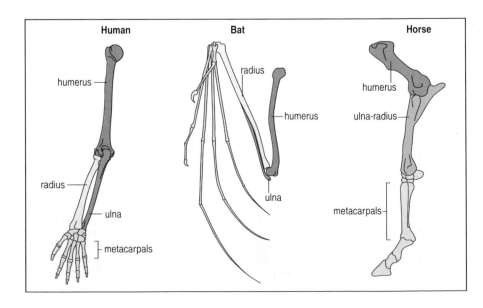

in the evolution of the horse's limb, and cases of limb reduction, as in snakes, are considered later.

A feature of limb evolution is that while reduction in digit number is common—there are only three in the chick wing and reduction is common in lizards—species with more than five digits are very rarely found. It seems that there is a developmental constraint on evolving more than five different kinds of digits. This may be due to the Hox genes providing only five discrete genetic programs for giving a digit an identity. In limbs with polydactyly, at least two of the digits are the same (see Section 10-4), and there are still only five different kinds of digits. This may be the reason why in an animal with an additional distinctive digit-like element, such as the giant panda's 'thumb', this digit is in fact a modified wrist bone.

15-3   Development of vertebrate and insect wings makes use of evolutionarily conserved mechanisms.

Vertebrate and insect wings have some superficial similarities and have similar functions, yet are very different in their structure. The insect wing is a double-layered epithelial structure, whereas the vertebrate limb develops mainly from a mesenchymal core surrounded by ectoderm. However, despite these great anatomical differences, there are striking similarities in the genes and signaling molecules involved in patterning insect legs, insect wings, and vertebrate limbs (Fig. 15.8, and see Chapter 10). Patterning along the antero-posterior axis of all these appendages uses signals encoded by *hedgehog*-related genes, and by members of the TGF-β family, such as *decapentaplegic* (in insects) and *BMP-2* (in vertebrates). It is remarkable that the dorsal surface of the insect wing is characterized by expression of the gene *apterous*, whereas the related gene *Lmx-1* is expressed in the dorsal mesenchyme of the vertebrate limb. A *fringe*-like gene is involved in the specification of the boundary between dorsal and ventral regions in both insect and bird wings.

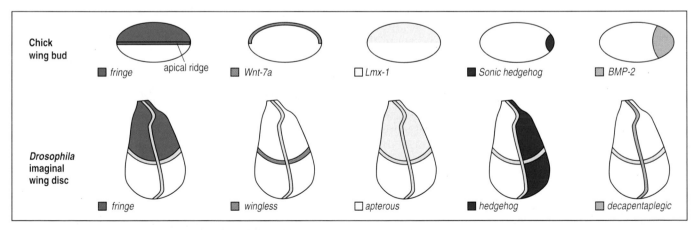

Fig. 15.8 Comparison of developmental signals in the chick wing bud and *Drosophila* wing imaginal disc. First column: the chick wing bud (top) is shown with the distal end facing. The double line bisecting it represents the apical ridge. The chick apical ectodermal ridge forms at the boundary between the dorsal cells, which express *fringe*, and the ventral cells, which do not. The dorso-ventral boundary in the insect wing disc also forms at the boundary between *fringe* and non-*fringe* cells (bottom). In the insect wing imaginal disc, the future dorsal and ventral regions are in the same plane. The vertical double lines represent the antero-posterior compartment border, and the horizontal double lines represent the dorso-ventral compartment border. Second column: the dorsal region of the chick wing is specified by *Wnt-7a* in the ectoderm, whereas *wingless* is expressed at the dorso-ventral insect wing margin. Third column: the gene *Lmx-1* is expressed in the dorsal region of chick wing bud mesoderm, whereas the *Drosophila* gene *apterous*, to which it is structurally related, specifies the dorsal region of the insect wing. Fourth and fifth columns: *Sonic hedgehog* in the chick wing and *hedgehog* in *Drosophila* are expressed in posterior regions and both induce expression of genes of the TGF-β family—*BMP-2* and *decapentaplegic*, respectively.

All these relationships suggest that, during evolution, a mechanism for patterning and setting up the axes of appendages appeared in some common ancestor of insects and vertebrates. Subsequently, the genes and signals involved acquired different downstream targets so that they could interact with different sets of genes, yet the same set of signals retain their organizing function in these very different appendages. The individual genes involved in specifying the limb axes are probably more ancient than either insect or vertebrate limbs.

Another example illustrating the conservation of the genetic machinery for appendages is provided by the gene *Distal-less*, which is expressed along the proximo-distal axis of a wide variety of developing append- ages, including annelid parapodia and the tube feet of sea urchins. It is also expressed during vertebrate limb outgrowth.

15-4	**Hox gene complexes have evolved through gene duplication.**

Hox genes play a key role in development both in vertebrates and insects. By comparing the organization and structure of the Hox genes in insects and vertebrates, we can determine how one set of important develop- mental genes has changed during evolution.

A major general mechanism of evolutionary change has been gene duplication. Tandem duplication of a gene, which can occur by a vari- ety of mechanisms during DNA replication, provides the embryo with an additional copy of the gene. This copy can diverge in its nucleotide sequence and acquire a new function and regulatory region, so changing its pattern of expression and downstream targets without depriving the organism of the function of the original gene. The process of gene dupli- cation has been fundamental in the evolution of new proteins and new patterns of gene expression; it is clear, for example, that the different hemoglobins in humans have arisen as a result of gene duplication (see Section 9-13).

One of the clearest examples of the importance of gene duplication in developmental evolution is provided by the Hox gene complexes. As we have seen, the Hox genes are members of the homeobox gene family, which is characterized by a short 180 base pair motif, the homeobox, which encodes a helix-turn-helix domain that is involved in transcrip- tional regulation (see Box 4A, page 104). Two features characterize all known Hox genes: the individual genes are organized into one or more gene clusters or complexes, and the order of expression of individual genes along the antero-posterior axis is usually the same as their sequential order in the gene complex.

Comparing the Hox genes of a variety of species, it is possible to reconstruct the way in which they are likely to have evolved from a simple set of six genes in a common ancestor of all species (Fig. 15.9). Amphioxus, which is a vertebrate-like chordate, has many features of a primitive vertebrate: it possesses a dorsal hollow nerve cord, a notochord, and segmental muscles that derive from somites. It has only one Hox gene cluster, and one can think of this cluster as most closely resembling the common ancestor of the four vertebrate Hox gene complexes—Hoxa, Hoxb, Hoxc, and Hoxd (see Fig. 15.9). It is possible that both the vertebrate and *Drosophila* Hox complexes evolved from a simpler ancestral complex by gene duplication. In *Drosophila*, the duplications could have given the *abd-A*, *Ubx*, and *Antp*. In vertebrates, the Hox genes are arranged in four separate clusters, each of which is on a different chromosome, and are not linked to each other. These separate clusters probably arose from duplications of whole chromosomal regions, in which new Hox genes had already been generated by tandem duplication. For example,

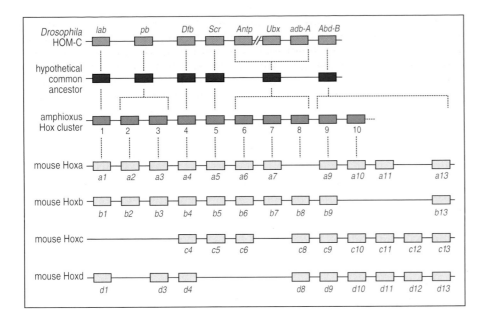

Fig. 15.9 Gene duplication and Hox gene evolution. A suggested evolutionary relationship between the Hox genes of a hypothetical common ancestor and *Drosophila* (an arthropod), amphioxus (a cephalochordate), and the mouse (a vertebrate). Duplications of genes of the ancestral set (red) could have given rise to the additional genes in *Drosophila* and amphioxus. Two duplications of the whole cluster in a chordate ancestor of the vertebrates could have given rise to the four separate Hox gene complexes in vertebrates. There has also been a loss of some of the duplicated genes in vertebrates. After Garcia-Fernandez, J., *et al.*: 1994.

sequence comparisons suggest that the multiple mouse Hox genes related to the *Drosophila* Abd-B gene complexes do not have direct homologs within the *Drosophila* bithorax complex (see Fig. 15.9); they probably arose by tandem duplication from an ancestral gene after the split of the insect and vertebrate lineages, but before duplication of the whole cluster in vertebrates. We now consider the role of these genes in evolution of the axial body plan.

15-5 Changes in specification and interpretation of positional identity have generated the elaboration of vertebrate and arthropod body plans.

Multicellular organisms are thought to have originated about 1500 million years ago, with the earliest generally accepted fossil of a multicellular animal dating from about 600 million years ago. There are currently about 35 extant animal phyla, each with its own distinctive basic body plan, and all of these had evolved by the end of the Cambrian period, around 500 million years ago. No new basic body plans have evolved since then, although body plans within the different phyla have been modified and elaborated to give organisms as different as fish and mammals within the chordates. In the following discussion on the elaboration of body plans within phyla, we focus on just two phyla —the chordates, which include the vertebrates, and the arthropods, which include the insects and the crustaceans—because we have a good understanding of the development of some of the members of both these phyla.

Hox genes are key genes in the control of development and are expressed regionally along the antero-posterior axis of the embryo. The apparent universality of Hox genes, and certain other genes, in animal development has led to the concept of the **zootype**. This defines the pattern of expression of these key genes along the antero-posterior axis of the embryo, which is present in all animals.

The role of the Hox genes is to specify positional identity in the embryo, rather than the development of any specific structure. These positional values are interpreted differently in different embryos to influence how the cells in a region develop into, for example, segments and append-ages. The Hox genes exert this influence by their action on the genes controlling the development of these structures. Changes in the downstream targets of the Hox genes can thus be a major source of change in evolution. In addition, changes in the pattern of Hox gene expression along the body can have important consequences. An example is a relatively minor modification of the body plan that has taken place within vertebrates. One easily distinguishable feature of pattern along the antero-posterior axis in vertebrates is the number and type of vertebrae in the main anatomical regions—cervical (neck), thoracic, lumbar, sacral, and caudal (see Fig. 4.10). The number of vertebrae in a particular region varies considerably among the different vertebrate classes —mammals have seven cervical vertebrae, whereas birds can have between 13 and 15. How does this difference arise? A comparison between the mouse and the chick shows that the domains of Hox gene expression have shifted in parallel with the change in number of vertebrae (see Section 4-3). For example, the anterior boundary of *Hoxc6* expression in the mesoderm in mice and chicks is always at the boundary of the cervical and thoracic regions. Moreover, the *Hoxc6* expression boundary is also at the cervical-thoracic boundary in geese, which have three more cervical vertebrae than chickens, and in frogs, which only have three or four cervical vertebrae in all. The changes in the spatial expression of *Hoxc6* correlate with the number of cervical vertebrae. Other Hox genes are also involved in the patterning of the antero-posterior axis, and their boundaries also shift with a change in anatomy.

Changes in patterns of Hox gene expression can also help explain the evolution of arthropod body plans. Insects and crustaceans are distinct groups of arthropods that have evolved from a common arthropod ancestor, which probably had a body composed of more or less uniform segments. A comparison of Hox gene expression in an insect, the grasshopper, with that in a crustacean, the brine shrimp *Artemia*, shows which body regions in these two present-day arthropods are homologous with each other, and how the different body plans might have evolved from the ancestral body plan. Such a comparison shows that both the pattern of Hox gene expression and the body regions to which particular Hox genes relate have changed during the evolution of these two groups (Fig. 15.10). The grasshopper has a pattern of Hox gene expression similar to that in *Drosophila*. As with *Drosophila*, the Hox genes *Antennapedia*, *Ultrabithorax*, and *abdominal-A* specify distinct segment types in the thorax and abdomen, but are expressed in overlapping domains, and segment types are defined by combinatorial expression (see Section 5-19). However, in the brine shrimp, these genes are all expressed together in a thoracic region that is composed of

Fig. 15.10 A comparison of body plans and Hox gene expression in two arthropods. A comparison of Hox gene expression in an insect, the grasshopper, with that in a crustacean, the brine shrimp *Artemia*, shows that both the pattern of Hox gene expression and the body regions to which particular Hox genes relate have changed during the evolution of these two groups of arthropods from their common ancestor. In *Artemia*, the three Hox genes *Antennapedia*, *Ultrabithorax*, and *abdominal-A* are all expressed throughout the thorax, where most of the segments are similar. However, the expression of these genes in the thorax and abdomen of the grass- hopper is different. They have overlapping and distinct patterns of expression which reflect the regional differences in the insect thorax. The gene *Abdominal-B* is expressed in the genital regions of both animals, indicating that these two regions are homologous. After Akam, M.: 1995.

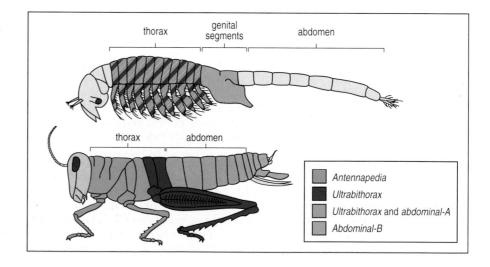

uniform segments. This suggests that the thorax of *Artemia* might be homologous to the whole insect thorax and much of the insect abdomen. Thus, in the evolution of the grasshopper and the brine shrimp from their common ancestor, the changes in body plan have resulted in part from changes in where particular Hox genes are expressed, although changes in their downstream targets have also been involved. These comparisons clearly show that Hox genes do not specify particular structures but simply provide a regional identity. How that identity is interpreted to produce a particular morphology is the role of genes acting downstream from the Hox genes. Further examples of the role of Hox genes are provided by arthropod appendages, which we will look at next.

15-6 The position and number of paired appendages in insects is dependent on Hox gene expression.

Insect fossils display a variety of patterns in the position and number of their paired appendages—principally the legs and wings. Some insect fossils have legs on every segment, whereas others only have legs in a distinct thoracic region. The number of abdominal segments bearing legs varies, as does the size and shape of the legs. Wings arose later than legs in insect evolution. Wing-like appendages are present on all the thoracic and abdominal segments of some insect fossils, but are restricted to the thorax in others. To understand how these different patterns of appendages arose during evolution we need to look at how the different patterns of appendages develop in two classes of modern insects, the Lepidoptera (the butterflies and moths) and the Diptera (the flies, including *Drosophila*).

The basic pattern of Hox gene expression along the antero-posterior axis is the same in all present day insect species that have been studied. Yet the larvae of Lepidoptera have legs on the abdomen as well as on the thorax, and the adults have two pairs of wings, whereas the Diptera, which evolved later, have no legs on the abdomen in the larva or adult, and only one pair of wings, with the second pair of wings having been modified into halteres. How are these differences related to differences in Hox gene activity in the two groups of insects?

In *Drosophila*, as discussed in Section 10-15, products of the bithorax gene complex suppress limb formation in the abdomen by repressing the expression of the *Distal-less* gene. This suggests that the potential for limb development is present in every segment, even in flies, and is actively repressed in the abdomen. It thus seems likely that the ancestral arthropod from which insects evolved had limbs on all its segments. During the embryonic development of Lepidoptera, the bithorax complex genes *Ultrabithorax* and *Abdominal-B* are turned off in the ventral parts of the abdominal segments and this results both in *Distal-less* being expressed and in limbs developing on the abdomen in the larva. The presence or absence of limbs on the abdomen is thus determined by whether or not a particular Hox gene is expressed there. This shows that changes in the pattern of Hox gene expression have played a key role in evolution. The Hox genes can also determine the nature of an appendage: we have seen how mutations can convert legs into antenna-like structures and an antenna into a leg (see Section 10-17).

It has been proposed that insect wings originated as outgrowths from the first leg segment. However, Hox genes do not seem to have been involved, as *Antennapedia*, which is expressed in the second thoracic wing-bearing segment of *Drosophila*, is not required for wing development. It is likely that wings were originally present on all thoracic and abdominal segments and that Hox genes have repressed and modified their development. For example, the differences between forewings and hindwings of insects with two pairs of wings, such as butterflies, are probably regulated by the *Ultrabithorax* gene.

It is clear that in the course of evolving, rather than new genes appearing, new regulatory interactions between the bithorax complex proteins and genes involved in limb and wing development have evolved.

15-7 | The body plan of arthropods and vertebrates is similar, but the dorso-ventral axis is inverted.

From paleontological, molecular, and cellular evidence, all multicellular animals are presumed to descend from a common ancestor, and the broad similarities in the pattern of Hox gene expression in vertebrates and arthropods, whose evolution diverged hundreds of millions of years ago, are taken as good supporting evidence for this concept. However, a comparison between the body plans of arthropods and chordates reveals an intriguing difference. In spite of many similarities in their basic body plan—both have an anterior head, a nerve cord running anterior to posterior, a gut, and appendages—the dorso-ventral axis of vertebrates is inverted, when compared to that of arthropods. The most obvious manifestation of this is that the nerve cord runs ventrally in arthropods, but dorsally in vertebrates (Fig. 15.11, left panels).

One explanation for this, first proposed in the 19th century, is that during the evolution of the vertebrates from their common ancestor with the arthropods, the dorso-ventral axis was turned upside-down, so that the ventral nerve cord of the ancestor became dorsal. This startling idea has recently found some support from molecular evidence showing that the same genes are expressed along the dorso-ventral axis in both insects and vertebrates, but in inverse directions. This inversion may have been dictated by the position of the mouth. The mouth defines the ventral side, and a change in the position of the mouth away from the side of the nerve cord would have resulted in the reversal of the dorso-ventral axis in relation to the mouth. The position of the mouth is specified during gastrulation, and it is not difficult to imagine how its position could have moved.

We have seen in Chapters 3 and 5 that the patterning of the dorso-ventral axis in vertebrates and insects involves intercellular signaling. In *Xenopus*, the protein chordin is one of the signals that specifies the dorsal region,

Fig. 15.11 The vertebrate and *Drosophila* dorso-ventral axes are related but inverted. In arthropods, the nerve cord is ventral whereas in vertebrates it is dorsal—dorsal and ventral being defined by the position of the mouth. In *Drosophila* and *Xenopus*, the signals specifying the dorso-ventral axis are similar, but are expressed in inverted positions. The protein chordin, a dorsal specifier in vertebrates, is related to sog, which is a ventral specifier in *Drosophila*, and the vertebrate ventral specifier BMP-4 is related to *Drosophila* decapentaplegic (dpp), which specifies dorsal. After Ferguson, E.L.: 1996.

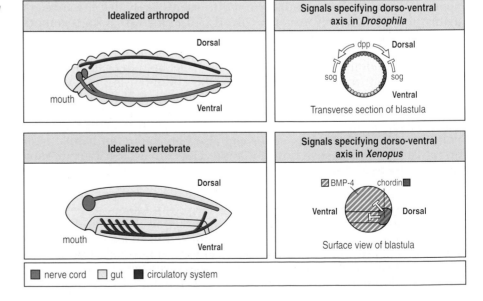

whereas the growth factor BMP-4 specifies a ventral fate. In *Drosophila*, the pattern of gene expression is reversed: the protein decapentaplegic, which is closely related to BMP-4, is the dorsal signal, and the protein short gastrulation (sog), which is related to chordin, is the ventral signal (see Fig. 15.11, right panels). These signaling molecules are experimentally interchangeable between insects and frogs. Chordin can promote ventral development in *Drosophila*, and decapentaplegic protein promotes ventral development in *Xenopus*. The molecules and mechanisms that set up the dorso-ventral axes in the two groups of animals are thus strikingly similar, strongly suggesting that the divergence in the body plans of the arthropods and vertebrates involved an inversion of this axis, by movement of the mouth during the evolution of the vertebrates.

Segmentation provides another example of similarity in the mechanisms involved in the body plan development of chordates and arthropods. The gene *engrailed*, which is expressed in the posterior compartment of each segment in *Drosophila*, is expressed in the posterior half of each of the first eight somites of amphioxus. Moreover, a homolog of the *Drosophila* pair-rule gene, *hairy*, is expressed in every other forming somite of the zebrafish.

Summary.

The development of an embryo provides insights into the evolutionary origin of the animal. Groups of animals that pass through a similar embryonic stage are descended from a common ancestor. During evolution, the development of structures can be altered so that they acquire new functions, as has happened in the evolution of the mammalian middle ear. The limbs of tetrapod vertebrates evolved from fins, with the digits as a novel feature. The development of vertebrate and insect limbs involves the same set of pattern-establishing genes, reflecting the evolution of limb development from an ancestral mechanism for specifying body appendages. Comparison of patterns of dorso-ventral gene expression suggests that during the evolution of the vertebrates the dorso-ventral axis of an invertebrate ancestor was inverted. The basic body plan of all animals is defined by patterns of Hox gene expression that provide positional identity, the interpretation of which has changed in evolution. The Hox genes themselves have undergone considerable evolution by gene duplication.

Summary: evolution of structures by developmental modification

articular and quadrate bones in reptile jaw   fins of ancestral fish
⇩   ⇩
malleus and incus bones of   limbs of tetrapods, with
mammalian middle ear   digits as novel structures

wings of chick and *Drosophila* use similar signals:
Sonic hedgehog—hedgehog; Lmx-1—apterous; Wnt-7a—wingless; BMP-2—decapentaplegic

Hox gene duplication, diversification, and interpretation
⇩
different body plans

signals for *Xenopus* and *Drosophila* dorso-ventral axes are similar but axes are inverted:
BMP-4—decapentaplegic; chordin—short gastrulation

Changes in the timing of developmental processes during evolution.

In the previous sections, we have focused on changes in spatial patterning that have occurred during evolution. But changes in the timing of developmental processes can also have major effects. In this section, we look at some examples of how changes in the timing of growth and sexual maturation can affect animal form, as well as behavior.

| 15-8 | Changes in relative growth rates can alter the shapes of organisms. |

Many of the changes that occur during evolution reflect changes in the relative dimensions of parts of the body. We have seen how growth can alter the proportions of the human baby after birth, as the head grows much less than the rest of the body (see Fig. 14.5). The variety of face shapes in the different breeds of dog, which are all members of the same species, also provides a good example of the effects of differential growth after birth. All dogs are born with rounded faces; some keep this shape, but in others the nasal regions and jaws elongate during growth. The elongated face of the baboon is also the result of growth of this region after birth.

Because structures can grow at different rates, the overall shape of an organism can be changed substantially during evolution by heritable changes in the duration of growth that leads to an increase in the overall size of the organism. In the horse, for example, the central digit of the ancestral horse grew faster than the digits on either side, so that it ended up longer than the lateral digits (Fig. 15.12). As horses continued to increase in overall size during evolution, this discrepancy in growth

Fig. 15.12 Evolution of the forelimb in horses. *Hyracotherium*, the first true equid, was about the size of a large dog. Its forefeet had four digits, of which one (the third digit in anatomical terms) was slightly longer, as a result of a faster growth rate. All digits were in contact with the ground. As equids increased in size, the lateral digits lost contact with the ground, as the result of the relatively greater increase in length of metacarpal 3 (the hand bone of the third digit). At a later stage, the transition from the lateral digits became even shorter, because of a separate genetic change. After Gregory, W.K.: 1957.

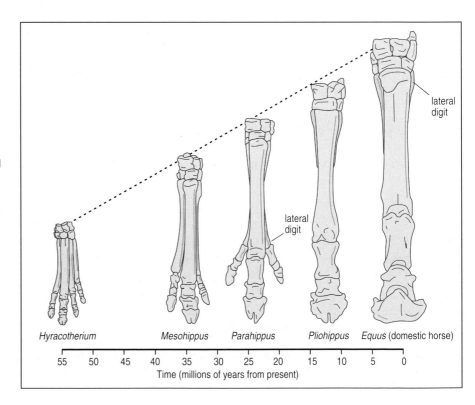

Hyracotherium   Mesohippus   Parahippus   Pliohippus   Equus (domestic horse)

55   50   45   40   35   30   25   20   15   10   5   0
Time (millions of years from present)

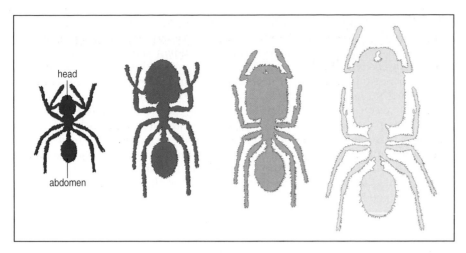

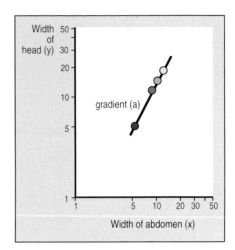

Fig. 15.13 Differential growth of body regions in the ant. As the ant grows, the width of the head (y) increases much faster than that of the abdomen (x) and so the head becomes proportionally larger. The relationship fits the equation $y = bx^a$, where a is the slope of the curve when log y is plotted against log x, and b is a constant.

rates resulted in the relatively smaller lateral digits no longer touching the ground because of the much greater length of the central digit. At a later stage in evolution, the now-redundant lateral digits became reduced even further in size because of a separate genetic change.

The mathematical analysis of the relative growth of parts of an organism during development is known as **allometry**. It has been found that the difference in growth rates of two structures in an organism can be determined from their final dimensions. There is often a mathematical relationship between two structures of lengths x and y, such that $y = bx^a$, where a and b are constants. So when one plots log y against log x, a straight line of slope a is obtained. The slope of the line indicates how much faster y grows than x. An example of this relationship is provided by the relative growth of the abdomen and head in an ant (Fig. 15.13). As the ant grows, the head becomes relatively much larger.

15-9 Evolution of life histories has implications for development.

Animals and plants have very diverse life histories: small birds breed in the spring following their birth, and continue to do so each year until their death; Pacific salmon breed in a suicidal burst at 3 years of age, and oak trees require 30 years of growth before producing acorns, which they eventually produce by the thousand. In order to understand the evolution of such life histories, evolutionary ecologists consider them in terms of probabilities of survival, rates of reproduction, and optimization of reproductive effort. These factors have important implications for the evolution of developmental strategies, particularly in relation to the rate of development. For example, a characteristic feature of many animal life histories is the presence of a feeding larval stage that is distinct from, and usually simpler in form, than the sexually mature adult, and which feeds in a different way. The evolutionary significance of this developmental strategy is not only that it provides the organism with both a means of dispersal and a means of obtaining nutrition before it becomes an adult, but that it also allows the organism to exploit different environments for feeding. Next, we consider two other issues in the evolution of life histories that impinge directly on development—selection for speed of development and selection for egg size.

Life histories help us to understand the evolution of long germ band insects, such as *Drosophila*, which are of more recent origin than short germ band insects. Compared with short germ band insects, such as grasshoppers, which have no larval stage and develop directly into small, immature adult-like forms, *Drosophila* develops very rapidly into a feeding larva. *Drosophila* develops into a feeding larva within 24 hours, whereas a grasshopper takes 5–6 days. It is easy to imagine conditions in which there would have been a selective advantage to insects whose larvae begin to feed as quickly as possible, and it is also likely that embryos are more vulnerable than adults. Thus it is likely that the evolution of the complex developmental mechanisms of long band insects—the system for setting up the whole antero-posterior axis in the egg (see Section 5-18)—resulted from selection pressure for rapid development.

Egg size can also be best understood within the context of life histories. If we assume that the parent has limited energy resources to put into reproduction, the question is how should these resources best be invested in making gametes, particularly eggs; is it more advantageous to make lots of small eggs or a few large ones? In general, it seems that the larger the egg, and thus the larger the offspring at birth, the better the chances of offspring survival, suggesting that in most circumstances an embryo needs to give rise to a hatchling as large as possible and as quickly as possible. Why then, do some species lay many small eggs capable of rapid development? The probable answer is that these species have evolved a strategy that is especially successful in variable conditions where populations can suddenly crash.

15-10   The timing of developmental events has changed during evolution.

Differences among species in the time at which developmental processes occur relative to one another, and relative to their timing in an ancestor, can have dramatic effects on both the structure and behavior of an organism. Differences in the feet of the members of a genus of tropical salamanders illustrate the effect of a change in developmental timing on both the morphology and ecology of different species. Many species of the salamander genus *Bolitoglossa* are arboreal (tree living), rather than typically terrestrial, and their feet are modified for climbing on smooth surfaces. The feet of the arboreal species are smaller and more webbed than those of terrestrial species, and their digits are shorter (Fig. 15.14). These differences seem to be mainly the result of the development and growth of the foot ceasing at an earlier stage in the arboreal species than in the terrestrial species. The term used to describe such differences in timing is **heterochrony**.

Some of the clearest examples of heterochrony come from alterations in the timing of onset of sexual maturity in organisms with larval stages. The acquisition of sexual maturity by an animal while still in the larval stage is a process that goes under the name **neoteny**. The development of the animal, although not its growth, is retarded in relation to the maturation of the reproductive organs. This occurs in the Mexican axolotl, a type of salamander; the larva grows in size and matures sexually, but does not undergo metamorphosis. The sexually mature form remains aquatic and looks like an overgrown larva. However, the axolotl can be induced to undergo metamorphosis by treatment with the hormone thyroxine (Fig. 15.15).

Larval stages may have evolved as a result of heterochrony. For example, if we assume that frog ancestors developed directly into adults, a change in the timing of events in the post-neurula stages, together with

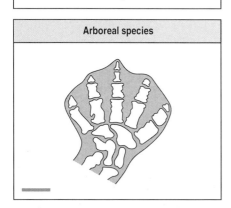

Fig. 15.14 Heterochrony in salamanders. In terrestrial species of the salamander *Bolitoglossa* (top panel), the foot is larger, has longer digits, and is less markedly webbed than in those that live in trees (bottom panel). This difference can be accounted for by foot growth ceasing at an earlier stage in the arboreal species. Scale bar = 1 mm.

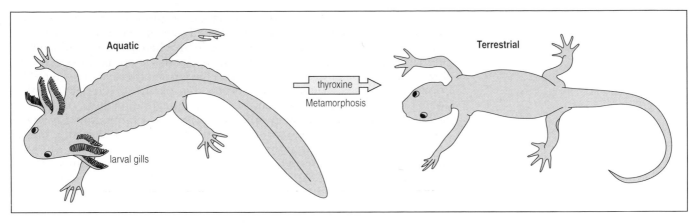

Fig. 15.15 Neoteny in salamanders. The sexually mature Mexican axolotl retains larval features, such as gills, and remains aquatic. This neotenic form metamorphoses into a typical terrestrial adult salamander if treated with thyroxine, which it does not produce, and which is the hormone that causes metamorphosis in other amphibians.

structural modifications, could have led to the interposition of a feeding tadpole stage. One of the changes required would have been a delay in the development of limbs.

Some modern frogs have evolved with direct development to the adult by a loss of the larval stage and the acceleration of the development of adult features. Frogs of the genus *Eleutherodactylus*, unlike the more typical amphibians, *Rana* and *Xenopus*, develop directly into an adult frog and there is no aquatic tadpole stage, the eggs being laid on land. Typical tadpole features, such as gills and cement glands, do not develop, and prominent limb buds appear shortly after the formation of the neural tube (Fig. 15.16). In the embryo, the tail is modified into a respiratory organ. Such direct development requires a large supply of yolk to the egg in order to support development through to an adult without a tadpole feeding stage. This increase in yolk may itself be an example of hetero-chrony, involving a longer or more rapid period of yolk synthesis in the development of the egg.

Most sea urchins have a larval stage that takes a month or more to become adult. But there are some species in which direct development has evolved so that they no longer go through a functional larval stage. Such species have large eggs and, as a result of their very rapid development, become juvenile urchins within 4 days. This has required changes in early development so that the directly developing embryo gives rise to a larval stage that lacks a gut and cannot feed, and which metamorphoses rapidly into the adult form.

Changes in timing were also involved during the evolution of the vertebrates. There are examples of the reduction or loss of limbs in several groups, notably in the whales and the snakes. In general, limb reduction occurs gradually during evolution, with limb structures being

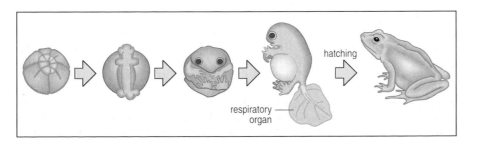

Fig. 15.16 Development of the frog *Eleutherodactylus*. Typical frogs, such as *Xenopus* or *Rana*, lay their eggs in water and develop through an aquatic tadpole stage, which undergoes meta-morphosis into the adult. Frogs of the genus *Eleutherodactylus* lay their eggs on land and the frog hatches from the egg as a miniature adult, without going through an aquatic free-living larval stage. The embryonic tail is modified as a respiratory organ.

lost in a distal to proximal sequence. In the evolution of the whales, the progressive loss of the hindlimbs occurred over many millions of years, as shown by fossil evidence. Some modern whales still retain internal remnants of a femur and tibia. Flightless birds, such as the kiwi, also show some limb reduction: kiwi wings have just one digit. These evolutionary changes can be partly understood in terms of changes in timing during limb development, such as a reduced rate of growth of the bud and early cessation of growth. The complete loss of the limbs in snakes and legless lizards can be understood in terms of a change in the timing of cell death and the disappearance of the apical ridge in the bud (see Section 10-3). Although the limb bud begins to develop in some legless species, cell death in the apical ridge starts at an early stage, followed by death of the cells in the bud itself, which generally results in the absence of limbs. But in the python, for example, the proximal elements of the limb still develop.

Summary.

Changes in the timing of developmental processes that have occurred during evolution can alter the form of the body, for if different regions grow faster than others, their size is proportionately increased if the animal gets larger. The decrease in size of the lateral digits of horses is partly due to this type of developmental change. Speed of development and egg size also have important evolutionary implications. Changing the time at which an animal becomes sexually mature can result in adults with larval characteristics. Some animals, such as frogs and sea urchins, which usually have a larval form, have evolved species that develop directly into the adult without a larval stage.

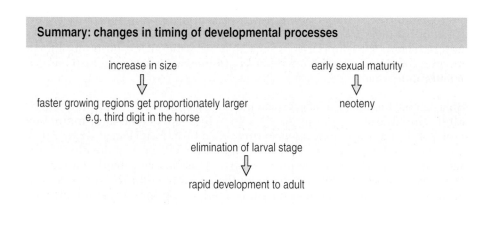

Summary: changes in timing of developmental processes

increase in size   early sexual maturity
⇩   ⇩

faster growing regions get proportionately larger   neoteny
e.g. third digit in the horse

elimination of larval stage
⇩
rapid development to adult

Summary to Chapter 15.

Many developmental processes have been conserved during evolution. While many questions relating to evolution and development remain unanswered, it is clear that development reflects the evolutionary history of ancestral embryos. All vertebrate embryos pass through a phylotypic developmental stage, although there can be considerable divergence both earlier and later in development. The signals involved in patterning of vertebrate and arthropod appendages, as well as of the dorso-ventral axis, show remarkable similarity and conservation. The pattern of Hox gene expression along the body axis of vertebrates and arthropods is conserved, and changes in body plan reflect both changes in Hox gene expression and their downstream targets. Changes in timing of developmental events have played an important role in evolution. Such changes can alter the overall form due to differences in growth rates of different structures, and can also result in sexual maturation at larval stages.

General references.

Raff, R.A.: *The Shape of Life.* University of Chicago Press, 1996.

Pennisi, E., Roush, W.: **Developing a new view of evolution.** *Science* 1997, **277**:34–37.

Section references.

15-1 Embryonic structures have acquired new functions during evolution.

Romer, A.S.: *The Vertebrate Body.* Philadelphia: W.B. Saunders, 1949.

15-2 Limbs evolved from fins.

Sordino, P., van der Hoeven, F., Duboule, D.: **Hox gene expression in teleost fins and the origin of vertebrate digits.** *Nature* 1995, **375**:678–681.

Tabin, C.J.: **Why we have (only) five fingers per hand: Hox genes and the evolution of paired limbs.** *Development* 1992, **116**:289–296.

15-3 Development of vertebrate and insect wings makes use of evolutionarily conserved mechanisms.

Panganiban, G., Irvine, S.M., Lowe, C., Roehl, H., Corley, L.S., Sherbon, B., Grenier, J.K., Fallon, J.F., Kimble, J., Walker, M., Wray, G.A., Swalla, B.J., Martindale, M.Q., Carroll, S.B.: **The origin and evolution of animal appendages.** *Proc. Natl. Acad. Sci.* 1997, **94**:5162–5166.

15-4 Hox gene complexes have evolved through gene duplication.

Holland, P.: **Homeobox genes in vertebrate evolution.** *BioEssays* 1992, **14**:267–273.

Krumlauf, R.: **Evolution of the vertebrate Hox homeobox genes.** *BioEssays* 1992, **14**:245–252.

Valentine, J.W., Erwin, D.H., Jablonski, D.: **Developmental evolution of metazoan bodyplans: the fossil evidence.** *Dev. Biol.* 1996, **173**:373–381.

15-5 Changes in specification and interpretation of positional identity have generated the elaboration of vertebrate and arthropod body plans.

Carroll, S.B.: **Homeotic genes and the evolution of arthropods and chordates.** *Nature* 1995, **376**:479–485.

Averof, M., Akam, M.: **Hox genes and the diversification of insect and crustacean body plans.** *Nature* 1995, **376**:420–423.

Akam, M.: **Hox genes and the evolution of diverse body plans.** *Phil. Trans. Roy. Soc. Lond. B* 1995, **349**:313–319.

Slack, J.M., Holland, P.W., Graham, C.F.: **The zootype and phylotypic stage.** *Nature* 1993, **361**:490–492.

15-6 The position and number of paired appendages in insects is dependent on Hox gene expression.

Carroll, S.B., Weatherbee, S.D., Langeland, J.A.: **Homeotic genes and the regulation and evolution of insect wing number.** *Nature* 1995, **375**:58–61.

15-7 The body plan of arthropods and vertebrates is similar, but the dorso-ventral axis is inverted.

Holland, L.Z., Kene, M., Williams, N.A., Holland, N.D.: **Sequence and embryonic expression of the amphioxus *engrailed* gene (*AmphiEn*): the metameric pattern of transcription resembles that of its segment-polarity homolog in *Drosophila*.** *Development* 1997, **124**:1723–1732.

Holley, S.A., Jackson, P.D., Sasai, Y., Lu, B., De Robertis, E., Hoffman, F.M., Ferguson, E.L.: **A conserved system for dorso-ventral patterning in insects and vertebrates involving *sog* and *chordin*.** *Nature* 1995, **376**:249–253.

Müller, M., von Weizsäcker, E., Campos-Ortega, J.A.: **Expression domains of a zebrafish homologue of the *Drosophila* pair-rule gene *hairy* correspond to primordia of alternating somites.** *Development* 1996, **122**:2071–2078.

Arendt, D., Nübler-Jung, K.: **Dorsal or ventral: similarities in fate maps and gastrulation patterns in annelids, arthropods and chordates.** *Mech. Dev.* 1997, **61**:7–21.

15-8 Changes in relative growth rates can alter the shapes of organisms.

Huxley, J.S.: *Problems of Relative Growth.* London: Methuen & Co. Ltd., 1932.

15-9 Evolution of life histories has implications for development.

Partridge, L., Harvey, P.: **The ecological context of life history evolution.** *Science* 1988, **241**:1449–1455.

15-10 The timing of developmental events has changed during evolution.

Alberch, P., Alberch, J.: **Heterochronic mechanisms of morphological diversification and evolutionary change in the neotropical salamander *Bolitoglossa occidentales* (Amphibia: Plethodontidae).** *J. Morphol.* 1981, **167**:249–264.

Lande, R.: **Evolutionary mechanisms of limb loss in tetrapods.** *Evolution* 1978, **32**:73–92.

Raynaud, A.: **Developmental mechanism involved in the embryonic reduction of limbs in reptiles.** *Intl. J. Dev. Biol.* 1990, **34**:233–243.

Wray, G.A., Raff, R.A.: **The evolution of developmental strategy in marine invertebrates.** *Trends Evol. Ecol.* 1991, **6**:45–56.

Glossary

The **acrosomal reaction** is the release of enzymes and other proteins from the acrosomal vesicle of the sperm head that occurs once a sperm has bound to the outer surface of the egg. It helps the sperm to penetrate the outer layers of the egg.

Adhesion molecules bind cells both to each other and the extracellular matrix. The main classes of adhesion molecules important in development are the cadherins, the immuno-globulin superfamily, and the integrins.

The **allantois** is a set of extra-embryonic membranes that develops in some vertebrate embryos. In bird and reptile embryos it acts as a respiratory surface, while in mammals its blood vessels carry blood to and from the placenta.

An **allele** is a particular form of a gene. In diploid organisms two alleles of each gene are present, which may or may not be the same.

Allometry is the mathematical study of the relative growth of parts of an organism.

The **amniotic sac** is an extra-embryonic membrane in birds, reptiles, and mammals, which forms a fluid-filled sac that encloses and protects the embryo. It is derived from extra-embryonic ectoderm and mesoderm.

An **androgenetic** embryo is an embryo in which the two sets of homologous chromosomes are both paternal in origin.

The **animal region** of an egg is the end of the egg where the nucleus resides, usually away from the yolk. The most terminal part of this region is the **animal pole**, which is directly opposite the vegetal pole at the other end of the egg. In *Xenopus* the pigmented animal half is called the **animal cap**.

The **animal-vegetal axis** passes through the animal and vegetal regions of an egg or early embryo.

The **antero-posterior axis** defines which is the 'head' end and which is the 'tail' end of an animal. The head is anterior and the tail posterior.

Anticlinal cell divisions are divisions in planes at right angles to the outer surface of a tissue.

The **apical ectodermal ridge** is a thickening of the ectoderm at the distal end of the developing chick and mammalian limb bud. Signals from the ridge specify the progress zone in the underlying mesoderm.

An **apical meristem** is the region of dividing cells at the tip of a growing shoot or root.

Apoptosis or programmed cell death is a type of cell death that occurs widely during development. In programmed cell death, a cell is induced to commit 'suicide', which involves fragmentation of the DNA and shrinkage of the cell. Apoptotic cells are removed by the body's scavenger cells and, unlike necrosis, their death does not cause damage to surrounding cells.

The **archenteron** is the cavity formed inside the embryo when the endoderm and mesoderm invaginate during gastrulation. It forms the gut.

The **area opaca** is the outer dark area of the chick blastoderm.

The **area pellucida** is the central clear area of the chick blastoderm.

Asymmetric divisions are cell divisions in which the daughter cells are different from each other because some cytoplasmic determinant(s) have been distributed unequally between them.

The **axes** of an organism define its polarity in various directions. In animals the two main axes are the antero-posterior axis, which runs from the head to the tail, and the dorso-ventral axis, which runs at right angles to this, from back to under side, with the mouth defining the ventral or underside. The main axis of a plant runs from shoot tip to root tip and is called the apical-basal axis.

Axons are long cell processes of neurons that conduct nerve impulses away from the cell body. The end of an axon forms contacts (synapses) with other neurons, muscle cells, or glandular cells.

The **basal lamina** is a sheet of extracellular matrix that separates an epithelial layer from the underlying tissues. For example, the epidermis of the skin is separated from the dermis by a basal lamina.

In amphibian limb regeneration, a **blastema** is formed from the dedifferentiation and proliferation of cells beneath the wound epidermis, and gives rise to the regenerated limb.

The **blastocoel** is the fluid-filled cavity that develops in the interior of a blastula.

The **blastocyst** stage of a mammalian embryo corresponds in form to the blastula stage of other animal embryos, and is the stage at which the embryo implants in the uterine wall.

A **blastoderm** is a post-cleavage embryo composed of a solid layer of cells rather than a spherical blastula, as in early chick and *Drosophila* embryos. The chick blastoderm is also known as the **blastodisc**.

Blastomeres are the cells derived from cleavage of the early embryo.

The **blastopore** is the slit-like or circular invagination on the surface of amphibian and sea urchin embryos where the mesoderm and endoderm move inside the embryo at gastrulation.

The **blastula** stage in animal development is the outcome of cleavage. The blastula is a hollow ball of cells, composed of an epithelial layer of small cells enclosing a fluid-filled cavity —the blastocoel.

The **body plan** describes the overall organization of an organism, for example the position of the head and tail, and the plane of bilateral symmetry, where it exists. The body plan of most animals is organized around two main axes, the antero-posterior axis and the dorso-ventral axis.

Catenins are intracellular proteins that interact with the cytoplasmic tails of cadherin molecules and form a link between the cadherins and the cell's cytoskeleton.

Cell adhesiveness is the property of cells that enables them to attach to each other and to a substratum.

The **cell cycle** is the sequence of events by which a cell duplicates itself and divides in two.

During **cell differentiation**, cells become functionally and structurally different from one another and become distinct cell types, such as muscle or blood cells.

Cell lineage restriction occurs when all the descendants of a particular group of cells remain within a 'boundary' and never mix with an adjacent group of cells. Compartment boundaries in insect development are boundaries of lineage restriction.

Cell motility refers to the ability of cells to change shape and move.

A **chimeric** animal is made up of cells from two or more different sources, and thus of different genetic constitutions.

The **chorion** is the outermost of the extra-embryonic membranes in birds, reptiles, and mammals. It is involved in respiratory gas exchange. In birds and reptiles it lies just beneath the shell. In mammals it is part of the placenta and is also involved in nutrition and waste removal. The chorion of insect eggs has a different structure.

Chromatin is the material of which chromosomes are made. It is composed of DNA and protein.

Cleavage occurs after fertilization and is a series of rapid cell divisions without growth that divides the embryo up into a number of small cells.

A **clone** is the genetically identical derivative of a single cell, or the genetically identical offspring of a single individual produced by asexual reproduction.

The **coding region** of a gene is that part of the DNA that encodes a polypeptide or functional RNA.

Induction of cell differentiation in some tissues depends on a **community effect**, in that there have to be a sufficient number of responding cells present for differentiation to occur.

Compaction of the mouse embryo occurs during early cleavage. The blastomeres flatten against each other and microvilli become confined to the outer surface of the ball of cells.

Compartments are discrete areas that contain all the descendants of a small group of founder cells and which show lineage restriction. Cells in compartments respect the compartment boundary and do not cross over into an adjacent compartment. Compartments tend to act as discrete developmental units.

Competence is the ability of a tissue to respond to an inducing signal. Embryonic tissues only remain competent for a limited period of time.

The **control region** of a gene is the region to which regulatory proteins bind and so determine whether or not the gene is transcribed.

Convergent extension is the process by which a sheet of cells changes shape by extending in one direction and narrowing —converging—in a direction at right angles to the extension.

Cortical rotation occurs immediately after an amphibian egg is fertilized. The egg cortex rotates with respect to the underlying cytoplasm, toward the point of sperm entry.

A **cotyledon** is the part of the plant embryo that acts as a storage organ for food.

Cyclins are proteins that periodically rise and fall in concentration during the cell cycle and are involved in controlling progression through the cycle.

Cytoplasmic localization is the nonuniform distribution of some factor or determinant in a cell's cytoplasm, so that when the cell divides, the determinant is unequally distributed to the daughter cells.

Dedifferentiation is loss of the structural characteristics of a differentiated cell, which may result in the cell then differentiating into a new cell type.

Dendrites are extensions from the body of a nerve cell that receive stimuli from other nerve cells.

The **dermamyotome** is the region of the somite that will give rise to both muscle and dermis.

The **dermatome** is the region of the somite that will give rise to the dermis.

The **dermis** of the skin is the connective tissue beneath the epidermis, from which it is separated by a basal lamina.

Desmosomes are specialized cell junctions between epithelial cells. Adhesion at these junctions is mediated by cadherins.

Determinants are cytoplasmic factors (e.g. proteins and RNAs) in the egg and in embryonic cells that can be asymmetrically distributed at cell division and so influence how the daughter cells develop.

Determination implies a stable change in the internal state of a cell such that its fate is now fixed, or **determined**. A determined cell will follow that fate when grafted into other regions of the embryo.

Diploid cells contain two sets of homologous chromosomes, one from each parent, and thus two copies of each gene.

Directed dilation is the extension of a tube-like structure at each end of a cell due to hydrostatic pressure, the direction of extension reflecting greater circumferential resistance to expansion.

A **dominant** allele is one that determines the phenotype even when present in only a single copy.

A **dominant-negative** mutation inactivates a particular cellular function by the production of a defective RNA or protein molecule that blocks the normal function of the gene product.

Dorsalized embryos develop much increased dorsal regions at the expense of ventral regions.

The **dorso-ventral** axis defines the relation of the upper surface or back (dorsal) to the under surface (ventral) of an organism or structure. The mouth is always on the ventral side.

Dosage compensation is the mechanism that ensures that although the number of X chromosomes in males and females is different, the level of expression of X-chromosome genes is the same in both sexes. Mammals, insects, and nematodes all have different dosage compensation mechanisms.

Ecdysis is a type of molting in arthropods in which the external cuticle is shed to allow for growth.

The **ectoderm** is the germ layer that gives rise to the epidermis and the nervous system.

Embryonic stem cells (ES cells) are derived from the inner cell mass of a mammalian embryo, usually mouse, and can be indefinitely maintained in culture. When injected into another blastocyst, they combine with the inner cell mass and contribute to the embryo.

Endochondral ossification is the replacement of cartilage with bone in the growth plate of vertebrate embryonic skeletal elements, such as those that give rise to the long bones of the limbs.

The **endoderm** is the germ layer that gives rise to the gut and associated organs, such as the lungs and liver in vertebrates.

The **endosperm** in higher plant seeds serves as a source of food for the embryo.

Enhancers are DNA sequences to which regulatory proteins bind to control the time and place of transcription of a gene. Enhancers can be many thousands of base pairs away from the gene's coding region.

The **epiblast** of mouse and chick embryos is a group of cells within the blastocyst or blastoderm, respectively, that gives rise to the embryo proper. In the mouse, it develops from cells of the inner cell mass.

Epiboly is the process during gastrulation in which the ectoderm extends to cover the whole of the embryo.

The **epidermis** in vertebrates and insects is the outer layer of cells that forms the interface between the organism and its environment.

Epimorphosis is a type of regeneration in which the regenerated structures are formed by new growth.

An **epithelium to mesenchymal transition** involves cells leaving an epithelial layer and becoming a loose mass of mesenchyme cells which can migrate individually.

Exogastrulas are artificially induced abnormal gastrulas in which the mesoderm does not enter the embryo but comes to lie outside it, connected to the ectoderm by a thin bridge.

Extra-embryonic ectoderm in mammals contributes to the formation of the placenta.

Extra-embryonic membranes are membranes external to the embryo proper that are involved in protection and nutrition of the embryo.

The **fate** of cells describes what they will normally develop into. By marking cells in the embryo, a **fate map** of embryonic regions can be constructed. However, having a particular normal fate does not imply that a cell could not develop differently if placed in a different environment.

A **floral meristem** is a region of dividing cells at the tip of a shoot that gives rise to a flower.

The individual parts of a flower develop from **floral organ primordia** generated by the floral meristem.

Follicle cells surround the oocyte and nurse cells during egg development in *Drosophila*.

The **gametes** are the cells that carry the genes to the next generation—in animals they are the eggs and sperm.

Gap genes are zygotic genes coding for transcription factors expressed in early *Drosophila* development that subdivide the embryo into regions along the antero-posterior axis.

The **gastrula** is the stage in animal development when the endoderm and mesoderm of the blastula move inside the embryo.

Gastrulation is the process in animal embryos in which the endoderm and mesoderm move from the outer surface of the embryo to the inside, where they give rise to internal organs.

Gene knock-out refers to the complete inactivation of a particular gene in a transgenic organism.

The **genital ridge** in vertebrates is the region of mesoderm lining the abdominal cavity from which the gonads develop.

The **genotype** is a description of the exact genetic constitution of a cell or organism in terms of the alleles it possesses for any given gene.

The **germ band** is the name given to the ventral blastoderm of the early *Drosophila* embryo, from which most of the embryo will eventually develop.

Germ cells are those cells that give rise to eggs and sperm.

The **germ layers** refer to the regions of the early animal embryo that will give rise to distinct types of tissue. Most animals have three germ layers—ectoderm, mesoderm, and endoderm.

Germ plasm is the special cytoplasm in some animal eggs, such as those of *Drosophila*, that is involved in the specification of germ cells.

Glia are supporting cells of the nervous system, such as Schwann cells.

The **gonads** are the reproductive organs of animals.

Axons of developing neurons extend by means of a **growth cone** at their tip. The growth cone both crawls forward on the substrate and senses its environment by means of filopodia.

Growth of vertebrate long bones occurs at the cartilaginous **growth plates**. The cartilage grows and is eventually replaced by bone by the process of endochondral ossification.

A **gynogenetic** embryo is one in which the two sets of homologous chromosomes are both maternal in origin.

Haploid cells are derived from diploid cells by meiosis and contain only one set of chromosomes (half the diploid number of chromosomes), and thus contain only one copy of each gene. In most animals the only haploid cells are the gametes—the sperm or egg.

A **haploid germ cell** is a germ cell that has undergone meiosis and so only has one set of chromosomes.

Hematopoiesis is the process by which all the blood cells are derived from a pluripotent stem cell. This occurs mainly in the bone marrow.

Hemi-desmosomes are specialized junctions between epithelial cells and the underlying basal lamina. They involve integrins.

Hensen's node is a condensation of cells at the anterior end of the primitive streak in chick and mouse embryos. Cells from the node give rise to the notochord. It corresponds to the Spemann organizer in amphibians.

Heterochromatin is the state of chromatin in regions of the chromosome that are so condensed that transcription is not possible.

Heterochrony is an evolutionary change in the timing of developmental events. A mutation that changes the timing of a developmental event is called a **heterochronic** mutation.

A diploid individual is **heterozygous** for a given gene when it carries two different alleles of that gene.

The **homeobox** is a region of DNA in homeotic genes that encodes a DNA-binding domain called the **homeodomain**. The homeodomain is present in a large number of transcription factors that are important in development, such as the products of the Hox genes and the Pax genes.

Homeosis is the phenomenon in which one structure is transformed into another, homologous structure. An example of a **homeotic transformation** is the development of legs in place of antennae in *Drosophila* as a result of mutation in a **homeotic gene**.

Homeotic selector genes in *Drosophila* are genes that specify the identity and developmental pathway of a group of cells. They encode homeodomain transcription factors and act by controlling the expression of other genes. Their expression is required throughout development. The *Drosophila* gene *engrailed* is an example of a homeotic selector gene.

Homologous genes share significant similarity in their nucleotide sequence and are derived from a common ancestral gene.

Homologous recombination is the recombination of two DNA molecules at a specific site of sequence similarity.

A diploid individual is **homozygous** for a given gene when it carries two identical alleles of that gene.

Hox genes are a particular family of homeobox-containing genes that are present in all animals (as far as is known) and are involved in patterning the antero-posterior axis. They are clustered on the chromosomes into one or more gene complexes.

The **hypoblast** in the early chick embryo is a sheet of cells that covers the yolk and gives rise to extra-embryonic structures such as the stalk of the yolk sac.

Imaginal discs are small sacs of epithelium present in the larva of *Drosophila* and other insects, which at metamorphosis give rise to adult structures such as wings, legs, antennae, eyes, and genitalia.

A gene is said to be **imprinted** when it is expressed differently (either active or inactive) in the embryo depending on whether it is derived from the mother or father.

Induction is the process whereby one group of cells signals to another group of cells in the embryo and so affects how they will develop.

An **inflorescence** in plants is a flowerhead—a flowering shoot. Shoots that can bear flowers develop as a result of the conversion of a vegetative apical meristem into an **inflorescence meristem**.

Ingression is the movement of individual cells from the outside of the embryo into the blastocoel.

Initials are cells in the meristems of plants that are able to divide continuously, giving rise both to dividing cells that stay within the meristem and to cells that leave the meristem and go on to differentiate.

The **inner cell mass** of the early mammalian embryo is derived from the inner cells of the morula, which form a discrete mass of cells in the blastocyst. Some of the cells of the inner cell mass give rise to the embryo proper.

In situ **hybridization** is a technique used to detect where in the embryo particular genes are being expressed. The mRNA that is being transcribed is detected by its hybridization to a labeled single-stranded complementary DNA probe.

In animals in which the larva goes through successive phases of growth and molting before developing into an adult, the phase between each molt is known as an **instar**.

Integrins are a class of cell adhesion molecules by which cells attach to the extracellular matrix.

Intercalary growth can occur in animals capable of epimorphic regeneration when two pieces of tissue with different positional values are placed next to each other. The intercalary growth replaces the intermediate positional values.

An **internode** is that portion of a plant stem between two nodes (sites at which a leaf or leaves form).

Invagination is the local inward movement of a sheet of embryonic epithelial cells to form a bulge-like structure, as in early gastrulation of the sea urchin embryo.

Involution is a type of cell movement that occurs at the beginning of amphibian gastrulation, when a sheet of cells enters the interior of the embryo by rolling in under itself.

Keratinocytes are differentiated epidermal skin cells that produce keratin, eventually die and are shed from the skin surface.

Knock-out, *see* **gene knock-out**.

The **lateral geniculate nucleus** is the region of the brain in mammals where the axons from the retina terminate.

The **lateral plate mesoderm** in vertebrate embryos lies lateral and ventral to the somites and gives rise to the tissues of the heart, kidney, gonads, and blood.

Lateral inhibition is the mechanism by which cells inhibit neighboring cells from developing in a similar way to themselves.

A cell's **lineage** is the sequence of cell divisions that give rise to that cell.

Lineage restriction, *see* **cell lineage restriction**.

In **long-germ development** the blastoderm gives rise to the whole of the future embryo, as in *Drosophila*.

The **marginal zone** of an amphibian embryo is the belt-like region of presumptive mesoderm at the equator of the late blastula.

Maternal-effect mutations are mutations in the genes of the mother which affect the development of the egg and later the embryo.

Maternal factors are proteins and RNAs that are deposited in the egg by the mother during oogenesis. The production of these maternal proteins and RNAs is under the control of so-called **maternal genes**.

Medio-lateral intercalation of cells occurs during convergent extension in amphibian gastrulation. The sheet of cells narrows and elongates by cells pushing in sideways between their neighbors.

Meiosis is a special type of cell division that occurs during formation of sperm and eggs, and in which the number of chromosomes is halved from diploid to haploid.

In **mericlinal chimeras** a genetically marked cell gives rise to a sector of an organ or of a whole plant.

Meristems are groups of undifferentiated, dividing cells that persist at the growing tips of plants. They give rise to all the adult structures—shoots, leaves, flowers, and roots.

Mesenchyme describes loose connective tissue, usually of mesodermal origin, whose cells are capable of migration; some epithelia of ectodermal origin, such as the neural crest, undergo an epithelial to mesenchymal transition.

A **mesenchyme to epithelium transition** occurs when loose mesenchyme cells aggregate and then form an epithelium like a tube, as in kidney development.

The **mesoderm** is the germ layer that gives rise to the skeleto-muscular system, connective tissues, the blood, and internal organs such as the kidney and heart.

The **mesonephros** in mammals is an embryonic kidney that contributes to the male and female reproductive organs.

Messenger RNA (mRNA) is the RNA molecule that specifies the sequence of amino acids in a protein. It is produced by transcription from DNA.

Metamorphosis is the process by which a larva is transformed into an adult. It often involves a radical change in form, and the development of new organs, such as wings in butterflies and limbs in frogs.

Metastasis is the movement of cancer cells from their site of origin to invade underlying tissues and to spread to other parts of the body. Such cells are said to **metastasize**.

Micromeres are small cells that result from unequal cleavage during early animal development.

The **midblastula transition** in amphibian embryos is when the embryo's own genes begin to be transcribed, cleavages become asynchronous, and the cells of the blastula become motile.

Mitosis is the division of the nucleus during cell division resulting in both daughter cells having the same diploid complement of chromosomes as the parent cell.

Molting is the shedding of an external cuticle when arthropods grow, and its replacement with a new one.

A **morphogen** is any substance active in pattern formation whose spatial concentration varies and to which cells respond differently at different threshold concentrations.

Morphallaxis is a type of regeneration that involves repatterning of existing tissues without growth.

Morphogenesis refers to the processes involved in bringing about changes in form in the developing embryo.

The **morphogenetic furrow** in *Drosophila* eye development moves across the eye disc and initiates the development of the ommatidia.

A **morula** is the very early stage in a mammalian embryo when cleavage has resulted in a solid ball of cells.

The **Müllerian duct** runs adjacent to the Wolffian duct in the mammalian embryo and becomes the oviduct in females.

Myoplasm is special cytoplasm in ascidian eggs involved in the specification of muscle cells.

The **myotome** is that part of the somite that gives rise to muscle.

Neoteny is the phenomenon in which an animal acquires sexual maturity while still in larval form.

The **neural crest cells** in vertebrates are derived from the edge of the neural plate. They migrate to different regions of the body and give rise to a wide variety of tissues, including the autonomic nervous system, the sensory nervous system, pigment cells, and some cartilage of the head.

Neural folds, *see* **neurulation.**

Neural plate, *see* **neurulation.**

Neural tube, *see* **neurulation.**

A **neuroblast** is an embryonic cell that will give rise to neural tissue (neurons and glia).

The **neuromuscular junction** is the specialized area of contact between a motor neuron and a muscle fiber, where the neuron can stimulate muscle activity.

Neurotrophins are proteins that are necessary for neuronal survival, such as nerve growth factor.

A **neurula** is the stage of vertebrate development at the end of gastrulation when the neural tube is forming.

Neurulation in vertebrates is the process in which the ectoderm of the future brain and spinal cord—the **neural plate**—develops folds (**neural folds**) and forms the **neural tube**.

The **Nieuwkoop center** is a signaling center on the dorsal side of the early *Xenopus* embryo. It forms in the vegetal region as a result of cortical rotation.

A **node** in a plant is that part of the stem at which leaves and lateral buds form.

The **notochord** in vertebrate embryos is a rod-like cellular structure that runs from head to tail and lies centrally beneath the future central nervous system. It is derived from mesoderm.

Nurse cells surround the developing oocyte in *Drosophila* and synthesize proteins and RNAs that are to be deposited in it.

Alternate **ocular dominance columns** in the visual cortex respond to the same visual stimulus from either the left or right eye.

Many of the genes involved in cell regulation can be mutated into **oncogenes**, which cause cells to become cancerous.

Ontogeny refers to the development of an individual organism.

An **oocyte** is an immature egg.

Oogenesis is the process of egg formation in the female.

The **optic tectum** is the region of the brain in amphibians and birds where the axons from the retina terminate.

An **organizer** or **organizing region** is a signaling center that directs the development of the whole embryo or of part of the embryo, such as a limb. In amphibians, the organizer usually refers to the Spemann organizer.

Organogenesis is the development of specific organs such as limbs, eyes, and heart.

The **oviduct** in female birds and mammals transports the eggs from the ovaries to the uterus.

The **pair-rule genes** in *Drosophila* are expressed in transverse stripes in the blastoderm, each pair-rule gene being expressed in alternate parasegments.

Genes within a species that have arisen by duplication and divergence are called **paralogs**. Examples are the Hox genes in vertebrates, which comprise several **paralogous subgroups** made up of **paralogous genes**.

Parasegments in the developing *Drosophila* embryo are independent developmental units that give rise to the segments of the larva and adult.

Pattern formation is the process by which cells in a developing embryo acquire identities that lead to a well ordered spatial pattern of cell activities.

Pax genes encode transcriptional regulatory proteins that contain both a homeodomain and another protein motif, the paired motif.

P elements are transposable DNA elements found in *Drosophila*. They are short sequences of DNA that can become inserted in different positions within a chromosome and can also move to other chromosomes.

Periclinal cell divisions are divisions in a plane parallel to the surface of the tissue.

In **periclinal chimeras** in plants, one of the three meristem layers has a genetic marker which distinguishes it from the other two.

The **phenotype** is the observable characters and features of a cell or an organism.

Phyllotaxy is the way the leaves are arranged along a shoot.

Its **phylogeny** is the evolutionary history of a species or group.

Vertebrate embryos pass through a developmental stage known as the **phylotypic stage** at which the embryos of the different vertebrate groups closely resemble each other. This is the stage at which the embryo possesses a distinct head, a neural tube, and somites.

A **plasmid** is a small circular DNA molecule that replicates independently of the bacterial genome.

The **pluteus** is the larval stage of the sea urchin.

Polar bodies are formed during meiosis in the developing egg. They are small cells containing a haploid nucleus and take no part in embryonic development.

In the developing chick and mouse limb buds, the **polarizing region** at the posterior margin of the bud produces a signal specifying position along the antero-posterior axis.

Pole cells give rise to the germ cells in *Drosophila* and are formed at the posterior end of the blastoderm.

Pole plasm is the cytoplasm at the posterior end of the *Drosophila* egg which is involved in specifying germ cells.

Polyspermy is the entry of more than one sperm into the egg.

Positional information in the form, for example, of a gradient of an extracellular signaling molecule, can provide the basis for pattern formation. Cells acquire a **positional value** that is related to their position with respect to the boundaries of the given field of positional information. The cells then interpret this positional value according to their genetic constitution and developmental history, and develop accordingly.

Posterior dominance or **posterior prevalence** is the process whereby the more posteriorly expressed Hox genes can inhibit the action of more anteriorly expressed Hox genes when they are expressed in the same region.

The **posterior marginal zone** of the chick embryo is a dense region of cells at the edge of the blastoderm, that will give rise to the primitive streak.

Post-translational modification of a protein involves changes in the protein after it has been synthesized. The protein can, for example, be enzymatically cleaved, phosphorylated, or glycosylated.

The **primitive ectoderm** or epiblast is the part of the inner cell mass in the mammalian blastocyst that gives rise to the embryo proper.

The **primitive endoderm** in mammalian embryos is that part of the inner cell mass that contributes to extra-embryonic membranes.

The **primitive streak** of the chick embryo is a strip of cells that extends inward from the posterior marginal zone and is the forerunner of the antero-posterior axis. During gastrulation, cells move through the streak into the interior of the blastoderm. The primitive streak in the mouse embryo has a similar function to that in the chick.

Programmed cell death, *see* **apoptosis**.

In chick and mouse limb buds the cells in the **progress zone** at the tip of the bud proliferate and acquire positional values.

The **promeristem** is the central region of the meristem that contains cells capable of continued division—the initials.

The **promoter** is a region of DNA close to the coding sequence to which RNA polymerase binds to begin transcription of a gene.

A **pronucleus** is the haploid nucleus of sperm or egg after fertilization but before nuclear fusion and the first mitotic division.

A **proto-oncogene** is a gene that is involved in regulation of cell proliferation which can, when mutated into an oncogene, or expressed under abnormal control, cause cancer.

Radial intercalation occurs in a multilayered ectoderm of an amphibian gastrula when cells intercalate in a direction perpendicular to the surface, so thinning and extending the cell sheet.

Reaction-diffusion mechanisms produce self-organizing patterns of chemical concentrations which could underlie periodic patterns.

A **recessive** mutation is a mutation in a gene that only changes the phenotype when both copies of the gene carry the mutation.

Redundancy refers to the apparent absence of an effect often seen when a gene that is normally active during development is inactivated. It is assumed that other pathways exist which can substitute for the missing gene action.

Regeneration is the ability of a fully developed organism to replace lost parts.

Regulation is the ability of the embryo to develop normally even when parts are removed or rearranged. Embryos that can regulate are called **regulative**.

The **rhombomeres** are a sequence of compartments of cell lineage restriction in the hindbrain of chick and mice embryos.

The **sclerotome** is that part of a somite that will give rise to the cartilage of the vertebrae.

Segmentation is the division of the body of an organism into a series of morphologically similar units or **segments**.

Segment-polarity genes in *Drosophila* are involved in patterning the parasegments and segments.

Selector genes in *Drosophila* determine the activity of a group of cells, and their continued expression is required to maintain that activity.

A **semi-dominant** mutation is a mutation which affects the phenotype when just one allele carries the mutation but where the effect on the phenotype is much greater when both alleles carry the mutation.

Senescence is the impairment of function associated with aging.

Short-germ development characterizes those insects in which most of the segments are formed sequentially by growth. The blastoderm itself only gives rise to the anterior segments of the embryo.

Signal transduction is the process by which a cell converts an extracellular signal, usually at the cell membrane, into a response, which is often a change in gene expression.

Somatic cells are any cells other than germ cells. In most animals, the somatic cells are diploid.

Somites in vertebrate embryos are segmented blocks of mesoderm lying on either side of the notochord. They give rise to body and limb muscles, the vertebral column, and the dermis.

A **specification map** shows how the tissues of an embryo will develop when placed in a simple culture medium.

A group of cells is called **specified** if when isolated and cultured in a neutral medium they develop according to their normal fate.

The **Spemann organizer** is a signaling center on the dorsal side of the amphibian embryo. Signals from this center can organize new antero-posterior and dorso-ventral axes.

Spermatogenesis is the production of sperm.

A **stem cell** is an undifferentiated type of cell found in certain adult tissues that is both self renewing and also gives rise to differentiated cell types.

A **synapse** is the specialized point of contact where a neuron communicates with another neuron or a muscle cell.

A **syncytium** is a cell with many nuclei in a common cytoplasm. Cell walls do not develop during nuclear division within the very early *Drosophila* embryo. This gives rise to the **syncytial blastoderm** in which the nuclei are arranged around the periphery of the embryo.

In leeches and other annelids, **teloblasts** give rise to the segmental structures.

The **teloplasm** is the cytoplasm in the eggs of leeches and other annelids which is involved in the specification of the blastomere that will give rise to the teloblasts.

Teratocarcinomas are solid tumors that arise from germ cells and which can contain a mixture of differentiated cell types.

A **threshold concentration** is that concentration of a chemical signal or morphogen that can elicit a particular response from a cell. Specific cellular responses to chemical signals often only occur above or below a particular threshold concentration of the signal.

Totipotency is the capacity of a cell to develop into any of the cell types found in that particular organism.

When a gene is active its DNA sequence is copied, or **transcribed**, into a complementary RNA sequence, a process known as **transcription**.

A **transcription factor** is a regulatory protein required to initiate or regulate transcription of a gene into RNA. Transcription factors act within the nucleus of a cell by binding to specific regulatory regions in the DNA.

Transdifferentiation is the process by which a differentiated cell can differentiate into a different cell type, such as pigment cell to lens.

Transfection is the technique by which mammalian and other animal cells are induced to take up foreign DNA molecules. The introduced DNA sometimes becomes inserted permanently into the host cell's DNA.

Transgenic organisms have had their genetic constitution deliberately changed by genetic engineering. A new gene may have been introduced or a particular gene may have been inactivated.

The process by which messenger RNA directs the order of amino acids in a protein during protein synthesis is known as **translation**.

A **transposon** is a DNA sequence that can become inserted into a different site on the chromosome, either by the insertion of a copy of the original sequence or by excision and reinsertion of the original sequence.

The **trophectoderm** is the outer layer of cells of the early mammalian embryo. It gives rise to extra-embryonic structures such as the placenta.

Tumor suppressor genes are genes that can cause a cell to become cancerous when both copies of the gene have been inactivated.

When an embryo is treated so that the amount of endoderm is increased at the expense of ectoderm it is **vegetalized**.

The **vegetal region** of an egg is the most yolky region, and is the region from which the endoderm will develop. The most terminal part of this region is called the **vegetal pole**, and is directly opposite the animal pole.

Embryos that are **ventralized** are deficient in dorsal regions and have much increased ventral regions.

The **visceral endoderm** is derived from the primitive endoderm that develops on the surface of the egg cylinder in the mammalian blastocyst and secretes proteins necessary for the embryo.

The **vitelline membrane** is an extracellular layer surrounding the eggs of the sea urchin and other animals. In the sea urchin it gives rise to the fertilization membrane.

Wolffian ducts are ducts associated with the mesonephros in mammalian embryos, They become the vas deferens in males.

The **yolk sac** is an extra-embryonic membrane in birds and mammals. In the chick embryo it surrounds the yolk.

The **zona pellucida** is a layer of glycoprotein surrounding the mammalian egg which serves to prevent polyspermy.

The **zone of polarizing activity** is another name for the polarizing region at the posterior margin of the limb bud.

The **zootype** refers to the pattern of expression of Hox genes (and certain other genes) along the antero-posterior axis of the embryo, which is characteristic of all animal embryos.

The **zygote** is the fertilized egg. It is diploid and contains the chromosomes of both the male and female parents.

Zygotic genes are those present in the fertilized egg and which are expressed in the embryo itself.

Index